54.00 100R

Neutron–Nucleus Collisions
A Probe of Nuclear Structure
(Burr Oak State Park, Ohio, 1984)

AIP Conference Proceedings
Series Editor: Hugh C. Wolfe
Number 124

Neutron–Nucleus Collisions
A Probe of Nuclear Structure
(Burr Oak State Park, Ohio, 1984)

Edited by
J. Rapaport, R. W. Finlay, S. M. Grimes
Ohio University
and
F. S. Dietrich
Lawrence Livermore National Laboratory

American Institute of Physics
New York   1985

Copying fees: The code at the bottom of the first page of each article in this volume gives the fee for each copy of the article made beyond the free copying permitted under the 1978 US Copyright Law. (See also the statement following "Copyright" below.) This fee can be paid to the American Institute of Physics through the Copyright Clearance Center, Inc., Box 765, Schenectady, N.Y. 12301.

Copyright © 1985 American Institute of Physics

Individual readers of this volume and non-profit libraries, acting for them, are permitted to make fair use of the material in it, such as copying an article for use in teaching or research. Permission is granted to quote from this volume in scientific work with the customary acknowledgment of the source. To reprint a figure, table or other excerpt requires the consent of one of the original authors and notification to AIP. Republication or systematic or multiple reproduction of any material in this volume is permitted only under license from AIP. Address inquiries to Series Editor, AIP Conference Proceedings, AIP, 335 E. 45th St., New York, N. Y. 10017.

L.C. Catalog Card No. 84-73216
ISBN 0-88318-323-4
DOE CONF- 840945

Proceedings of the

Conference on Neutron Nucleus Collisions--

A Probe of Nuclear Structure

Burr Oak State Park, Glouster, Ohio

September 5-8, 1984

PREFACE

The desire to hold this topical conference grew in part out of the success of informal workshops sponsored by Lawrence Livermore National Laboratory and held in Asilomar, California, in 1981, 1982 and 1983. The purpose was to discuss how neutrons as a probe can yield information on contemporary issues in nuclear reaction and in nuclear structure. The emphasis of the program was on reactions and scattering in the energy range up to 100 MeV, an energy range where we have or expect soon to have experimental data. However, the discussion also spanned the intermediate energy range, especially in charge exchange reactions.

The Conference was organized in six sessions including a session of Applications and Techniques and a session on Astrophysics. Another session was devoted to contributed papers. One special session was dedicated to the memory of Peter Moldauer.

The theoretical interpretation of nucleon-nucleus collision processes may be considered to be at a transition point between traditional phenomenological models and a more fundamental microscopic description. The latter has been very successful at intermediate energies where the impulse approximation may be used and the t-matrix is replaced by the free nucleon-nucleon interaction. Below 100 MeV, effects of the nuclear medium become much more important and the need of a density dependent effective interaction is clearly manifested.

Relative contributions of target neutrons and target protons to "collective" nuclear excitations has been one of the traditional problems in nuclear structure. Recent progress in this area, obtained by simultaneous analysis of nucleon inelastic scattering and from electromagnetic excitations, has been very rewarding.

In the recent years perhaps one of the most exciting results in nuclear physics has been the clear observation of spin excitations in nuclei. The most dramatic example is the observation of Gamow-Teller and M1 giant resonances in medium and heavy mass nuclei via (p,n) and (p,p') reactions at or near zero degrees with 100-200 MeV incident protons. The above data are used to obtain information on β^- strength distribution in nuclei. At present there is no information on β^+ strength distribution, which is badly needed in astrophysical problems and double beta decay.

The purpose of the Conference was to address the above questions as well as other topics in which neutron plays an important role and to elicit exchange of ideas between the participants.

One important aspect of the Conference was to review the available neutron facilities and to discuss future development plans. A workshop in experimental techniques was held to address these questions and to indicate priorities on future new facilities.

The major credit for the success of the Conference goes to all the speakers for their stimulating talks and to the participants for their lively contributions in the discussion periods. From the early planning stages of the Conference to the editing of the proceedings, Ms. Cindy White, secretary for the Ohio University Accelerator Laboratory, dedicated a sizable fraction of her time in the organization of the Conference. We certainly appreciate and commend her for her enthusiasm and competence. A large part of the Conference arrangements were done by the Ohio University Office of Workshops and Conferences. Here we acknowledge the invaluable assistance of Andrew Chonko and Steve Orth of the Workshops Office.

In these proceedings, contributions are presented as they were presented at the Conference. Photographs of the Conference here reproduced were taken by R.O. Lane who deserves all the credit and our thanks for his efforts.

Thanks are also due to the Harshaw-Filtrol and Tennelec Corporations not only for their interesting displays but also for their economic help towards the expenses of the social functions.

J. Rapaport
Athens, Ohio
November 1984

SPONSORS

U.S. National Science Foundation

Argonne Universities Association Trust Fund

Lawrence Livermore National Laboratory

Ohio University 1804 Alumni Fund

Ohio University Physics Department

ADVISORY COMMITTEE

S.M. Austin, Michigan State University, East Lansing, MI
H. Barschall, University of Wisconsin, Madison, WI
J.C. Browne, Los Alamos National Laboratory, Los Alamos, NM
G. Bertsch, University of Tennessee, Knoxville, TN
H. Feshbach, Massachusetts Institute of Technology, Cambridge, MA
H. von Geramb, Universitat Hamburg, Hamburg, West Germany
J. Harvey, Oak Ridge National Laboratory, Oak Ridge, TN
C. Mahaux, University of Liege, Liege, Belgium
A. Michaudon, Institut Laue-Langevin, France
H. Orihara, Tohoku University, Sendai, Japan
D. Seeliger, Tech. University, Dresden, East Germany
J. Speth, Inst. fur Kernphysik, KFA, Julich, West Germany
C. Zafiratos, University of Colorado, Boulder, CO

PROGRAM COMMITTEE

B. Clark, Ohio State University, Columbus, OH
P. Brady, University of California--Davis, Davis, CA
F. Dietrich, Lawrence Livermore National Laboratory, Livermore, CA
M. McEllistrem, University of Kentucky, Lexington, KY
A. Kerman, Massachusetts Institute of Technology, Cambridge, MA
F. Petrovich, Florida State University, Tallahassee, FL
J. Rapaport, Ohio University, Athens, OH
R.L. Walter, Duke University, Durham, NC

ORGANIZING COMMITTEE

F.S. Dietrich, Lawrence Livermore National Lab., Livermore, CA
R.W. Finlay, Ohio University, Athens, OH
S.M. Grimes, Ohio University, Athens, OH
C. Poppe, Lawrence Livermore National Laboratory, Livermore, CA
J. Rapaport (chair), Ohio University, Athens, OH

CONFERENCE PARTICIPANTS

M. AHMAD, OHIO UNIVERSITY, ATHENS OH 45701

R. ALARCON, OHIO UNIVERSITY, ATHENS OH 45701

S. AUSTIN, MICHIGAN STATE UNIVERSITY, E. LANSING MI 48824

N. BACK, LAWRENCE LIVERMORE NAT. LAB., LIVERMORE CA 94550

H.H. BARSCHALL, UNIVERSITY OF WISCONSIN, MADISON WI 53706

S. BLOOM, LAWRENCE LIVERMORE NAT. LAB., LIVERMORE CA 94550

C.D. BOWMAN, LOS ALAMOS NATIONAL LAB., LOS ALAMOS NM 87544

P. BRADY, UNIV. OF CALIFORNIA-DAVIS, DAVIS CA 95616

C. BRIENT, OHIO UNIVERSITY, ATHENS OH 45701

B.A. BROWN, MICHIGAN STATE UNIVERSITY, E. LANSING MI 48824

R. BYRD, INDIANA UNIV. CYCLOTRON FAC., BLOOMINGTON IN 47401

R.F. CARLTON, MIDDLE TENNESSEE STATE UNIV., MURFREESBORO TN 37132

J.A. CARR, FLORIDA STATE UNIVERSITY, TALLAHASSEE FL 32306

Q. CHEN, INDIANA UNIV. CYCLOTRON FAC., BLOOMINGTON IN 47405

B. CLARK, OHIO STATE UNIVERSITY, COLUMBUS OH 43210

T.B. CLEGG, UNIVERSITY OF NORTH CAROLINA, CHAPEL HILL NC 27514

G. CODDENS, UNIVERSITY OF ANTWERPEN, ANTWERP BELGIUM

H. CONDE, UPPSALA UNIVERSITY, UPPSALA SWEDEN

J.C. DAVIS, LAWRENCE LIVERMORE NAT. LAB., LIVERMORE CA 94550

J.P. DELAROCHE, SERVICE DE PHYSIQUE NUCLEAIRE, BRUYERES-LE-CHATEL FRANCE

F.S. DIETRICH, LAWRENCE LIVERMORE NAT. LAB., LIVERMORE CA 94550

P. EGUN, OHIO UNIVERSITY, ATHENS OH 45701

H. FESHBACH, M.I.T., CAMBRIDGE MA 02139

R. FINLAY, OHIO UNIVERSITY, ATHENS OH 45701

T. FORD, UNIV. OF CALIFORNIA-DAVIS, DAVIS CA 95616

C.C. FOSTER, INDIANA UNIV. CYCLOTRON FAC., BLOOMINGTON IN 47405

A. FRASCA, WITTENBERG UNIVERSITY, SPRINGFIELD OH 45501

D. FRIESEL, INDIANA UNIV. CYCLOTRON FAC., BLOOMINGTON IN 47405

F. GABBARD, UNIVERSITY OF KENTUCKY, LEXINGTON KY 40553

C.D. GOODMAN, INDIANA UNIV. CYCLOTRON FAC., BLOOMINGTON IN 47405

G.C. GOSWAMI, UNIVERSITY OF LOWELL, LOWELL MA 01854

C. GOULD, NORTH CAROLINA STATE UNIV., RALEIGH NC 27695

S. GRAHAM, OHIO UNIVERSITY, ATHENS OH 45701

CONFERENCE PARTICIPANTS

S. GRIMES, OHIO UNIVERSITY, ATHENS OH 45701

R. HAIGHT, LAWRENCE LIVERMORE NAT. LAB., LIVERMORE CA 94550

J.M. HANLY, UNIVERSITY OF KENTUCKY, LEXINGTON KY 40506

L.F. HANSEN, LAWRENCE LIVERMORE NAT. LAB., LIVERMORE CA 94550

G. HAOUAT, CENTRE D'ETUDES B.P. 12, C.E.A. - SERVICE P2N FRANCE

J.A. HARVEY, OAK RIDGE NATIONAL LAB., OAK RIDGE TN 37831

R. HERSHBERGER, UNIVERITY OF KENTUCKY, LEXINGTON KY 40506

S. HICKS, UNIVERSITY OF KENTUCKY, LEXINGTON KY 40506

P.E. HODGSON, OXFORD UNIVERSITY, OXFORD ENGLAND

N. HOLDEN, BROOKHAVEN NATIONAL LAB., UPTON NY 11973

G. HONORE, DUKE UNIVERSITY, DURHAM NC 27706

C. HOWELL, DUKE UNIVERSITY, DURHAM NC 27706

S. ISLAM, OHIO UNIVERSITY, ATHENS OH 45701

C. JOHNSON, OAK RIDGE NATIONAL LAB., OAK RIDGE TN 37830

C. KALBACH-WALKER DUKE UNIVERSITY, DURHAM NC 27706

H. KLAGES, KFK, KARLSRUHE W. GER D7500

H. KNOX, OHIO UNIVERSITY, ATHENS OH 45701

P. KOEHLER, OHIO UNIVERSITY, ATHENS OH 45701

R. KOHLER, CBNM GEEL, GEEL BELGIUM

CH. LAGRANCE, SERVICE DE PHYSIQUE NUCLEAIRE, BRUYERES-LE-CHATEL FRANCE

R. LANE, OHIO UNIVERSITY, ATHENS OH 45701

R. LAWSON, ARGONNE NATIONAL LAB., ARGONNE IL 60439

A. LEJEUNE, UNIVERSITY OF LIEGE, LIEGE 1 BELGIUM B4000

W.G. LOVE, UNIVERSITY OF GEORGIA, ATHENS GA 30602

J.E. LYNN, AERE HARWELL, OXFORDSHIRE ENGLAND

A.D. MacKELLAR, UNIVERSITY OF KENTUCKY, LEXINGTON KY 40506

D. MADLAND, LOS ALAMOS NATIONAL LAB., LOS ALAMOS NM 87545

V. MADSEN, LAWRENCE LIVERMORE NAT. LAB., LIVERMORE CA 94550

C. MAHAUX, UNIVERSITY OF LIEGE, B-4000 LIEGE 1 BELGIUM

G. MATHEWS, LAWRENCE LIVERMORE NAT. LAB., LIVERMORE CA 94550

M.T. McELLISTREM, UNIVERSITY OF KENTUCKY, LEXINGTON KY 40506

A.S. MEIGOONI, OHIO UNIVERSITY, ATHENS OH 45701

C. MEITZLER, OHIO UNIVERSITY, ATHENS OH 45701

CONFERENCE PARTICIPANTS

S. MELLEMA, UNIV. OF WISCONSIN-MADISON, MADISON WI, 53706

K., MURPHY, DUKE UNIVERSITY, DURHAM NC 27706

N. OLSSON, STUDSVIK SCIENCE RESEARCH LAB, 82 NYKOPING SWEDEN

D. ONLEY, OHIO UNIVERSITY, ATHENS OH 45701

H. ORIHARA, TOHOKU UNIVERSITY, SENDAI 980 JAPAN

F. OSTERFELD, JKP,KFA JUELICH, JUELICH W. GER

R. PEDRONI, DUKE UNIVERSITY, DURHAM NC 27706

J. PETLER, OHIO UNIVERSITY, ATHENS OH 45701

F. PETROVICH, FLORIDA STATE UNIVERSITY, TALLAHASSEE FL 32306

H. PFUTZNER, DUKE UNIVERSITY, DURHAM NC 27706

C.H. POPPE, LAWRENCE LIVERMORE NAT. LAB., LIVERMORE CA 94550

J. RAPAPORT, OHIO UNIVERSITY, ATHENS OH 45701

G. RAWITSCHER, UNIVERSITY OF CONNECTICUT, STORRS CT 06268

D. RESLER, OHIO UNIVERSITY, ATHENS OH 45701

W. RODNEY, NATIONAL SCIENCE FOUNDATION, WASHINGTON D.C. 20550

J.L. ROMERO, UNIV. OF CALIFORNIA-DAVIS, DAVIS CA 95616

E. SADOWSKI, OHIO UNIVERSITY, ATHENS OH 45701

S. SARAF, OHIO UNIVERSITY, ATHENS OH 45701

H. SATYANARAYANA, OHIO UNIVERSITY, ATHENS OH 45701

D. SCHRAMM, UNIVERSITY OF CHICAGO, CHICAGO IL 60637

P. SCHWANDT, INDIANA UNIV. CYCLOTRON FAC., BLOOMINGTON IN 47405

J.Q. SHAO, UNIVERSITY OF LOWELL, LOWELL MA 01854

J.R. SHEPARD, UNIVERSITY OF COLORADO, BOULDER CO 80309

A.B. SMITH, ARGONNE NATIONAL LAB., ARGONNE IL 60439

E.R. SUGARBAKER, OHIO STATE UNIVERSITY, COLUMBUS OH 43212

T.N. TADDEUCCI, INDIANA UNIV. CYCLOTRON FAC., BLOOMINGTON IN 47401

N. TAMIMI, OHIO UNIVERSITY, ATHENS OH 45701

W. TORNOW, UNIVERSITAT TUBINGEN, TUBINGEN W. GER.

P.A. TREADO, GEORGETOWN UNIVERSITY, WASHINGTON D.C. 20057

J. TRURAN, UNIVERSITY OF ILLINOIS, URBANA IL 61801

I.J. VAN HEERDEN, INDIANA UNIV. CYCLOTRON FAC., BLOOMINGTON IN 47405

R.L. VARNER JR., UNIVERSITY OF NORTH CAROLINA, CHAPEL HILL NC 27514

H.V. VON GERAMB, UNIVERSITAT HAMBURG. HAMBURG 2000 GERMANY

CONFERENCE PARTICIPANTS

R. WALTER, DUKE UNIVERSITY, DURHAM NC 27706

J. WAMBACH, UNIV. OF ILLINOIS AT URBANA, URBANA IL 61801

D. WANG, OHIO UNIVERSITY, ATHENS OH 45701

Y. WANG, OHIO UNIVERSITY, ATHENS OH 45701

J.L. WEIL, UNIVERSITY OF KENTUCKY, LEXINGTON KY 40506

S. WENDER, LOS ALAMOS NATIONAL LAB., LOS ALAMOS NM 87545

R. WHITE, LAWRENCE LIVERMORE NAT. LAB., LIVERMORE CA 94550

R.R. WINTERS, DENISON UNIVERSITY, GRANVILLE OH 43023

C.D. ZAFIRATOS, UNIVERSITY OF COLORADO, BOULDER CO 80309

Z. ZHOU, UNIVERSITY OF KENTUCKY, LEXINGTON KY 40506

M. ZIRNBAUER, MAX PLANCK INSTITUT, 6900 HEIDELBERG 1 W. GER.

TABLE OF CONTENTS

SESSION A

THE OPTICAL MODEL POTENTIAL

The phenomenological neutron optical potential --
P.E. Hodgson 1

Implication of microscopic reaction analyses below
100 MeV -- H.V. von Geramb 14

Nuclear structure approach to the imaginary optical
potential and inelastic form factor -- F. Osterfeld
and V.A. Madsen 26

Contribution:

Momentum and frequency dependence of the self-energy
in nuclear matter: A path to study the effective
nucleon-nucleon interaction -- A. Lejeune,
V. Bernard, P. Grangé and M. Martzolff 50

SESSION B

OBSERVABLES AND EFFECTIVE INTERACTIONS

Analyzing power measurements for neutron-nucleus scattering
and the spin-orbit potential -- R.L. Walter 53

Scattering of the polarized and unpolarized neutrons from
^{28}Si and ^{32}S -- C.R. Howell 72

Test of effective interactions for nucleon scattering
and charge exchange below 60 MeV -- F.S. Dietrich and
F. Petrovich 90

Isovector nucleon-nucleon effective interactions --
J.R. Shepard 107

Dirac phenomenology at low energies -- B.C. Clark, S. Hama,
S.G. Kälbermann, E.D. Cooper and R.L. Mercer 123

Contributions:

Interpretation of the Perey-Buck nonlocality in terms of
the relativistic optical model formalism --
G.H. Rawitscher 135

Few-nucleon experiments with fast polarized neutrons --
H.O. Klages, R. Aures, F.P. Brady, P. Doll,
E. Finckh, J. Hansmeyer, W. Heeringa, J.C.
Hiebert, K. Hofmann, H. Krupp, Ch. Maier, W. Nitz,
P. Plischke, J. Wilczynski. H. Zankel and
B. Zeitnitz 137

Isovector "$\Delta J^{\pi} = 0^-$" transitions observed in the
charge-exchange (p,n) reactions on ^{13,14}C and
^{16}O -- H. Orihara 139

The (n,p) reaction at zero degrees from ^{12}C, ^{13}C, ^{90}Zr
and ^{209}Bi at 65 MeV -- T.D. Ford, F.P. Brady,
C.M. Castaneda, J.L. Romero, J.R. Drummond
B. McEachern, M.L. Webb, N.S.P. King, J. Martoff
and K. Wang 141

The analog of the ^{40}Ca G.D.R. through the (n,p)
reaction at 65.5 MeV -- C.M. Castaneda, J.L. Romero,
F.P. Brady, J.R. Drummond, T.D. Ford and
B. McEachern 143

SESSION C

NUCLEAR STRUCTURE WITH NEUTRONS

Isovector vibrational modes in heavy nuclei --
J. Wambach 146

Isospin effects in nuclear vibrations -- V.A. Madsen
and V.R. Brown 171

New aspects of nuclear structure from the viewpoint of
neutron induced reactions -- B.A. Brown 183

Sensitivity of neutron scattering to low energy
collective excitations -- M.T. McEllistrem 208

Determination of nuclear transition densities with
various probes -- J.A. Carr, F. Petrovich and
J.J. Kelly 230

Contribution:

Nucleon scattering from ^{34}S -- R. Alarcon, J. Rapaport,
R.T. Kouzes, W.H. Moore and B.A. Brown 256

SESSION DA

APPLICATIONS AND TECHNIQUES

Nuclear physics in the 10-300 MeV energy range using a
  pulsed white neutron source -- C.D. Bowman,
  S.A. Wender and G.F. Auchampaugh 259

Neutron scattering above 25 MeV with monoenergetic
  neutrons -- R.W. Finlay 274

Neutron measurements for biomedical and fusion technology
  applications -- H.H. Barschall 286

  Contributions:

  Can an atomic beam polarized source be improved for
    neutron emission experiments by using an ECR
    ionizer? -- T.B. Clegg 297

  On performing (p,n), (d,n) and (n,n) measurements with
    a rotating magnet beam swinger -- F.S. Dietrich and
    T.B. Clegg 299

SESSION DB

CONTRIBUTED PAPERS

  Contributions:

  Neutron, alpha and total widths and spin assignments
    for resonances in ^{33}S+n from 10-400 keV --
    G.P. Coddens, M. Salah, J.A. Harvey and N.W. Hill . . 302

  $d_{5/2}$-single particle strength in ^{48}Ca+n -- J.A. Harvey,
    C.H. Johnson, R.F. Carlton and B. Castel 304

  High resolution neutron resonance spectroscopy --
    R. Köhler, L. Mewissen, F. Poortmans, I. Van Parys
    and H. Weigmann 306

  Optical model scattering functions for low energy
    neutrons on ^{86}Kr -- R.F. Carlton, J.A. Harvey and
    C.H. Johnson 308

  Energy dependence of the local optical potential for
    neutron-nucleus scattering -- J.P. Delaroche,
    P.P. Guss, G.M. Honore, C.R. Howell and R.L.
    Walter 310

The imaginary part of the spin-orbit interaction for
neutron-nucleus scattering -- R.L. Walter,
W. Tornow and P.P. Guss 312

Neutron scattering on deformed nuclei -- L.F. Hansen,
R.C. Haight, B.A. Pohl, C. Wong and Ch. Lagrange . . . 314

Semi-microscopic calculations of elastic, inelastic
and total neutron scattering by ^{239}Pu --
Ch. Lagrange, D.G. Madland and M. Girod 318

Semi-microscopic interpretation of fast neutron
scattering from ^{208}Pb -- G. Haouat, Ch. Lagrange,
J.C. Brient, Y. Patin, R. de Swiniarski and
F. Dietrich 320

Phenomenological mapping of the Fermi-surface anomaly
with neutron-nucleus collisions -- R.W. Finlay,
J.R.M. Annand, J.S. Petler and F.S. Dietrich 322

Nucleon induced excitation of $K^\pi = 0^+$, 0_2^+, 1^- and 3^-
bands in ^{12}C -- A.S. Meigooni, R.W. Finlay, J.S.
Petler and J.P. Delaroche 324

SESSION DC

WORKSHOP IN EXPERIMENTAL TECHNIQUES

Summary of facilities for experimental studies of neutron-
induced reactions -- C.D. Zafiratos 327

An accelerator system for the production of an intense
neutron beam for research -- D.L. Friesel 361

A beam swinger for neutron scattering -- C.D. Goodman . . . 375

Facilities for neutron induced reactions -- F.P. Brady . . . 382

Neutron spin transfer measurements: Why and how --
T.N. Taddeucci 394

A facility for neutron scattering measurements at 22 MeV --
N. Olsson and B. Trostell 401

Planning of neutron physics at the rebuilt cyclotron of
the Gustaf Werner Institute, Uppsala University --
H. Condé 403

SESSION E

IN MEMORY OF PETER MOLDAUER

Contribution:

Peter Arnold Moldauer: Memorial Session Introductory
   Remarks -- A.B. Smith 410

Some comments on the theory of nuclear reactions --
   H. Feshbach 412

Statistical theories of neutron cross sections of the
   actinides -- J.E. Lynn 427

Optical models for low-energy s-, p- and d-wave cross
   sections -- C.H. Johnson 446

Level density calculations: Past, present and future --
   S.M. Grimes 463

Contribution:

Complete solution of a model in statistical nuclear
   reaction theory -- M.R. Zirnbauer 481

SESSION FA

ASTROPHYSICS

r and s-processes: Chronometers, thermometers and neutron
   dosimeters -- R.R. Winters 484

Neutron capture processes of heavy element synthesis --
   J.W. Truran 504

Contribution:

A parametric study of dynamic s-process neutron-capture
   nucleonsyntheses: Nuclear data needs -- G.J.
   Mathews, W.M. Howard, K. Takahashi and R.A. Ward . . . 511

Neutron capture processes in astrophysics -- B.S. Meyer
   and D.N. Schramm 515

SESSION FB

SUMMARY

Summary and outlook--experimental -- S.M. Austin 527

Theoretical contributions--hindsight and outlook --
   C. Mahaux 540

SESSION A

THE OPTICAL MODEL POTENTIAL

THE PHENOMENOLOGICAL NEUTRON OPTICAL POTENTIAL

P.E. Hodgson
Nuclear Physics Laboratory, Oxford, U.K.

ABSTRACT

The present status of the phenomenological neutron optical potential is briefly reviewed, with particular attention to the dependence on the nuclear asymmetry. It is shown that the conventional parametrisation is inadequate and some possible improvements are discussed. These improvements are of two types: firstly a more flexible global parametrisation and secondly some ways of incorporating data relating to each particular nucleus. Two ways of doing this are proposed, using the RMS matter radius to improve the real potential and the nuclear level density to improve the imaginary potential.

1. INTRODUCTION

Many studies of neutron scattering by nuclei have shown that it is possible to reproduce the cross-sections and polarisations to a high degree of accuracy by a suitably parametrised one-body optical potential. It is also found that the parameters of this potential are very similar for a wide range of energies and nuclei, and that the variations from the average values have a dependence on E, A and Z that can be easily represented by simple expressions. Several parameter sets have been proposed that give a good global fit to a wide range of neutron elastic scattering data. Such global potentials usually refer to nuclei at or near closed shells, since most of the available data are for such nuclei. In other regions of the periodic table the strong coupling to low-lying collective states affects the elastic scattering and hence the optical potential used to describe it. Sometimes the coupling is quite similar for a range of nuclei, such as the actinides, and it is then possible to find a regional potential that fits the scattering for these nuclei.

Many potentials of these types have been reviewed recently (Hodgson, [1]), and will not be discussed again here. Instead, attention will be directed to a new aspect of the optical potential that could well become important as the new and accurate neutron data are analysed. This concerns the deviations from smooth global behaviour that are attributable to the details of the structure of the nuclei concerned, and will be referred to as the fine structure of the nucleon optical potential.

2. THE FINE STRUCTURE OF THE OPTICAL POTENTIAL

The scattering of a neutron by a nucleus is the result of a very complicated series of interactions between the neutron and the nucleons of the target nucleus. It is rather surprising that this can be represented at all by a simple optical potential, and as we increase the accuracy of our measurements we expect the inadequacies of a global potential to become increasingly evident. In each case these inadequacies can to some extent be rectified by further adjustment of parameters but then we lose the advantages of a global potential.

The question we want to examine is whether these divergences from a global potential can themselves be parametrised in terms of some feature of the target nucleus. If this proves possible, then we have recovered the advantages of a global potential and increased the accuracy of its fits to the data although this must be paid for by some additional complexity.

It is usual in optical model analyses to fix the form factor parameters to average values, and to fit the data by adjusting the potential depths U, W and U_s. These may then be parametrised to allow for energy, Coulomb and asymmetry effects. For the real part of the potential this is usually done by the expressions

$$U_p = U_o - \varepsilon E + \alpha U_1 + U_c \qquad (2.1)$$

$$U_n = U_o - \varepsilon E - \alpha U_1$$

where the subscripts refer to protons and neutrons, $\alpha = (N-Z)/A$ is the nuclear asymmetry, U_1 is the asymmetry potential and $U_c \approx 0.4Z/A^{1/3}$ the Coulomb correction. Similar expressions may be used for W and U_s. It should be particularly noted that U_1 includes geometrical as well as isospin effects.

The question just posed can be specified more precisely in the framework of the relations (2.1). If, for example, we plot U_p or U_n as a function of α at a fixed energy, then do the points corresponding to analysis of data relating to particular nuclei fall in a straight line as required by (2.1)? Or do they fall on a more complicated family of curves that may themselves be parametrised? Finally, may the deviations from these curves themselves be related to the structure of the nuclei?

Some progress has already been made towards answering these questions, and this work is described in sections 3 and 4. It will become apparent that the equations (2.1) are perhaps not the best way of parametrising the optical potential, and that the whole question needs to be radically re-thought in a more fundamental way. In particular, it seems desirable to allow the radius and surface diffuseness parameters to depend on N and Z, and some preliminary work on this is described in section 5.

It is important to recognise the existence of this fine structure for several reasons. Firstly the use of a parametrisation of the potential that includes the fine structure will make possible more accurate fits to neutron data. Secondly it strengthens the links between the nucleon optical potentials and nuclear structure, and thus puts them on a sounder physical basis. Finally it is relevant to global optical potentials and in particular shows how to obtain the neutron potential from the proton potential. It is not correct to do this simply by reversing the sign of the asymmetry term and omitting the Coulomb term U_c.

Several studies of nucleon optical potentials have concentrated attention on the volume integrals of the potential, as these are believed to be less sensitive to the details of the form factor. However the presence of fine structure in the volume integrals does not necessarily imply fine structure in the potential depths. This may be seen by considering the expression for the volume integral

$$J = U(A,Z,N-Z) \int f(R,a,r) d\underline{r} \qquad (2.2)$$

In most optical model calculations, the radius parameter R is taken proportional to $A^{1/3}$. However if the A dependence of U is separable, then it is possible in principle to allow R, and indeed the diffuseness parameter a as well, to depend on A in such a way that the A-dependence of U is removed. If this is done, then U shows no fine structure, while fine structure remains in the volume integral J. A particular example of this will be discussed in section 4. Such potentials have the great advantage that the isospin and geometrical contributions to the asymmetry dependence of the potential are clearly separated.

3. EVIDENCE FOR FINE STRUCTURE OF PROTON POTENTIAL

Proton data are generally more accurate than neutron data, and so evidence for fine structure is more likely to be found for protons. Many analyses of proton elastic scattering data have been made, and the resulting potentials fitted by the expression (2.1). It is found in many cases that the potentials vary linearly with asymmetry parameter and the slope gives the potential U_1 (Perey, [2]).

There are however two analyses that show a more complicated dependence of the potential on the asymmetry factor, namely that of Perey and Perey [3] at 11 MeV and that of Noro et al [4] at 65 MeV. Perey and Perey found that the values of the real potential depth obtained from analysis with the same form factor lie on a family of lines of constant isospin $T=\frac{1}{2}(N-Z)$ when plotted as a function of nuclear asymmetry $\alpha=(N-Z)/A$. Noro et al found that the volume integrals of the real potentials lie on a family of curves of constant Z when plotted as a function of α.

Examination of these two plots, shown in figures 3.1 and 3.2, shows that the points of Perey and Perey also lie on lines of constant Z, and that the points of Noro et al also lie on lines of constant T. Since the volume integral of the real potential is proportional to the potential depth, these two investigations reach essentially the same conclusion, namely that the real potential depth is a function of T, Z and α.

$$U_p(T,Z,\alpha) \equiv U_p(\tfrac{1}{2}(N-Z),Z,\tfrac{N-Z}{A}) \equiv U_p(A,Z,(N-Z)) \qquad (3.1)$$

It is useful to represent this dependence by a convenient parametrised formula. The potentials of Perey and Perey may be represented by the expression (Hodgson[5])

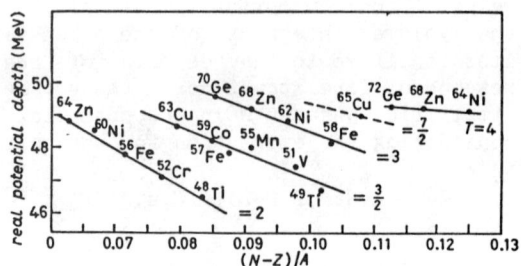

Figure 3.1. Real potential depths as a function of the nuclear-symmetry parameter α = (N-Z)/A for 11 MeV protons elastically scattered by a range of nuclei. The results indicated by solid points correspond to the geometry r_U = 1.285, a = 0.65 (Perey and Perey 3).

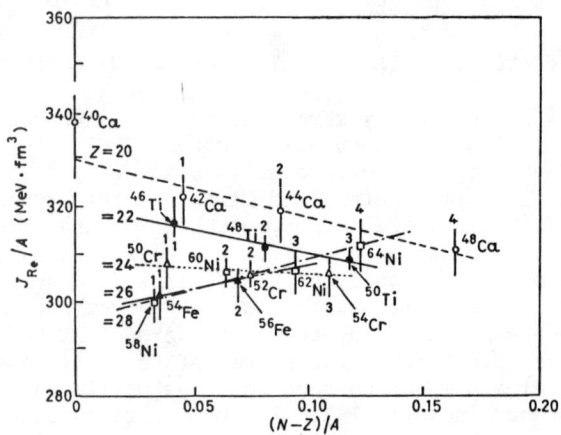

Figure 3.2. The volume integral per nucleon of the real part of the optical potential as a function of (N-Z)/A. Each line represents the linear fit to the points for nuclei with the same Z. The T-values 1, 2, 3, 4 are indicated above and below the bars (Noro et al [4]).

$$U_{p1} = \frac{106.7A}{R^3(1+\pi^2 a^2/R^2)} + 27\frac{N-Z}{A} + 0.4Z/A^{1/3} \qquad (3.2)$$

where

$$R = 1.204A^{1/3} + 0.305 \qquad (3.3)$$

In this expression the first term ensures the approximate constancy of the volume integral, the second is the isospin term and the third the Coulomb correction. As it is preferable to use a simpler expression, a fit was attempted with

$$U_{p2} = U_o + \beta A^{-1/3} + U_1 \frac{N-Z}{A} + 0.4Z/A^{1/3} \qquad (3.4)$$

which is obtained from (3.2) and (3.3) by expanding in powers of $A^{-1/3}$ and ignoring higher order terms. The fit obtained with this expression is shown in Figure 3.3. Fits of essentially the same quality can be obtained with similar expressions with terms proportional to $A^{1/3}$ or to A. It is desirable to find an expression with a firmer physical basis, and this is discussed further in section 5.

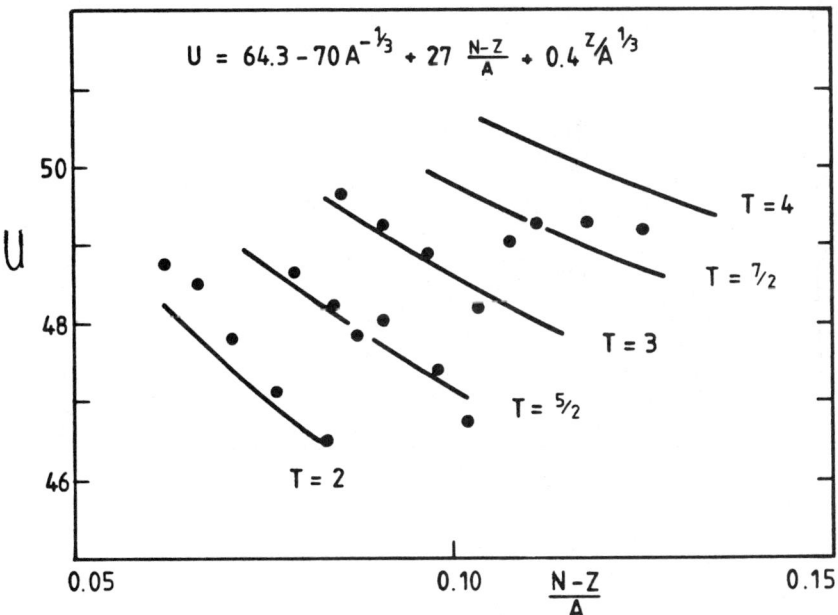

Figure 3.3. Real potential depths of Perey and Perey 3 as a function of nuclear asymmetry parameter for 11 MeV protons compared with a simple formula with a term proportional to $A^{-1/3}$.

4. FINE STRUCTURE OF THE NEUTRON POTENTIAL

Since fine structure is certainly found in the proton optical potential, it is very likely that it also occurs in the neutron potential, but has not yet been found due to the lower statistical accuracy of the neutron data.

The first attempt to find evidence for fine structure in the neutron optical potential used data on total cross-sections, since these are unaffected by the energy averaging. This ensures that the analysis is not affected by the presence of the compound nucleus contributions to the cross-section that are often important at energies of a few MeV. Unfortunately the total cross-sections fluctuate appreciably even at around 5 MeV so that the experimental energy-averaged total cross-sections cannot be obtained with sufficient accuracy to enable the fine structure in the potentials to be detected. An additional complication is that most experimental total cross-sections are available only for the natural isotopic mixtures.

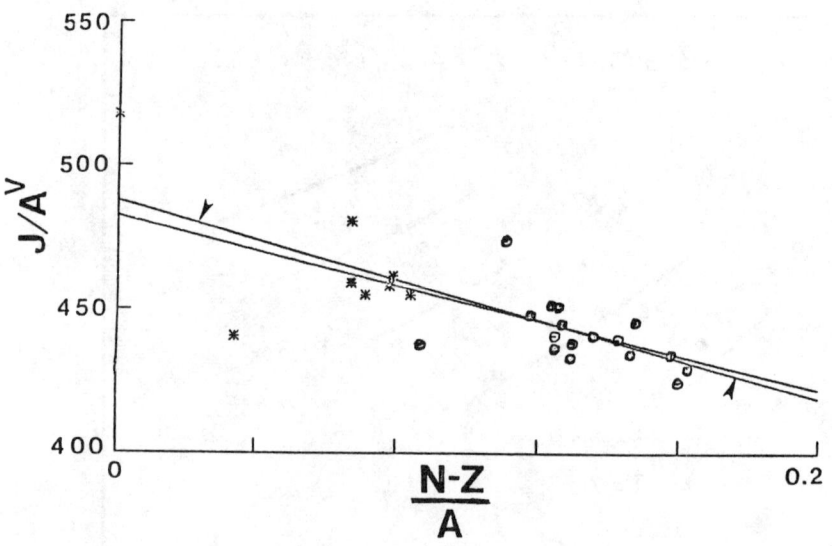

Figure 4.1. Volume integrals per nucleon of the neutron optical potentials as a function of (N-Z)/A. The symbols are .=present work, θ and *=previous work. The lines are least-squares fits to the data points (Smith et al [6]).

The second attempt was based on some very detailed analyses of 1.5 to 4 MeV neutron elastic scattering by Smith et al [6]. The potentials they obtained are shown in Figure 4.1; they vary overall linearly with the asymmetry parameter, with some deviations. The question is whether such deviations are attributable to experimental uncertainties or are indications of fine structure.

Smith et al fitted their data with a regional optical potential:

$$U = 52.58 - 0.3E - 30 \frac{N-Z}{A}; \quad r_u = 1.131 - 0.00107A;$$

$$a_u = 1.203 - 0.00511A$$

$$W = 11.7 - 25\frac{N-Z}{A} - 1.8\cos\{2\pi(A-90)/29\}; \quad r_W = 2.028 - 0.00683A;$$

$$a_W = 0.00551A - 0.1061$$

The real potential has the Saxon-Woods form and the imaginary potential the Saxon-Woods derivative form, with $R=r_i A^{1/3}$. The spin-orbit potential is real, with depth 6 MeV and Thomas form. This potential is applicable only for nuclei with $A\approx 85-125$ and $E_n < 5$ MeV.

There are several notable features of this potential. Firstly the radius and diffuseness parameters depend on A. This automatically introduces a fine structure into the plot of J against (N-Z)/A, as shown in Figure 4.2. There is however no fine structure in the real potential depth U. This shows that the form taken by the fine structure depends on the parametrisation chosen, and raises the important question whether there is an optimum parametrisation and if so how can it be found. This will be further discussed in section 5.

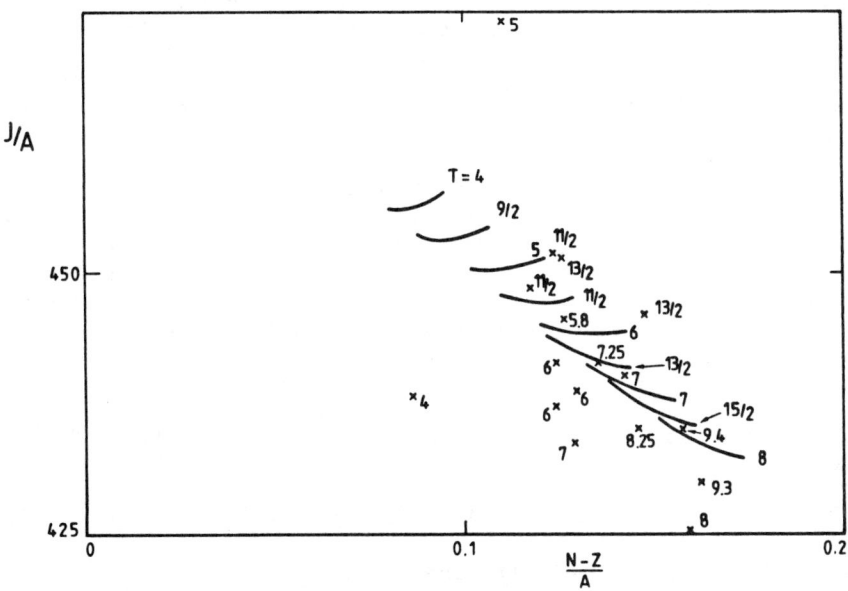

Figure 4.2. Volume integrals per nucleon of the neutron optical potentials of Smith et al [5] compared with values calculated from the global potential.

Examination of Figure 4.2 shows that indeed the global formula is in overall agreement with the data but the scatter of the individual points obscures the fine structure. At preseent one must conclude that the presence of fine structure in the neutron optical potential is not established.

The second notable feature of the potential is the cosine term appearing in the expression for W. This represents in an approximate fashion the variation of W in the region of the shell closure at A=90, and is an example of a parametrisation of the optical potential, valid for a limited range of nuclei, that relates the potential to a feature of nuclear structure. The question whether this can be done in more detail, with consequential improvement of the potential, is also considered in section 5.

5. PHYSICAL BASIS FOR FINE STRUCTURE

The results presented in the preceding sections show that it is desirable to choose the parametrisation of the optical potential to simplify as far as possible the fine structure. A particular example of such a potential is that of Smith et al in which the radius and diffuseness parameters depend on A in such a way that there is no fine structure in the real potential depth although it does, of course, appear in the volume integral.

The expressions of Smith are purely phenomenological and are valid in a limited region of the periodic table. The question addressed in this section is whether more general relations can be obtained on physical grounds, in particular by using the relation between the optical potential V(r) and the nuclear matter distribution $\rho(r)$

$$V(r) = \int \rho(r')v(r,r')dr' \qquad (5.1)$$

We also use the result of Myers [7] that the equivalent sharp radius R_V of the real part of the optical potential extends a fixed amount Δ beyond the equivalent sharp radius R_ρ of the nuclear density distribution. Then, using the results of Srivastava et al [8]

$$R_V = R_\rho + \Delta \approx 1.13A^{1/3} + 0.55 \qquad (5.2)$$

This difference is a result of the density dependence of the effective interaction

$$v(r,r') = v(|\underline{r}-\underline{r}'|)\{1-\beta\rho^{2/3}[\tfrac{1}{2}(\underline{r}+\underline{r}')]\} \qquad (5.3)$$

If the optical potential has the form

$$U(r) = \frac{U_o}{1+\exp(\frac{r-R}{a})} \qquad (5.4)$$

then the radius R_V and mean square radius Q_V of the equivalent sharp distribution are given by

$$R_v^2 = R^2(1 + \frac{2}{3}\frac{\pi^2 a^2}{R^2}) \tag{5.5}$$

and

$$Q_v^2 = R^2(1 + \frac{7}{3}\frac{\pi^2 a^2}{R^2}) \tag{5.6}$$

Hence

$$R^2 = \frac{7}{5}R_v^2 - \frac{2}{5}Q_v^2 \tag{5.7}$$

and

$$a^2 = \frac{3}{5\pi^2}(Q_v^2 - R_v^2) \tag{5.8}$$

These relations can be used to express R and a in terms of A.
To do this, we use (5.2) together with the relations

$$Q_v^2 = \frac{5}{3}\langle r^2 \rangle_v = \frac{5}{3}[\langle r^2 \rangle_\rho + \langle r^2 \rangle_{eff}] \tag{5.9}$$

and

$$\langle r^2 \rangle_{eff} = C_o + C_1 A^{1/3} + C_2 A^{-1/3} \tag{5.10}$$

The mean square radius $\langle r^2 \rangle_\rho$ of the nuclear matter distribution is expressed in terms of its equivalent sharp radius R_ρ and diffuseness a_ρ

$$\langle r^2 \rangle_\rho = \frac{3}{5}Q_\rho^2 = \frac{3}{5}R_\rho^2 + \pi^2 a_\rho^2 \tag{5.11}$$

Making use of the numerical values of R_p, a_p, C_o, C_1 and C_2 given by Srivastava et al [9] gives the required A-dependence of the radius and diffuseness of the optical potential

$$R^2 = 1.277 A^{2/3} - 3.18 + 1.24 A^{1/3} - 0.38 A^{-1/3} \tag{5.12}$$

$$a^2 = 0.5295 + 0.057 A^{-1/3} \tag{5.13}$$

These are plotted in Figure 5.1.

These expressions for R and a give the overall variation of the form factor with A, so if a global optical model analysis were made using them it is possible that the fine structure of the type described in sections 2-4 would not appear. Further elaboration of these expressions for the form factor parameters could be made by also allowing them to depend on N or Z as well as A.

It is however unlikely that such elaboration would be worthwhile because such parametrisations take no account of shell effects and other variations of nuclear structure. They are therefore likely to be of use only in very limited regions of the periodic table for groups of nuclei of similar properties. To illustrate the importance

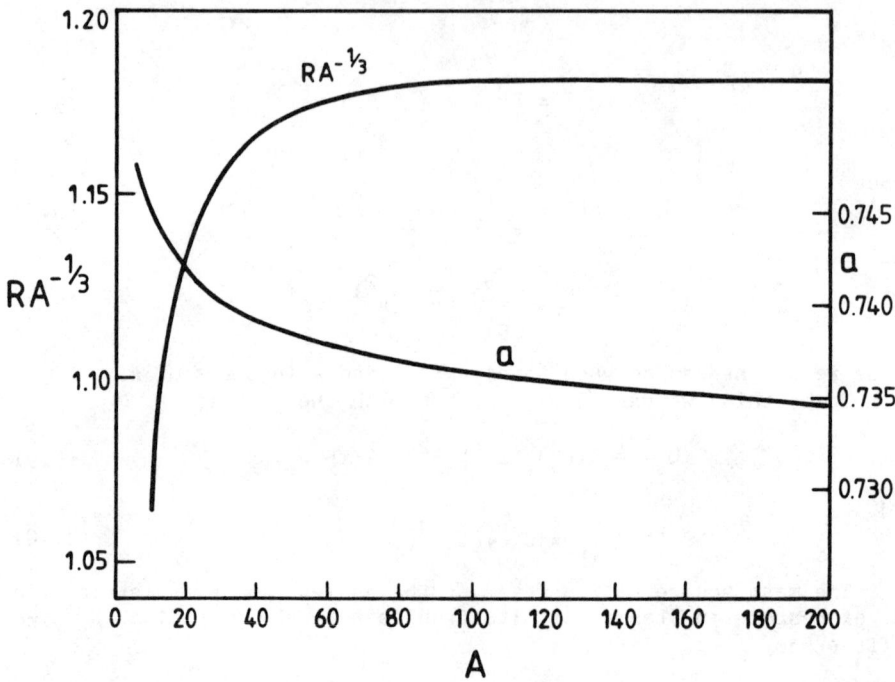

Figure 5.1. Variation of $RA^{-1/3}$ and a with A according to (5.12) and (5.13).

of shell effects Figure 5.2 shows the differences between experimental RMS charge radii and those calculated from the single-particle potential model using a Saxon-Woods potential with parameters adjusted to fit the charged distributions of selected nuclei and varying smoothly with A (Brown, Bronk and Hodgson, [10]). While the deviations are quite small, it is clearly impracticable to parametrize them in a manageable way. We therefore seek a way of defining the optical potential that makes use of some experimental data for the nucleus of interest.

This may be done by expressing the radius R in terms of the equivalent sharp radius of the matter distribution R_ρ. Using the above relations gives

$$R^2 = R_\rho^2 + 1.54 R_\rho - 3.1805 - 0.4973 A^{1/3} - 0.3720 A^{-1/3} \quad (5.14)$$

Unfortunately the matter radius is not so well known as the charge radius, but it should be possible to overcome this difficulty by using the charge radius, which is accurately known for many

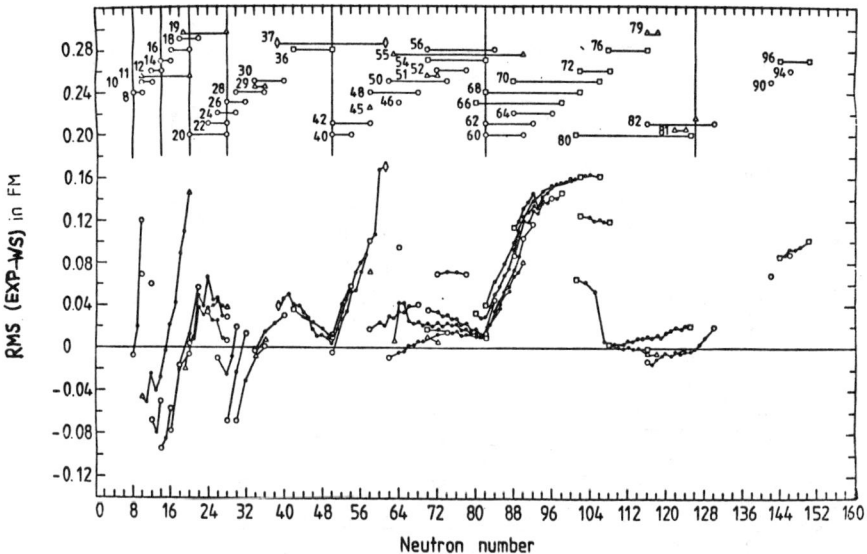

Figure 5.2. Differences between experimental RMS charge radii and theoretical values calculated from the Saxon-Woods potential model. Isotopic chains are connected by lines with special symbols at the ends. Circles are used for the even-Z isotopes with experimental absolute normalisation, squares for the even Z isotopes with arbitrary normalisation, triangles for the odd Z isotopes with experimental absolute normalisation and diamonds for the odd Z isotopes with arbitrary normalisation (Brown, Bronk and Hodgson, [10]).

nuclei, and correcting for the difference between the charge and matter radii using the single-particle potential model as parametrised by Streets et al [11], or Skyrme Hartree-Fock calculations with parameters adjusted to give the nuclear single-particle energies. Calculations along these lines are in progress.

A more fundamental way of tackling this problem is to calculate the optical potential from first order Brueckner theory. This method has been used by Haider et al [12] to analyse the data of Noro et al [4], with considerable success. Such calculations are however very time-consuming, so the simpler methods discussed here may still be useful.

The imaginary part of the optical potential is even more strongly related to the shell structure of the target nucleus. This is shown, for example, by the abnormally low values of the absorption cross-sections for low energy neutrons in the regions of shell closure (Perey and Buck, [13]). This is essentially a consequence of the reduced level density in such nuclei, so it is suggested that it might be possible to find a way of obtaining improved values of the imaginary part of the optical potential by using the appropriate nuclear level densities. Such a method has already been used with some success by Quesada et al [14] to determine the imaginary potential for heavy-ion scattering.

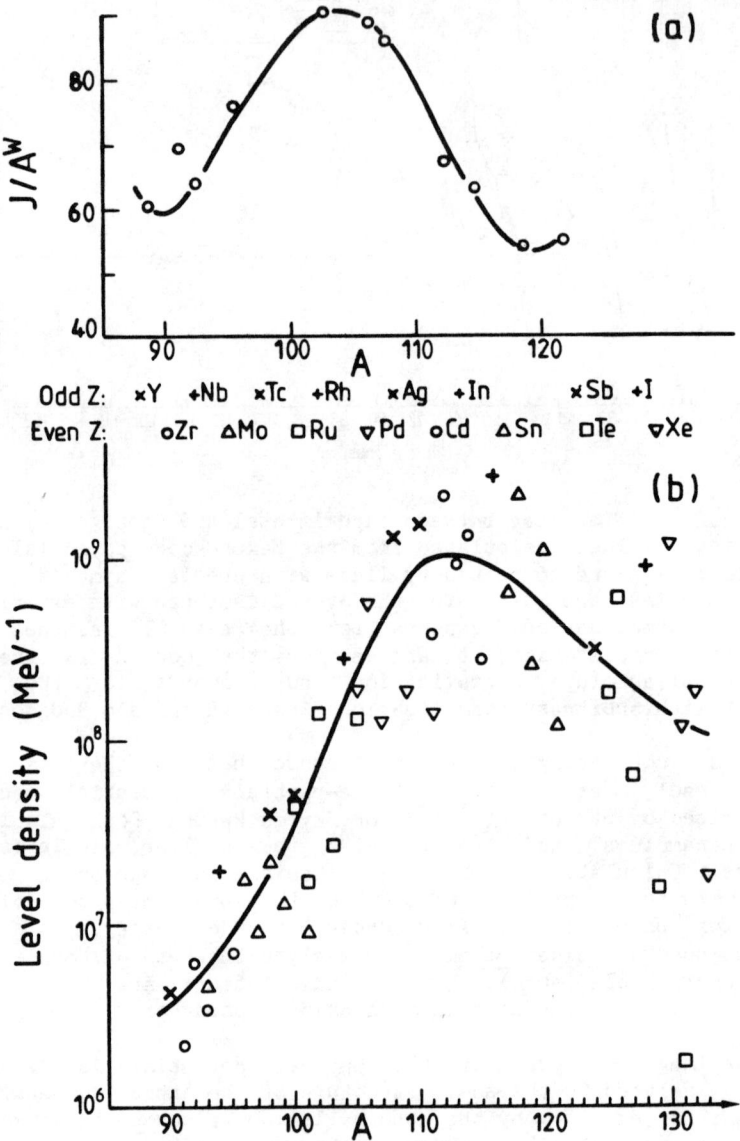

Figure 5.3. (a) Volume integrals of imaginary potential as a function of A (Smith et al [6]. (b) Nuclear level density at 4 MeV as a function of A calculated from the formula of Brancazio and Cameron [16]. The line separates the points corresponding to nuclei with odd Z and even Z (Madurga [15]).

To see if this approach might be applicable to neutron scattering, Madurga [15] has calculated the level densities of the nuclei studied by Smith et al, and the results are compared with the imaginary potentials of these authors in Figure 5.3. In these calculations Madurga used the level density formula of Brancazio and Cameron [16] which contains shell corrections and evaluated them for an incident neutron energy of 4 MeV. It is notable that the two curves peak at approximately the same value of A, so it seems likely that it will prove possible to obtain improved values of the imaginary neutron potential from the nuclear level density parameter.

ACKNOWLEDGEMENTS

I am grateful to Miss G.M. Field for the calculations relating to the fine structure of the proton and neutron potentials, to Dr. D.K. Srivastava for help in the derivation of the expressions (5.12) and (5.13), to Professor G. Madurga for permitting me to quote his work on the imaginary potential, and to Mrs J. Long for some numerical calculations.

REFERENCES

1. P.E. Hodgson, Rep.Prog.Phys. 47, 613, (1984).
2. F.G. Perey, Phys.Rev. 131, 745, (1963).
3. C.M. Perey and F.G. Perey, Phys.Lett. B26, 123, (1968).
4. T. Noro, H. Sakaguchi, M. Nakamura, K. Hatamaka, F. Ohtami, H. Sakamoto and S. Kobayashi, Nucl.Phys. A366, 189, (1981).
5. P.E. Hodgson, Nucl.Phys. A150, 1, (1970).
6. A.B. Smith, P.T. Guenther and J.F. Whalen, Argonne Preprint, (1983).
7. W.D. Myers, Nucl.Phys. A204, 465, (1973).
8. D.K. Srivastava, N.K. Ganguly and P.E. Hodgson, Phys.Lett. 51B, 439, (1974).
9. D.K. Srivastava, D.N. Basu and N.K. Ganguly, Phys.Lett. 124B, 6, (1983).
10. B.A. Brown, C. Bronk and P.E. Hodgson, J.Phys.G.(in press), (1984).
11. J. Streets, B.A. Brown and P.E. Hodgson, J.Phys. G8, 839, (1982).
12. W. Haider, A.M. Kobos, J.R. Rook and K.F. Pal, Nucl.Phys. A419, 521, (1983).
13. F.G. Perey and B. Buck, Nucl.Phys. 32, 353, (1962).
14. J.M. Quesnda, M. Lozana and G. Madurga, Phys.Lett. 125, 14, (1983).
15. G. Madurga, (private communication), 1984.
16. P.J. Brancazio and A.G.W. Cameron, Can.J.Phys. 47, 1029, (1969).

IMPLICATION OF MICROSCOPIC REACTION ANALYSES BELOW 1oo MEV

H.V. von Geramb
Theoretische Kernphysik
Universität Hamburg
Luruper Chaussee 149, 2ooo Hamburg 5o, W.-Germany

ABSTRACT

Elastic scattering of neutrons, protons and antiprotons from complex nuclei is microscopically analysed with Brückner's reaction matrix as effective interaction. This interaction is generated with the Paris (NN) and (N$\bar{\text{N}}$) potentials as fundamental input and it houses medium effects such as density and energy dependencies. Here we address ourselves to energies between 25 and 1oo MeV and discuss the relevant aspects for the three different projectiles. Properties of the effective interaction play thereby a central role.

INTRODUCTION

The non-relativistic *nuclear matter approach* is now generally accepted as a method to describe the global features of the complex optical model potential (OMP) for nucleon-nucleus scattering [1]. For the energy range 1oo-4oo MeV, this approach fits over 8 orders of magnitude the experimental data without adjustable parameters.[2] Towards lower energies fits are less impressive or rely on some adjustments [3,4]. The literature which accumulated in the last years, however, lends full support to the usefulness of this theoretical approach.[5]

During the last year the antiproton storage ring LEAR at CERN went in operation and this allow us to analyse consistently neutron, proton and antiproton data within the same theoretical framework.[6]

The nuclear matter approach identifies the local scattering in a nucleus with the situation in an infinite medium and traces the bulk properties of the OMP back to the features of the two body reaction matrix. This reflects the properties of the two nucleon potential with its energy dependence and spin/isospin dependencies in a wide sense (central, tensor, spin orbit etc. interaction) as well as medium effects.

With this talk I shall concentrate on features associated with the energy dependence of the parametrized reaction matrix below 1oo MeV for nucleons and antinucleons. The aim of this exposé is to find an energy region which may be favoured for a comparison between the three projectiles. Since antiprotons are distinguished from nucleons, antisymmetrization of the projectile - target wavefunction is not needed. Contrary to antiproton scattering, for nucleon scattering antisymmetrization is very important and the exchange amplitudes are the source of non-locality in the microscopic OMP. Its minimization is the problem.

THEORY AND DISCUSSION

The theoretical basis of the microscopic OMP was laid by Brueckner and has been reviewed by several authors.[3,5] The approach identifies the single particle mass operator with the OMP in infinite nuclear matter in lowest order of the hole line expansion. Driving operator is the Brueckner reaction matrix which we compute with the Paris NN and NN̄ potential.[7] The Pauli blocking operator limits nucleons (but not antinucleons) to propagate above the Fermi sea. Technical details how to solve the Bethe-Goldstone equation and how to generate the effective complex density and energy dependent operator in the form of a two body potential with central, spin orbit and tensor components

$$V(r) = \sum_{ST} V_0^{ST}(r) P^S P^T + \sum_T V_1^T(r)(L \cdot S) + V_2^T(r) S_{12} \quad (1)$$

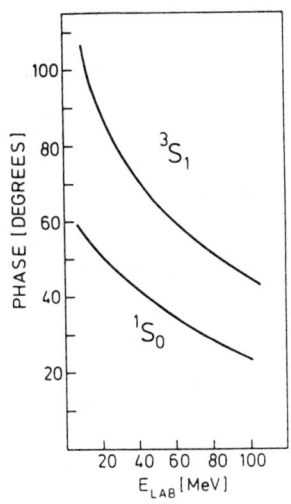

Fig. 1

can be found elsewhere. Tables of these interactions are available for (NN) and (NN̄) systems for energies up to 4oo MeV.[8]

In the low energy region (< 1oo MeV) the $^3S_1$ and $^1S_0$, Fig. 1 phase shifts vary most rapidly. This is the source of a rapid variation of the triplet even (S=1, T=o) and singlet even (S=o, T=1) component of the central interaction. In Fig. 2 this can be seen, where the volume integrals are shown for the real and imaginary parts.

$$4\pi \int V_0^{ST}(r) r^2 dr \quad (2)$$

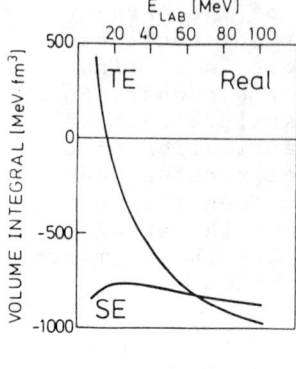

Fig. 2: Energy dependence of the volume integrals of the effective interactions in the limiting case $k_F \to 0$.

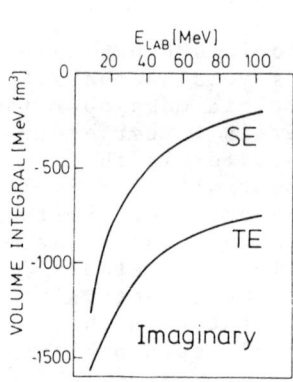

Figs. 3 and 4 display the importance of density dependence. The solid line is computed with the free t-matrix, whereas the dashed line is based on the energy and density dependent interaction. The shaded area indicates medium corrections.

Fig. 4

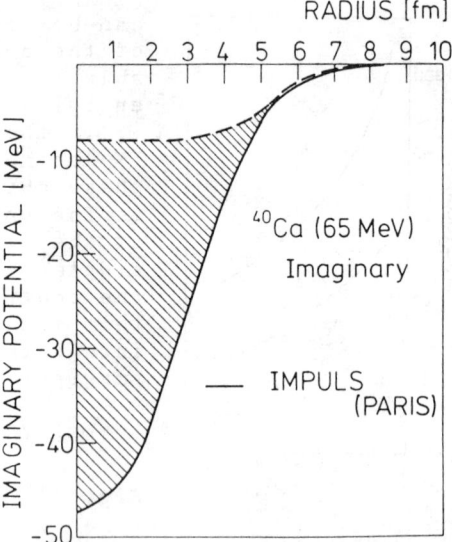

Fig. 3

The energy dependence is severe with the exception of the SE real potential. To appreciate the density effects of the medium upon the two body scattering mechanism Figs. 3 and 4 show the theoretical OMP computed (see expression for the local optical model potential below) with the free ($k_F = 0$) scattering amplitude (solid line) and with the reaction matrix (dashed line). The real part of the OMP changes comparably little up to approximately 80% density whereas the imaginary potential already starts to change within the radius of 10% nuclear density. This 65 MeV example remains qualitatively unaltered towards lower energies, with the restriction that the imaginary potential is even more strongly suppressed yielding the well known surface absorption. These features are in accordance with phenomenological OMP and they establish the success of the microscopic OMP in the nuclear matter approach. Quantitatively, the choice of the nucleon-nucleon potential matters. Furthermore, the self-consistent choice of the single particle energies in the Bethe-Goldstone propagator is important and towards lower energies essential. Numerically this is not a trivial task and we are aware that present low energy calculations of the OMP hinge on this lack. Improvements are anticipated but for the present discussion this should be of no importance. Before we continue with the discussion of the properties of the effective interaction some relevant expressions for the folded OMP are recalled.[2,3]

The complex OMP is a convolution of the central and spin orbit effective interaction respectively. It contains the local direct and the localized knock-on exchange potentials which may be computed in coordinate space or momentum space. The following equations hold for the neutron optical potentials. For protons interchange $p \leftrightarrows n$. The antiproton OMP is obtained with the expression for the neutron OMP but without knock-on exchange contributions.
The following abbreviations for standard Fourier transforms are used

$$\tilde{\rho}(k) = 4\pi \int_0^\infty \rho(x) j_0(kx) x^2 dx \qquad (3)$$

$$\tilde{V}_{nn}^D = \tilde{V}_{pp}^D(k) = \pi \int_0^\infty (V^{SE}(x) + 3V^{TO}(x)) j_0(kx) x^2 dx \qquad (4)$$

$$\tilde{V}_{np}^D = \tilde{V}_{pn}^D(k) = \frac{\pi}{2} \int_0^\infty (V^{SO}(x) + 3V^{TE}(x) + V^{SE}(x) + 3V^{TO}(x)) j_0(kx) x^2 dx \qquad (5)$$

$$\tilde{V}_{nn}^E = \tilde{V}_{pp}^E(k) = \pi \int_0^\infty (V^{SE}(x) - 3V^{TO}(x)) Q(x, \kappa, k_F) j_0(kx/2) x^2 dx \quad (6)$$

$$\tilde{V}_{np}^E = \tilde{V}_{pn}^E(k) = \frac{\pi}{2} \int_0^\infty (3V^{TE}(x) - V^{SO}(x) + V^{SE}(x) - 3V^{TO}(x)) \times \\ Q(x, \kappa, k_F) j_0(kx/2) x^2 dx \quad (7)$$

$$Q(x, \kappa, k_F) = j_0(x\kappa(r)) \times \frac{3j_1(xk_F(r))}{xk_F(r)} \quad (8)$$

$$k_F(r) = \left[\frac{3\pi^2}{2}(\rho_p(r) + \rho_n(r))\right]^{1/3} \quad (9)$$

$$\kappa(r) = \sqrt{2m(E - U(r) - V_C(r))}/\hbar \quad , \quad \varepsilon(r) = \mathrm{Re}(E - U(r) - V_C(r)) \quad (10)$$

$$U_n(r, E) = \int d^3s \{V_{nn}^D(s, k_F(r), \varepsilon(r)) \rho_n(|\vec{r} - \vec{s}|) \\ + V_{np}^D(s, k_F(r), \varepsilon(r)) \rho_p(|\vec{r} - \vec{s}|)\} \quad (11)$$

$$+ \int d^3s \{V_{nn}^E(s, k_F(r), \varepsilon(r)) \rho_n(|\vec{r} + \vec{s}/2|) \\ + V_{np}^E(s, k_F(r), \varepsilon(r)) \rho_p(|\vec{r} + \vec{s}/2|)\}$$

$$\times \frac{3j_1(k_F(r)s)}{k_F(r)s} \cdot j_0(\kappa(r)s)$$

$$= \frac{1}{2\pi^2} \int_0^\infty k^2 dk \, j_0(kr) \{(\tilde{V}_{nn}^D(k) + \tilde{V}_{nn}^E(k)) \tilde{\rho}_n(k) \quad (11a) \\ + (\tilde{V}_{np}^D(k) + \tilde{V}_{np}^E(k)) \tilde{\rho}_p(k)\}$$

In momentum space, this expression shows a direct product of the effective interaction with the density and the OMP is a weighted form of the effective interaction. As weights act the nuclear point densities in momentum space. They are smooth functions with ρ_p extracted from electron scattering and ρ_n fitted to nucleon nucleus scattering. Indeed, many elastic scattering experiments are often motivated with the aim to determine matter distributions. However, the qualitiy of this result depends crucially on the effective interaction and used approximations. In comparing neutrons, protons and anti-

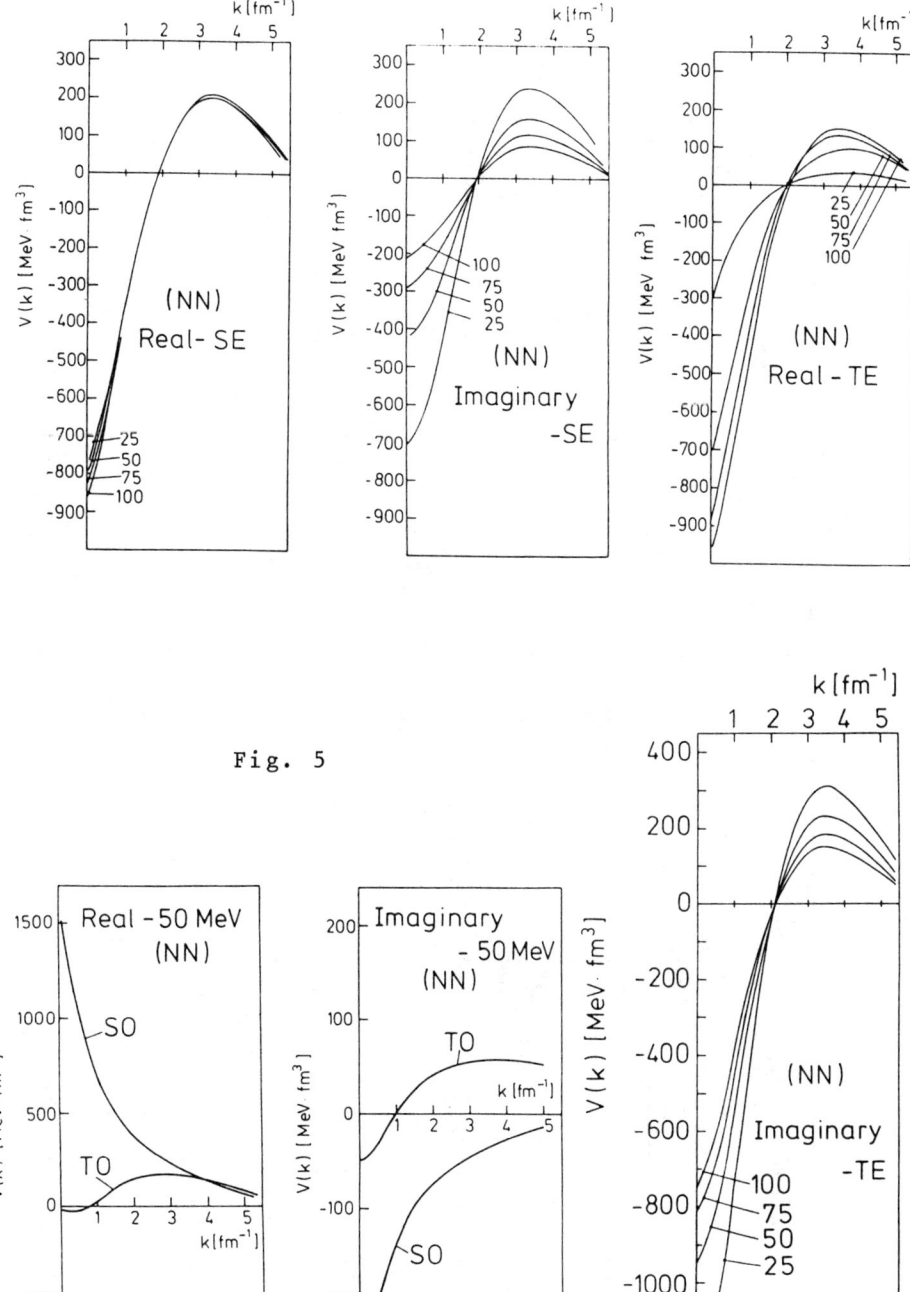

Fig. 5

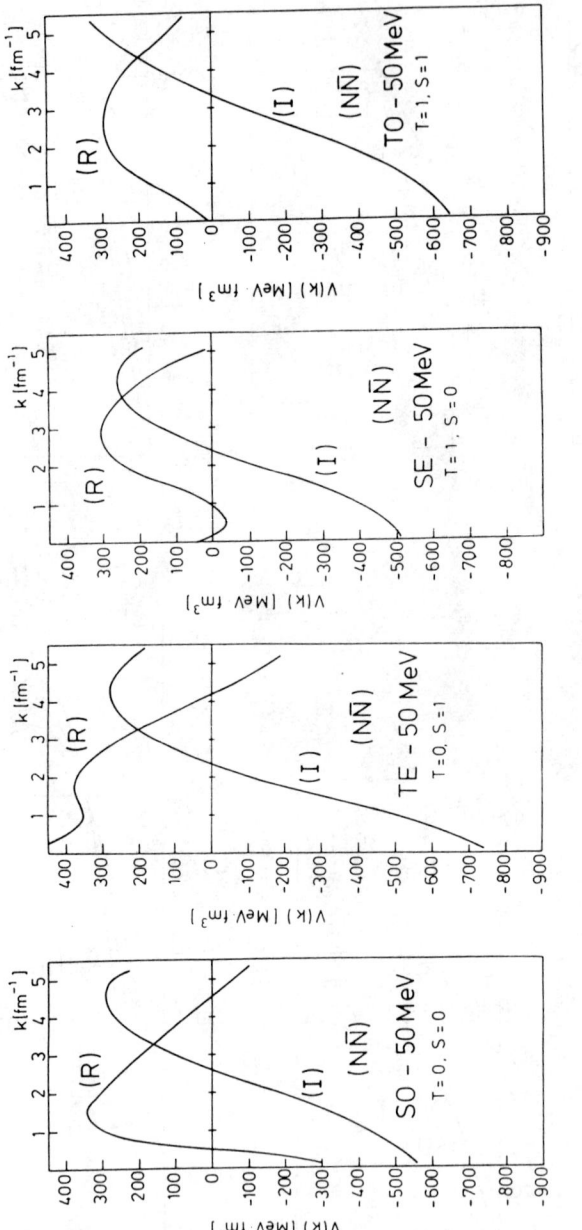

Fig. 6

protons it appears therefore obvious to find kinematic regions where the knock-on exchange contributions are minimal.

To discuss this aspect the effective interactions are shown in Figs. 5 and 6 split into real/imaginary parts and various (NN)/(N$\bar{\text{N}}$) spin/isospin components. For some components the energy dependence is explicitly included. As a unique feature we observe the node, $k \sim 2\text{fm}^{-1}$, in the SE and TE channel for the real and imaginary parts. The (N$\bar{\text{N}}$) interaction shows for all channels a quite similar shape for the imaginary potentials. With these features in mind and using Eqs. 11, 11a we found the energy window, 6o - 65 MeV, where the (NN) SE and TE exchange contributions are minimal. The density dependence of the interaction hinders it from being exactly zero. The odd state interaction, SO and TO, is comparable short ranged and direct/exchange contributions have opposite signs. In the limit of a δ-function interaction in R-space (constant in momentum space) they cancel to zero. On the basis of these results the 6o - 65 MeV energy window is optimal. All arguments put forward for studies of elastic scattering hold invariantly for inelastic scattering to normal parity states. Non normal parity transitions are dominated by the non central forces such as spin orbit and tensor forces. However, with the use of microscopic DWTA distorted waves may be generated optimally in the above mentioned energy window and therefore have a smaller uncertainty.

A few months ago the first differential cross sections for antiproton scattering from carbon was obtained with the LEAR facility, CERN.[6] This represents a challenge to compare the theory of nucleon nucleus scattering. The semi-microscopic (N$\bar{\text{N}}$) potential from the Paris group is obtained by G-parity operation on the Paris (NN) potential together with a readjustment of the phenomenological core parameters and the introduction of an energy and state dependent absorption to account for the prevailing annihilation. As a first calculation, a low density approximation is used in which Pauli blocking for the valence nucleon is included and self-consistency neglected.

$$Q(q, K, k_F) = \begin{cases} 1 & \text{for } |K/2-q| > k_F \\ 0 & \text{for } K/2+q < k_F \end{cases}, \text{ else } ((K/2+q)^2 - k_F^2)/2Kq \quad (12)$$

The OMP uses Eq. 11 without exchange. Fig. 7 gives an impression of the radial shapes relevant for antiproton scattering. The imaginary potential is much larger as for nucleon scattering. The real central potential is repulsive in the centre and shows an attractive pocket in the surface region which will be enhanced by the addition of the attractive Coulomb potential. The imaginary potential is very absorptive and follows roughly the underlying density distribution of here ^{40}Ca. The spin orbit potentials have the same signs as those known from nucleon nucleus scattering and the imaginary part is substantially larger.

The results for the real potential are in contradiction to phenomenological analyses of atomic antiproton states and OMP analyses but they are in agreement with other t-matrix calculations in the non relativistic and relativistic frameworks.[9] The repulsive potential illustrates an important aspect of the nuclear matter approach since it is very sensitive to the essential difference between folding the free ($N\bar{N}$) potential or the reaction matrix. Its behaviour as well as similar features in the other potentials can be associated with the strong short ranged annihilation potential. It is easily shown that a purely (absorptive) imaginary potential generates a repulsive real OMP. Conversely taking only the real part of the Paris ($N\bar{N}$) potential as input results in an attractive real potential. This latter result was expected and underlies phenomenological fits. In Fig. 8 the differential cross section for 47 MeV scattering from carbon is compared with the microscopic OMP result. The imaginary OMP was thereby enhanced by 2o% which appears to be a short-coming of the Paris ($N\bar{N}$) potential which uses an extremely short range annihilation potential. In terms of the momentum space behaviour of the interaction shown in Fig. 6 a longer ranged annihilation potential is suggested with a node around $1.8f^{-1}$. For completeness a table of the effective interactions for nucleons and antinucleons (5o MeV) is included.

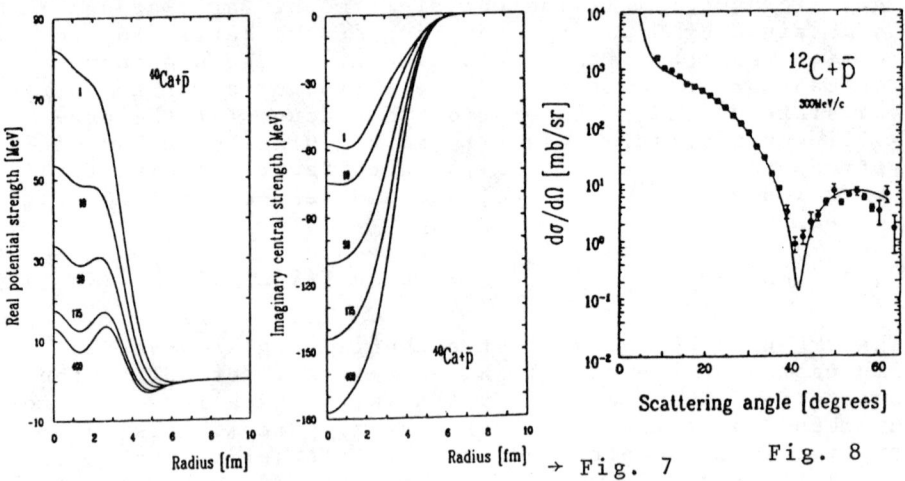

→ Fig. 7   Fig. 8

23

ENERGY = 50.000(MEV)   DENSITY = 0.0000(FM**-3)   KF = 0.0246(FM**-1)
COMPLEX NUCLEON NUCLEON EFFECTIVE INTERACTION HAMBURG 1984

	<<< T = 0 >>>		<<< T = 1 >>>	
	< REAL >	< IMAG >	< REAL >	< IMAG >

CENTRAL SINGLET INTERACTION

MASS	POTENTIAL STRENGTH, U		POTENTIAL STRENGTH, U	
0.69	0.422231280+02	0.557324150+01	0.142192950+02	0.724626080+00
3.10	0.739127270+02	-0.175924310+03	-0.341291430+03	-0.306391110+03
5.40	0.711369080+04	0.122897510+04	0.538149020+05	0.474541800+05
5.90	-0.770717370+04	-0.112466410+04	-0.534425540+05	-0.467673330+05

VOLUME INTEGRAL 1494.4   -253.49   -795.32   -420.14

CENTRAL TRIPLET INTERACTION

MASS	POTENTIAL STRENGTH, U		POTENTIAL STRENGTH, U	
0.69	-0.180865510+02	-0.554954000+01	-0.182144220+01	-0.247394370+01
3.10	0.201694360+04	0.495782080+04	-0.134805950+04	-0.119895060+03
5.40	0.307185640+05	0.726271830+05	0.198022510+05	0.164322570+04
5.90	-0.300762550+05	-0.712424280+05	-0.196584310+05	-0.164533110+05

VOLUME INTEGRAL -727.25   -919.81   218.91   -40.992

L-S INTERACTION

MASS	POTENTIAL STRENGTH, U		POTENTIAL STRENGTH, U	
0.76	-0.160888570+01	-0.202853260-01	0.163604460-01	0.113144910+01
1.60	-0.400675630+02	0.441880030+02	0.238693770-01	0.168537980+02
3.80	-0.411788310+04	0.101288620+04	0.239380570-03	-0.388255720+03
5.70	0.113564740+05	-0.199508420+04	0.609187710+04	0.665526540+03

VOLUME INTEGRAL 609.48   325.89   -166.93   35.282

TENSOR INTERACTION

MASS	POTENTIAL STRENGTH, U		POTENTIAL STRENGTH, U	
0.76	-0.268809480+00	0.557558850+00	0.395454350+00	0.956701720+00
1.60	-0.623843000+02	0.225177700+02	0.151490820+02	0.448612530+02
3.80	-0.498733830+04	-0.360064850+04	0.172788400+04	0.571033170+04
5.70	-0.262326820+05	0.139804930+05	-0.885317040+04	-0.363698810+05

VOLUME INTEGRAL -833.82   39.187   255.92   -819.45

CENTRAL, SPIN ORBIT = U*EXPT-T*MSS*R)/R   TENSOR = U*R*R*EXPT-T*MSS*R)/R

ENERGY (MEV) = 50.000   RHO (FM**-3) = 0.0001   KF (FM**-1) = 0.114
COMPLEX ANTINUCLEON NUCLEON EFFECTIVE INTERACTION

	<<< T = 0 >>>		<<< T = 1 >>>	
	< REAL >	< IMAG >	< REAL >	< IMAG >

CENTRAL SINGLET INTERACTION

MASS	POTENTIAL STRENGTH, U		POTENTIAL STRENGTH, U	
0.70	0.254288310+01	0.409729340+02	0.353708340+02	0.114251380+02
1.20	-0.516719660+02	-0.318678580+02	-0.169825000+03	-0.635268330+02
6.60	0.646634630+04	0.259924530+05	0.406623180+04	-0.220381820+05
8.40	-0.111844630+05	-0.306341550+06	0.416287190+05	0.346460280+05
	-0.167823210+05			

VOLUME INTEGRAL 415.97   -547.66   41.314   -498.66

CENTRAL TRIPLET INTERACTION

MASS	POTENTIAL STRENGTH, U		POTENTIAL STRENGTH, U	
0.90	-0.115840750+00	0.105360220+00	0.433762730-02	0.237427310-01
1.20	-0.133195750+02	-0.308801710-01	0.180528850+02	-0.235704630-01
6.60	0.345726690+05	-0.185541920+03	-0.772187000+03	0.241635810+05
8.40	-0.509454930+05	-0.327932000+04	-0.127193700+05	-0.204994580+05
		0.700593030+04	0.169391230+05	0.364747760+05

VOLUME INTEGRAL -333.93   -842.75   152.38   -603.30

L-S INTERACTION

MASS	POTENTIAL STRENGTH, U		POTENTIAL STRENGTH, U	
0.90	-0.433228180+01	0.308301170+01	0.230762730-02	-0.424767750-01
1.20	-0.208116550+02	-0.184152800+02	0.198415280+02	0.235109400+02
6.60	-0.345885270+05	0.449882770+05	0.243175200+05	0.204994580+05
8.40	-0.572141200+05	0.132750+06	0.175895500+05	-0.798080400+05
			0.421519120+05	0.310567620+05

VOLUME INTEGRAL -1141.0   3541.0   1203.2   -93.528

TENSOR INTERACTION

MASS	POTENTIAL STRENGTH, U		POTENTIAL STRENGTH, U	
0.90	-0.115840750+00	0.105360220+00	0.151770290+00	0.572177910+02
1.20	-0.796787830+00	0.388065710-02	-0.190799430+02	-0.267470120+02
6.60	-0.998675940+05	0.177659500+05	-0.174844930+05	-0.798080400+05
8.40	0.346602730+06	-0.682437300+05	0.178910610+05	0.298440010+06
			-0.688019970+05	

VOLUME INTEGRAL 2305.5   -3784.1   110.23   1520.3

Table: Example of (NN) and (NN̄) effective interactions based on the Paris potential.

REFERENCES

+ work supported by Bundesministerium für Forschung und Technologie

1. J. Hüfner and C. Mahaux, Ann.Phys. (N.Y.) 73, 525 (1972)
   J.P. Jeukenne, A. Lejeune and C. Mahaux, Phys.Rev. C25, 83 (1976) and Phys.Rep. 25C, 85 (1976)

2. H.V. von Geramb, *The interaction between medium energy nucleons in nuclei* - 1982, ed. H.O. Meyer, Americ. Inst. Phys. Conf. Proc., 97 (1983)
   L. Rikus, K. Nakano and H.V. von Geramb, Nucl.Phys. A414, 413 (1984)
   L. Rikus, H.V. von Geramb, Nucl.Phys. A426, 496 (1984)

3. F.A. Brieva and J.R. Rook, Nucl.Phys. A291, 299 (1977); ibid. A297, 2o6 (1978); ibid. A3o7, 493 (1978)
   W. Haider, A.M. Kobos and J.R. Rook, Nucl.Phys. A419, 521 (1984)

4. F.S. Dietrich, R.W. Finlay, S. Mellema, R. Randers-Pehrson, F. Petrovich, Phys.Rev.Lett. 51, 1629 (1983)
   S. Mellema, R.W. Finlay, F.S. Dietrich and F. Petrovich, Phys.Rev. C28, 2267 (1983); ibid. Phys.Rev. C29, 2385 (1984)

5. C. Mahaux, *Internat. conf. on Nucl. Structure 1982*, Nucl.Phys. A396 9c (1983) and *Lectures given at the Winter College on Fundamental Nuclear Physics*, Trieste 1984, to be published
   R. Brugger, M.K. Weigel and J. Winter, Univ. München, preprint (1984)

6. D. Garreta et al., Phys.Lett. 135B, 266 (1984); and see *Proc. of Tenth Internat. Conf. Particles and Nuclei*, Heidelberg (1984)

7. M. Lacombe et al., Phys.Rev. C21, 861 (198o);
   J. Cote, M. Lacombe, B. Loiseau, B. Moussalam and R. Vinh Mau, Phys.Rev.Lett. 48, 1319 (1982)

8. H.V. von Geramb and K. Nakano, *Workshop on the Interaction Between Medium Energy Nucleons in Nuclei 1982*, ed. H.O. Meyer, Americ. Inst.Phys. Conf. Proc. 97 (1983) and internal reports.

9. C.Y. Wong et al., Phys.Rev. C29, 574 (1984);
   C.J. Batty, E. Friedman and J. Lichtenstadt, Phys.Lett. 142B, 241 (1984);

9. A.M. Green and S. Wycech, Nucl.Phys. $\underline{A377}$, 441 (1982); T. Suzuki and H. Narumi, Phys.Lett. $\underline{125B}$, 251 (1983; B.C. Clark et al., Proc. of Tenth Int.Conf. Particles and Nuclei, PANIC 1984, C4o, and preprint Ohio University (1984)

NUCLEAR STRUCTURE APPROACH TO THE IMAGINARY OPTICAL POTENTIAL AND INELASTIC FORM FACTOR

Franz Osterfeld
Institut für Kernphysik, KFA Jülich, D-5170 Jülich, West Germany
and
Victor A. Madsen
Oregon State University, Corvallis, OR 97331, USA, and
Institut für Kernphysik, KFA Jülich, D-5170 Jülich, West Germany

ABSTRACT

Microscopic calculations of the second order imaginary optical potential are presented for neutron and proton scattering from spherical target nuclei using random phase approximation transition densities for intermediate excited states. Both inelastic and charge-exchange intermediate states are considered. The contribution of the deuteron channel to the absorption is estimated. The sensitivity of the surface absorption on details of the nuclear structure wave functions is discussed. Calculations are presented for the isoscalar term W_0 of the optical potential, the isovector term W_1 (Lane-term), and the Coulomb correction term ΔW_c. Elastic scattering cross sections are calculated from the microscopic nonlocal potentials and are compared to cross sections which were obtained from the equivalent local potentials. The validity of the local approximation to the nonlocal potentials is investigated. The differences and similarities between the nuclear matter approach and the nuclear structure approach to the calculation of the imaginary potential are discussed. The nuclear structure approach is also applied to the calculation of the imaginary inelastic form factor used in inelastic scattering. First results are presented for the ^{48}Ca(p,n)^{48}Sc(0^+, E_x = 6.8 MeV) reaction at an incident energy of 25 MeV.

I. INTRODUCTION

Although the optical model has been widely used in analysing nucleon-nucleus elastic scattering data[1] the theoretical problem of calculating the optical potential from some basic nucleon-nucleon interaction is not yet completely solved. Basically two different approaches have been formulated so far for such a microscopic calculation of the optical potential. One is the so-called "nuclear matter approach" of Jeukenne, Lejeune and Mahaux[2,3] and Brieva and Rook[4,5] in which they consider the large target limit (A→∞, A = nucleon mass number) and calculate either the optical potential[2,3] or the two-nucleon t-matrix[4,5] in nuclear matter.

They then obtain the optical potential for finite nuclei by making either a local density approximation on the potential itself[2,3] or on the two-nucleon t-matrix[4,5] which is then folded into the target ground state density distribution to get the optical potential. This approach has the advantage that it starts from a realistic nucleon-nucleon interaction and allows for a more or less parameter-free calculation of the optical potential in nuclear matter. It has also proven to be accurate enough to account for many bulk properties of the optical model like its energy dependence and the development of a wine bottle shaped real central potential for medium projectile energies (E ~ 100 MeV)[5]. But, on the other hand, apart from a reduction of density at the nuclear surface this approach does not take into account any specific effects which are due to the finiteness of the nucleus such as the possibility of collective shape oscillations, rearrangement collisions, etc.

Another approach, which we adopt here, is the so-called nuclear structure approach of N. Vinh Mau[6] and other authors[7-11] in which one assumes that the effective nucleon-nucleon interaction is basically known, but in which the inelastic excitations of the finite nucleus are taken into account. In some sense this approach is complementary to the first one, although it is not understood to what extent they overlap. While the nuclear matter approach includes inelastic excitation effects of a finite nucleus only in an average way (via the local density approximation), the nuclear structure approach treats them explicitly and therefore includes specific features of the target nucleus like its shell structure, collectivity etc. in detail. The correct treatment of the energetically open reaction channels, however, should be especially important in determining the imaginary part of the optical potential since the absorption should be quite sensitive to the nature and number of the energetically open intermediate states.

In the following we shall mainly summarize results on the imaginary part of the nucleon-nucleus optical potentials as obtained by the Jülich group[9-11] using the nuclear structure approach of N. Vinh Mau and collaborators[6]. Our results corroborate to a large extent those of Refs. 6-8 calculated with similar approximations. Then we present an extension of these calculations to the imaginary inelastic form factor. First results are shown for the imaginary Lane potential W_1 in the ^{48}Ca(p,n)-reaction at E = 25 MeV.

In section II we briefly describe the formulations and methods used in the nuclear structure approach. In section III we discuss the results obtained for the microscopic optical potentials, and in section IV we make a suggestion how one could possibly unify the nuclear matter and nuclear structure approach. In section V we discuss the microscopic calculation of the imaginary inelastic transition form factor. Section VI contains a summary and conclusions.

II. THEORY

Following Feshbach[12] we write the generalized optical potential for elastic scattering at energy E as

$$U_{opt}(E) = (\Psi_0|V|\Psi_0) + (\Psi_0|V Q \frac{1}{E-H_{QQ}+i\varepsilon} Q V|\Psi_0)$$
$$= V_{oo} + \Delta U \quad (II.1)$$

where it is understood that we take the limit $\varepsilon \to 0$. In eq. (II.1), Ψ_0 denotes the exact target nucleus ground state, Q is a projection operator which projects off the ground state: $Q \equiv I-|\Psi_0)(\Psi_0|$ (I: = identity operator); H is the total Hamiltonian of the scattering system, and V is the interaction between the projectile and the target. Exchange effects are not exhibited explicitly in eq. (II.1) but will be included in the calculations. If we start with a real effective interaction V then V_{oo} is a real and nonlocal folding potential. It gives the dominant contribution to the real part of the optical potential and was found to be in good agreement with the real parts of phenomenological potentials[13].

In this talk we are mainly concerned with an approximate evaluation of the second term ΔU of eq. (II.1) and in particular with its imaginary part, $Im(\Delta U)$. The evaluation of this term requires a detailed knowledge of the dynamics of the nuclear system, which in general is very complicated. Therefore we are forced to make some drastic approximations and assumptions. In the nuclear structure approach one approximates $Im(\Delta U)$ by the second order expression

$$W(E,\vec{r},\vec{r}') = Im \sum_N{}' \langle 0|V|N\rangle_{\vec{r}} \, g_N(\vec{r},\vec{r}') \langle N|V|0\rangle_{\vec{r}'} \quad (II.2)$$

where the matrix elements are over the nuclear coordinates only, and g_N is a projectile optical Greens function evaluated at an energy $E-E_N$, E_N being the excitation energy of the target nuclear state N. Obviously W is nonlocal. Because of the restriction to second order, only the intermediate particle-hole strength is directly excited. Since the random phase approximation (RPA) gives a good description of the particle-hole strength including collective effects, it is advantageous to use available RPA transition densities for the intermediate excitations. These RPA-states act as "doorways" for the absorption into more complicated states. From a physical point of view we expect the most important "doorways" to be direct inelastic scattering, direct rearrangement collisions and capture into bound states where the incoming nucleon is caught into a bound state which deexcites by γ-emission, for instance. At sufficiently high incident energies (E ~ 30 MeV) the absorption due to capture into bound states should be small in comparison to the absorption into the other reaction channels. The most important rearrangement channels will be the (p,d) or (n,d) pickup channels with a deuteron propagating in the intermediate states. These channels are not mutually orthogonal with the inelastic one-particle-one-hole (1p1h) channels and lead therefore to double counting. How serious the double counting problem really is depends on the truncation of the inelastic basis. For proton scattering the excited nucleon has to be a neutron and highly correlated in position and momentum with the incident proton in order to correspond to a deuteron. The probability of this occurring for a given inelastic 1p1h-excitation seems to be small. However, when summing over many intermediate inelastic states, the double counting could also be appreciable. Therefore we ignore the deuteron channel in most of our calculations which therefore provide a <u>lower limit</u> for the magnitude of the absorbing potential.

The second order expression for W in eq. (II.2) has been derived for a <u>non-identical</u> projectile. Particle identity, however, introduces some essential complications, which have been treated by several authors[6,9,15] using the Greens function formalism, in which the optical potential is identified with the mass or proper self energy operator[14] $S^*(x,x')$. The operator S^* is the sum of all proper self energy graphs connected to the projectile particle line.

Using RPA wave functions for the intermediate excited states $|\Psi_N\rangle$ in eq. (II.2) and considering the consequences of the particle identity as required by the mass operator of the Greens function formalism one obtains the following expression for W (in momentum space)[6,9,15]:

$$W(\vec{k}',\vec{k}) = \text{Im} \sum_{N,q>f}' [\langle\Psi_0|J(\vec{k}',\vec{q})|\Psi_N\rangle \frac{1}{E-\epsilon_q-E_N+i\eta} \langle\Psi_N|J(\vec{q},\vec{k})|\Psi_0\rangle$$

$$- \frac{1}{2}\langle\Psi_0|J(\vec{k}',\vec{q})|\mathcal{S}_N\rangle \frac{1}{E-\epsilon_q-E_N^{(0)}+i\eta} \langle\mathcal{S}_N|J(\vec{q},\vec{k})|\Psi_0\rangle] , \quad \text{(II.3a)}$$

where

$$J(q,k) = \sum_{\beta\gamma} a_\beta^+ \langle\beta|V|k\gamma\rangle a_\gamma \quad \text{(II.3b)}$$

is a Hermitian one-body operator representing the effect of the scattering of the projectile, including exchange, on the target nucleus. In eq. (II.3) Ψ_0 and Ψ_N denote the exact ground state and the exact excited states of the A-nucleon system, respectively, while Ψ_0 and Ψ_N denote the corresponding states of the uncorrelated system. The second (uncorrelated) term in eq. (II.3) represents a double counting correction which occurs due to the identity of the projectile and target nucleon in the leading order (noninteracting particle and hole) RPA graphs[9]. The functions $\langle \Psi_N | J(\vec{q},\vec{k}) | \Psi_0 \rangle$ and $\langle \varphi_N | J(\vec{q},\vec{k}) | \varphi_0 \rangle$ of eq. (II.3) are calculated according to the formulation of Ref. 9, approximating Ψ_N by RPA wave functions[16] and φ_N by unperturbed particle-hole states, respectively. Since the detailed calculation technique is rather lengthy we don't repeat it here and refer the interested reader to Ref. 9.

For the Green's function g_N in eq. (II.2) we have used either an optical or, equivalently, an energy averaged Green's function. Both choices take care of the fact that there is absorption in intermediate channels which is not explicitly included in our calculations. In the first case we achieve this by adding a small imaginary part ($W_0 \sim 1$ MeV) to the real potential V_{00} or, better, to its local equivalent $\tilde{V}_{00}$ which we use to generate the intermediate Green's function. In the second case we simply calculate g_N at the complex energy $(E-E_N)+iI/2$ where I is the width of a Lorentzian energy distribution function used in the energy averaging process. Inside the nucleus, the wave equation for this complex energy is identical to that with an absorptive potential $W = \Gamma/2$.

The calculated microscopic imaginary potential $W(\vec{r},\vec{r}')$ is nonlocal. In order to compare it with empirical potentials we first transform it into an "equivalent" local potential $\tilde{W}(R)$ where $\vec{R} = \frac{1}{2}(\vec{r}+\vec{r}')$. We obtain $\tilde{W}(R)$ by applying the local approximation of Perey and Saxon[17] to the nonlocal potential $W(\vec{r},\vec{r}')$.

III. THE IMAGINARY PART OF THE NUCLEON-NUCLEUS OPTICAL POTENTIAL

In the Lane model[18] the nuclear part of the optical potential is given by

$$U_N(E,r) = U_0(E,r) \pm \frac{N-Z}{4A} U_1(E,r) + \Delta U_C(E,r) \qquad (III.1)$$

where U_0 is an isospin independent potential, U_1 is the Lane potential, and ΔU_C is a Coulomb correction made for protons. Both of the latter two are charge dependent, but the Lane potential term comes from a $\vec{T}\cdot\vec{t}$ term, which can also produce charge exchange while the Coulomb correction term cannot. Each term in eq. (III.1) consists of a real part and an imaginary part

$$U_k(E,r) = V_k(E,r) + iW_k(E,r); \quad k=0,1 \qquad (III.2a)$$

$$\Delta U_C(E,r) = \Delta V_C(E,r) + i\Delta W_C(E,r) \qquad (III.2b)$$

In the following we shall present microscopic calculations for the different imaginary terms $(W_k)_{k=0,1}$ and for the imaginary Coulomb correction term ΔW_c.

III.1 THE ISOSCALAR TERM W_0

Calculations of the isoscalar part W_0 of the imaginary optical potential have been performed for ^{40}Ca(n,n) at various incident energies E using eq. (II.3) as basis for the calculations. For the effective projectile-target nucleon interaction V we used the finite range Eikemeier-Hackenbroich (EH) t-operator[19,20] and the RPA wave functions of Krewald and Speth[16] for the intermediate states Ψ_N. The Green's function was generated from an optical potential, the real part of which was obtained by folding the EH-interaction into the shell model ground state density distribution. The exchange contribution to the folding potential was included by using the zero-range pseudopotential approximation, but the magnitude was adjusted downward to agree roughly with the ratio of exchange to direct strengths from exact exchange calculations[21]. We include all energetically open intermediate states in the calculations, i.e., all inelastically excited natural and unnatural parity states and also the corresponding natural and unnatural charge exchange states. In a doubly closed shell nucleus such as ^{40}Ca most of the states in the low excitation energy region are of natural parity, while the first unnatural parity states open up at higher excitation energies (~10 MeV). Therefore a large fraction

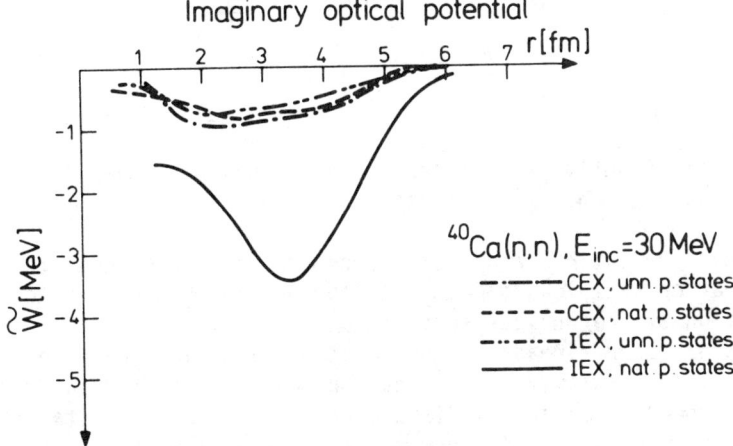

Fig. 1. Contributions to the imaginary optical potential from various types of intermediate states: unnatural and natural parity, charge exchange (CEX) and inelastic (IEX) scattering intermediate states.

of the absorption is due to the natural parity states, particularly the low-lying 3^- and 5^- states. At an incident energy of 30 MeV, however, the giant dipole, the isoscalar giant quadrupole, and also an appreciable fraction of the hexadecapole strength are excited and contribute substantially to the absorption. On the other hand, the unnatural parity states are less collective and therefore contribute less. In Fig. 1 we show the contributions to the imaginary potential for $^{40}Ca(n,n)$ which arise from the different classes of intermediate states. It is somewhat surprising that charge exchange gives such a large contribution to the absorption. This results from the fact that the charge exchange states in (n,p) reactions are shifted down in energy by the Coulomb energy differences of the nuclei (Z,N) and (Z-1,N+1) and to the greater (x2) projectile matrix element for charge exchange compared to inelastic scattering. In Fig. 2 we show the total contributions of

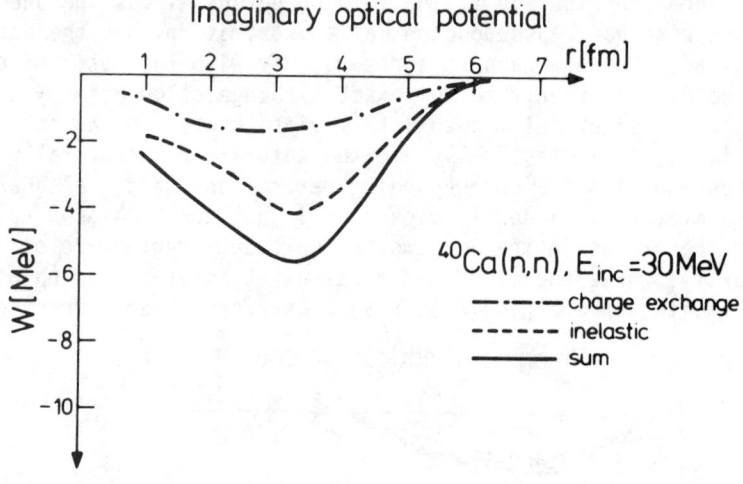

Fig. 2. Contribution to the imaginary optical potential from all charge exchange and all inelastic intermediate states.

inelastic excitations and charge exchange reactions separately and find that charge exchange accounts for nearly 30 % of the total absorption at the nuclear surface.

As has been already discussed in section II we have to correct the RPA results for the contributions coming from the excitation of unperturbed intermediate particle-hole doorway states, since otherwise we would count these contributions twice. Simultaneously this particle-hole doorway contribution is also the result which one would obtain with uncorrelated particle-hole intermediate states only. It can be seen from Fig. 3 that the particle-hole correction is rather small. This means that a calculation

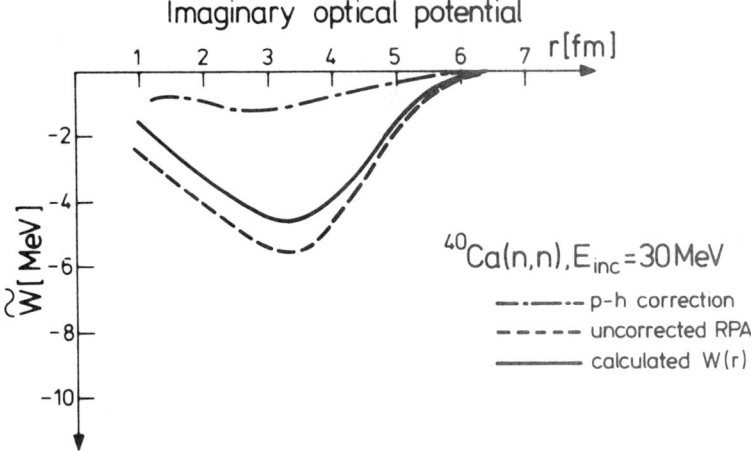

Fig. 3. Effect of the double counting (ph) correction to the imaginary optical potential. The dot-dashed curve is also the optical potential which is obtained for 30 MeV neutrons using all open unperturbed particle-hole intermediate states. (It is itself already corrected for double counting.)

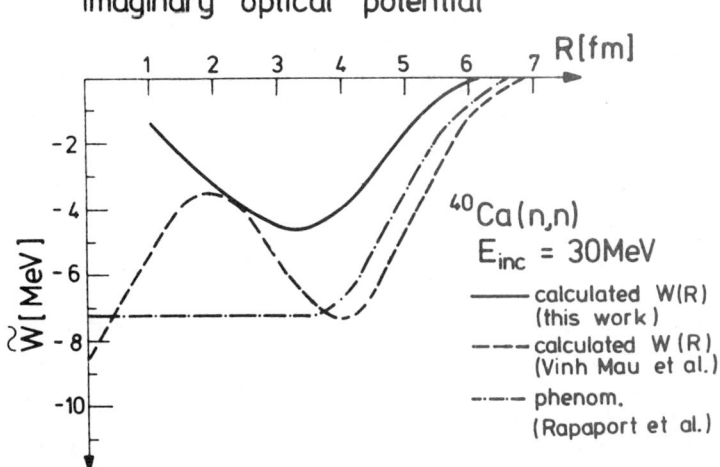

Fig. 4. Comparison of our optical potential with a similar calculation by Vinh Mau (Ref. 6) and a phenomenological potential (Ref. 24) at 30 MeV neutron energy.

with pure particle-hole intermediate states would greatly underestimate the absorption. Thus a large fraction of the absorption and surface peaked character must be attributed to the collective effects of a nucleus, which are reasonably well described by the RPA.

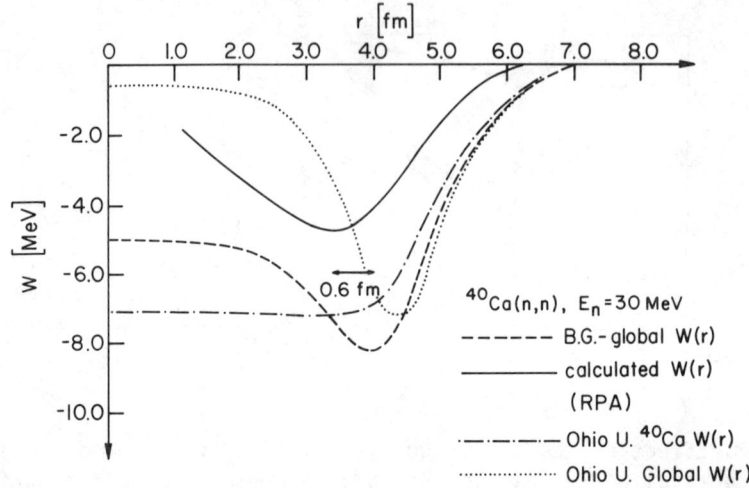

Fig. 5. Comparison of our calculated potential with phenomenological potentials of Becchetti-Greenlees (Ref. 25) and Ohio University (Ref. 24).

In Figs. 4 and 5 we compare our calculated potential with various phenomenological potentials and with the calculated potential of Vinh Mau[6]. Our imaginary potential is surface peaked, but the peak occurs at a smaller radius and is weaker than the phenomenological potentials by about 2 MeV. The underestimate in depth is perhaps not surprising since we include only the open inelastic channels explicitly but no rearrangement channels, which would make an added contribution to the absorption. A comparison between our calculation and that of Vinh Mau shows that our potential peaks at a smaller radius and is also not as deep. Clarification of the difference between our result and that of Vinh Mau requires consideration of the following points.

(i) We use the RPA-transition densities of Krewald and Speth[16], while Vinh Mau uses those of Gillet and Sanderson[22]. A comparison of the RPA-transition densities for the low-lying 3⁻ collective state, for instance, shows that the Gillet-Sanderson transition density peaks at a larger radius (by about 0.3 fm) than that of Ref. 16.

(ii) The difference in depth, however, is harder to understand. The RPA calculations of Krewald and Speth have been performed in a model space which is appreciably larger (nearly all $3\hbar\omega$ excitations are included) than that in the calculations of Gillet and Sanderson. Therefore a larger fraction of the total transition strength especially for states of higher multipolarity is included, which should also lead to a more strongly absorbing potential.

There is, however, an additional difference between the two calculations which concerns the effective projectile-target nucleon interaction used. In our calculations we use the EH-force[19,20] which on the average reproduces inelastic scattering cross sections for the low-lying collective states. This we have checked by performing microscopic, antisymmetrized DWBA calculations for these states using the RPA wave functions of Ref 16. Therefore we may say that we have calibrated the effective projectile-target nucleon interaction to inelastic scattering before using it in the calculation of the optical potential. Vinh Mau, on the other hand, uses the Reichstein-Tang interaction[23] which gives inelastic cross sections which are larger by a factor of 1.5 than that obtained with the EH-force. Therefore we would also obtain a 50 % deeper imaginary optical potential if we used the Reichstein-Tang interaction. It would also, however, give too large an inelastic cross section. The effective projectile-target nucleon interaction is ambiguous and has to be calibrated in some way, for instance in inelastic scattering, before it can be used in the calculation of the optical potential.

We have mentioned already that our calculated potential for ^{40}Ca(n,n) peaks somewhat below the nuclear radius, compared with most of the phenomenological potentials, like that of Refs. 24,25. There are various possible explanations for this shift: one is that the EH interaction lacks a density dependence. A density-dependent force has the tendency to shift the absorptive potential to large radii[9]. Another possibility is that the RPA transition densities to the intermediate collective states in ^{40}Ca have not the right shape and peak at too small a radius. Then also the absorptive potential will show this property since it is mainly the collective states which determine the shape of the imaginary potential at low projectile energies, E < 50 MeV. Finally, the shift could also be due to the neglect of noninelastic intermediate states like transfer channels which are not included in our calculations.

In order to clarify these questions we have performed a study[26] of the microscopic imaginary optical potential $W_N = Im(U_N)$, eq. (III.1), for proton scattering from ^{208}Pb. We chose this target since the transition densities to many collective states in ^{208}Pb have been measured in inelastic electron scattering experiments[27]. Furthermore, the RPA calculations of Speth and collaborators[28] work best for this nucleus and describe excitation energies, $B(E\lambda)$ values, and transition densities of many low- and high-lying collective states in a quantitative way.

Therefore we may assume that in the case of ^{208}Pb the nuclear structure input to our microscopic potential calculations is known, and consequently we can find out whether the shift of $\tilde{W}$ in ^{40}Ca to smaller radii is a real physical effect or whether it is simply due to a defect in the RPA transition densities used.

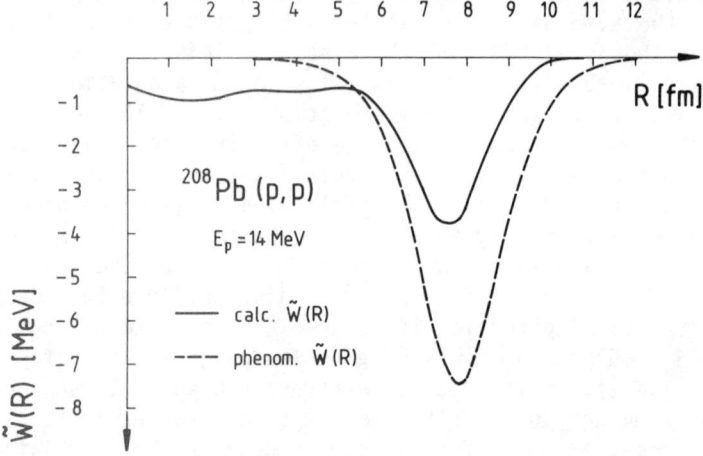

Fig. 6. Comparison of the microscopic local equivalent imaginary potential with the phenomenological potential of Ref. 29.

In Fig. 6 the equivalent local potential for ^{208}Pb(p,p) is compared with the phenomenological potential of Ref. 29. The calculated potential is surface peaked and, in contrast to ^{40}Ca(n,n), the surface peak occurs now in the case of ^{208}Pb(p,p) nearly at the right position. The microscopic $\tilde{W}$ peaks at r = 7.6 fm in good agreement with the phenomenological potential which peaks at r = 7.8 fm. The shift in the peak position by about 0.2 fm is much smaller than in the case of nucleon-^{40}Ca scattering (0.6 fm, see Fig. 5) and can easily be accounted for by a density-dependent effective projectile-target nucleon interaction[9]. This result clearly shows that good structure wave functions are needed for the description of the intermediate excited states in order to obtain the right surface peaking of the microscopic absorptive potential. The volume integral per nucleon J_W of the microscopic potential amounts to 33 MeV fm^3 and that of the phenomenological potential to 73 MeV fm^3. So, the volume integral of $\tilde{W}$ is about a factor of 2 smaller than that of the phenomenological potential. A similar factor of 2 was also found already in our studies of the microscopic imaginary potential for ^{40}Ca(p,p) and ^{40}Ca(n,n) scattering[11].

III.2 THE ISOVECTOR TERM W_1

From a microscropic point of view the isovector potential W_1 is produced in $N \neq Z$ nuclei by second and higher order inelastic transitions involving neutrons (protons) of the last occupied shells which form the neutron (proton) excess. Apart from isospin factors the potential W_1 is identical with the imaginary part of the charge-exchange transition potential to the isobaric analogue state (IAS) of the target nucleus ground state. Therefore we choose this transition to determine W_1. There are four types of intermediate particle-hole excitations which will contribute in second order to the excitation of the IAS. These involve intermediate inelastic excitations and intermediate charge-exchange excitation each with either the intermediate particle or intermediate hole scattered in the second step of the excitation as shown in

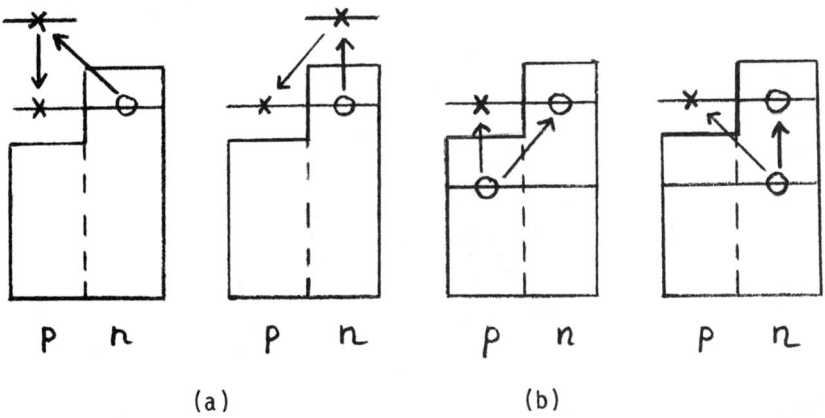

Fig. 7. Particle scattering (a) and hole scattering processes (b) contributing in second order to the imaginary W_1-potential.

Fig. 7. The summation of the contributions from all these diagrams gives a second order charge-exchange operator which is proportional to $U_1 = V_1 + iW_1$.

For the isoscalar part W_0, collectivity in the intermediate inelastic transitions is very important. The residual interaction pushes isoscalar strength downwards into the open channel energy region, thereby enhancing the absorption by inelastic channels and increasing the imaginary optical potential. The strong isoscalar collective states, however, don't have strong matrix elements for charge exchange to the IAS. Isovector states which couple strongly to the IAS, on the other hand, are pushed up in energy and will therefore tend to diminish rather than to increase the absorption. For this reason we choose to calculate W_1 using unperturbed particle-hole states knowing that we are not losing a large fraction of the charge-exchange strength. The excited particle is scattered either into bound states or into the continuum which we treat exactly.

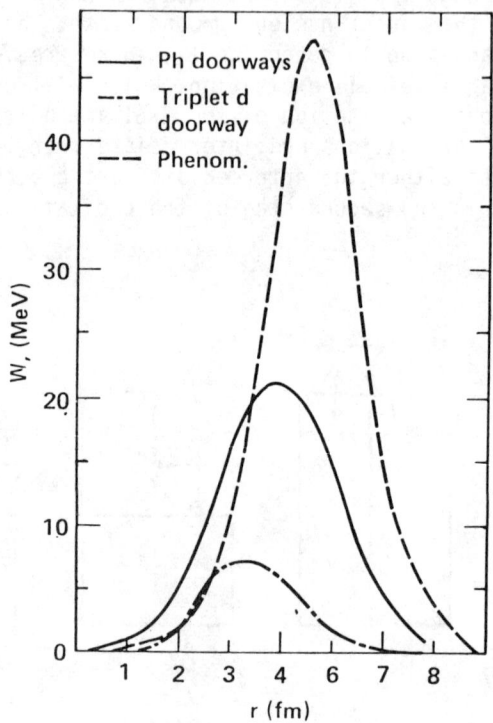

Fig. 8. Comparison of the calculated imaginary isovector potential W_1 with the phenomenological potential (W_1) of Becchetti-Greenlees[25]. Also the contribution of the intermediate deuteron channel to W_1 is shown.

The calculated imaginary potential W_1 is shown in Fig. 8 along with a separate calculation of the contribution for a single intermediate ground state deuteron channel. Since the intermediate deuteron is also a projectile-plus-particle-hole state of the nucleus, it is nonorthogonal to inelastic and charge-exchange intermediate states. We have therefore not added the contributions from these two types of doorways but have displayed them both to give an estimate of the strength of this one strong intermediate channel. The sum of the deuteron doorway and ph doorways therefore represents an upper limit on the contribution from both kinds of doorways. Also shown in Fig. 8 is the Becchetti-Greenlees (BG)[25] phenomenological potential W_1. The volume integral of the BG-potential is a litte greater than double of that of the theoretical particle-hole potential.

THE COULOMB CORRECTION TERM ΔW_c

The Coulomb correction is usually made in the real part of the proton optical potential, accounting for the difference in kinetic energy of neutrons and protons in the region of the nucleus. This correction is necessary because the nuclear part of the optical potential is energy dependent and the presence of a Coulomb potential means that a proton senses the nuclear potential at an energy reduced by the average Coulomb potential in the nucleus.

It has recently been recognized that a Coulomb correction should also be made for the imaginary part of the optical potential. This recognition has come about not only from comparison of neutron and proton empirical optical potentials[30], but also from theoretical calculations of the imaginary optical potential using either the nuclear matter[31] or the nuclear structure approach[10]. In the former the Coulomb correction arises from the dependence of the imaginary potential for nuclear matter on the Fermi energy, which, in turn, depends on the background Coulomb potential while in the latter the Coulomb effect is directly related to the different absorption felt by a proton or neutron scattered at given energy E from a finite nucleus. The nuclear structure approach seems actually ideally suited to study the Coulomb dependence of the potential, since all the Coulomb effects are rather realistically included in the different phases of the calculations. Likewise, it has the advantage that the spectrum of intermediate particle-hole strength is fairly realistically described by the RPA transition densities for both proton and neutron scattering.

In the case of nucleon scattering from ^{40}Ca, the parameter $\varepsilon = (N-Z)/A$ in eq. (III.1) is zero, which leaves us with

$$U_p = U_0(E,r) + \Delta U_c(E,r) \qquad (III.3a)$$

$$U_n = U_0(E,r) \qquad (III.3b)$$

Therefore the Coulomb correction can be best determined by a comparison of neutron and proton optical potentials for N=Z nuclei. Essentially there are two important Coulomb effects which lead to different absorption in neutron and proton scattering. The first comes from charge-exchange intermediate states. To explain this effect we consider a self-conjugate target nucleus like ^{40}Ca, so the corresponding members of the T=1 multiplet are available for the ^{40}Ca(p,n)^{40}Sc and the ^{40}Ca(n,p)^{40}K reactions. The states in ^{40}K, however, are shifted down in energy by twice the Coulomb energy E_C compared to their double analog states in ^{40}Sc. Thus, while many states are energetically open in (n,p), their double analog states are not open in (p,n) and will therefore not contribute to the imaginary optical potential for protons. This effect of excluding intermediate states as absorbing channels from (p,p) scattering as compared to (n,n) scattering is partly compensated for by the fact that in (n,p)(p,n) one has a Coulomb propagator for the intermediate proton, which will effectively close out some of the high-lying states.

The second Coulomb effect comes from the intermediate Greens function which is different for protons and neutrons due to the presence of the Coulomb potential in the proton propagator. In nuclear matter many intermediate states are excluded due to the constant, repulsive, background Coulomb potential, which effectively lowers the energy of protons and closes out high-lying states. For scattering from finite nuclei, however, no states are actually excluded asymptotically since the Coulomb potential goes to zero at r=∞. At finite radii high-lying intermediate states may be effectively closed for protons because of the Coulomb repulsion. Therefore one would expect that the imaginary potential for protons would be smaller than for neutrons. The opposite, however, is the case. At larger radii near the nuclear surface the proton propagator is typically larger than that of the neutrons reflecting the larger amount of time spent by the slower moving proton in the repulsive Coulomb field. (In the WKB approximation this property is expressed by way of a $[k(x)]^{-1/2}$ factor in the wave function.) This effect is accentuated upon calculation of the local-equivalent imaginary potential, for which smaller local wave numbers are favored, leading to a larger absorbing potential for protons. These two effects are so great that, if it were not for charge exchange, the absorption would always be greater for protons than for neutrons.

In Fig. 9 we show the calculated local equivalent imaginary optical potentials for both ^{40}Ca(n,n) and ^{40}Ca(p,p) at projectile energies of 17.7 MeV. The calculated potentials show surface absorption with their peak somewhat inside and their strengths somewhat lower than that of phenomenological potentials. Because the calculated forms are different in radial shape, we follow Rapaport[30] and present in Table I volume integrals of the various calculated potentials as well as ratios between the neutron and proton volume integrals. We see that the neutron volume integral at 17.7 MeV is greater than that of protons at the same bombarding energy, but smaller than that of 25 MeV protons, which have the

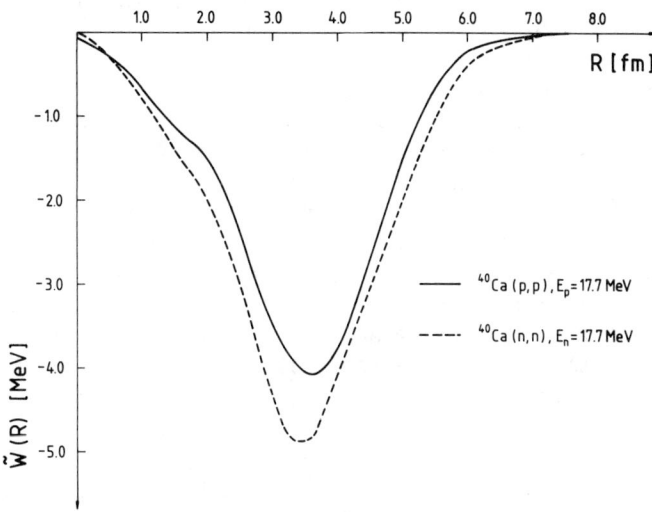

Fig. 9. Local equivalent neutron and proton imaginary potentials for ^{40}Ca(n,n) and ^{40}Ca(p,p) scattering at 17.7 MeV.

Table I. Volume integrals /A.

Projectile	E (MeV)	J/A (MeV fm^3)	J_n/J_p[a]	J_n/J_p[b]
n	17.7	56.32		
p	17.7	45.38	1.24	1.26
p	25.0	65.92	0.85	

[a] This work. [b] Ref. 24.

same average kinetic energies as the 17.7 MeV neutrons. The usual Coulomb correction made for the real proton potential is based on the assumption that neutron and proton potentials at the same kinetic energy should be equal. Such an approach to the imaginary potential, although in the right direction, would give too little absorption. In our calculations, it is fairly clear that the difference between the absorption of neutrons at 17.7 MeV and that of protons at 25 MeV is owing to opening of a strong 2$^+$ giant resonance at 19.6 MeV. The fact that protons have an average Coulomb potential energy of (25-17.7) MeV is irrelevant as far as their ability to excite these resonances. The agreement of the calculated and empirical ratios in Table 1 is reasonably good.

III.4 ELASTIC CROSS SECTION CALCULATIONS FROM THE MICROSCOPIC NONLOCAL POTENTIALS

The calculated microscopic potentials $W(\vec{r},\vec{r}')$ are nonlocal. In section III.1-3 we transformed them into "equivalent" local potentials $\tilde{W}$ by using the local approximation of Perey and Saxon[17]. We then compared these equivalent local potentials to the phenomenological ones and discussed differences and similarities between both. A further stringent test of the quality of the microscopic potentials is their direct use in elastic cross section calculations. Recently we have developed a new program which can solve the potential scattering problem also for nonlocal potentials[11]. Using this program we have calculated elastic scattering cross sections from the microscopic nonlocal and the "equivalent" local potentials. The nonlocal potential used has the form

$$V_\ell(r,r') = [V_f(r)+V_C(r)]\delta^3(\vec{r}-\vec{r}') + iW_\ell^{(2)}(r,r') , \quad (III.4)$$

where V_f is the folding potential described above, V_C is the Coulomb potential, and $W_\ell^{(2)}(r,r')$ is our second-order microscopic nonlocal imaginary potential for partial wave ℓ. The "equivalent" local potential is given by

$$V(R) = V_f(R) + V_C(R) + \tilde{W}(R) , \quad (III.5)$$

where $\tilde{W}(R)$ is the local equivalent to the nonlocal potential $\tilde{W}(r,r')$. The results obtained with these potentials are compared with experimental elastic scattering data in Fig. 10 for

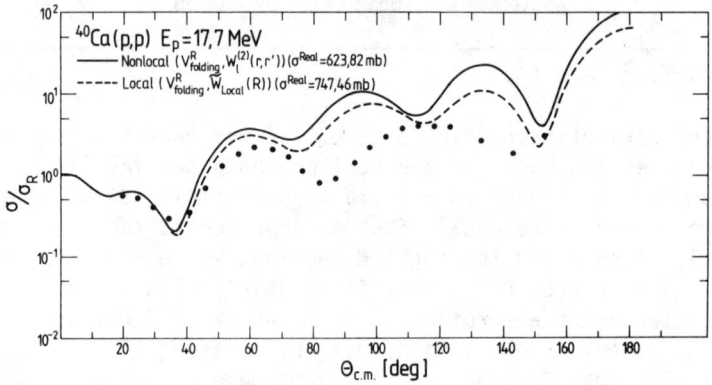

Fig. 10. Comparison of calculated ^{40}Ca(p,p) differential cross sections at 17.7 MeV to the experimental data of Ref. 32. The full line is the exact nonlocal calculation. The dashed line is the calculation with the equivalent local imaginary potential.

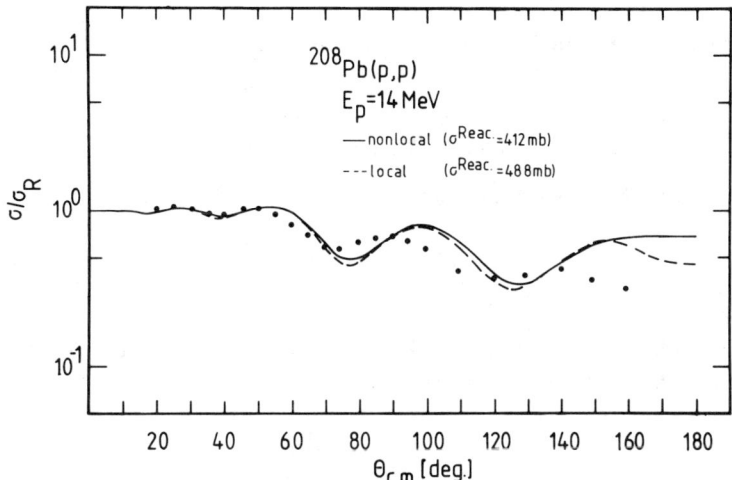

Fig. 11. Comparison of calculated ^{208}Pb(p,p) differential cross sections at 14 MeV to the experimental data of Ref. 29. The full line is the exact nonlocal calculation. The dashed line is the calculation with the equivalent local imaginary potential.

^{40}Ca(p,p) and in Fig. 11 for ^{208}Pb(p,p). The calculations for ^{40}Ca(p,p) agree well with the data up to 60°, but at larger angles the calculated cross sections are too high, indicating too little absorption. Although the behaviour of the two angular distributions calculated with the nonlocal and equivalent local potentials is very similar, there are also differences up to 100 % at large scattering angles, indicating too little absorption. The reaction cross section of the nonlocal calculation amounts to 624 mb and that of the local equivalent calculation to 748 mb. A similar result is obtained for ^{208}Pb(p,p) (see Fig. 11). Although the behaviour of the two calculated angular distributions is very similar in this case, there are again differences in the reaction cross section by about 20 %. This again indicates a lack of accuracy of the local approximation.

IV. UNIFICATION OF THE NUCLEAR MATTER AND NUCLEAR STRUCTURE APPROACH

The advantage of the nuclear matter approach is that it starts from a realistic free nucleon-nucleon interaction V_{NN} like the Paris potential[33], for instance, and calculates the energy dependent t-matrix in infinite nuclear matter by solving the Bethe-Goldstone equation

$$t(E) = V_{NN} + V_{NN} G^{(+)}(E) t(E) . \qquad (IV.1)$$

The Green's function $G_Q^{(+)}(E)$ is defined by

$$G_Q^{(+)}(E) = \sum_{|\vec{b}|,|\vec{c}|>k_F}{}' \frac{|\Phi_{bc}\rangle\langle\Phi_{bc}|}{E-e(b)-e(c)+i\varepsilon} \qquad (IV.2)$$

and involves the Pauli operator

$$Q = \sum_{|\vec{b}|,|\vec{c}|>k_F}{}' |\Phi_{bc}\rangle\langle\Phi_{bc}| \qquad (IV.3)$$

which projects onto unoccupied intermediate states $|\Phi_{bc}\rangle$. The starting energy E is identified with

$$E = e(a) + e(m) \qquad (IV.4)$$

using the notation that the particle labels a,b,c,d denote particles above the Fermi sea while m,n,... refer to states within the Fermi sea. From eq. (IV.1) one obtains the t-matrix for finite nuclei by means of a local density approximation. Such an energy and density dependent t-matrix could now also be used in the nuclear structure approach to the optical potential, but, of course, one has to take care that there occurs no double counting. The latter can be avoided in the following way[34]: One chooses the Pauli operator in eq. (IV.2) such that it projects onto those intermediate states only which are orthogonal to the model space used in the nuclear structure calculations of the finite nucleus. This can be achieved, for instance, by replacing in eq. (IV.2) k_F by a certain k_F' with $K_F' > K_F$. In this way one would construct an energy dependent and density dependent t-matrix which could be used as effective projectile-target nucleon interaction in the nuclear structure approach to the optical potential. An optical potential calculation carried out in this way would include energy and density dependent effects as well as the particular properties of the finite nucleus.

V. THE IMAGINARY INELASTIC TRANSITION FORM FACTOR

It has long been recognized[35-37] that a complex transition form factor is needed to describe the inelastic scattering to collective final nucleus states. In the collective model one obtains the inelastic transition from factor as the derivative of the optical potential

$$F(\vec{r}) = -\beta_L R \frac{d}{dr}[V(r)+iW(r)]Y_{LM}(r) . \qquad (V.1)$$

Such a form factor is naturally complex. In the microscopic model one normally calculates a form factor from perturbation theory by folding wave functions for the nuclear states with the two-nucleon interaction

$$F(r) = \langle \Psi_f(\xi) | \sum_{n=1}^{A} V(r,r_n) | \Psi_i(\xi) \rangle , \qquad (V.2)$$

which gives a real form factor. However, the data tend to agree better[35-37] with eq. (V.1) than with eq. (V.2), so a hybrid theory is often used in which the real part of the form factor is calculated from the microscopic model and the imaginary part from the macroscopic model.

The leading contribution to the imaginary interaction for inelastic scattering can be written (see eq. (II.2))

$$\text{Im } F(\vec{r},\vec{r}') = \text{Im} \sum_N \langle \Psi_f | V | \Psi_N \rangle \langle \Psi_N | V | \Psi_i \rangle G_N(r,r') \qquad (V.3)$$

The imaginary part of the projectile Green function $G_N(r,r')$ is always negative. Since for elastic scattering, $\Psi_f = \Psi_i$, an absolute square of the matrix element appears, all the terms of eq. (V.3) add constructively. For inelastic scattering, the different terms need not have the same sign. They might even be random and therefore destructive. On the other hand the collective model eq. (V.1) gives an imaginary interaction simply related through the parameter β to the strength of the real form factor.

In Ref. 38 it was proven that the interference between the different terms of eq. (V.3) is, indeed, partly destructive, however, not in a random but in a systematic way. The proof was based on the following two assumptions: (i) The initial and final nuclear states were assumed to be related by creation and destruction operators as follows

$$\Psi_f = \sum_{\underline{mn}} A_{\underline{mn}} a_{\underline{m}}^+ a_{\underline{n}} \Psi_i , \qquad (V.4)$$

where m refers to an unoccupied state (particle) and n an occupied state (hole) based on the target. Eq. (V.4) holds for TDA and RPA collective states as well as simple single-particle transitions. (ii) The intermediate states can be taken as simple particle-hole doorway states based on the ground state or as linear combinations of those. Both assumptions are perfectly fulfilled for our case of the isobaric analog transition $^{48}\text{Ca}(p,n)^{48}\text{Sc}(0^+, E_x = 6.68 \text{ MeV})$. In this case the imaginary form factor Im $F(\vec{r},\vec{r}')$ of eq. (V.3) is identical with $W_1(\vec{r},\vec{r}')$ of eq. (II.2). We have learned already that the calculated microscopic $\widetilde{W}_1$ shows indeed the same behaviour

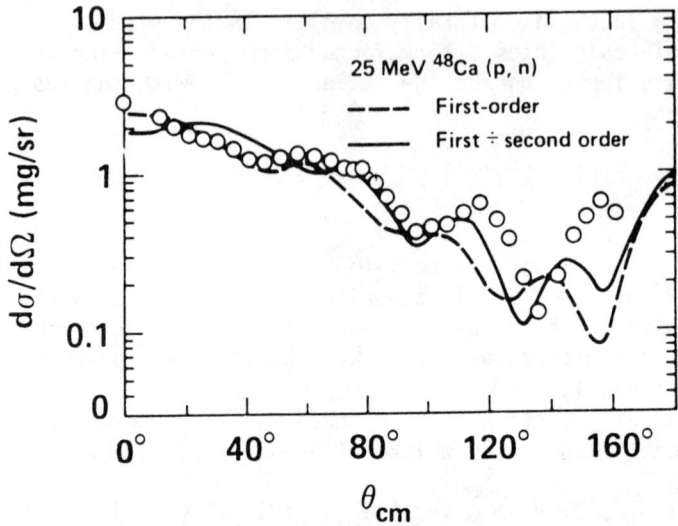

Fig. 12. Angular distributions for the ^{48}Ca(p,n)^{48}Sc(0^+, E_x = 6.68 MeV)-reaction at 25 MeV incident energy. The data were taken from Ref. 39. The dashed curve is the result of a microscopic, antisymmetric first order DWBA-calculation and the full curve represents a fully antisymmetric first and second order calculation including the nonlocal imaginary transition potential $W_1(\vec{r},\vec{r}\,')$.

as suggested by phenomenological potentials and as predicted in Ref. 38. In Fig. 12 we show angular distributions for the ^{48}Ca(p,n)^{48}Sc(0^+, E_x = 6.68 MeV) reaction at 25 MeV incident energy. Two different calculated angular distributions are compared to the data[39]. One is the result of a microscopic, antisymmetric first order DWBA calculation using the EH t-operator[19,20] for the effective projectile-target nucleon interaction (dashed curve) and the second is a fully antisymmetric first plus second order DWBA calculation (full curve) which includes the nonlocal imaginary transition potential $W_1(\vec{r},\vec{r}\,')$. As one can see from Fig. 12 there seems to be an improvement in the description of the experimental data at large scattering angles by performing a complete second order DWBA calculation.

VI. SUMMARY AND CONCLUSIONS

The calculation of the imaginary part of the nuclear optical potential W has been made to second order using a finite range effective projectile-target nucleon interaction and an optical Greens function for the intermediate-state propagation including exchange and all open inelastic and charge exchange RPA intermediate states. For neutron scattering charge exchange intermediate states are rather important, accounting for more than 1/3 of W. The calculated W tends to be surface peaked due largely to surface-peaked transition densities of the important intermediate collective states. The position of the peak is very sensitive to the details of the nuclear structure wave functions. Especially the transition densities to the collective intermediate states have to be in agreement with experiment in order to obtain the surface peak of the imaginary potential at the right position. A strongly surface-peaked optical potential does not occur when unperturbed particle-hole intermediate states are used. In all cases considered the volume integrals of the calculated microscopic potentials are about a factor of 2 weaker than those of phenomenological potentials. One can think of several reasons for this underestimate in absorption. An obvious one is that the inclusion of only inelastic channels as doorways in the potential calculations is not sufficient in order to obtain the right amount of absorption. Other reaction channels like transfer channels would have to be included which, on the other hand, however, lead to double counting problems. Another possibility is that the coupling between the excited doorways is of importance. This effect is not included in our calculations. But there are also simpler possibilities. The EH-interaction which we use for the effective projectile-target nucleon interaction is density independent. From the nuclear matter calculations we know, however, that the density dependence of the force is very important. Therefore it would be interesting to perform a calculation of the microscopic optical potential within the nuclear structure approach, but with density dependent finite range effective interactions. Finally, since a deviation by a factor of 2 between the volume integrals of the microscopically calculated and the phenomenological potentials is not very large, it could also well be that a consideration of all three effects, inclusion of rearrangement channels, coupling between doorways, and the use of a density dependent effective projectile-target nucleon interaction, is needed to obtain better agreement between the calculated and phenomenological potentials.

Acknowledgement: The work of VAM was supported in part by the U.S. Department of Energy under grant number DE-AT06-79ER-10405.

REFERENCES

1. For a review of phenomenological optical potentials see for example the book: P.E. Hodgson, Nuclear reactions and nuclear structure (Clarendon Press, Oxford, 1971);
P.E. Hodgson, invited paper at this conference.

2. J.P. Jeukenne, A. Lejeune, and C. Mahaux, Phys. Rep. 25, 83 (1976).
3. C. Mahaux, Microscopic optical potentials, Lecture Notes in Physics, edited by H.V. Geramb (Springer, Berlin, 1979), p. 1; IUCF Workshop 1982, edited by H.O. Meyer, AIP Conference Proc. No. 97 (American Inst. of Physics, New York, 1983), p. 20.
4. F. Brieva and J.R. Rook, Nucl. Phys. A291, 299 (1977); A291, 317 (1977); A307, 493 (1978).
5. F.A. Brieva, H.V. Geramb, J.R. Rook, Phys. Lett. 79B, 177 (1978);
H.V. Geramb, IUCF Workshop 1982, edited by H.O. Meyer, AIP Conference Proc. No. 97 (American Inst. of Physics, New ork, 1983);
H.V. Geramb, invited paper at this conference.
6. N. Vinh Mau, Theory of nuclear structure (IAEA, Vienna, 1970), p. 931;
N. Vinh Mau and A. Bouyssy, Nucl. Phys. A257, 189 (1976);
N. Vinh Mau, Phys. Lett. 71B, 5 (1977);
in Microscopic optical potentials, edited by H.V. Geramb, Lecture Notes in Physics,Vol. 89 (Springer, Berlin, 1979), p. 40;
A. Bouyssy, H. Ngo, and N. Vinh Mau, Nucl. Phys. A371, 173 (1983).
7. C.L. Rao, M. Reeves III, and G.R. Satchler, Nucl. Phys. A207, 182 (1973).
8. P.W. Coulter and G.R. Satchler, Nucl. Phys. A293, 269 (1977).
9. F. Osterfeld, J. Wambach, and V.A. Madsen, Phys. Rev. C23, 179 (1981).
10. F. Osterfeld and V.A. Madsen, Phys. Rev. C24, 2468 (1981).
11. H. Dermawan, F. Osterfeld, and V.A. Madsen, Phys. Rev. C27, 1474 (1983); Phys. Rev. C25, 180 (1982).
12. H. Feshbach, Ann. Phys. (N.Y.) 5, 257 (1958); 19, 287 (1962).
13. D.A. Slanina and H. McManus, Nucl. Phys. A116, 27 (1968);
L.W. Owen and G.R. Satchler, Phys. Rev. Lett. 25, 1720 (1970);
C.B. Dover and N. Van Giai, Nucl. Phys. A190, 373 (1972).
14. A.L. Fetter and J.D. Walecka, Quantum theory of many-particle systems (McGraw-Hill, New York, 1976).
15. F. Villars, in Fundamentals of nuclear theory, edited by A. de Shalit and C. Villi (IAEA, Vienna, 1967).
16. S. Krewald, J. Speth, Phys. Lett. 74B, 295 (1974).
17. F.G. Perey and D.S. Saxon, Phys. Lett. 10, 107 (1964).
18. A.M. Lane, Nucl. Phys. 35, 676 (1962);
19. H. Eikemeier and H. Hackenbroich, Nucl. Phys. A169, 407 (1971).
20. The effective interaction used has been obtained from the Eikemeier-Hackenbroich (EH) two-nucleon force[19] by taking the long range part of the interaction according to the Scott-

Moszkowski approximation (S.A. Moszkowski and B.L.Scott, Ann. Phys. 11, 65 (1960)) to the G-matrix. For further details see Ref. 10 and the paper of K.A. Amos, H.V. Geramb, R. Sprickmann, J. Arvieux, M. Buenerd, and G. Perrin, Phys. Lett. 52B, 138 (1974).
21. J. Atkinson and V.A. Madsen, Phys. Rev. C1, 1377 (1970)
22. V. Gillet and E.A. Sanderson, Nucl. Phys. 54, 472 (1964); A91, 292 (1967).
23. I. Reichstein and Y.C. Tang, Nucl. Phys. A139, 144 (1969).
24. J. Rapaport, V. Kulkarni, and R.W. Finlay, Nucl. Phys. A330, 15 (1979);
J. Rapaport, J.D. Carlson, D. Bainum, T.S. Cheema, and R.W. Finlay, ibid. A286, 232 (1977).
25. F.D. Becchetti and G.W. Greenlees, Phys. Rev. 182, 1190 (1969).
26. H. Dermawan, F. Osterfeld, V.A. Madsen, Phys. Rev. C29, 1075 (1984).
27. J. Heisenberg, Nucl. Phys. A396, 171 (1983), and references cited therein.
28. J. Speth and A. van der Woude, Rep. Prog. Phys. 44, 719 (1981), and references cited therein.
29. J.S. Eck and W.J. Thompson, Nucl. Phys. A237, 83 (1975).
30. J. Rapaport, Phys. Lett. 92B, 233 (1980); Phys. Rep. 87, 27 (1982).
31. J.P. Jeukenne, A. Lejeune, and C. Mahaux, Phys. Rev. C15, 10 (1977); 16, 80 (1977).
32. W.T.H. Van Oers, Phys. Rev. C3, 1550 (1971).
33. M. Lacombe, B. Loiseau, J.M. Richard, R. Vinh Mau, J. Coté, P. Pirès, and R. de Tourreil, Phys. Rev. C21, 861 (1980).
34. V.A. Madsen, F. Osterfeld, and J. Wambach, in Microscopic Optical Potentials, Lecture Notes in Physics, edited by H.V. Geramb (Springer, Berlin, 1979), p. 151.
35. G.R. Satchler, Phys. Lett. 35B, 279 (1971).
36. G.R. Satchler, Z. Phys. 260, 209 (1973).
37. H.V. Geramb and P. Hodgson, Nucl. Phys. A246, 173 (1975).
38. G. Baur, V.A. Madsen, F. Osterfeld, Phys. Rev. C17, 819 (1978).
39. D.M. Patterson, R.R. Doering, A. Galonsky, Nucl. Phys. A263 (1976) 261

MOMENTUM AND FREQUENCY DEPENDENCE OF THE SELF-ENERGY IN NUCLEAR MATTER : A PATH TO STUDY THE EFFECTIVE NUCLEON-NUCLEON INTERACTION

A. Lejeune,
Institut de Physique B5, Université de Liège, Sart Tilman,
B-4000 Liège 1, Belgium

V. Bernard, P. Grangé and M. Martzolff,
Physique Nucléaire Théorique, C.R.N., B.P. 20,
F-67037 Strasbourg Cedex, France.

The properties of nuclear matter with the Paris N-N potential have been already investigated[1,2] in the BHF approximation with the continuous choice of the single particle field at the Fermi surfaces. In Figs. 1 and 2, the continuous curves show our results at $k_F = 1.36$ fm^{-1} for the real and imaginary parts of the optical potential and for the effective mass, respectively. We focus here on the investigation of the momentum k and frequency ω dependence of the mass operator $\Sigma(k,\omega)$ not considered in earlier studies.[3] This is believed a relevant way to understand the true energy dependence of the effective interaction. To investigate this dependence we envisage a model N-N interaction acting in S-state only, with a short range momentum dependent repulsion in keeping with the main feature of the Paris N-N interaction. This model allows for an analytic determination of both the polarization and correlation contributions to Imag $\Sigma(k,\omega)$. The real parts are obtained via dispersion relations.

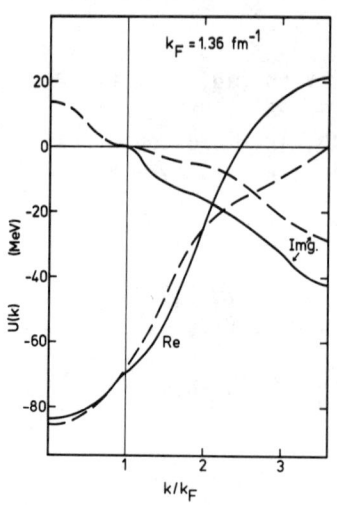

Fig. 1.

Real and imaginary parts of the optical potential versus k/k_F (see text).

In Fig. 1 the dashed curves represent the Hartree-Fock plus the polarization contribution to the real part of the optical potential, the correlation and the polarization contributions to the imaginary part. In its present form, the model gives a good qualitative agreement with the BHF contribution to $\Sigma(k,\omega)$ up to energies of 200 MeV. We are then encouraged to consider the (k,ω) dependence of the whole mass operator as resulting from the contribution of the model interaction. In Fig. 2, the short dashed curve (right hand scale) shows the variations with k/k_F of the "momentum" mass $\tilde{m}/m$ and the long dashed curve the variations of m^*/m for the model interaction. The lack of short range repulsion of the model interaction above 200 MeV shown in Fig. 1 accounts for the shift in scale[4] of the different effective masses m^*/m.

Through the comparison with the BHF approximation of the mass operator for the Paris potential we have thus constrained our model in a way such that the matrix elements

$$\langle k|G^{Model}(\omega = \omega_k \pm \Delta)|k\rangle = \langle k|V^{Model} + V^{Model}\frac{Q}{e}V^{Model}|k\rangle$$

$$\simeq \langle k|G^{Paris}(\omega = \omega_k \pm \Delta)|k\rangle \quad , \quad (1)$$

with $\Delta \simeq 8$ MeV and up to values of $k \approx 2k_F$. The model determines the function $F(k,k'; \omega - \omega_k \mp \Delta)$ such that

$$\langle k'|G^{Model}(\omega)|k\rangle = \langle k'|G^{Model}(\omega = \omega_k \pm \Delta)|k\rangle + F(k,k';\omega - \omega_k \mp \Delta). \quad (2)$$

Taking advantage of relations (1) and (2) we infer that the far off-shell energy dependence of the Paris potential G-matrix is determined by

$$\langle k'|G^{Paris}(\omega)|k\rangle = \langle k'|G^{Paris}(\omega = \omega_k \pm \Delta)|k\rangle + F(k,k';\omega - \omega_k \mp \Delta). \quad (3)$$

The energy dependent effective interaction $\langle k'|V_{eff}(\omega)|k\rangle$ may then be constructed from this G-matrix.

This work has been carried out under support from NATO Scientific Affairs Division, Research grant n° 025.81.

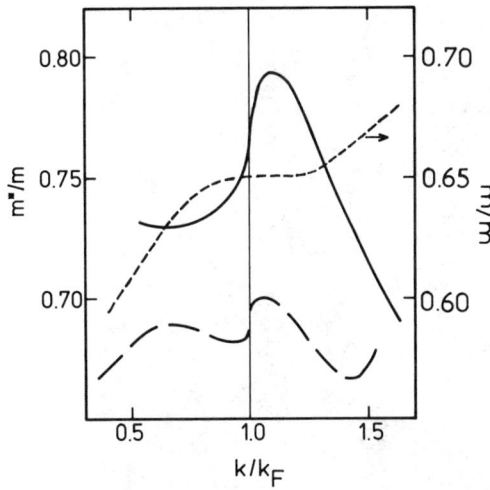

Fig. 2.

Effective masses versus k/k_F (see text).

References

1. A. Lejeune & P. Grangé, in "Recent Progress in Many-Body Theories" ed. by H. Kümmel & M.L. Ristig (Springer-Verlag, Berlin 1984), (Lecture Notes in Physics, vol. 198).
2. M.A. Matin & Dey, Phys. Rev. 27C, 2356 (1983).
3. J.P. Jeukenne, A. Lejeune & C. Mahaux, Phys. Rep. 25C, 83 (1976).
4. V. Bernard & C. Mahaux, Phys. Rev. 23C, 888 (1981).

SESSION B

OBSERVABLES AND

EFFECTIVE INTERACTIONS

ANALYZING POWER MEASUREMENTS FOR NEUTRON-NUCLEUS
SCATTERING AND THE SPIN-ORBIT POTENTIAL

Richard L. Walter
Department of Physics, Duke University, Durham, NC 27706
and
Triangle Universities Nuclear Laboratory,* Durham, NC 27706

ABSTRACT

Analyzing power $A_y(\theta)$ and cross section measurements have been obtained from 10 to 17 MeV for 20 isotopes ranging from ^{6}Li to ^{208}Pb. These combined data sets provide a unique data base for nuclear model development. The experimental method for the $A_y(\theta)$ measurements and comparisons to coupled-channels and spherical optical model calculations are given.

INTRODUCTION

Measurements of the analyzing power $A_y(\theta)$ for proton elastic and inelastic scattering have been crucial in determining the nature of the spin-dependent terms of the nuclear potential and, more generally, in creating a fruitful basis for nuclear model development. Until recently, most of the information regarding the details of the neutron spin-dependent interaction in neutron-nucleus collisions has been at least implicitly taken from the findings of proton-nucleus scattering. The reason, of course, is that very little polarization data for neutron scattering has been obtained in an energy region where the nuclear reactions are not strongly effected by compound-nucleus scattering or resonance reactions. Until a few years ago, about the only useful $A_y(\theta)$ data for neutron scattering for nuclei with Z>2 and neutron energies E_n>6 MeV were from two experiments[1,2] on 6 nuclei at 24 MeV and 2 nuclei at 10.4 MeV, each measurement having been performed with only moderate statistical accuracy. Recently though, elastic scattering has been investigated for a wide range of targets at 7.75 MeV at Stuttgart[3] where a high-current 4 MeV accelerator is used in conjunction with the ^{9}Be(α,n) reaction to provide a source of highly polarized neutrons.

Alternatively, it has been known for about 15 years that the ^{2}H($\vec{d},\vec{n}$) and ^{3}H($\vec{d},\vec{n}$) polarization transfer reactions can provide highly polarized neutron beams at 0° where the differential cross section peaks. Furthermore, these polarized neutron sources are continuously variable in energy, a feature which allows for studying the energy dependence of $A_y(\theta)$ for neutron-nucleus scattering. Only recently though has the intensity of polarized deuteron beams been sufficient to permit high accuracy neutron data to be obtained. Presently, two laboratories, Erlangen and TUNL, have exploited the ^{2}H($\vec{d},\vec{n}$) source of neutrons for the case where the source reaction was induced with direct-current deuteron beams. In addition, at TUNL we have developed[4] a method for producing pulsed polarized deuteron beams. This technique thereby permits studies that employ neutron time-of-flight spectroscopy. The present paper concentrates on the program at TUNL and gives a brief outline of the method, the quality of the data and a few of the model calculations.

EXPERIMENTAL METHOD

Two features of our technique are responsible for our ability to make $A_y(\theta)$ measurements to the typical accuracy of ± 0.04 that we typically report for elastic scattering of polarized neutrons. The first is in the nuclear physics of the $^2H(d,n)^3He_{g.s.}$ reaction. This source reaction produces the largest yield of monoenergetic neutrons per unit energy spread of any source reaction that is variable across the neutron energy range of 8 to 17 MeV. The neutron flux emitted at 0° is much more intense than the flux emitted at other angles. Furthermore, the polarization transfer coefficient is among the highest ever observed, permitting 90% of the vector polarization of the incident deuteron beam to be "transferred" to the neutrons emitted into the cone near 0°. In our case at TUNL the deuteron beam typically is about 65% polarized, so we can obtain an intense, clean beam of about 60% polarization. Of course, a distinct advantage of using the polarization transfer method for a neutron source is that instrumental asymmetries can be kept to a minimum because the orientation of the neutron polarization vector can be interchanged from down to up merely by changing the deuteron polarization axis at the ion source. This allows for keeping the detectors fixed throughout the $A_y(\theta)$ measurement, and thus, to probably second order, the neutron background from the surrounding room is held constant.

The second feature is integral to the design of Lamb-shift polarization ion sources. For example, in our source the deuteron beam drifts for nearly 2 meters at the relatively low energy of 1 keV. We have taken advantage of this feature, which is unique to Lamb-shift sources. By superposing a sawtooth voltage of about 100 V at the gap which accelerates the beam to the 1 keV energy, we introduce a 3% dispersion in the velocity of the deuteron beam. This causes the beam (normally d.c.) to arrive at the end of the ion source in bursts about 60 ns wide, sufficiently narrow to fit into the phase acceptance of the two-stage buncher at the entrance to the tandem accelerator. This buncher compresses these beam pulses into 2 ns wide bursts at the $^2H(d,n)$ target. The overall efficiency of the system permits us typically to have beams of about 150 nA on target, which is about 80% of the normal d.c. polarized beam.[4]

The scattering setup is fairly standard, so only a few pertinent details are given here. As the counting rates are low because of the 150 nA beam, several compromises had to be made. A relatively tight source-to-scatterer geometry is employed and somewhat large scatterers are used. Hence, corrections to the data for multiple scattering and geometry effects must be treated carefully. In addition, the neutron flux has been increased by using a fairly thick deuterium gas target, usually giving a neutron energy spread between 300 to 400 keV. Until recently, we had chosen a survey of $A_y(\theta)$ for elastic scattering for a range of targets and energies, and usually obtained the inelastic data more as a byproduct. In fact, the flight paths to the neutron detectors (up to 6 m for one detector and 4 m for the other) are only modest and resolution has been a problem in some instances. Yet, with the 2 ns resolution we have been able to obtain some information on $A_y(\theta)$ for inelastic scattering to low-lying states separated by 800 keV. A sample time-of-flight spectrum for n+^{28}Si scattering, which study is reported in this volume in a contribution by C.R. Howell, is illustrated in fig. 1. The elastic and inelastic scattering peaks rest well above the residual background, and are well resolved for this nucleus.

MEASUREMENTS AND ANALYSES

The targets studied to date at TUNL are ^{1,2}H, ^{3,4}He, ^{6}Li, ^{9}Be, ^{10,11}B, ^{12}C, ^{16}O, ^{27}Al, ^{28}Si, ^{32}S, ^{40}Ca, ^{54,56}Fe, ^{58,60}Ni, ^{65}Cu, ^{89}Y, ^{93}Nb, ^{116,120}Sn, and ^{208}Pb. The data for n-p, n-d and n-He scattering were obtained with a d.c. beam and are of high statistical accuracy (±0.002 to ±0.02). They will not be discussed here. For the 1p-shell nuclei typically the energy range is 8 to 17. For the heavier nuclei the measurements were usually performed at 10, 14 and 17 MeV. Much of the data is still being interpreted by students for their theses, and, as such, has not been released yet. Spectra obtained in the Li, O and Al measurements and measurements at 17 MeV are still being processed.

All the analyses include $\sigma(\theta)$ data, most of which were obtained at TUNL and some, usually at 11 and 26 MeV, have been taken from

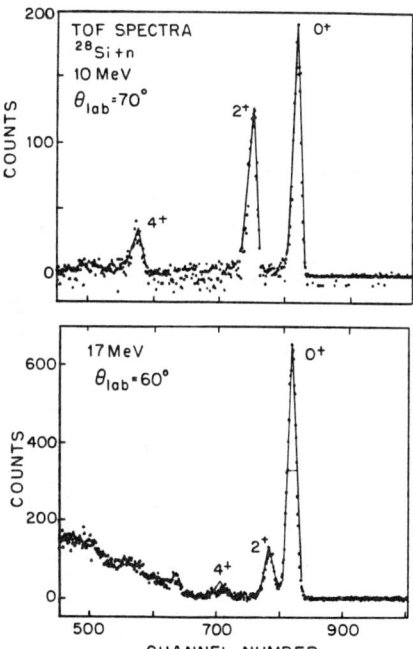

Fig. 1. Time-of-flight spectra for the scattering of polarized neutrons from ^{28}Si. At 17 MeV the residual background from neutrons originating in the ^{2}H(d,np) reactions becomes a problem for resolving inelastic neutron scattering peaks. (Note that the vertical scale is offset from zero counts.)

publications of the Ohio University neutron group. The analyses have followed several paths. The light nuclei have been approached through the Lane model, weighing heavily on self-consistency with (p,n) and (p,p) cross sections and $A_y(\theta)$, where available. The other nuclei are studied with both a spherical and a deformed optical model approach, paying heed to the total cross section σ_T, the s- and p-wave strength functions, and some of the (p,p) data. Several measurements and analyses will now be singled out. In addition, the papers in this volume by Delaroche et al., Walter et al., and Howell report on some other aspects of our analyses.

LANE MODEL FOR NUCLEON + ^{11}B

Our most recent Lane model calculations for 1p-shell nuclei involve nucleon interactions with ^{11}B, i.e., ^{11}B(n,n), ^{11}B(p,n), and ^{11}B(p,p). For this study $\sigma(\theta)$ and $A_y(\theta)$ data were measured for the (n,n) and (p,n) channels and $\sigma(\theta)$ data for the (p,p) channel were obtained from the literature. Unfortunately, no $A_y(\theta)$ for the (p,p) channel are available in our energy range, to our knowledge. In our model the neutron elastic scattering potential U_{nn} and the symmetry potential U_1 were used as a basis for calculating the observables for (n,n) and (p,p) elastic scattering and the (p,n) quasi-elastic channel through the Lane model equations:

$$U_{nn} = U_{nn}$$

$$U_{pp} = U_{nn} + 4TU_1/A$$

$$U_{na} = U_{nn} + 2U_1/A$$

$$U_{pn} = -2(2T)^{1/2}U_1/A$$

The U_{na} is the potential for neutron scattering from the isobaric analog state of the target nucleus. Here U_{nn} and U_1 are of the form $U = V + iW + (V^{so}+iW^{so})\vec{\sigma}\cdot\vec{\ell}$ and the form factors are the commonly used Woods-Saxon for V and W and the Thomas-type for V^{so} and W^{so}, except for V_1^{so}, in which the $\frac{1}{r}$ factor is replaced by a constant.

The graphs of the potentials for neutron scattering, V_{nn} and $W_{nn} = (W_d)_{nn}$, are shown in fig. 2. The data points for V_{nn} and W_{nn} correspond to values obtained in single energy searches on ^{11}B(n,n), except for 15 MeV for which the W_{nn} was forced to lie on a smooth curve for $W_{nn}(E)$. The value for the spin-orbit term V_{nn}^{so} was held fixed at 5.9 MeV following initial searches on the (n,n) and p,n) sets.

For describing the ^{11}B(p,n) data, it was important to introduce a correction for the difference in the kinetic energies of the neutron and proton inside the nuclear region. If the energy dependence of the Lane equations is displayed, they take the following form:

$$U_{nn}(E_{inc}) = U_{nn}(E_{inc})$$
$$U_{pp}(E_{inc}) = U_{nn}(E_{inc}-\Delta E) + 4TU_1(E_{inc})$$
$$U_{nA}(E_{inc}) = U_{nn}(E_{inc}-\Delta E) + 2U_1(E_{inc})$$
$$U_{pn}(E_{inc}) = \qquad\qquad -2(2T)^{1/2}U_1(E_{inc})$$

Here ΔE is the difference in the kinetic energies inside the interior of the nucleus and was estimated from the combined effects of the Coulomb potential in the nuclear interior and the symmetry potential. Values for U_{pp} and U_{nA} were then obtained by estimating from the plot of $U_{nn}(E_{inc})$ in fig. 2 its value at $U_{nn}(E_{inc}-\Delta E)$. The values obtained for the corrections to the real well depth, $\Delta V = V_{nn}(E_{inc}-\Delta E) - V_{nn}(E_{inc})$, are about -0.1 MeV below 14 MeV and about $+1.5$ MeV at 17 MeV. The value $0.4Z/A^{1/3}$, which is commonly used for this correction, yields a value of about $+1$ MeV. The corrections $\Delta W = W_{nn}(E_{inc}-\Delta E) - W_{nn}(E_{inc})$ to the imaginary well are about -2 MeV below an E_{inc} of about 14 MeV and $+2$ MeV at 17 MeV, reflecting the change in slope of W_{nn} of fig. 2.

Calculations using the potential values in fig. 2 are shown in fig. 3 alongside the $\sigma(\theta)$ data for (n,n). In fact, for all three channels the $\sigma(\theta)$ data have been fairly well reproduced, particularly at the higher end of the energy range. Likewise, $A_y(\theta)$ for $^{11}B(n,n)$ is successfully reproduced, as shown in fig. 4. The calculations for $A_y(\theta)$ for $^{11}B(p,n)$ are not shown here. The agreement is quite good except that for back angles $A_y(\theta)$ is overestimated in magnitude.

The real part of the symmetry potential U_1 is not surface peaked as has been found for heavier nuclei, but rather has a shape similar to that found in our previous studies[5] for the mirror nuclei ^9Be, ^{13}C, and ^{15}N in this mass range. This is consistent with the view of surface peaking in U_1 arising from the concentration of excess neutrons on the surface, since for these nuclei the neutron excess is very small.

The sensitivity of the $A_y(\theta)$ data allows one to investigate the need for the isospin spin-orbit interaction $V_1^{SO}(r)$. In the ^{11}B analysis, we find a consistent preference for a positive strength of 3 to 5 MeV. No need was established for a W_1^{SO} (nor a W_{nn}^{SO}) in this work.

In a reanalysis of nucleon scattering from ^9Be, which now incorporates our recently available $A_y(\theta)$ data for $^9Be(n,n)$, our fits are of somewhat better quality than those shown here. We are finding a need for a V_1^{SO} strength of 2.5 to 3.0 MeV, similar to the strength found for ^{11}B. However, for ^9Be in which $A_y(\theta)$ data exists for all three channels, fitting the data requires a large strength for W_{nn}^{SO} (+1.9 to 0.7 MeV) although the proton channel prefers $W_{pp}^{SO} = 0$. That is, the only way we can find to fit all the data sets simultaneously is to introduce an imaginary part for the U_1^{SO}, i.e., a $W_1^{SO} \neq 0$.

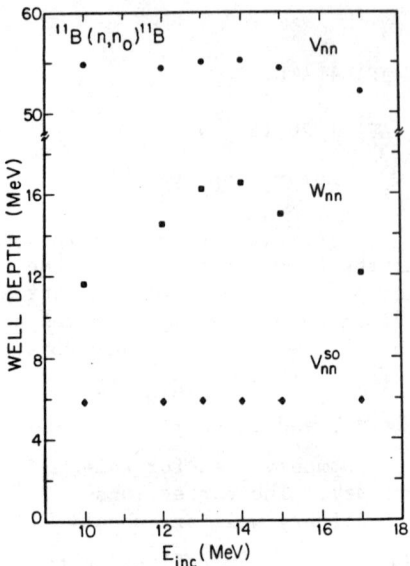

Fig. 2. Potentials for ^{11}B(n,n).

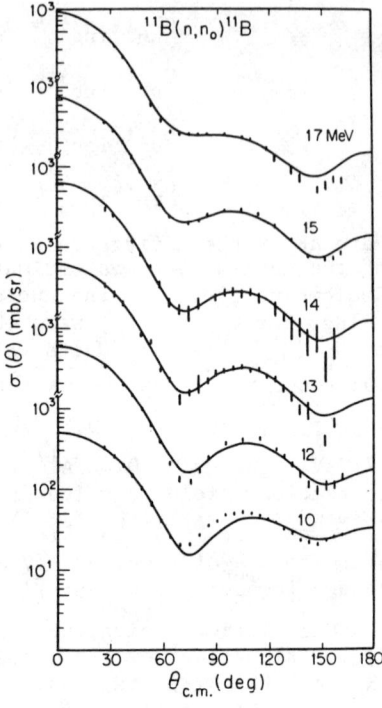

Fig. 3. Data and Lane model fits for $\sigma(\theta)$ for ^{11}B(n,n).

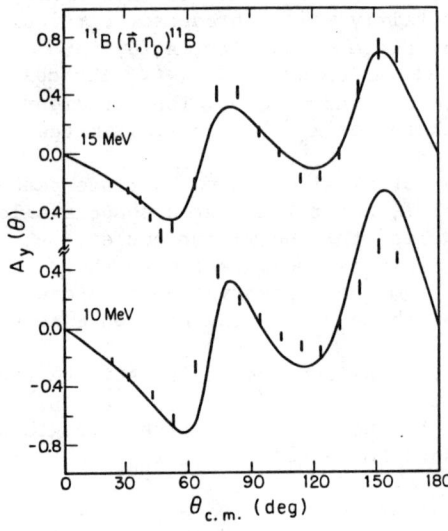

Fig. 4. Data and Lane model fits for $A_y(\theta)$ for ^{11}B(n,n).

Deformed Potential Calculations

Elastic and inelastic neutron scattering data for many of our nuclei have been analyzed through coupled-channels (CC) calculations. Much of this phase of our program has been strongly influenced by the experiences of our collaborator J. P. Delaroche. In fact, the contribution to this volume by Howell on ^{28}Si and ^{32}S illustrates the thoroughness of the CC analyses for individual nuclei, and by Delaroche et al. on the total cross section σ_T at high energies presents another aspect of the results of our constrained CC analyses.

Two cases that we have carefully studied are the pairs ^{58,60}Ni and ^{116,120}Sn. The results of the calculations, which are based on fixed geometry parameters and smooth energy dependencies for the potential strengths, are shown in figs. 5-7. A similar set of data for $\sigma(\theta)$ for ^{60}Ni and ^{116}Sn was also obtained and is described equally well in each case by using the same potential (which includes the symmetry term to account for the different neutron numbers) for each pair of isotopes. In the analyses we looked for consistency with the s- and p-wave strength functions (when available), the total cross section $\sigma_T(E)$ and proton scattering data. Also, in the CC calculations, the full Thomas form[6] of the spin-orbit interaction was used; this form allowed the s.o. potential to be deformed. We permitted the s.o. deformation parameter β_{so} to take on values which were different from that of the central wells, β_c.

To illustrate the type of parameterization we obtained, the potential parameters for Sn are given here:

$V = 51.97 - 20.83\varepsilon - 0.22E$ (0.0 < E < 100 MeV)

$W_D = 3.29 - 15.97\varepsilon + 1.71E^{1/2}$ (0.0 < E < 14 MeV)

$W_D = 9.69 - 15.97\varepsilon - 0.15(E-14)$ (14 MeV < E < 100 MeV)

$W_V = 0.0$ (0.0 < E < 14 MeV)

$W_V = 0.14(E-14)$ (14 < E < 100 MeV)

$V_{so} = 6.50 - 0.030E$ (0.0 < E < 100 MeV)

$r_V = 1.23 \quad r_D = 1.25 \quad r_{so} = 1.12$

$a_V = 0.66 \quad r_D = 0.54 \quad a_{so} = 0.50$

Here, ε is the symmetry parameter $(N-Z)/A$. The coupling basis in this work is $(0^+, 2^+, 3^-, 4^+, 5^-)$. The deformation parameters for ^{120}Sn, for example, are $\beta_{2^+} = 0.100$, $\beta_{3^-} = 0.150$ and $\beta_{so}r_{so} \sim \beta_c r_c$. The β_c values for neutron scattering from the Sn isotopes (as well as for Ni) are about 10% smaller than the corresponding values that we extracted for proton scattering, a finding which is consistent

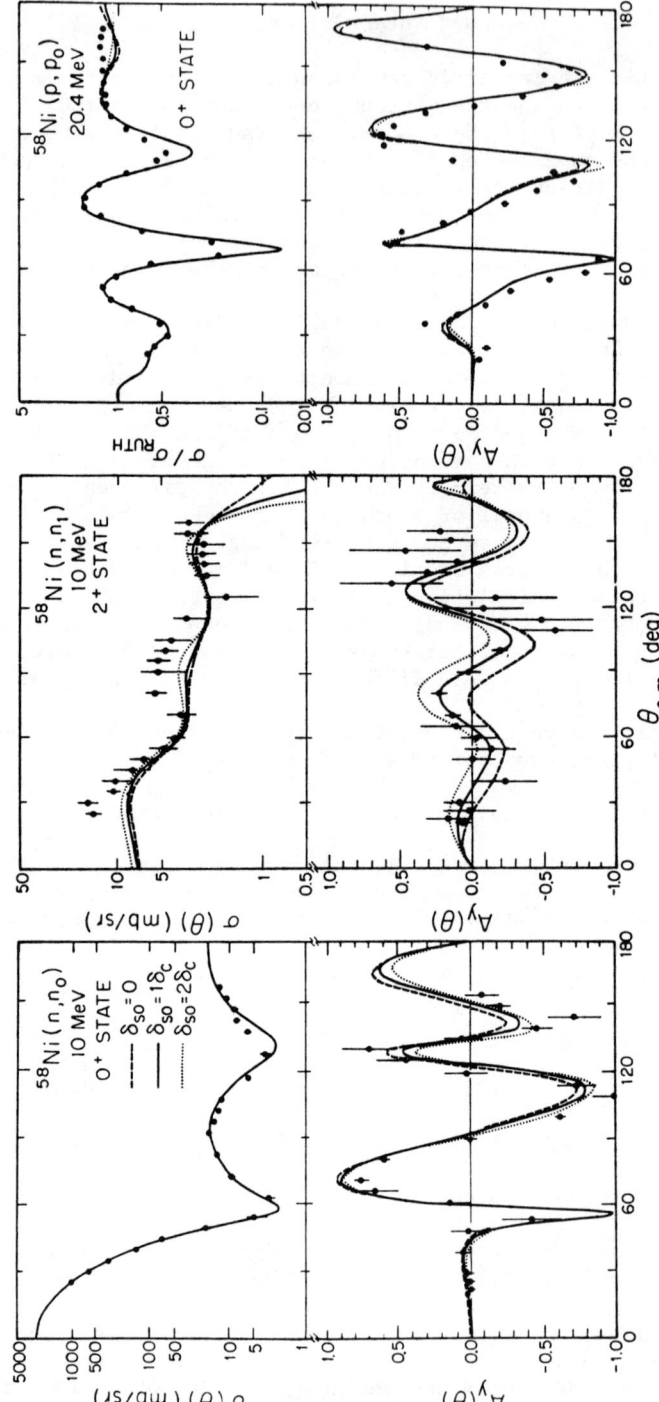

Fig. 5. Coupled channels calculations for neutron and proton scattering from ^{58}Ni.

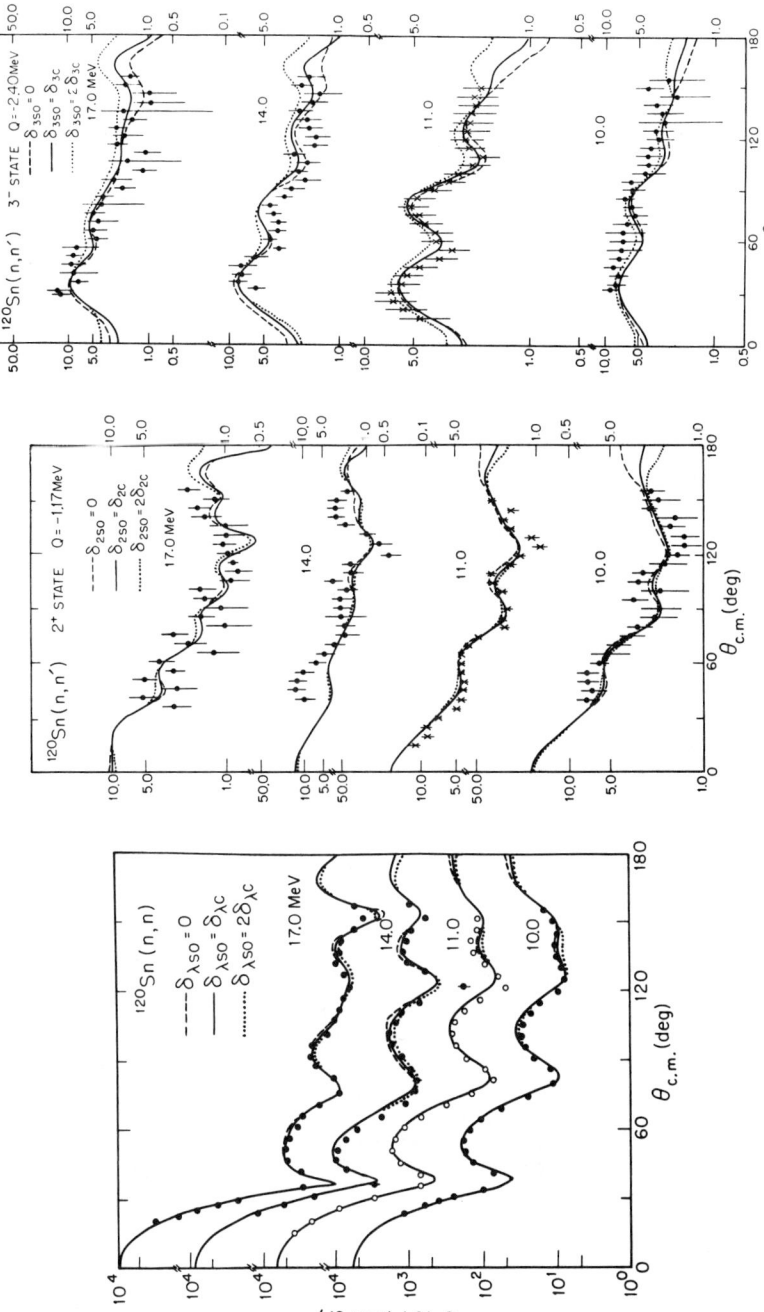

Fig. 6. Coupled channels calculations compared to (n,n) and (n,n') data for ^{120}Sn.

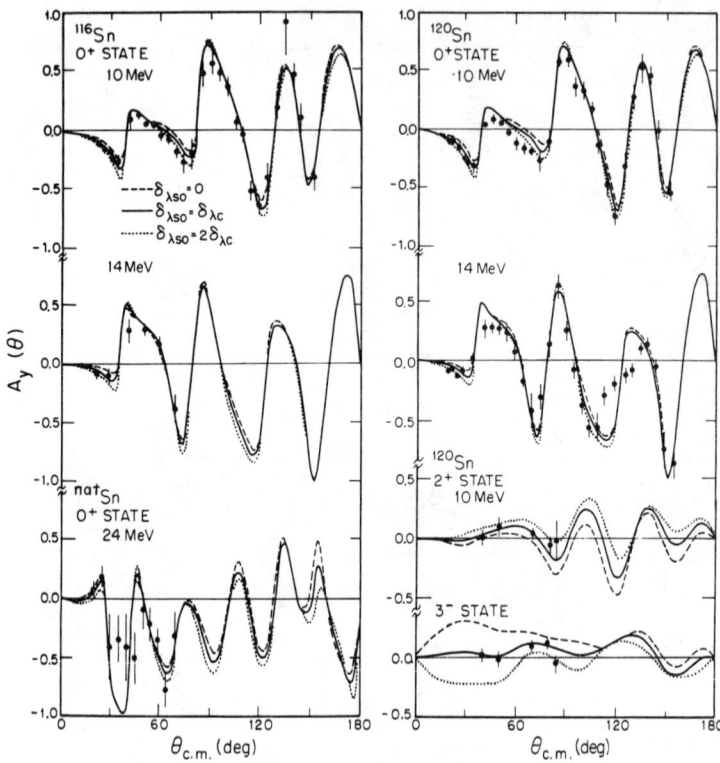

Fig. 7. Coupled channels calculations compared to (n,n) and (n,n') data.

with the core-polarization model of Brown and Madsen.[7]

The sensitivity of $A_y(\theta)$ for inelastic scattering to the size of the deformation length $\delta_{so} = \beta_{so} r_{so}$ is shown in fig. 8, which is taken from a recent publication.[8] Since that report, we have obtained more (n,n') data at 10, 14 and 17 MeV and for ^{56}Fe and ^{60}Ni in order to determine if there is an energy dependence for β_{so} and, if so, whether it corresponds to that seen for (p,p') scattering. In all the calculations shown in fig. 8, and in figs. 6 and 7 for Sn, we had set the imaginary part of U_{so} equal to zero. We know now that the fit to the Sn data for $A_y(\theta)$ would have been slightly improved with a $W_{so} \sim +0.7$ MeV. Furthermore, we are now aware that there is an ambiguity[9] between the magnitudes of δ_{so} and W_{so}, as far as the calculations of $A_y(\theta)$ for (n,n') is concerned. It is important to have comprehensive data sets for these analyses in order to draw suitable conclusions.

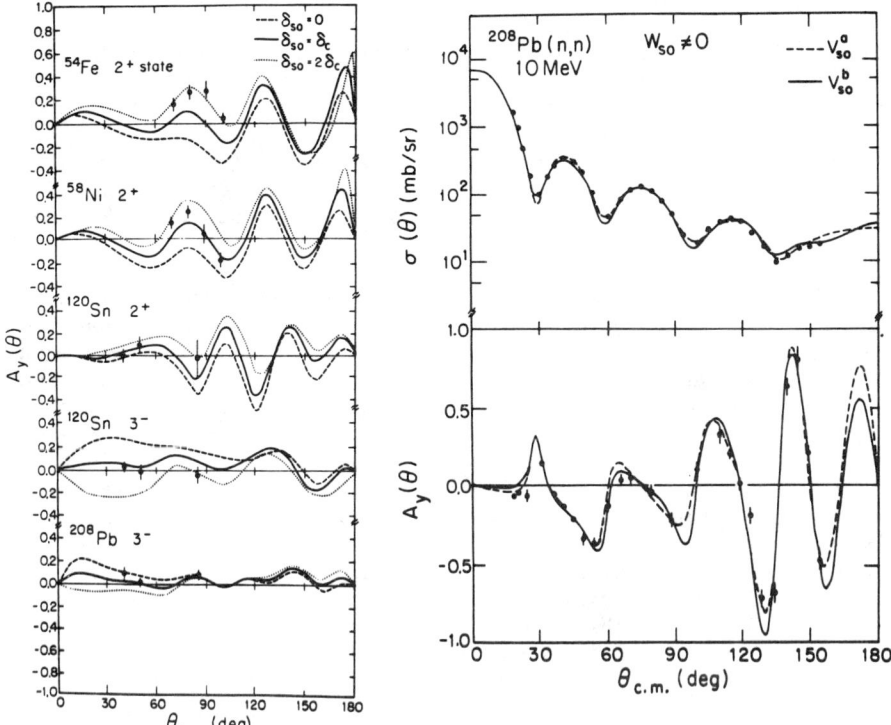

Fig. 8. The $A_y(\theta)$ measurements for (n,n') compared to CC calculations using several values for δ_{so}. Except for ^{54}Fe, the data favor $\delta_{so} \sim \delta_c$.

Fig. 9. The $\sigma(\theta)$ and $A_y(\theta)$ for ^{208}Pb(n,n) compared to SOM calculations using a conventional Thomas form V_{so}^a and microscopic form V_{so}^b.

Spherical Optical Models

Although spherical optical models (SOM) give a poorer representation of the nuclear interaction than do the deformed potential models, it is more convenient to make an initial survey of the data base for elastic scattering with an SOM. Furthermore, the parameters derived in the SOM are useful for i) studying the systematics of nuclei, ii) predicting scattering for neighboring nuclei, iii) comparing proton scattering to neutron scattering and determining Coulomb corrections and isospin parameters, and iv) providing optical potentials for DWBA calculations and for applied purposes. In addition, the parameters can be used as a starting set for the more complex and time consuming CC calculations.

For all the nuclei that we have studied, we have looked individually or in pairs for a suitable SOM description of the $\sigma(\theta)$ and $A_y(\theta)$ for elastic scattering. Here again, available data up to either 26 or 40 MeV are also included in the data base. The individual SOM models have been quite successful in describing the data for most nuclei. The parameters for the real and central wells fall among those commonly accepted for protons, making proper allowances for the symmetry and the Coulomb correction terms. For U_{SO} a real part with $5.0 < V_{SO} < 7.5$ MeV, $1.05 < r_{SO} < 1.21$ fm and $a_{SO} \simeq 0.5$ fm and an imaginary part with $+0.2 < W_{SO} < +1.1$ MeV has been found for medium-weight to heavy nuclei. For light nuclei (analyzed with an SOM) we have found that the data prefer an a_{SO} under 0.4 fm, and ^{40}Ca and ^{9}Be require a positive W_{SO} to achieve good fits. Whether our data can establish the size and sign of the symmetry spin-orbit interaction is still undecided.

Several specific examples of SOM studies will now be discussed. In the first case, we found in the combined CC and SOM analyses for the Ni and Sn isotopes that we were able to achieve almost as good fits to the elastic scattering data with an SOM which used all the geometries and strengths of the CC potentials, except for an anticipated increase of about 2 MeV in the strength of W_D. For example, for Sn we found that changing the CC equation for W_D given in the previous section to the relation $W_D = 11.43 - 16.0\varepsilon - 0.15(E-14)$ for $14 < E < 100$ MeV gives a good description of the elastic scattering data. This approach of deriving an SOM from a starting point of such a highly constrained CC calculation is the reverse of the traditional method--however, it yielded a good SOM representation. Because of the SOM ambiguities, it probably would have been fortuitous if the opposite procedure would have yielded such good CC results. One conclusion of this work regarding the spin-orbit potential is that coupling in the additional channels does not introduce a need for changing V_{SO} from the one which gives good CC fits. However, our data are probably still too inaccurate to test an analogous statement for the imaginary part of U_{SO}.

In fig. 9 two calculations are shown for ^{208}Pb(n,n) at 10 MeV. The purpose of this study[10] was to see if the data favored a semi-microscopic form for $V_{SO}(r)$ over the conventional Thomas-type derivative Wood-Saxon form. In the microscopic case neutron and proton densities, which are determined in Hartree-Fock calculations based on an effective internucleon force, are related to the $V_{SO}(r)$ through Scheerbaum's prescription.[11] Both calculations agree equally well, although the Thomas-type had twelve free parameters, two more than the microscopic method. (In order to obtain the good fits shown in fig. 9, it was necessary to introduce a $W_{SO} = +0.7$ MeV for both calculations.)

Before moving to our global analyses, it seems worthwhile to illustrate the significance of the spin-orbit interaction on the angular distribution for the scattering of unpolarized neutrons. Likewise, the illustration also shows how analyses of accurate differential cross sections alone can indeed be a sensitive probe for defining the spin-orbit strength. To permit this, we define the differential cross section for a completely polarized spin-up beam as $\sigma_\uparrow$ and for a completely polarized spin-down beam as $\sigma_\downarrow$. Then the analyzing power will be given by $(\sigma_\uparrow - \sigma_\downarrow)/(\sigma_\uparrow + \sigma_\downarrow)$ where $(\sigma_\uparrow + \sigma_\downarrow)$ is the usual cross section $\sigma(\theta)$ that is observed if the incident beams were unpolarized. Because the spin-orbit interaction either decreases or increases the strength of the nuclear interaction, depending on the relative orientation of $\vec{\ell}$ and $\vec{s}$, the diffraction maxima (and minima) shift either toward or away from the central maximum (at 0°) for the spin-up or spin-down beams. In fig. 10 a series of calculations (similar to those shown previously for higher energy protons) are shown for 10-, 14- and 17-MeV neutrons using an SOM which describes our $\sigma(\theta)$ and $A_y(\theta)$ data for Pb. The results dramatically illustrate how the complex structure of $\sigma(\theta)$ at back angles is intimately tied to the cross sections for the spin-up and spin-down components of an unpolarized beam (equal parts spin-up and spin-down). To indicate that a spin-orbit sensitivity also can occur in the forward hemisphere and for nuclei lighter than Pb in our energy range, in fig. 11 we have plotted the results of a calculation for ^{65}Cu which also describes our $\sigma(\theta)$ and $A_y(\theta)$ measurement. In the region from 45° to 60°, the spin-down cross section is responsible for nearly 80% of the normal differential cross section $\sigma(\theta)$ and is the source of the second diffraction peak. (In these curves and all the SOM calculations shown in this paper, the Mott-Schwinger interaction has been included. This interaction causes the large analyzing power at scattering angles near 1°.)

Lastly, we present a preview of one of the global optical models P. P. Guss and I have been developing as the $A_y(\theta)$ data is is becoming available to us. We have created a large data base which incorporates much of the TUNL $A_y(\theta)$ and $\sigma(\theta)$ results. We have also included the new σ_T data from ORELA, $\sigma(\theta)$ and $A_y(\theta)$ data near 24 MeV from Ohio University and Livermore, respectively, and $\sigma(\theta)$ data at 30 and 40 MeV from Michigan State University. Of course, our goal is to find an SOM parametrization which will produce the best global representation of the data base, and then to investigate the significance of the potential terms. Our model is based on the conventional Woods-Saxon form factors and the Thomas-type spin-orbit interaction. We allow for an imaginary part of the spin-orbit potential as well.

Starting values are guided by our previous searches on individual nuclei, but the geometry is kept essentially independent of A, except for the diffuseness of W_V. After the initial attempts, it was decided to heavily constrain the energy dependence of V_{SO} and W_{SO} by including the 80 MeV (p,p) data for $A_y(\theta)$ that Schwandt et al.[12] measured for ^{40}Ca, ^{90}Zr and ^{208}Pb. For neutrons we also split off the data for nuclei lighter than ^{54}Fe--these lighter nuclei require freeing up the geometries and the energy dependencies more than we are willing to permit in our initial model.

The status of our work to date is indicated by the fits shown in figs. 12-14. The 80 MeV (p,p) $A_y(\theta)$ data are shown along with the neutron data in fig. 13. This is the first time such a large data base has been used in an SOM search above 10 MeV. Many of the data sets are quite well described, including the 80 MeV (p,p) data sets. As seen in fig. 15, the energy dependencies for the volume and surface absorption terms are linear, and the surface term goes to zero around 80 MeV; the symmetry term, which is proportional to (N-Z)/A and which has been introduced into the surface term, causes

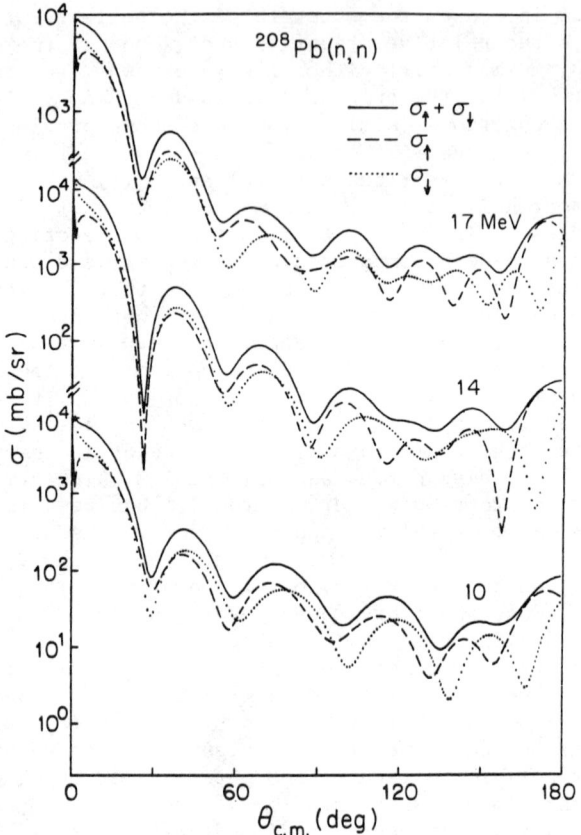

Fig. 10. The SOM calculations of the cross sections for 100% polarized beams ($\sigma_\uparrow$ or $\sigma_\downarrow$) and the sum ($\sigma_\uparrow + \sigma_\downarrow$) which corresponds to an unpolarized beam condition.

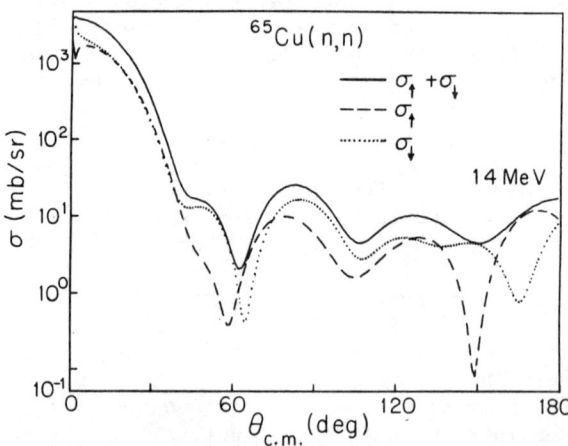

Fig. 11. See caption for fig. 10.

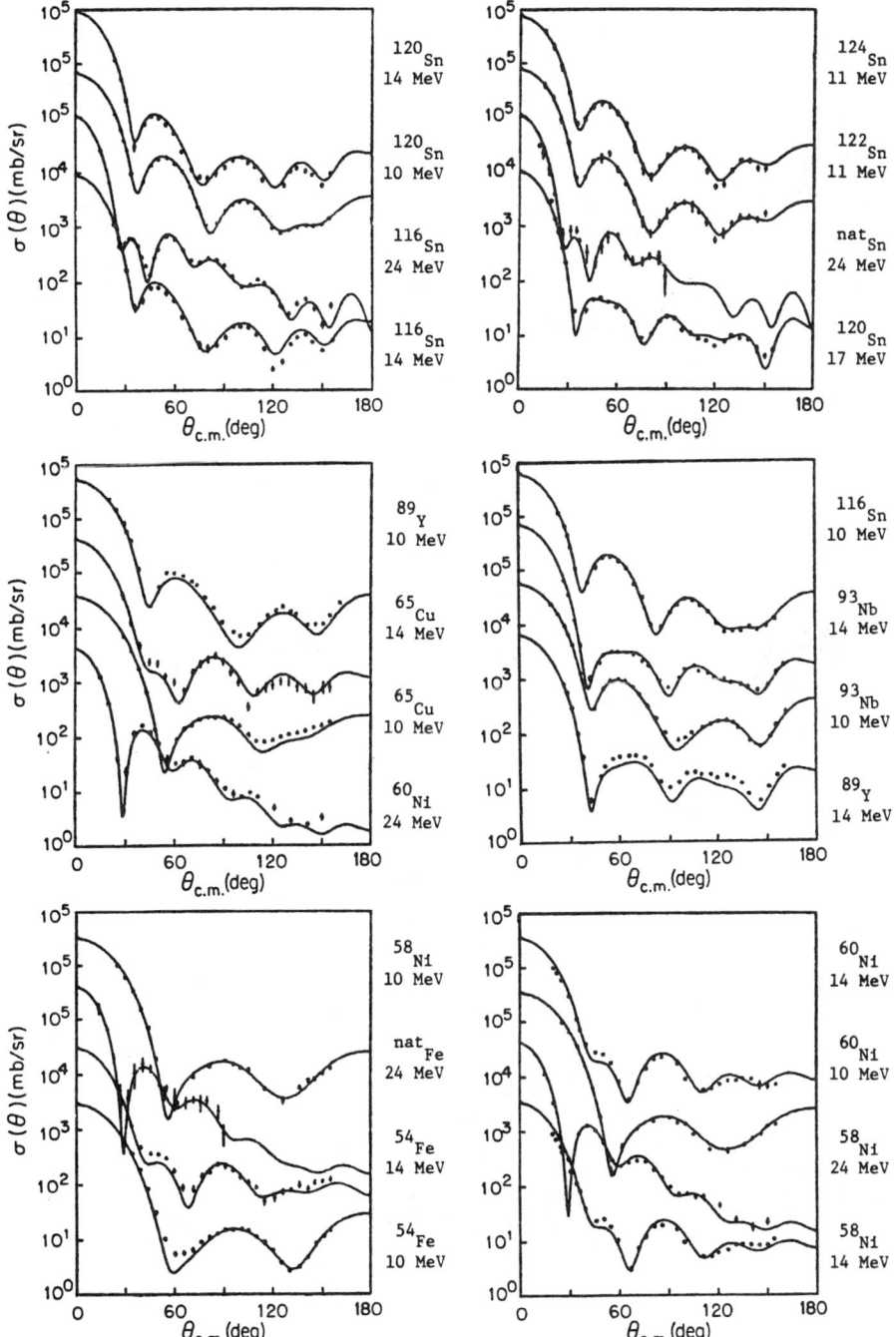

Fig. 12. Predictions of global SOM for (n,n) scattering.

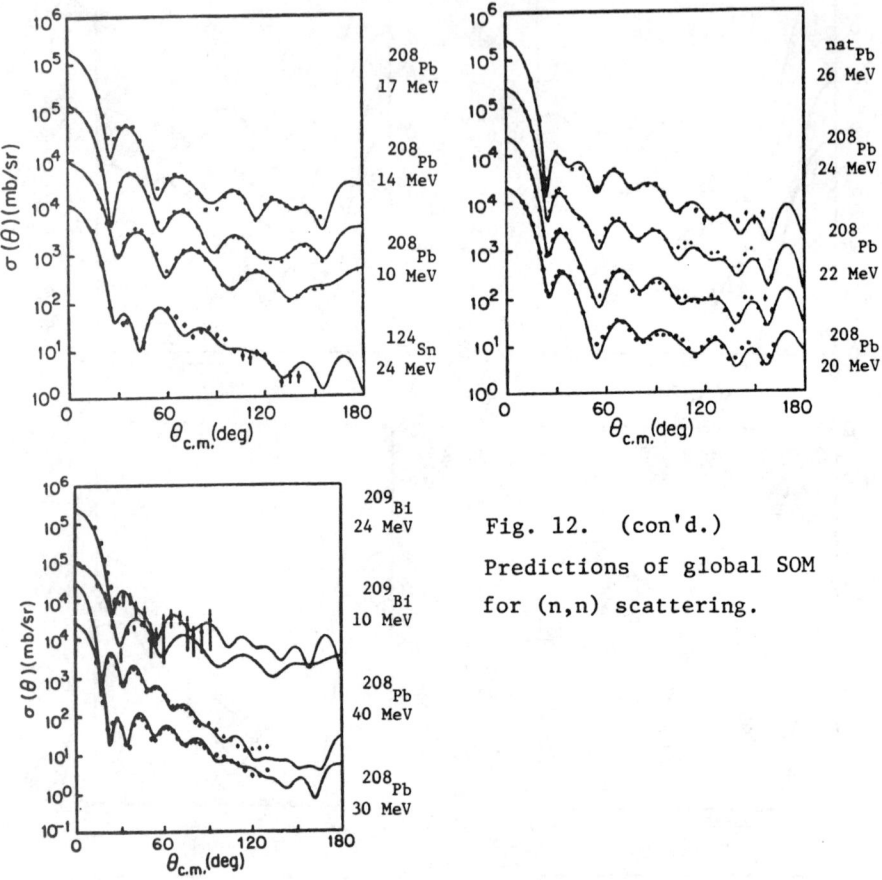

Fig. 12. (con'd.) Predictions of global SOM for (n,n) scattering.

the slopes to be different for ^{54}Fe and ^{208}Pb. Likewise, V_{SO} and W_{SO} are linear. The real central potential V is linear up to 60 MeV and becomes logarithmic thereafter. All of the potentials tie onto those of Schwandt et al. very well, except for W_V, indicating that we may have to decrease our slope somewhat. However, the global search seems to prefer an even larger slope. There are other puzzles out there in χ^2 space, and some might be blamed on our simple model. As we do not have relativistic corrections in the global search code yet, there will need to be some slight adjustment to our potentials. Therefore, we will refrain from presenting the parametrization until we can deal with this problem. However, since this paper has been focusing on the spin-orbit potential, we will state that our present position gives V_{SO}= (6.36-0.0174E) MeV, r_{SO}= 1.15, a_{SO}= 0.612, W_{SO}= (1.11 - 0.0203E) MeV, r_{WSO}= 1.19, $a_{SO} \approx 0.80$.

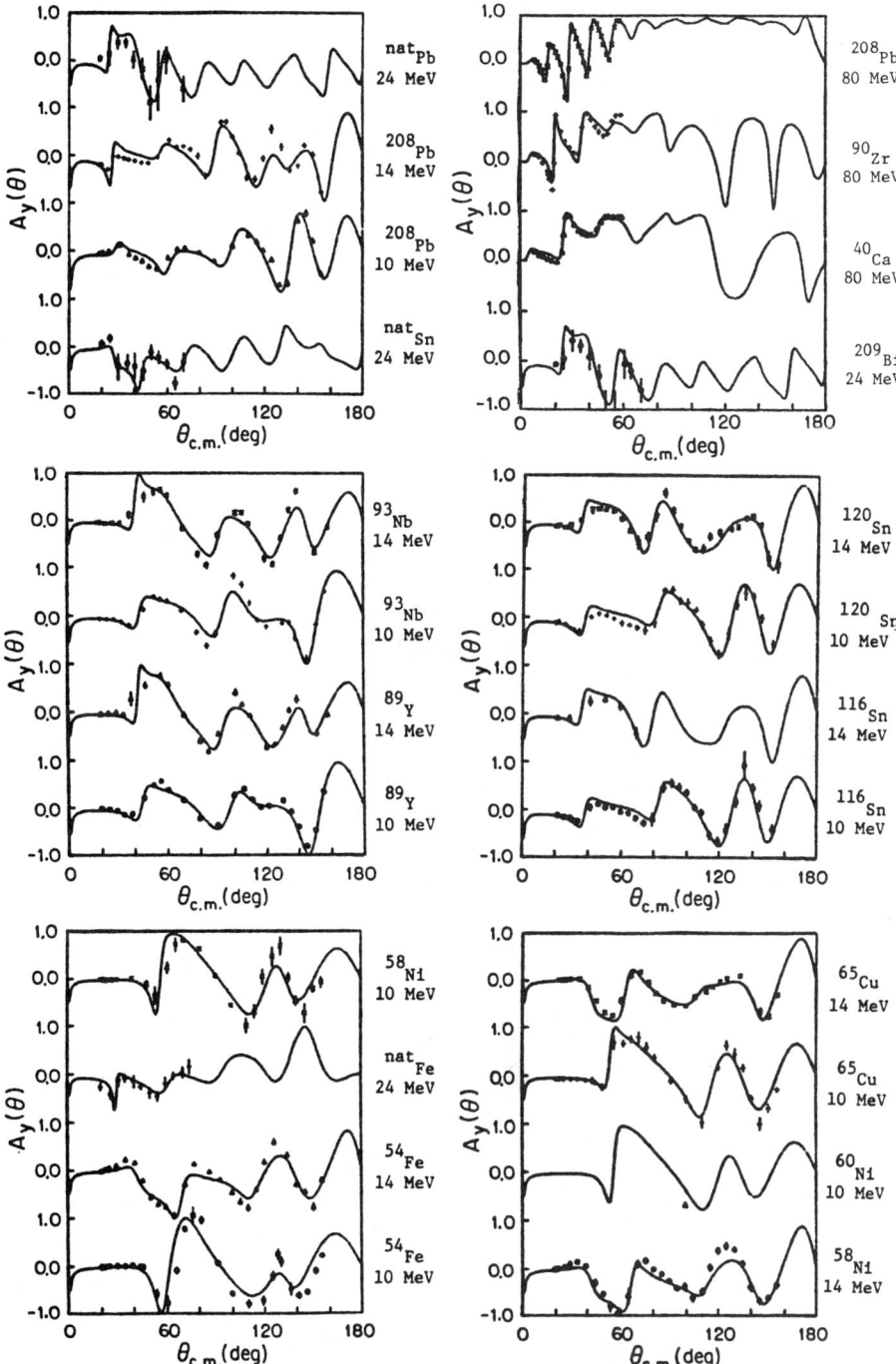

Fig. 13. Predictions of global SOM for (n,n) and (p,p) scattering.

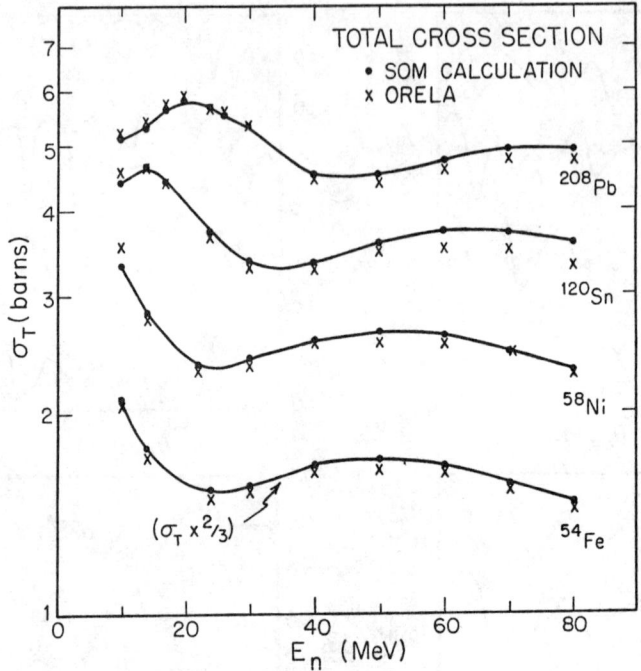

Fig. 14. The global SOM calculations (circles) compared to the recent ORELA measurements (crosses). The solid curve is a guide to the eye through the circles.

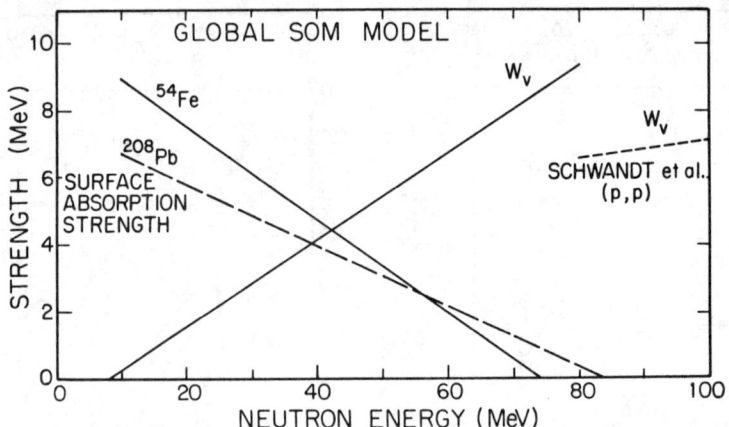

Fig. 15. The strengths of the imaginary surface and volume terms derived in the global SOM search with the data shown in the previous figures.

In summary, our measurements of $A_y(\theta)$ and the associated $\sigma(\theta)$ are becoming accurate enough to examine some details of the neutron-nucleus interaction that have not been accessible before. Furthermore, the measurements are producing greater constraints on the nuclear models, and because the total quantity of data is becoming quite large now in the energy region above 10 MeV, global properties of the neutron-nucleus interaction are becoming better defined. But even with the $A_y(\theta)$ data that we have provided, it will be important to obtain such data with similar accuracy at higher energies where the spin-dependent interaction for neutrons has not been well tested.

I would like to kindly acknowledge the students and fellows at TUNL whose diligent contributions are reviewed here: R.C. Bryd, P.P. Guss, C.E. Floyd, K. Murphy, C.R. Howell, R. Pedroni, M. Roberts, G. Honore, H. Pfutzner, G. Tungate, Anli Li, and S.A. Wender. Drs. J.P. Delaroche and W. Tornow have contributed effectively in interpreting the data and Prof. T.B. Clegg has carried the major responsibility for the upkeep of the polarized ion source and the design for pulsing the polarized deuteron beam.

*Supported in part by the U. S. Department of Energy Director of Energy Research, Office of High Energy and Nuclear Physics, under Contract No. DE-AC05-76ER01067.

1. C. Wong, J. D. Anderson, J. W. McClure and B. D. Walker, Phys. Rev. 128, 2339 (1962).
2. A. H. Hussein et al., Phys. Rev. C 15, 233 (1955).
3. G. Bulski et al., in Nuclear Data for Science and Technology, ed. by K. H. Bockhoff, D. Reidel Publ. Co., Holland (1983), p. 783.
4. S. A. Wender, C. E. Floyd, T. B. Clegg and W. R. Wylie, Nucl. Instrum. Methods 174, 341 (1980).
5. R. C. Byrd et al., Nucl. Phys. A399, 94 (1983); Nucl. Phys. A404, 29 (1983); Nucl. Phys. A410, 29 (1983).
6. H. Sherif and J. S. Blair, Phys. Lett. 26B, 489 (1969).
7. V. R. Brown and V. A. Madsen, Phys. Rev. C11, 1298 (1975); Phys. Rev. C12, 1205 (1975).
8. P. P. Guss et al., Phys. Rev. C 25, 2854 (1982).
9. C. E. Floyd et al., Phys. Rev. C 28, 1498 (1983).
10. J. P. Delaroche et al., Phys. Rev. C 28, 1410 (1983).
11. R. R. Scheerbaum, Nucl. Phys. A257, 77 (1976).
12. P. Schwandt et al., Phys. Rev. C 26, 55 (1982).

SCATTERING OF POLARIZED AND UNPOLARIZED NEUTRONS FROM ^{28}Si and ^{32}S

C. R. Howell
Duke University and Triangle Universities Nuclear Laboratory*,
Durham, N.C. 27705

ABSTRACT

The most complete set of differential cross sections $\sigma(\theta)$ and vector analyzing powers $A_y(\theta)$ for neutron scattering from ^{28}Si and ^{32}S over the 8 to 40 MeV energy range has been collected and evaluated. These data have been successfully described with the spherical optical model (SOM) and with phenomenological coupled-channels (CC) calculations. For the case of ^{28}Si, various calculations were performed to determine the sensitivity of the data and calculations to the sign of the potential deformation parameters β_2 and β_4. Comparisons of the present analyses to the corresponding analyses of proton-scattering data have yielded interesting results concerning the magnitudes of the Coulomb corrections to the OM potentials for ^{28}Si and ^{32}S.

INTRODUCTION

The even-even T=0 nuclei of ^{28}Si and ^{32}S have received much theoretical and experimental consideration. These nuclei are particularly intriguing because of the strong collective nature of their excited states. The investigation of these nuclei using nucleons as a probe has progressed in the usual manner with the proton-scattering data and analyses preceding that of the neutron work. Analyses of ^{28}Si(p,p') data over the energy range from 14 to 40 MeV with coupled-channels (CC) calculations have yielded much information concerning the deformation parameters β_2 and β_4 of the central potential and β_{so} of the spin-orbit potential.[1,2] Similar analyses of ^{32}S(p,p') data have been conducted over the energy range from 15 to 35 MeV yielding information about the collective properties of the low-lying states in ^{32}S.[1,3,4] Only recently have neutron-scattering data of comparable accuracy to the proton data become available for these nuclei.[5] Because optical model (OM) analyses of these high accuracy neutron data have been conducted only at single energies and over narrow energy ranges spanning 10 MeV at most, the trends of the OM parameters for the neutron-nucleus interaction have not been studied for these nuclei over a broad energy range.

The objectives of the present investigation were: 1) to study the collective properties of the low-lying states in ^{28}Si and ^{32}S using neutrons as a probe, 2) to determine the energy dependencies of the OM parameters for these nuclei over a broad energy range, 3) to parametrize for the first time the spin-orbit (SO) part of the optical model potential (OMP) for the neutron-nucleus interaction, and 4) to extract accurate values for the Coulomb corrections to the real and imaginary parts of the OMP. The fulfillment of these objectives required very accurate differential cross sections $\sigma(\theta)$, vector analyzing powers $A_y(\theta)$, and neutron total cross sections σ_T over a wide energy range. To obtain the necessary data a literature search was performed which resulted in the collection of much high accuracy $\sigma(\theta)$ data for elastic and inelastic neutron scattering from 10 to 40 MeV. The neutron facilities at TUNL were used to fill the gaps in the existing $\sigma(\theta)$ data and to measure a complete set of $A_y(\theta)$ data from 8 to 17 MeV. Prior to the present measurements no $A_y(\theta)$ data for $^{32}S+n$ existed above 8 MeV and only one $A_y(\theta)$ angular distribution for $^{28}Si+n$ had been reported above 8 MeV, that of Böttcher et al. at 14.1 MeV.[5]

THE DATA SET

To reduce the ambiguities in the OM parameters the analyses were extended over a wide energy range by including data from several references. High quality neutron-scattering data from recent measurements were collected and evaluated to form the most complete sets of $\sigma(\theta)$ and $A_y(\theta)$ data for neutron scattering from ^{28}Si and ^{32}S in existence. At energies where several independent measurements existed, only the highest accuracy data were included in the data sets. Both data sets spanned an energy range from 8 to 40 MeV and consisted of 22 $\sigma(\theta)$ distributions and 6 $A_y(\theta)$ distributions. Differential cross sections for elastic scattering and inelastic scattering to the first excited state were measured for both nuclei at incident energies of 8, 10, 12, 14, and 17 MeV using the neutron time-of-flight facilities at TUNL. These facilities and the techniques applied to measure the $\sigma(\theta)$ data have been described in detail in reference 6 and therefore will not be presented here. The data at 9.8 and 14.8 MeV were obtained by private communications with members of the neutron group at Bruyeres-le-Châtel (BRC). The data at 11, 20, and 26 MeV were measured at Ohio University and were reported in R. C. Tailor's Ph.D. dissertation.[5] The $\sigma(\theta)$ for elastic scattering at 30.3 and 40 MeV were measured at Michigan State University by DeVito et al.[5] The relative uncertainties in the elastic distributions ranged from 2% to 5% and those in the inelastic data were 5% to 10%. All distributions had a 5% normalization uncertainty. All $A_y(\theta)$ data included in the data sets were measured at TUNL using the pulsed polarized neutron facilities described in references 6 and 7. The uncertainties in the $A_y(\theta)$ data ranged from 3% to 8% for the elastic data and from 6% to 12% for the inelastic distributions. There was a 2% normalization uncertainty in these data due to the uncertainty in the value for K_y. Total cross sections were taken from ENDF/B-5 for neutron energies below 20 MeV and from the ORELA data at higher energies.

A comparison of the present data to the data reported by Böttcher et al., hereafter referred to as the Erlangen data, is presented in fig. 1. With the exception of the apparent discrepancies in the elastic $A_y(\theta)$ around 40° and 110° c.m., the two data sets are in good agreement. These discrepancies have been attributed to the inability of Böttcher et al. to completely resolve the elastically scattered neutrons from those that inelastically scattered to the first excited state. The Erlangen data are not included in the $A_y(\theta)$ data set because of these apparent discrepancies. Because the model calculations were very sensitive to $A_y(\theta)$ in the angular region around the discrepancies, the Erlangen point near 40° c.m. was plotted along with the TUNL data in comparison to the calculations.

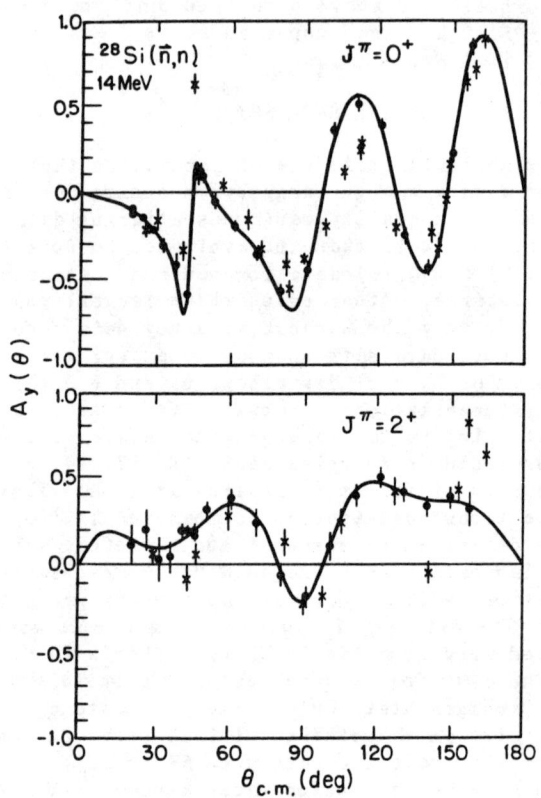

Fig. 1. The $A_y(\theta)$ for elastic and inelastic neutron scattering from ^{28}Si at 14 MeV. The circles are the present data and the crosses represent the data reported by Böttcher et al.[5] The curves are fits derived from Legendre polynomial fits to $A_y(\theta) \cdot \sigma(\theta)$ of the present data.

THE SPHERICAL OPTICAL MODEL ANALYSES

In the spherical optical model (SOM) analyses a complex potential with Woods-Saxon form factors was used with the conventional form of the spin-orbit interaction. The imaginary part of the central potential consisted of both surface and volume absorption terms. From preliminary calculations it was concluded that the description of the $A_y(\theta)$ data did not require the inclusion of an imaginary spin-orbit term. The calculations were performed with a modified version of Perey's search code GENOA. This modified version of the code included a correction term for the Mott-Schwinger interaction.[8] Compound nucleus (CN) contributions were added to both $\sigma(\theta)$ and $A_y(\theta)$ calculations at energies below 14 MeV. The CN cross sections were calculated with the computer code HAUSER*5 which was developed by the nuclear analysis group at the Hanford Engineering Development Laboratory.[10]

The mean values of the geometrical parameters resulting from "best" fits at each energy provided energy-averaged sets of parameters for both nuclei. At all energies the available $\sigma(\theta)$ and $A_y(\theta)$ distributions along with the σ_T were simultaneously fitted to reduce ambiguities in the parameters. The starting parameters for the best fits were those of Böttcher et al.[5] for ^{28}Si and those of Rapaport et al.[9] for ^{32}S. In order to define the SO parameters early in the analyses, the energies at which both $\sigma(\theta)$ and $A_y(\theta)$ data were available were considered first. The potential strengths and geometrical parameters were simultaneously varied, with care being taken to avoid the notorious $V_R r_R$, $W_V r_I$, and $W_D a_I$ ambiguities. The rigid constraints that $r_I \geq r_R > r_{so}$ and $a_R \geq a_I > a_{so}$ imposed by Van Oers et al.[11] were applied along with the requirement that all radius and diffuseness parameters be greater than 1.0 fm and 0.5 fm, respectively. The latter constraint was imposed because the $A_y(\theta)$ data consistently favored $r_{so} < 1.0$ fm and $a_{so} < 0.5$ fm. The best fits are represented by the dashed curves in figures 2 and 3.

After best fits had been obtained at each energy the data were then described using the energy-averaged geometries. The energy-averaged geometrical parameters for silicon were

$r_R = 1.16$ fm $a_R = 0.689$ fm $r_I = 1.31$ fm $a_I = 0.574$ fm

$r_{so} = 1.01$ fm $a_{so} = 0.500$ fm   and $V_{so} = 6.60$ MeV .

The volume absorption strength W_V was parametrized as $W_V = 0$ for $E \leq 13.3$ MeV and $W_V = -2.0 + 0.15E$ for $E > 13.3$ MeV. For the case of sulfur,

$r_R = 1.184$ fm $a_R = 0.703$ fm $r_I = 1.260$ fm $a_I = 0.525$ fm

$r_{so} = 1.01$ fm $a_{so} = 0.50$ fm   and $V_{so} = 6.31$ MeV .

The volume absorption strength was $W_V = 0$ for $E \leq 12.5$ MeV and

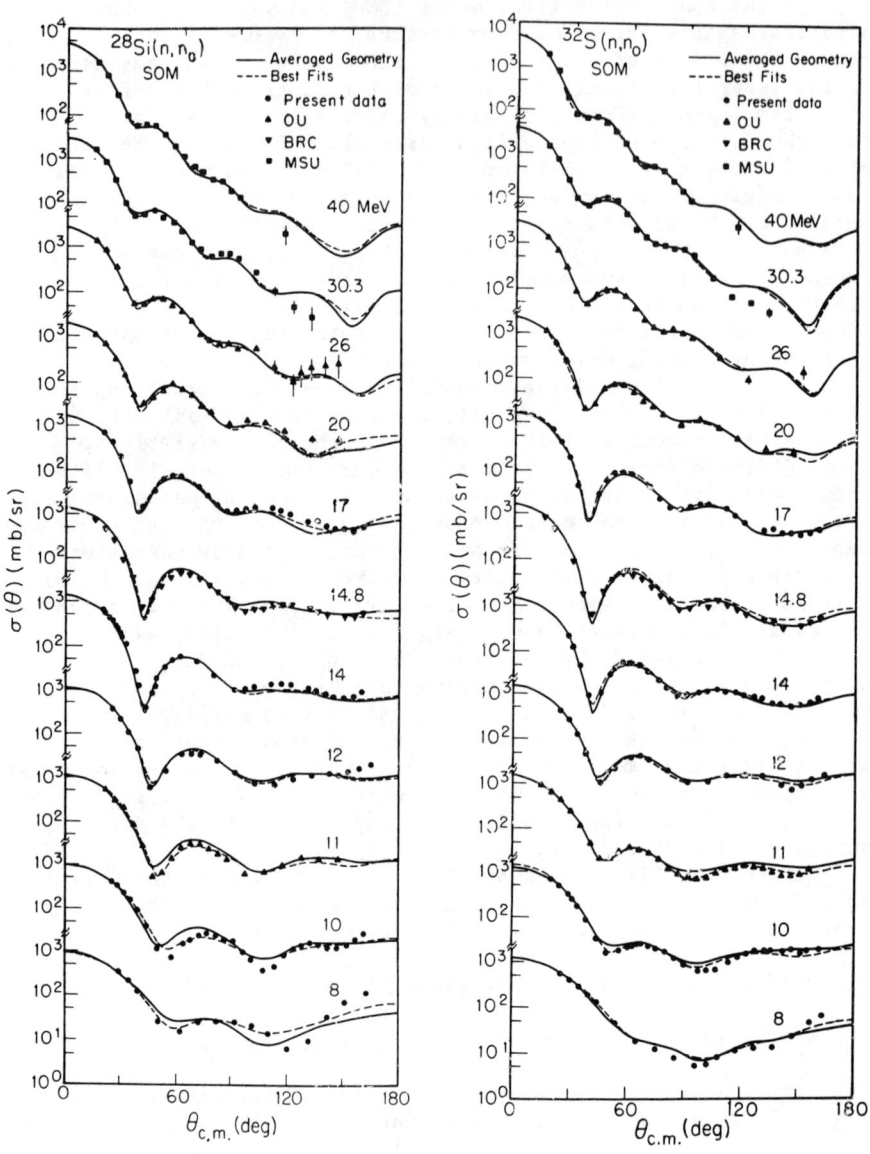

Fig. 2. SOM calculations compared to elastic $\sigma(\theta)$ data for ^{28}Si+n and ^{32}S+n. The solid curves represent the fits with constant geometry, and the dashed curves are the best fits. All curves include the calculated contributions for CN elastic scattering.

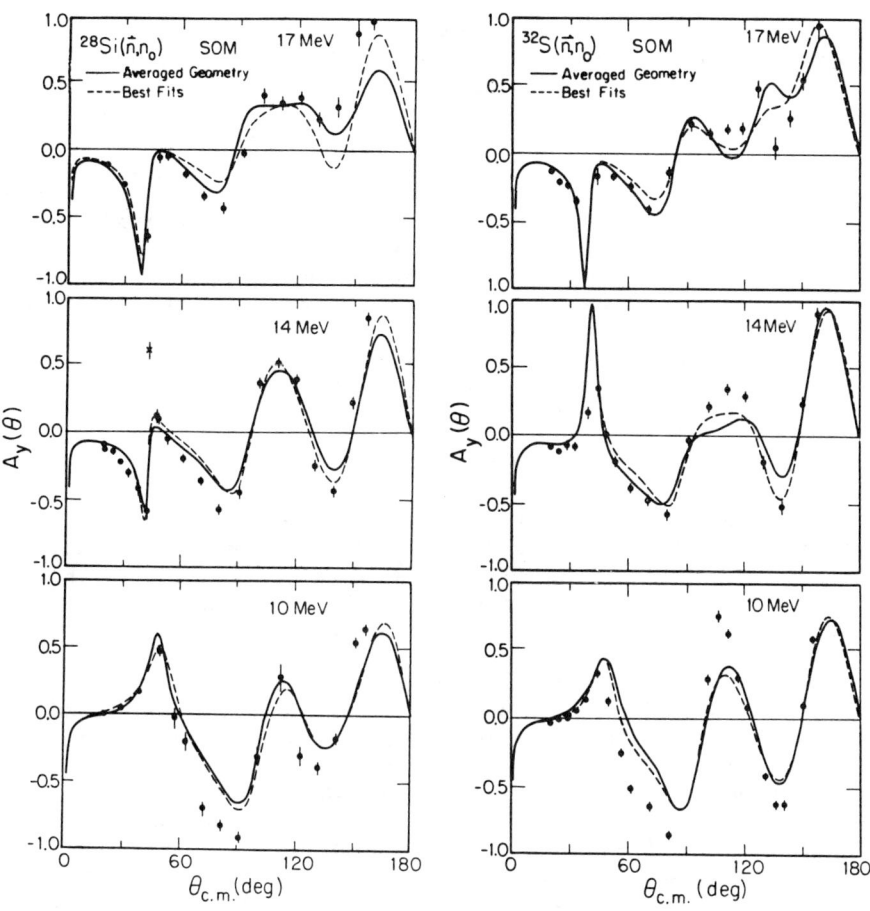

Fig. 3. SOM calculations compared to elastic $A_y(\theta)$ data for $^{28}Si+\vec{n}$ and $^{32}S+\vec{n}$. The solid curves represent the fits with constant geometry, and the dashed curves are the best fits. All curves include the calculated contributions for CN elastic scattering.

$W_V = -2.09 + 0.17E$ for $E > 12.5$ MeV. The fits were optimized at each energy by varying the strengths of the real potential V_R and the surface absorption W_D. The resulting fits are illustrated with the solid curves in figures 2 and 3. The resultant potential strengths and corresponding volume integrals per nucleon are plotted in figure 4. Assuming simple linear energy dependencies and fitting the resultant potential strengths with a least-square fitting routine, the strength parameters V_R and W_D were parametrized as shown in figure 4.

In conclusion, the SOM was capable of describing the differential cross sections very well and the analyzing powers reasonably well for elastically scattered neutrons from both nuclei at incident energies between 8 and 40 MeV. The geometrical parameters were held constant over the entire energy range and simple linear energy dependencies for the potential strengths were deduced. Because of the good agreement between the SOM calculations and the data, it was concluded that the available data were consistent with each other (no normalization discrepancies between the $\sigma(\theta)$ measured by the various contributors to the data set) and with the trends of the OM.

COUPLED-CHANNELS ANALYSES

The coupled-channels calculations were performed with Raynal's search code ECIS79.[12] The $\sigma(\theta)$ and $A_y(\theta)$ data for elastic and inelastic scattering from ^{28}Si and ^{32}S were successfully described over the 8 to 40 MeV energy range using CC calculations with a constant geometry. The first energy-averaged parametrization of the spin-orbit potential (SOP) for neutron scattering from these nuclei were deduced in the present analyses. For the purposes of extracting values for the Coulomb corrections to the OMP the data were fit a second time using geometrical parameters that were derived solely from the analyses of proton-scattering data. These latter parameters will be denoted as set B, while the parameters derived from the present analyses of the neutron data will be referred to as set A. Because both nuclei are strongly deformed, the transmission coefficients used in the CN calculations were computed using the CC code ECIS79 so that the nuclear deformations could be explicitly included in the calculations. The CN contributions were added to the CC calculations for elastic scattering and inelastic scattering to the first excited state at incident energies below 14 MeV. For higher excitations the CN contributions were added to the CC calculations up to incident energies of 20 MeV.

In these analyses the 0^+, 2^+, and 4^+ states in ^{28}Si were modeled as the lower levels of a $K^{\pi} = 0^+$ rotational band. Starting with the average geometrical parameters deduced in the SOM analysis, both geometrical parameters and potential strengths were varied to minimize the chi square of the fits to the elastic and inelastic data. Again the $A_y(\theta)$ data preferred $r_{so} < 1.0$ fm and $a_{so} < 0.5$ fm, therefore these parameters were fixed to the more conventional values of $r_{so} = 1.01$ fm and $a_{so} = 0.5$ fm. The deformation length

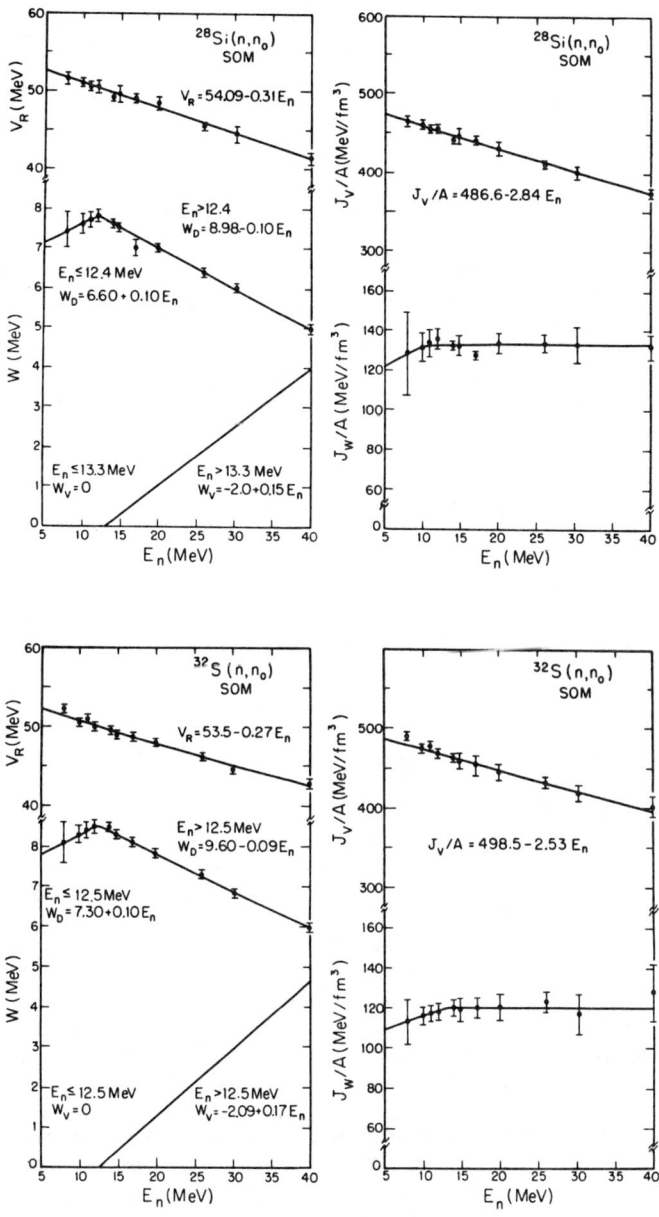

Fig. 4. Plots of the potential strengths and corresponding volume integrals per nucleon resulting from the present SOM analyses of ^{28}Si+n and ^{32}S+n. The lines through the points represent least-square fits. For both nuclei the parametrization of W_V was adjusted to optimize the fits to the data above 20 MeV.

$\delta_{SO} = \beta_{SO}R_{SO}$ of the SOP was adjusted to optimize the fits to the 2^+ $A_y(\theta)$ data. After several iterations a set of energy-averaged geometrical and deformation parameters was derived. The resulting parameters were denoted as set A and are listed below,

$\beta_2 = -0.36$ $\qquad \beta_4 = +0.20 \qquad \delta_{SO} = 1.2\delta_c$

$r_R = 1.170$ fm $\quad a_R = 0.654$ fm $\quad r_I = 1.278$ fm $\quad a_I = 0.580$ fm

$r_{SO} = 1.010$ fm $\quad a_{SO} = 0.500$ fm $\quad$ and $\quad V_{SO} = 6.0$ MeV .

The volume absorption was fixed to the line derived in the SOM analysis $W_v = 0$ for $E < 13.3$ MeV and $W_v = -2.0 + 0.15E$ for $E \geq 13.3$ MeV. The fits illustrated with the solid curves in figure 5 were obtained by fixing the geometrical parameters to the average values and searching upon V_R and W_D. The resultant values of V_R and W_D were parametrized with the simple linear energy dependencies of

and
$$V_R = 54.09 - 0.29E$$

$W_D = 00.90 + 0.46E \quad$ for $E \leq 11.0$ MeV

$ = 6.81 - 0.08E \quad$ for $E > 11.0$ MeV .

Because the predictions of vector analyzing powers for neutron scattering from ^{28}Si have been shown to be very sensitive to the sign of the quadrupole deformation parameter β_2, various calculations were performed to test the sensitivity of the present data and calculations to the signs of the potential deformation parameters β_2 and β_4. The $A_y(\theta)$ data at 10 MeV were not included in these calculations because of the large CN contributions in these data. In each case the geometrical parameters of set A were used, W_v was set to zero, and V_R and W_D were searched upon to optimize the fits. The results of these tests are shown in fig. 6.

The fits to the data at 14 and 17 MeV preferred an oblate shaped potential ($\beta_2 < 0$) with $\beta_4 > 0$. The descriptions of the elastic $\sigma(\theta)$ at back angles were very sensitive to the sign of β_2 with slightly better fits resulting with $\beta_2 < 0$. The quality of the fits to the $\sigma(\theta)$ data for inelastic scattering to the 2^+ state were about the same for both signs of β_2, however the calculations with $\beta_4 < 0$ were substantially lower than the data. The calculations of $\sigma(\theta)$ for scattering to the 4^+ state showed little sensitivity to the signs of β_2 and β_4.

The $A_y(\theta)$ data of the elastically scattered neutrons showed only little sensitivity to the sign of β_2. However, at 14 MeV, the calculated $A_y(\theta)$ for elastic scattering around $\theta_{c.m.} = 40°$ was extremely sensitive to the sign of β_4. Similarly, Böttcher et al. cited a sensitivity around $\theta_{c.m.} = 40°$ to the values of the SOP parameters, especially to a_{SO}. In the present analysis, it was also found that the $A_y(\theta)$ in this narrow angular region was very sensitive to the parametrization of the central well. In light of this finding, it was concluded that the high sensitivity of the

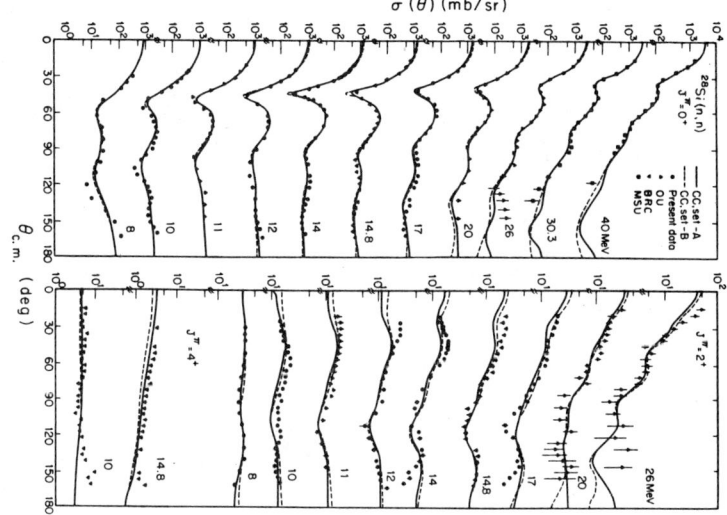
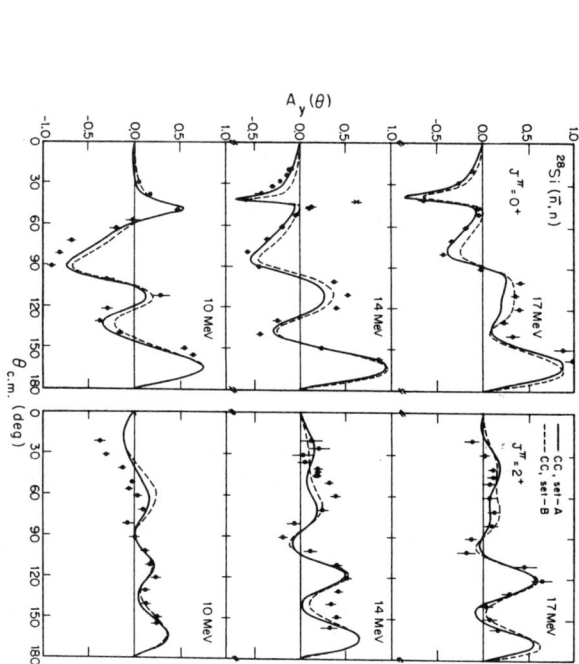

Fig. 5. The σ(θ) and $A_y(\theta)$ for elastic and inelastic scattering of neutrons for 28Si. The curves through the data are the results of CC calculations which model 28Si as a rotational nucleus with the 0^+, 2^+, and 4^+ states forming the coupling basis.

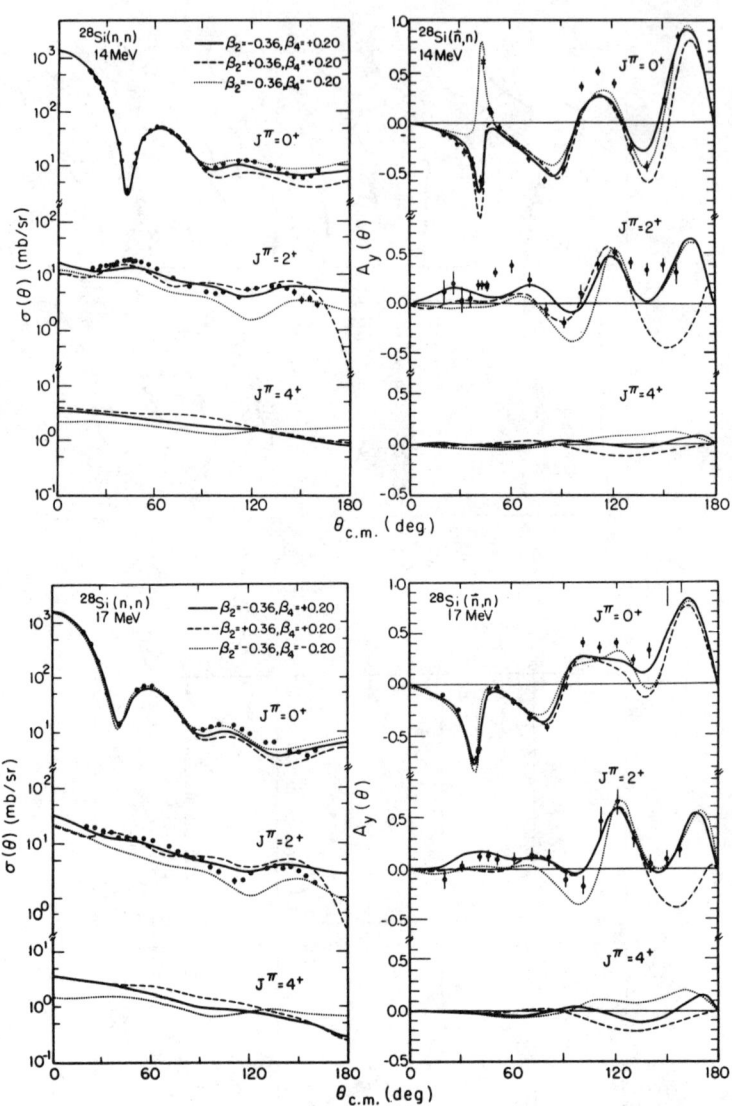

Fig. 6. CC calculations of ^{28}Si+n at 14 and 17 MeV to illustrate the sensitivity of the data and calculations to the signs of β_2 and β_4. In all cases, $\delta_{so} = 1.2\delta_c$, $W_V = 0$, and V_R and W_D were searched upon to optimize the fits.

$A_y(\theta)$ around $\theta_{c.m.} = 40°$ to the OM parameters was probably due to the combination of partial waves necessary to produce the deep minimum in the elastic $\sigma(\theta)$ around $\theta_{c.m.} = 40°$, and therefore, the description of the $A_y(\theta)$ in this narrow angular region did not provide a valid test for determining the sign of β_4.

One pronounced characteristic was that the data and calculations for back angle scattering showed a significantly higher sensitivity to the shape of the nuclear potential than did the forward angle scattering. For instance, the most sensitive observables to the signs of β_2 and β_4 were the back angle $A_y(\theta)$ data for inelastic scattering to the 2^+ and 4^+ states, respectively. In all cases the data preferred $\beta_2 < 0$ and $\beta_4 > 0$. These findings were consistent with the accepted signs of β_2 and β_4.[13,14]

Because the vibrational character of the excited states in ^{32}S has been well established by the various studies, the present analysis concentrated on the vibrational model.[15] As in the case of the ^{28}Si analysis, these calculations were performed with the CC code ECIS79. The 0^+, 2_1^+, 2_2^+, 4^+, and 3^- states formed the coupling basis with the 2_1^+ and 2_2^+ states being described as admixtures of one-phonon and two-phonon states. The wave functions for the 2_1^+ and 2_2^+ states were described according to the method introduced by Tamura[16]

$$|I\rangle = \cos(\phi)|\text{one phonon}\rangle + \sin(\phi)|\text{two phonon}\rangle.$$

The magnitudes of the coupling parameter β_2 and the phonon-admixing angle ϕ_2 for the 2_2^+ state were adjusted to simultaneously describe the $\sigma(\theta)$ and $A_y(\theta)$ of the 2_1^+ state and the $\sigma(\theta)$ of the 2_2^+ state. The calculations for the inelastic data showed strong sensitivity to the symmetry of the wave functions for the 2_1^+ and 2_2^+ states with the inelastic $A_y(\theta)$ at back angles being the most sensitive observable. In fact, significantly better fits to all data were obtained with a symmetric 2_1^+ state (i.e., $\sin(\phi_1) > 0$) and an anti-symmetric 2_2^+ state. The resulting parameters were $\beta_2 = +0.31$ and $\phi_2 = -56°$. Because the wave functions for the 2_1^+ and 2_2^+ states must be orthogonal, the phonon-admixing angle ϕ_1 was computed as $\phi_1 = \phi_2 + 90° = 34°$. The coupling parameter β_3 was adjusted to fit the $\sigma(\theta)$ of the 3^- state at 10 and 26 MeV. A set of average geometrical parameters was derived from the parameters resulting from best fits at each energy. This parameter set will be referred to as set A and is listed below,

$r_R = 1.18$ fm $a_r = 0.683$ fm $r_I = 1.21$ fm $a_I = 0.612$ fm

$r_{so} = 1.04$ fm $a_{so} = 0.542$ fm $V_{so} = 5.30$ MeV and $\beta_{so} = \beta_c$.

The fits at each energy were optimized by searching only upon V_R and W_D, fixing W_V to the values derived in the SOM analysis. The resultant potential strengths were parametrized as

$$V_R = 52.40 - 0.25E$$

$$W_D = 1.71 + 0.33E \quad \text{for } E \leq 13.0 \text{ MeV}$$

$$= 7.22 - 0.09E \quad \text{for } E > 13.0 \text{ MeV}.$$

The resulting fits are represented by the solid curves in figures 7 through 9.

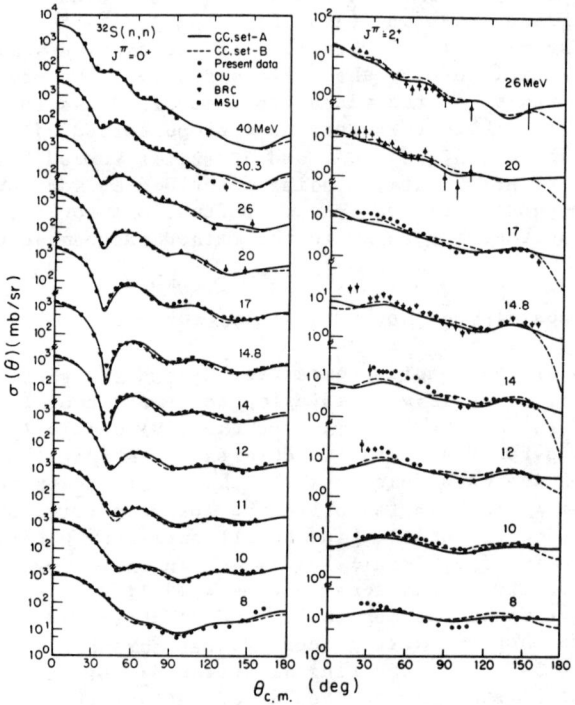

Fig. 7. The $\sigma(\theta)$ for elastic and inelastic scattering of neutrons from ^{32}S. The solid curves through the data are the results of CC calculations which model ^{32}S as a vibrator with admixing of one-phonon and two-phonon states. The dashed curves result from an axially symmetric rotational model.

As cited by Rapaport et al.[17], the determination of the Coulomb correction to the OMP can be made less dependent upon the choice of geometrical parameters by comparing volume integrals instead of potential strengths. Assuming linear energy dependencies, the volume integrals per nucleon for the real part of the OMP for incident protons and incident neutrons can be written as

85

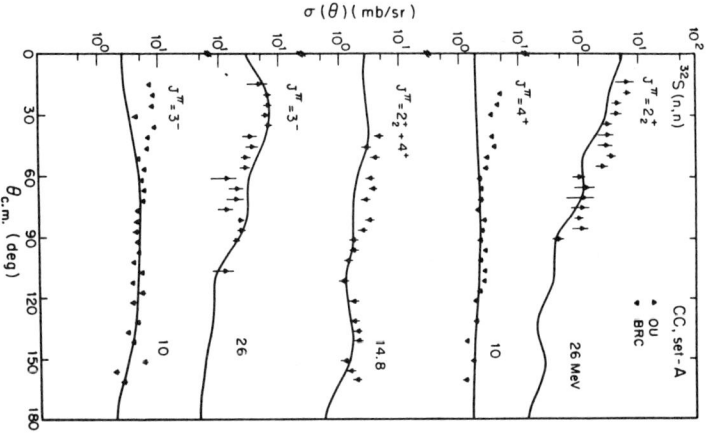

Fig. 8. The σ(θ) for inelastic scattering to the 2_2^+, 4^+, and 3^- states in 32S. The curves through the data are the results of CC calculations which model 32S as a vibrator with admixing of one-phonon and two-phonon states.

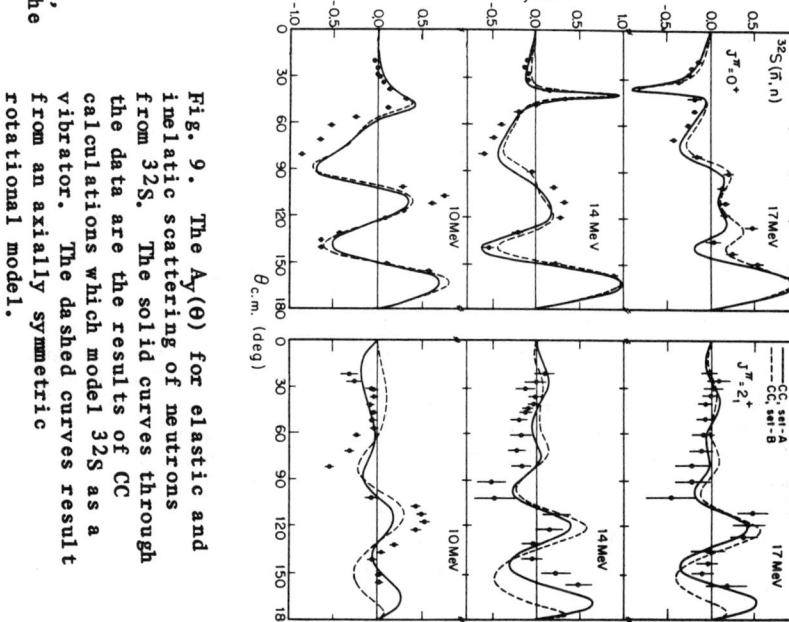

Fig. 9. The $A_y(\theta)$ for elastic and inelastic scattering of neutrons from 32S. The solid curves through the data are the results of CC calculations which model 32S as a vibrator. The dashed curves result from an axially symmetric rotational model.

$$(J_v/A)_p = (J_{v_0}/A) - \alpha E + (\Delta J_v/A)_c$$

and

$$(J_v/A)_n = (J_{v_0}/A) - \alpha E .$$

And similarly for the imaginary potential

$$(J_w/A)_p = J_{w_0}(E)/A + (\Delta J_w/A)_c$$

and

$$(J_w/A)_n = J_{w_0}(E)/A .$$

Thus, $(\Delta J_w/A)_c$ may be written as

$$-(\Delta J_w/A)_c = (J_w/A)_n - (J_w/A)_p .$$

To extract values of the Coulomb corrections, the neutron data were re-analyzed with CC calculations using the OM parameters reported by Tailor et al. in their analyses of the corresponding proton data.[5] Both nuclei were modeled as symmetric rotors. The geometrical parameters used in these calculations are referred to as set B and are listed below for ^{28}Si

$\beta_2 = -0.39$ $\beta_4 = +0.16$ $\delta_{so} = 1.3\delta_c$

$r_R = 1.148$ fm $a_R = 0.663$ fm $r_I = 1.330$ fm $a_I = 0.600$ fm

$r_{so} = 1.020$ fm $a_{so} = 0.680$ fm $V_{so} = 6.3$ MeV ,

and for ^{32}S

$\beta_2 = +0.283$ $\delta_{so} = \delta_c$

$r_R = 1.158$ fm $a_R = 0.703$ fm $r_I = 1.215$ fm $a_I = 0.640$ fm

$r_{so} = 1.029$ fm $a_{so} = 0.661$ fm   and $V_{so} = 6.08$ MeV .

The fits were optimized at each energy by searching upon V_R and W_D. The resulting fits are represented by the dashed curves in figures 5, 7, 8, and 9. As can be seen, the descriptions of the angular distributions using the proton parameters were of comparable quality to those obtained with the neutron parameters. However, the predictions of the total neutron cross sections were consistently about 5% higher than the data.

Figure 10 contains plots of volume integrals per nucleon resulting from the above analyses of the neutron data in comparison to those derived from Tailor et al.'s analyses of the corresponding proton data. For the real potential, the dashed line and shaded region represent the mean and standard deviation of the neutron

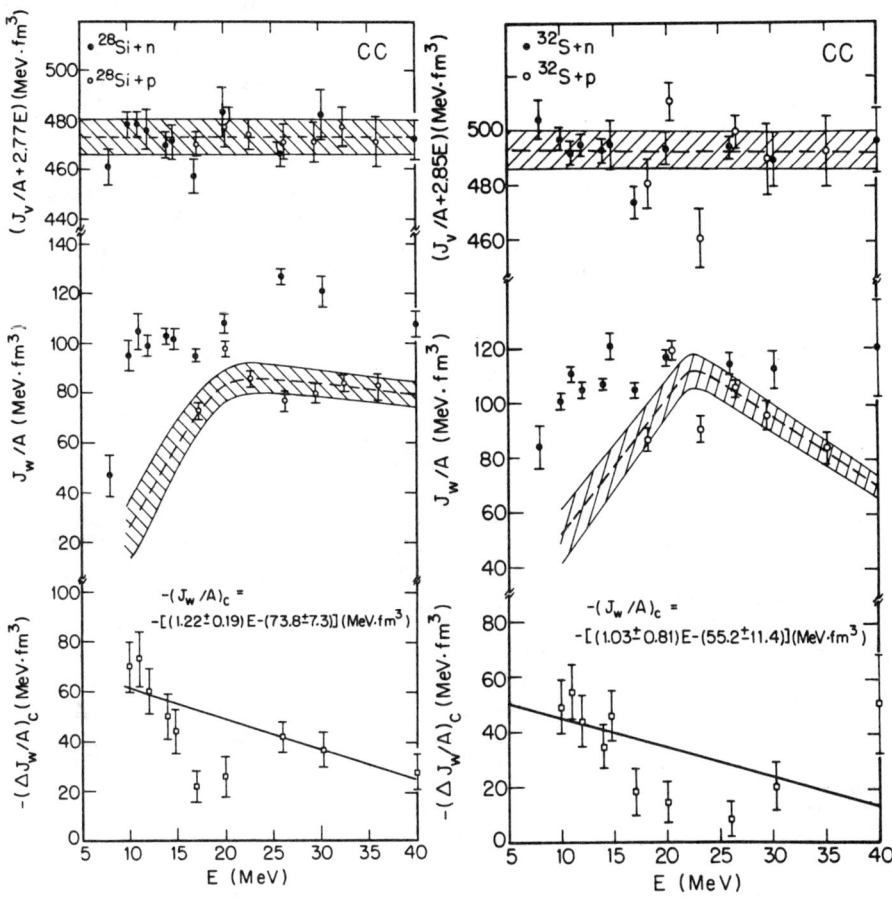

Fig. 10. Volume integrals per nucleon versus incident projectile energy. Comparison of (n,n) and (p,p) analyses to deduce the Coulomb corrections to the real and imaginary parts of the OMP.

data. Because the proton points were statistically indistinguishable from the neutron data, reliable values for the Coulomb corrections to the real part of the OMP could not be extracted from the present analyses. However, this was not the case for the absorptive potential. The neutron data for the absorptive potential were consistently higher than the corresponding proton points, therefore permitting the extraction of values for the Coulomb corrections to the absorptive potential for both nuclei.

The following people are recognized for their contributions to the the present work: R. S. Pedroni, G. M. Honoré, H. G. Pfützner, G. Tungate, R. C. Byrd, and R. L. Walter. I would like to thank Dr. R. DeVito for sharing his $\sigma(\theta)$ data for neutron scattering from silicon and sulfur at 30.3 and 40 MeV prior to it being published. Similarly, I am grateful of the $\sigma(\theta)$ data at 9.8 and 14.8 MeV that were contributed by Dr. G. Haouat. Dr. D. C. Larson is recognized for allowing us to use the ORELA data for the neutron total cross section for silicon.

*Work partially supported by the U.S. Department of Energy Director of Energy Research, Office of High Energy and Nuclear Physics, under Contract No. DE-AC05-76ER01067.

REFERENCES

1. R. de Swiniarski and D. L. Pham, Lett. Al Nuovo Cimento 16 (1976) 391; R. de Swiniarski et al., Nucl. Phys. A261 (1976) 111; R. de Swiniarski et al., Can. J. of Phys. 51 (1973) 1293.
2. R. de Leo et al., Phys. Rev. C19 (1979) 646.
3. R. M. Lombard and J. Raynal, Phys. Rev. Lett. 31 (1973) 1015.
4. R. de Leo et al., Il Nuovo Cimento 59A (1980) 101.
5. R. P. DeVito et al., Phys. Rev. C28 (1983) 2530 and private communications; J. Böttcher et al., Phys. G: Nucl. Phys. 9 (1983)L65.; G. Haouat et al., Procedings of the International Conference on Nuclear Data for Science and Technology, (Boston: D. Reidel Publishing Company, 1982) 796 and private communications; R. C. Tailor, Ph.D. Dissertation, Ohio University (1982); S. Kliczewski and Z. Lewandowski, Nucl. Phys. A304 (1978) 269; A. W. Obst and J. L. Weil, Phys. Rev. C7 (1972) 1076.
6. C. R. Howell and R. L. Walter, IEEE Trans. Nucl. Sci., NS-30 No. 2 (1983)1132.
7. C. R. Howell, Ph.D. Dissertation, Duke University (1984); To be published.
8. C. E. Floyd, R. L. Walter, and R. G. Seyler, Bull. Amer. Phys. Soc. 27 (1982)722.
9. J. Rapaport et al., Nucl. Phys. A286(1977)232.
10. F. M. Mann, HEDL-TME 78-83 UC-79d, Hanford Engineering Development Laboratory, Richland, WA (1979).
11. W. T. H. Van Oers et al., Phys. Rev. C10(1974)307.
12. J. Raynal, code ECIS79 (unpublished).
13. H. C. Lee and R. Y. Cusson, Annals of Phys. 72(1972)353.
14. W. J. Thompson and J. S. Eck, Phys. Lett. 67B(1977)151.

15. O. Häusser et al., Nucl. Phys. $\underline{A175}$(1971)593; G. T. Garvey et al., Phys. Lett. $\underline{29B}$(1969)108; F. Ingebretsen et al., Nucl. Phys. $\underline{A161}$(1971)433; W. F. Coetzee, M. A. Meyer, and D. Reitmann, Nucl. Phys. $\underline{A185}$(1974)644; A. Olin et al., Nucl. Phys. $\underline{A221}$(1974)555; R. M. Lombard and J. Raynal, Phys. Rev. Lett. $\underline{31}$(1973)1015.
16. Taro Tamura, Supp. of Prog. of Theor. Phys. $\underline{37}$ and $\underline{38}$(1966)383.
17. J. Rapaport, V. Kulkarni, and R. W. Finlay, Nucl. Phys. $\underline{A330}$(1979)15.

TESTS OF EFFECTIVE INTERACTIONS FOR NUCLEON SCATTERING AND CHARGE EXCHANGE BELOW 60 MeV

F. S. Dietrich
Lawrence Livermore National Laboratory, Livermore, CA 94550

F. Petrovich
Florida State University, Tallahassee, FL 32306

ABSTRACT

Significant progress has been made over the past several years in developing energy- and density-dependent effective interactions and optical potentials based on free nucleon-nucleon potentials. This leads to the hope that the nucleon-scattering reaction mechanism may be well enough known to probe details of nuclear spectroscopy (i.e., transition densities) more accurately than with purely phenomenological approaches. Of particular interest to this conference is the possibility of separating neutron and proton transition densities by comparing proton and neutron scattering, which can be done reliably only if the isovector parts of the effective interaction are well known. This paper attempts to assess the accuracy of presently available interactions through comparisons with elastic and inelastic scattering over a wide mass range, and also with the (p,n) isobaric-analog reaction. Particular emphasis has been placed on the isovector parts of the interaction and on Coulomb corrections by comparing proton and neutron scattering on the same targets. In a number of cases, precise neutron data have been measured to facilitate these comparisons.

I. INTRODUCTION

An important goal of the study of nuclear reaction mechanisms is to understand scattering cross sections in terms of the interaction between the nucleons of the projectile and those of the target. The work described in this paper employs a one-step folding model to describe the potentials for nucleon scattering (optical potential for elastic scattering, or form factor for inelastic and charge-exchange reactions). Because the free nucleon-nucleon interaction has a short-distance behavior that is too strong to be used in a perturbative framework, this approach requires an effective nucleon-nucleon interaction. Until a few years ago, the properties of the effective interaction were determined primarily from phenomenological parameterizations of simple interaction forms (e.g., a real interaction with a single Yukawa form factor, and with several strength parameters representing the various components of the interaction).

A typical study of such an interaction has been given by Austin.[1] Although the phenomenological interactions yielded useful comparisons with data in a limited range of energies and angular-momentum transfer, particularly when the calculated form factors were supplemented with phenomenological imaginary parts, the wide range of data presently available from low through intermediate energies has demanded an effective interaction that is so complicated that it requires further theoretical guidance for its parameterization. The most important characteristics of the effective interaction that must be described are its exchange mixture, range structure, and dependence on energy and density. A major step during the past decade has been the development of effective interactions[2-5] and nuclear-matter optical potentials[6] based on free nucleon-nucleon interactions and the use of the techniques of nuclear many-body theory. These calculations address the characteristics of an effective interaction listed above. Additionally, most of them[3-6] yield estimates of the imaginary components of the interaction (or potential). The strong calculated density dependence requires the use of a local density approximation (LDA) in applying the results to finite nuclei.

This work presents results from a systematic comparison of several theoretical interactions, together with the necessary local-density and exchange approximations, with experimental results for elastic, inelastic, and isobaric-analog charge-exchange reactions. It is expected that the merits and deficiencies of the theoretical interactions will give guidance for the development of a more general phenomenological interaction.

In principle, when a DWBA description of an inelastic or charge-exchange reaction is appropriate, the same effective interaction should be used to describe the distorting optical potentials and the transition form factors. We present several examples of these types of reactions, to assess whether such a consistent treatment is within reach of the currently available interactions and the approximations used to implement them. A further goal is to understand whether the interactions are sufficiently reliable to yield accurate spectroscopic information, such as the reduced transition strengths B(EL) and ratios M_n/M_p of neutron to proton multipole matrix elements for a given transition. Extracting such information depends on the accuracy of the isospin terms in the interaction, and accordingly neutron and proton scattering on the same target nucleus has been analyzed in nearly all cases.

Since the purpose of the present work is to test the reaction mechanism, we have concentrated on cases for which there is already experimental knowledge of the relevant ground-state or transition densities. The proton densities are accurately known from electron scattering, and the neutron densities have either been inferred from the results of 800-MeV proton scattering experiments, or assumed (in the case of self-conjugate nuclei) to be equal to the proton densities. For a particular nucleus and effective interaction, the procedure that has been followed is to first calculate the real and imaginary central potentials and the spin-orbit potential in a folding model. Details of the calculations may be found in Refs. 7,8. Elastic scattering is calculated by inserting these potentials into a spherical optical-model code, and comparison with experimental data is made by least-squares adjustment of two normalizing parameters λ_V and λ_W for the real and imaginary central parts of the optical potential U:

$$U = \lambda_V V + i \lambda_W W + V_{s.o.} \qquad (1)$$

In the energy range considered here, it has not been found necessary to alter the spin-orbit potential.

Comparison of the normalizing parameters for neutron and proton scattering constitutes a first test of the accuracy of the isospin content of the microscopic reaction mechanism. Once the energy dependence of the normalizing parameters has been established by the analysis of elastic scattering, normal-parity inelastic scattering and charge exchange are then calculated in DWBA with no further parameter variation. The same normalizing factors are applied to the real and imaginary central parts of the form factors that are used for the elastic distorting potentials.

Achieving consistency between proton and neutron scattering requires not only an adequate description of the isovector parts of the effective interaction, but also a correct treatment of Coulomb effects on proton scattering. To understand whether the common practice of replacing the incident energy E by $E-V_{Coul}$ everywhere in a local optical potential is reasonable, it is necessary to clarify the sources of energy dependence in the optical potential. In the microscopic folding model we identify two separate energy dependences. One of these is the intrinsic E-dependence of the effective interaction, which is weak for the real part, but strong for the imaginary part. The other is an energy dependence associated with the local-momentum approximation for exchange, in which conservation of energy is employed to eliminate the momentum dependence arising from the exchange approximation:

$$p^2/2m = E - V_{nuc} - V_{Coul}. \qquad (2)$$

The combination $E-V_{Coul}$ explicitly appears in the exchange approximation; this is well understood to be responsible for the major part of the energy dependence of the real optical potential. We have argued[7] that the energy dependence of the effective interaction should be measured from the Fermi energy, and that the near equality of neutron and proton Fermi energies in a heavy nucleus implies that effective interactions for neutron and proton scattering at the same incident energy should be evaluated at the same energy, rather than at energies shifted by the Coulomb potential. As will be illustrated below, this choice is necessary to achieve neutron-proton consistency for the imaginary potential.

II. INTERACTIONS AND CALCULATIONAL DETAILS

We have tested the central effective interactions of Brieva and Rook[3] and Yamaguchi et al.[5], both based on the Hamada-Johnston[9] potential, but calculated with different numerical techniques; Yamaguchi et al. also required self-consistency in the energies appearing in the Bethe-Goldstone equation. We have also tested the potential of Jeukenne, Lejeune, and Mahaux[6] (JLM), based on the Reid hard-core potential[10]. The JLM nuclear-matter optical potential has been treated as an effective interaction by dividing it by the density, and accounting for the range of the interaction a posteriori by a multiplicative Gaussian form factor with an empirically determined range parameter. This procedure, originally discussed by JLM for elastic scattering, has also been applied previously to inelastic scattering and charge exchange[11]. Since exchange has already been included in the JLM nuclear-matter optical potential, which is local, the distinction between intrinsic and exchange energy dependences may not be utilized in comparing neutron and proton scattering. However, JLM have concluded that proton scattering should be calculated with the substitution $E \to E-V_{Coul}$ in both the real and imaginary parts of the potential. Calculations of an effective interaction based on the Paris potential[12] in the range 25-100 MeV have been reported by von Geramb at this conference. Since this interaction has shown promise in the range 100-400 MeV[4], it will be interesting to compare its predictions with those of the above interactions.

In the present energy range, we have found that the real, density- and energy-independent spin-orbit interaction of Bertsch et al.[2] (Elliott matrix elements) provides a remarkably successful description of elastic-scattering analyzing-power angular distributions, without requiring renormalization. This interaction is very similar to that of Brieva and Rook[13]. The spin-orbit interaction of Yamaguchi et al. is also very similar in shape (or in its q-dependence), but its strength is lower than that of Bertsch et al. The spin-orbit interaction of Bertsch et al. has been used for all the calculations herein, except those involving the Yamaguchi et al. potential, in which case the spin-orbit interaction of Yamaguchi et al. has been used, with its strength normalized upward by 30%.

The shape and strength of the calculated form factors and potentials depend on the prescription used to evaluate the density dependence of the effective interaction. We have consistently employed a "midpoint LDA," in which the density is calculated at the arithmetic mean of the projectile and target coordinates. Comparisons of potentials calculated with various prescriptions may be found in Ref. 8. The choice of prescription mainly affects the normalization of the imaginary central potential, at the level of $\approx 15\%$; the shape differences, when the potentials are viewed in q-space, appear at values of momentum transfer too large to have very important effects on the scattering angular distributions.

A further technical point concerns the use of effective-mass factors to modify parts of the calculated potentials. It has been pointed out by Negele and Yazaki[14] that the imaginary part of the JLM potential should be multiplied by a factor m_k/m, which is the part of the effective mass associated with nonlocality (or, alternatively, momentum dependence); a similar discussion has been

given by Fantoni et al.[15]. To understand whether a similar
factor should be applied to optical potentials based on effective
interactions requiring an explicit approximation for exchange, we
have examined the reduction of the folding potential from a
nonlocal to a local form by extending the technique of
Fiedeldey[16], who studied the Perey-Buck transformation. We
conclude[17] that if the real potential is used in eq. (2) to
eliminate the momentum dependence, then an effective mass factor
should multiply the imaginary potential; the correct factor
depends on the radial and exchange structure of the interaction,
and is calculated along with the optical potential. On the other
hand, if the potential V_{nuc} in eq. (2) is the full, complex
optical potential, which yields a complex wave number, no
effective-mass correction is required. The near equivalence of
these two procedures has been verified by calculation. The
consequence of either approach is a reduction of the
imaginary-potential volume integral by 20-30%, and a slight shift
in the shape of the imaginary potential toward a more
surface-peaked form, which is in the direction of improved
agreement with phenomenological potentials. Extending the same
derivation to reactions (i.e., inelastic scattering and charge
exchange) shows that no effective-mass corrections should be made
to the DWBA form factor, but that the initial- and final-state
wave functions should each be multiplied by a Perey factor, which
is just the square root of the momentum-dependent effective-mass

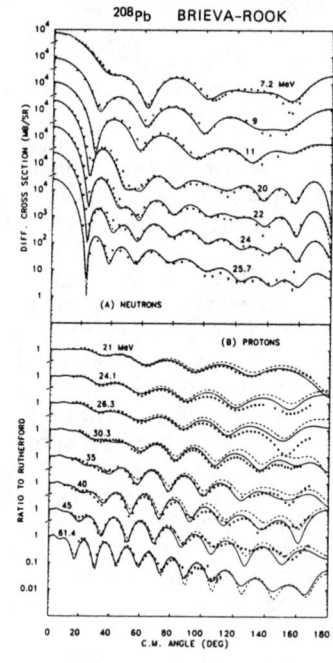

Figure 1

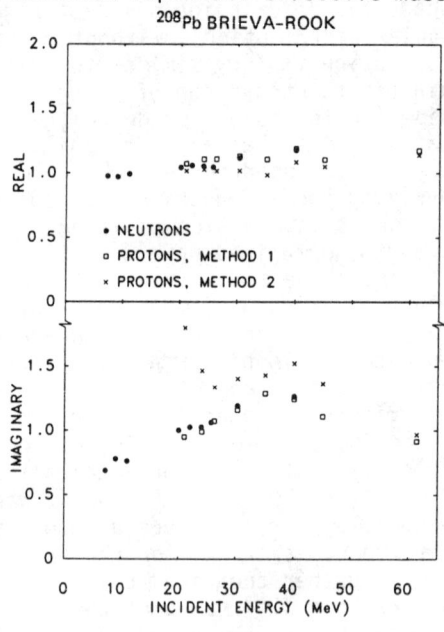

Figure 2

factor $\sqrt{(m_k/m)}$. The Perey factors, which are typically in the range 0.8-0.85, reduce the reaction cross section by 20-30%, and thus should be included when consistency between elastic scattering and reactions is sought.

III. ELASTIC SCATTERING

Fig. 1 shows calculations with the Brieva-Rook interaction of proton and neutron elastic scattering from ^{208}Pb, and Fig. 2 shows the corresponding normalizing coefficients. The sources of the data are listed in Ref. 7. The dashed curves were calculated with a neutron density proportional to the proton density[18]. The solid curves, calculated with a neutron density inferred from 800-MeV neutron scattering[19], show that low-energy nucleon scattering is sensitive to small differences in the shape of the densities (in this case, to the slope of the density in the surface region). The normalizing parameters show that consistency between neutron and proton scattering is achieved only if Method I is used for evaluating the energy dependence of the interaction; in Method I this energy dependence is measured from the Fermi surface, whereas in Method II the substitution $E \rightarrow E-V_{Coul}$ is used. Angular distributions calculated with the JLM and Yamaguchi et al. potentials are very similar, and are not shown; however, the normalizing coefficients are shown in Figs. 3 and 4. Consistency between neutron and proton scattering is achieved in the JLM calculations if the Coulomb correction (CC) suggested by JLM is applied only to the real part of the proton potentials. Calculations with the Yamaguchi et al. interaction

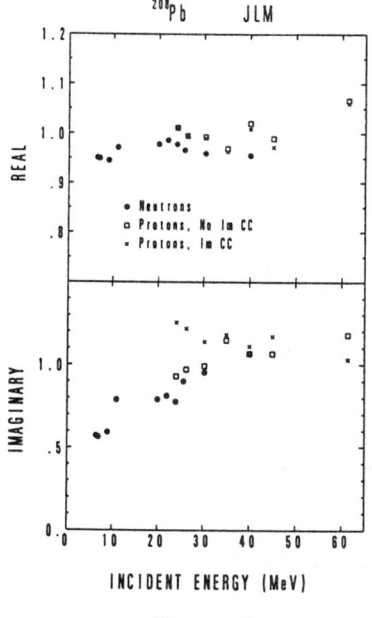

Figure 3

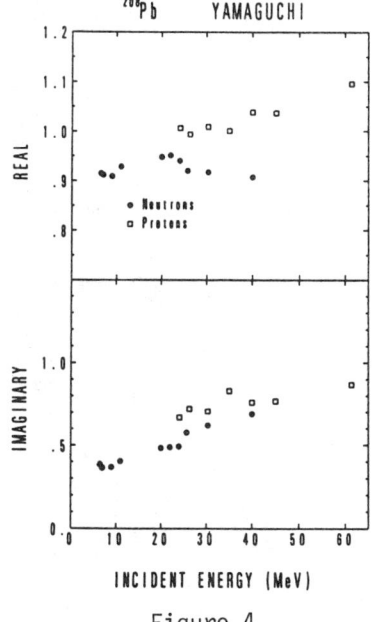

Figure 4

were made only with Method I. We note that the real normalizing factors for all three interactions are within ≈10% of unity. On the other hand, the imaginary factors show a systematic decrease with decreasing energy below 30 MeV, which may reflect the inability of nuclear-matter based potentials to account for the lowered density of states at low energies near a doubly-magic nucleus. Neutron analyzing-power data[20] are compared with calculations using the BBML[2] spin-orbit interaction and the JLM and Brieva-Rook central potentials in Fig. 5; the agreement is excellent in both cases.

Results of a similar study[8] of the Brieva-Rook and JLM potentials for a medium-weight nucleus (^{54}Fe) are shown in Figs. 6-10. Only the neutron data[8,21-23] are shown; the 24-MeV analyzing-power data are for natural Fe, and were analyzed as ^{56}Fe. The JLM potentials yield a better reproduction of the angular distributions and less energy dependence in the normalizing parameters than the Brieva-Rook. By examination of the q-space behavior of the potentials, we have concluded[8] that the systematic overshoot of the low-angle cross sections and the total cross sections with the Brieva-Rook potentials is correlated with an anomalous "bump" at the nuclear surface in the real potential; as will be shown below, this feature is absent in optical potentials generated from the interaction of Yamaguchi et al., which is also based on the Hamada-Johnston potential.

As a further systematic test of the mass dependence of the potentials, angular distributions of 14.6-MeV elastically scattered neutrons have been measured on a series of targets from A = 9 to 209, and compared with the scattering calculated from the

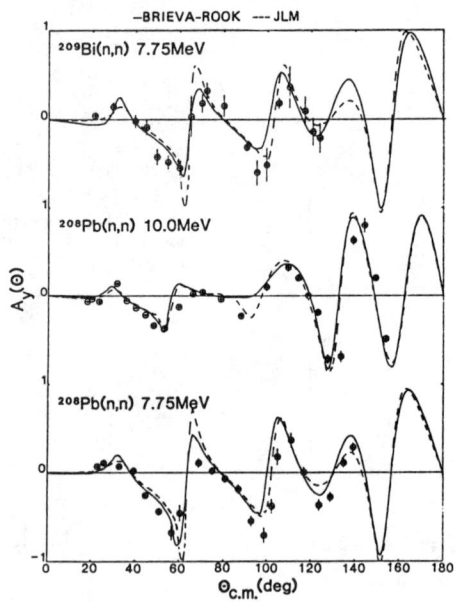

Figure 5

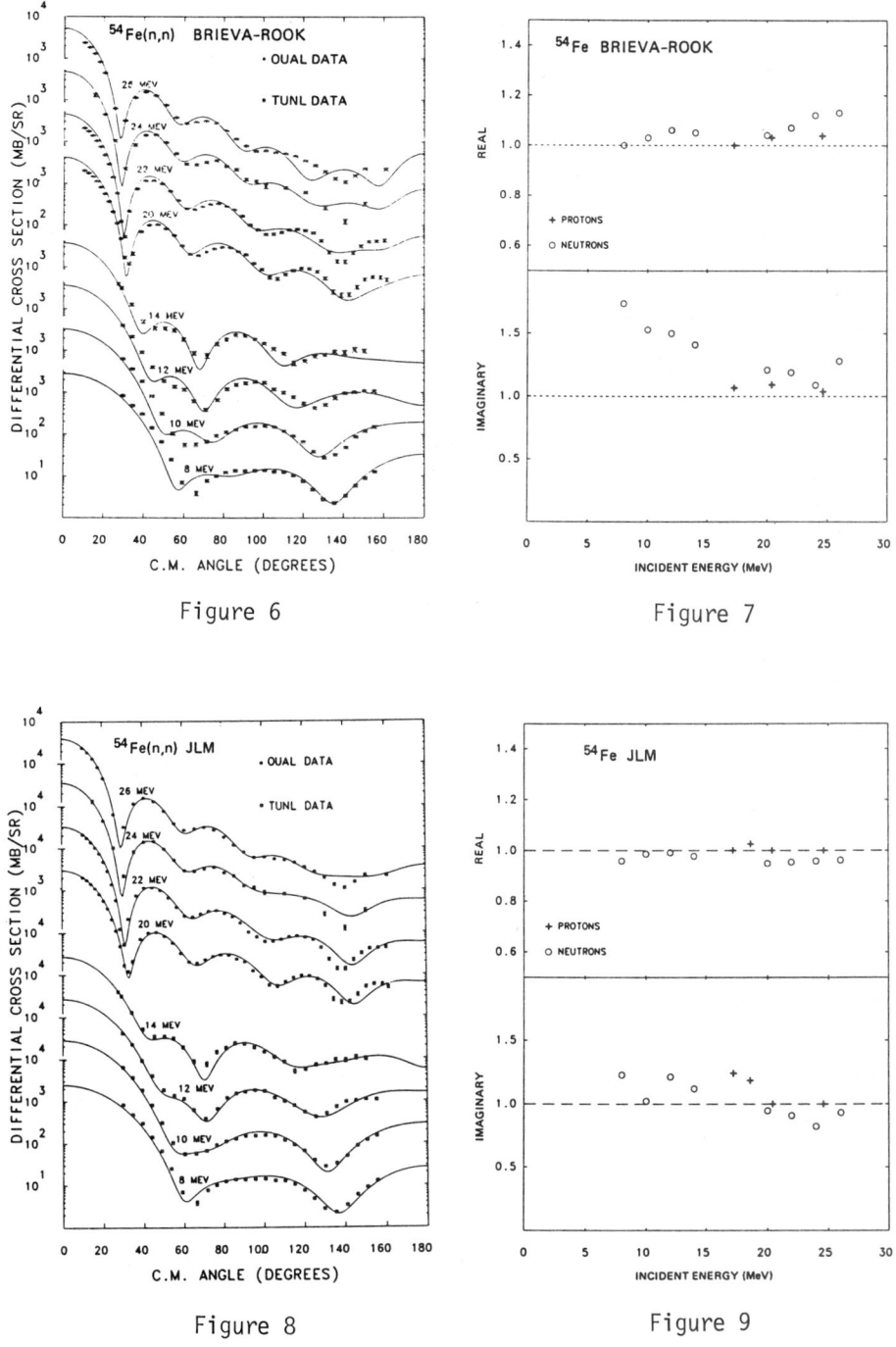

Figure 6

Figure 7

Figure 8

Figure 9

Brieva-Rook and JLM potentials[24]. The results from A = 9 to 59, shown in Fig. 11, illustrate the worsening of the agreement with decreasing target mass at both forward and backward angles for the Brieva-Rook potentials, whereas the JLM potentials yield reasonable agreement for nuclei as light as ^{12}C. The poorer agreement for ^{9}Be is probably due largely to the spin (3/2) of this nucleus, which allows the probe to sense the nuclear deformation directly in a one-step process.

The JLM potentials have proven to yield quite stable results for the light nuclei. Proton and neutron scattering from ^{12}C, ^{13}C, ^{14}C, and ^{16}O in the range 20-50 MeV appears to be consistent with constant real and imaginary normalizing parameters of 1.0 and 0.8, respectively. Figs. 12 and 13 show comparisons with angular distributions for neutron[25] and proton[26] scattering on ^{12}C. We note the well-known difficulty of reproducing the back-angle behavior of proton-scattering data on A=4N nuclei. Nevertheless, the JLM potentials yield agreement with the data that is at least as good as obtained[28] with the phenomenological proton potential of Fabrici et al.[27], who have made a systematic study of proton scattering on light nuclei. Calculations with the Brieva-Rook interaction are compared with neutron scattering, while results with the Yamaguchi et al. interaction are shown for the proton data. The interaction of Yamaguchi et al. yields results that are qualitatively very similar to those for JLM; however, the imaginary normalization is rather small (≈ 0.5).

As a final example, Fig. 14 shows neutron total cross sections for ^{140}Ce in the range 4-60 MeV, compared with

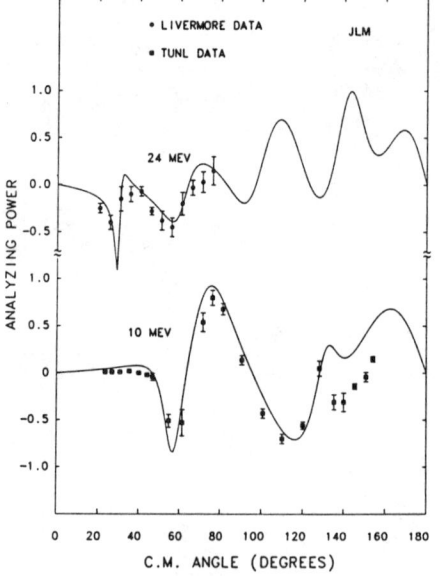

Figure 10

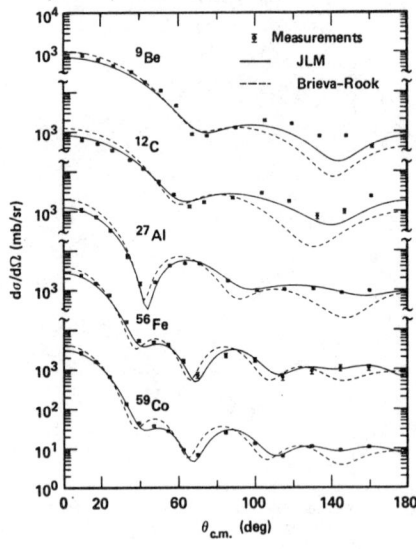

Figure 11

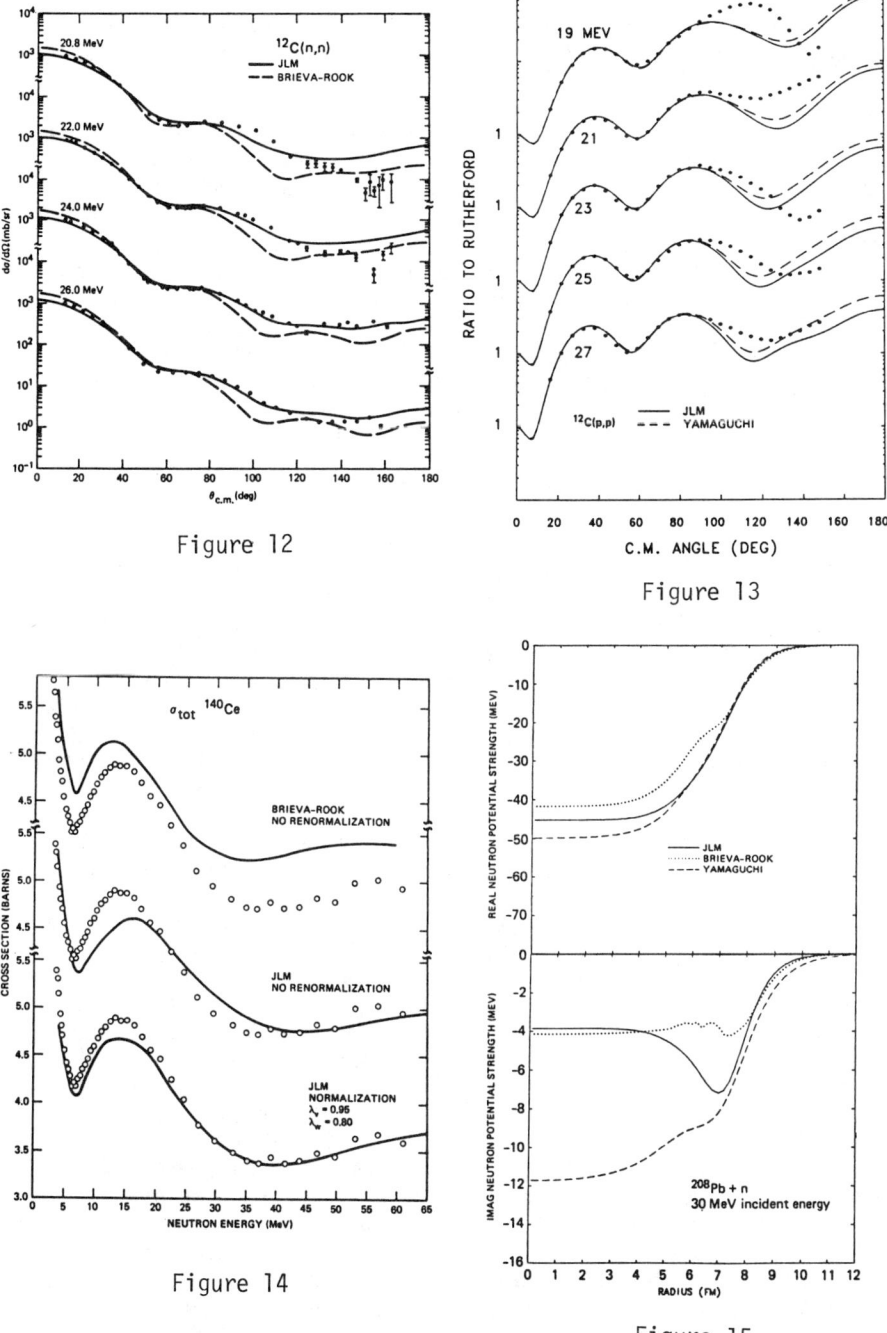

Figure 12

Figure 13

Figure 14

Figure 15

calculations using the Brieva-Rook and JLM potentials. The total cross sections with the Brieva-Rook interaction are systematically high over the entire energy range, whereas reasonable agreement with the JLM potentials is obtainable with only slight, energy-independent normalizations.

To illuminate partially the different results obtained with the various interactions, the real and imaginary central potentials calculated with each of the interactions are shown in Fig. 15 for 30 MeV neutron scattering on ^{208}Pb. The differences between the potentials are quite pronounced, with the JLM resembling most closely the phenomenological forms. The absence of the surface anomaly in the Yamaguchi et al. potential suggests that this feature of the Brieva-Rook potential is associated with the treatment of the many-body problem, rather than an intrinsic feature of the underlying Hamada-Johnston interaction. We also note that the Yamaguchi et al. imaginary potential appears to be insufficiently damped in the nuclear interior.

IV. CHARGE EXCHANGE

Two apparent problems arise in applying the folding model to the (p,n) reaction to isobaric-analog states. The first is that the Q-value, which is large in heavy nuclei, seems to render ambiguous the choice of energy at which the quantities in the form factor are evaluated. The second concerns the use of the Fermi energy as the basis for evaluating the energy dependence of the effective interaction. In the present case, there are two Fermi energies, corresponding closely to the lowest T-1/2 and T+1/2 levels in the compound system; these are separated by a symmetry energy $\Delta\varepsilon_F$, which is similar in magnitude to the Coulomb displacement energy (or -Q).

We have investigated[17] these problems by writing the (p,n) reaction in nonlocal, coupled isospin channels, requiring isospin conservation for the nuclear interaction, identifying the potential functions that appear with quantities calculated in a folding model, and then making the usual local-momentum approximation for exchange. This procedure fully prescribes the energy dependence. In a heavy nucleus, we find that the form factor is proportional to the difference of proton-target and neutron-target optical potentials $U_p(E_p,k_p)-U_n(E_n,k_n)$. In this expression, the E's represent the energy at which the interaction is to be evaluated; E_p is the incident proton energy, and $E_n=E_p-\Delta\varepsilon_F$. The local momenta k_p and k_n correspond to proton and neutron elastic scattering with incident energies E_p and E_p+Q, respectively. If the effective interaction were independent of energy, this procedure would correspond closely to that of Brieva and Lovas[30]; however, the energy dependence induces an extra contribution to the form factor, whose form is that of an isoscalar density folded with the difference of isoscalar effective interactions evaluated at E_p and E_n.

We have calculated the (p,n) charge-exchange reaction to the isobaric-analog state in the reaction ^{208}Pb(p,n)^{208}Bi using the DWBA. Results with the Brieva-Rook and JLM interactions are compared with data[31] at 25, 35, and 45 MeV incident energy in Figs. 16 and 17. The experimentally-determined proton and neutron densities underlying these calculations are the same as those used earlier for elastic scattering.

The solid curves in Fig. 16 represent a fully self-consistent treatment using the Brieva-Rook interaction, whereas the dashed curve was calculated with the microscopic form factor, but with phenomenological distorting potentials[32]. It is clear that the goal of a satisfactory self-consistent calculation has not been attained. However, the reasonable agreement when the phenomenological distortion is used suggests that the main problems may lie with the details of the effective interaction (e.g., distorting effects of the potential-surface anomaly), rather than with the approach.

Self-consistent calculations with the JLM interaction are shown in Fig. 17. The form factors were obtained by subtracting proton and neutron potentials at incident energies separated by $\Delta\epsilon_F$. For consistency with the results of Fig. 3, a Coulomb correction was applied only to the real part of the proton potential. Although the angular-distribution structure is reproduced, the calculations are deficient in magnitude by a factor typically in the range 4-6. The origin of a discrepancy in magnitude with a similar calculation in Ref. 11 is not understood.

The problems in reproducing the magnitude of the (p,n) reaction focus attention on the fact that the ratio of real isovector to isoscalar terms in the nuclear-matter based microscopic optical potentials is significantly smaller than that

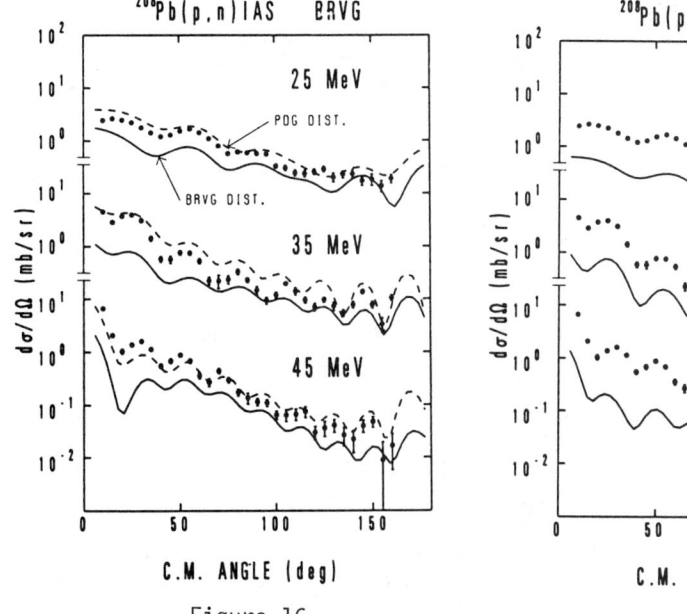

Figure 16

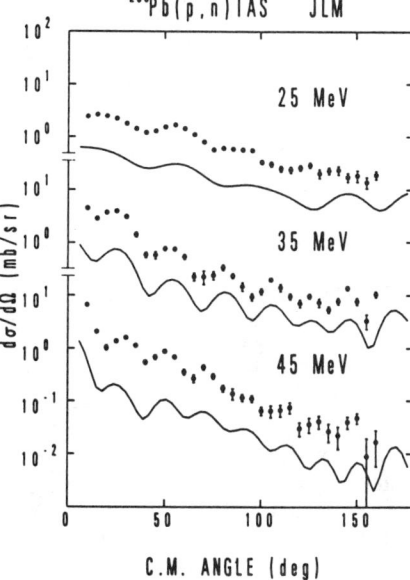

Figure 17

found in phenomenological potentials and effective interactions. The ratio is density dependent, and decreases with increasing density for all interactions tested. To illustrate this, we have calculated the real potentials for 30-MeV neutron scattering on ^{208}Pb, parametrized them as $V = -V_0+V_1(N-Z)/A$, and find the following results for the ratio V_1/V_0 near the origin and at 7 fm, which is approximately half maximum density:

	R = 0 fm	R = 7 fm
Brieva-Rook	0.21	0.33
Yamaguchi et al.	0.28	0.37
JLM	0.19	0.24

As examples of phenomenological results, V_1/V_0 = 0.5 for the Becchetti-Greenlees potential[33], and for the empirical interaction of Austin[1], v_τ/v_0 = 0.54. The discrepancy in the magnitude of the cross sections calculated with the JLM potentials is very likely to be correlated with the very low isovector strength shown above, and the systematically small values of the ratio raise the question as to whether the nuclear-matter approach can yield an accurate description of the isovector interaction.

V. INELASTIC SCATTERING

We show examples of self-consistent inelastic scattering calculations for the first 2^+ level in ^{54}Fe (E_x = 1.41 MeV), and the first 3^- levels in ^{208}Pb (2.62 MeV) and ^{16}O (6.13 MeV). Further details of the calculational techniques that are specific to inelastic scattering may be found in Refs. 34,35.

Calculations[36] with the Brieva-Rook interaction for neutron and proton scattering are compared with data[36,37] in Figs. 18 and 19. The proton transition density was a Tassie-model form, normalized to the B(E2) determined from Coulomb excitation[38]. The solid curves show the excellent agreement obtained with a neutron transition density 0.8 times that for protons, which is in accord with the core-polarization model of Brown and Madsen[39]. The dashed curve in Fig. 19, calculated with the hydrodynamic assumption (N/Z) for the ratio of neutron to proton transition densities, is clearly inconsistent with the combined study of neutron and proton scattering.

Fig. 20 shows calculations using the Brieva-Rook and JLM interactions in comparison with data for neutron[40] and proton[41,42] scattering to the first 3^- state of ^{208}Pb. The proton and neutron transition densities were taken from analyses of electron[43] and 800-MeV proton[44] scattering measurements, respectively. The 61.4-MeV proton data are acceptably reproduced with both interactions. It should be noted that the neutron/proton transition-density ratio is 1.73, which is slightly larger than the hydrodynamic value 1.54. The JLM results are in qualitative accord with a calculation[11] using theoretical (RPA) transition densities. On the other hand, the results at lower energies, where both proton and neutron data are available,

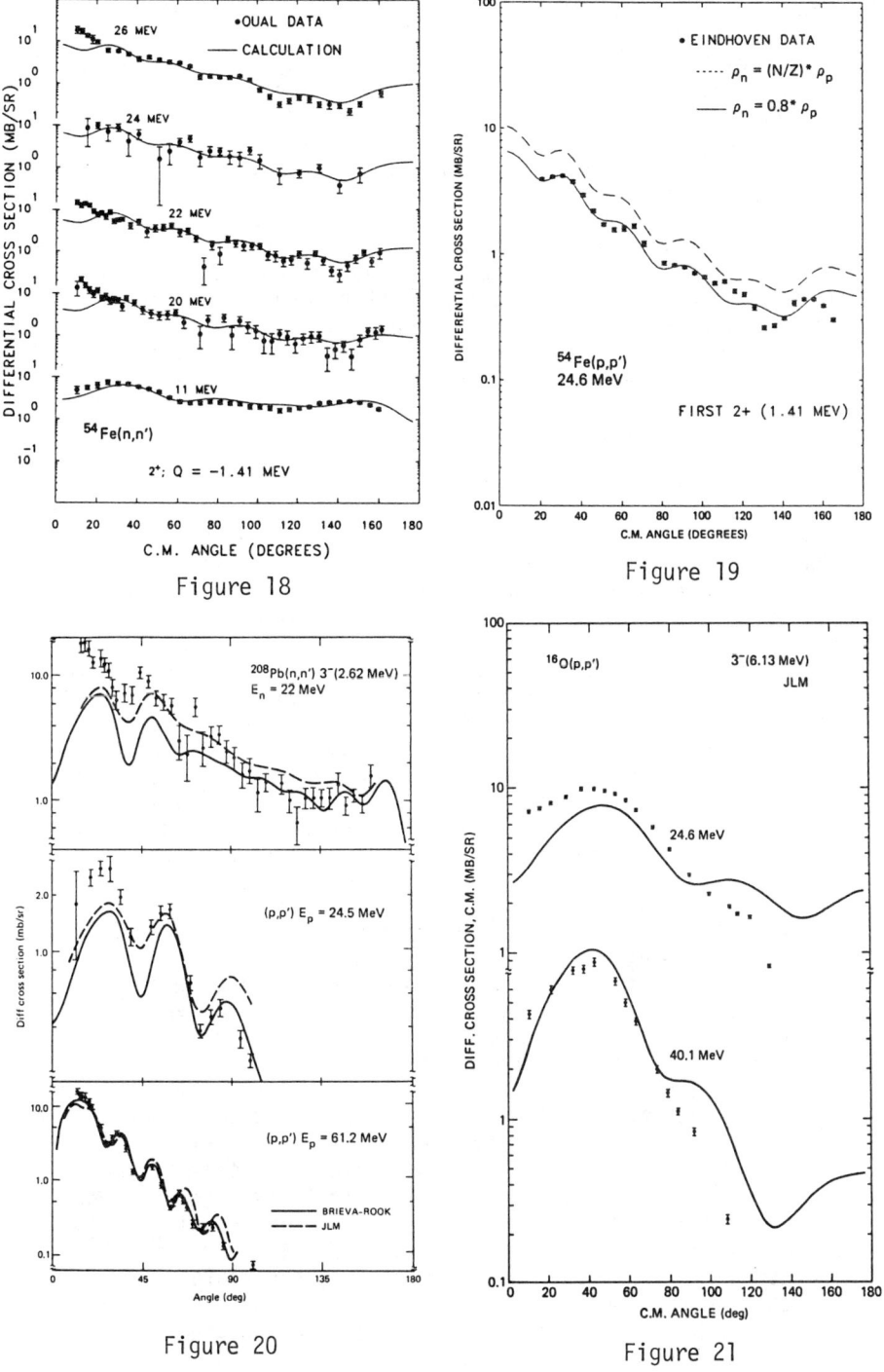

Figure 18

Figure 19

Figure 20

Figure 21

are much less satisfactory. Both interactions, particularly the Brieva-Rook, underpredict the forward cross sections. This characteristic of the Brieva-Rook calculations is reminiscent of that for the (p,n) reaction. Although the behavior of the calculations is similar for both protons and neutrons in the 20-25 MeV range, it is evident that in this nucleus no reliable conclusion may be drawn about the ratio of neutron to proton transition densities by comparison of neutron and proton probes with the interactions presently available in this energy range.

As a final example, Fig. 21 shows calculations with the JLM interaction for proton scattering[45] near 25 and 40 MeV to the first 3- level of ^{16}O, using a transition density[46] derived from electron scattering. Even though these parameter-free calculations are deficient at back angles, they are comparable in quality to phenomenological treatments.

VI. SUMMARY AND ACKNOWLEDGMENTS

The degree of success obtained in the systematic analyses of the previous sections shows that the folding-model approach with nuclear-matter based effective interactions is very useful, even though the results are not universally perfect. The model is quite successful in reproducing elastic-scattering differential cross sections and analyzing powers, particularly when the JLM interaction is used for the central potential. The model works well over a wide mass range, without requiring (or, in fact, allowing) arbitrary changes in geometrical parameters. The model provides a suitable framework for addressing the problems of Coulomb effects and the consistent treatment of elastic scattering and reactions, since the sources of energy dependence and of nonlocality in the reaction model are explicit. In some cases, such as the ^{54}Fe inelastic-scattering example, the model appears adequate for the extraction of spectroscopic information on neutron-proton transition density differences by comparison of neutron and proton scattering. On the negative side, there is significant variation of the predictions of the various effective interactions, even of those based on the same underlying nucleon-nucleon potential. The isovector spin-independent parts of the interactions appear to be too small, particularly for the JLM calculations, which otherwise are very successful.

Our conclusion from the present work is that the folding-model approach shows the promise of providing a self-consistent treatment of elastic scattering and reactions, and that efforts should be made to further refine the effective interactions. These efforts should include not only further development of theoretically-based interactions, such as those investigated here, but also the development and testing of phenomenological interactions guided by the features of the theoretical interactions that appear to be successful. Neutron-scattering data, in comparison with proton data, will continue to be of great importance in evaluating the isospin-dependent parts of these interactions.

We wish to acknowledge the participation of numerous colleagues in various aspects of this work, including R. W. Finlay, S. H. Mellema, and J. S. Petler of Ohio University; N. L. Back, L. F. Hansen and T. W. Phillips of Lawrence Livermore National Laboratory; J. J. Kelly of MIT; and A. Carpenter and J. A. Carr of Florida State University. This work was performed under the auspices of the U.S. Department of Energy by the Lawrence Livermore National Laboratory under contract number W-7405-ENG-48 and the National Science Foundation under Grant No. PHY-8122131.

REFERENCES

1. S. M. Austin, in The (p,n) Reaction and the Nucleon-Nucleon Force, C. D. Goodman et al., eds., Plenum, N. Y., 1980, p. 203.
2. G. Bertsch, J. Borysowicz, H. McManus, and W. G. Love, Nucl. Phys. A284, 399 (1977).
3. F. A. Brieva and J. R. Rook, Nucl. Phys. A291, 299, 317 (1977); Yukawa parameterization of interaction by H. V. von Geramb.
4. L. Rikus, K. Nakano, and H. V. von Geramb, Nucl. Phys. A414, 413 (1984), and references therein.
5. N. Yamaguchi, S. Nagata, and T. Matsuda, Prog. Theor. Phys. (Japan) 70, 459 (1983).
6. J. P. Jeukenne, A. Lejeune, and C. Mahaux, Phys. Rev. C16, 80 (1977); A. Lejeune, Phys. Rev. C21, 1107 (1980).
7. F. S. Dietrich et al., Phys. Rev. Lett. 51, 1629 (1983).
8. S. H. Mellema, R. W. Finlay, F. S. Dietrich, and F. Petrovich, Phys. Rev. C28, 2267 (1983).
9. T. Hamada and I. D. Johnston, Nucl. Phys. 34, 382 (1962).
10. R. V. Reid, Ann. Phys. (N.Y.) 50, 411 (1968).
11. Ch. Lagrange and J. C. Brient, J. Phys. (Paris) 44, 27 (1983).
12. M. Lacombe et al., Phys. Rev. C21, 861 (1980).
13. F. A. Brieva and J. R. Rook, Nucl. Phys. A297, 206 (1978).
14. J. W. Negele and K. Yazaki, Phys. Rev. Lett. 47, 71 (1981).
15. S. Fantoni, B. L. Friman, and V. R. Pandharipande, Phys. Lett. 104B, 89 (1981).
16. H. Fiedeldey, Nucl. Phys. 77, 149 (1966).
17. F. S. Dietrich and F. Petrovich, to be published.
18. B. Frois et al., Phys. Rev. Lett. 38, 152 (1977).
19. G. W. Hoffmann et al., Phys. Rev. C21, 1488 (1980).
20. G. Bulski et al., Proc. Int. Conf. Nucl. Data, Antwerp, 1982; J. P.Delaroche et al., Phys. Rev. C28, 1410 (1983).
21. S. M. El-Kadi et al., Nucl. Phys. A390, 509 (1982).
22. C. E. Floyd in Polarization Phenomena in Nuclear Physics -- 1980, AIP Conf. Proc. No. 69, G. G. Ohlson et al., eds. (AIP, New York, 1981).
23. C. Wong, J. D. Anderson, J. W. McClure, and B. D. Walker, Phys. Rev. 128, 2339 (1962).
24. L. F. Hansen et al., to be published in Phys. Rev. C, Jan., 1985.

25. A. S. Meigooni, J. S. Petler, and R. W. Finlay, Phys. Med. Biol. $\underline{29}$, 643 (1984).
26. N. L. Back, F. S. Dietrich, I. Proctor, and J. Woodworth, to be published.
27. E. Fabrici et al., Phys. Rev. $\underline{C21}$, 830 (1980).
28. N. L. Back, private communication.
29. H. C. Camarda, T. W. Phillips, and R. M. White, Phys. Rev. $\underline{C29}$, 2106 (1984).
30. F. A. Brieva and R. G. Lovas, Nucl. Phys. $\underline{A341}$, 377 (1980).
31. R. R. Doering, D. M. Patterson, and A. Galonsky, Phys. Rev. $\underline{C12}$, 378 (1975).
32. D. M. Patterson, R. R. Doering, and A. Galonsky, Nucl. Phys. $\underline{A263}$, 261 (1976).
33. F. D. Becchetti and G. W. Greenlees, Phys. Rev. $\underline{182}$, 1190 (1969).
34. F. Petrovich and W. G. Love, Nucl. Phys. $\underline{A354}$, 499 (1981).
35. J. J. Kelly et al., Phys. Rev. Lett. $\underline{45}$, $\overline{2012}$ (1980).
36. S. H. Mellema, R. W. Finlay, F. S. Dietrich, and F. Petrovich, Phys. Rev. $\underline{C29}$, 2385 (1984).
37. P. J. Van Hall et al., Nucl. Phys. $\underline{A291}$, 63 (1977).
38. M. J. LeVine, E. K. Warburton, and D. Schwalm, Phys. Rev. $\underline{C23}$, 244 (1981).
39. V. R. Brown and V. A. Madsen, Phys. Rev. $\underline{C11}$, 1298 (1975), and V. A. Madsen, private communication.
40. R. W. Finlay et al., Phys. Rev. $\underline{C30}$, 796 (1984).
41. J. Saudinos et al., Phys. Lett. $\underline{22}$, 492 (1966).
42. A. Scott, N. P. Mathur, and F. Petrovich, Nucl. Phys. $\underline{A285}$, 222 (1977).
43. D. Goutte et al., Phys. Rev. Lett. $\underline{45}$, 1618 (1980).
44. L. Ray and G. W. Hoffmann, Phys. Rev. $\underline{C27}$, 2133 (1983).
45. D. Bayer, thesis, Michigan State University (1971), unpublished.
46. J. J. Kelly, thesis, Massachusetts Institute of Technology (1981), unpublished.

ISOVECTOR NUCLEON-NUCLEON EFFECTIVE INTERACTIONS

J. R. Shepard
Department of Physics, University of Colorado, Boulder, Co. 80309

Experimental capabilities have recently been developed which have provided data allowing us to test in new ways our models of the effective interacions which drive nucleon-nucleon scattering processes. Some of these new measurements have been made at low energies ($T_N < 100$ MeV) where effects of the nuclear medium must be incorporated via the G-matrix formalism. Others have been made at higher energies where the impulse approximation is felt to be well-justified allowing the effective interaction to be identified with the free NN t-matrix. In the following I will address the relevance of the new data to our understanding of the effective interaction, with emphasis on the isovector channel.

Rather than try to present a comprehensive picture of the present state of knowledge concerning the low-energy isovector effective interaction, I would like to discuss the interpretation of a few specific experimental results in terms of one, well-known effective interaction, namely M3Y.[1] At the Osaka Meeting in the spring of 1983, Orihara et al.[2] reported data for the ^{12}C, ^{14}C, ^{16}O, ^{24}Mg, ^{28}Si(p,n) reactions at $T_p = 35$ MeV. In these experiments, energy resolution of $\Delta T_n = 30 - 170$ keV was achieved. Of particular interest in this study are the angular distributions for the low-lying 0^-, 1^-, 2^-, 3^- quartet in the ^{16}O(p,n) ^{16}F reaction, the identification of which was made possible by the excellent energy resolution. DWBA calculations for these transitions using M3Y for the effective interaction do not reproduce well either the shapes or the relative magnitudes of the angular distributions. Similar results were obtained using the interaction of Anataraman et al.[3] based on the Paris potential. The agreement is much improved by dropping the odd components of the central and spin orbit pieces of M3Y. However, reproduction of the 0^- angular distribution -- the weakest and most difficult to measure -- requires that the odd components of the tensor force be retained. This phenomenological modification of M3Y yields a gratifyingly consistent description of not only the ^{16}O(p,n) data, but also the data for many other "magnetic" transitions measured in this study.

Sakai et al.[4] presented unique $(\vec{p},\vec{n})$ spin transfer data at the recent (Nov. 1983) Tokyo Symposium on the Electromagnetic Properties of Atomic Nuclei. They measured the spin transfer observable D_{nn} for the ^{12}C(p,n)^{12}N(1^+ g.s.) transition at 65 MeV. The found that the theoretical value of this quantity is quite sensitive to the high-q ($q > 1$ fm^{-1}) behavior of the tensor component of the effective interaction through the knock-on exchange process. Sakai et al. found that, while an overall reduction of the tensor interaction as was found necessary by Love[5] and Petrovich[6] to describe cross sections worsens agreement with D_{nn}, a reduction of the high q-portion alone results in improved agreement. Sakai et al. were guided in choosing the form of this modification by considering the meson-exchange origin of the tensor force.

While the two studies disussed above cannot be said to have resulted in definitive conclusions about the effective interaction, they do serve to illustrate the usefulness of new and more complete kinds of measurements. This in turn justifies the difficulty they entail. It seems certain that experimentalists will increasingly concern themselves with such experiments in the future, to the benefit of reaction theorists.

In the high energy domain ($T_p > 100$ MeV), as stated above, the impule approximation is typically employed; i.e., the effective interaction is assumed to be equal to the free NN interaction. (Note that Rikus, Nakano and von Geramb[7] have recently published a G-matrix analysis of the $^{12}C(p,p')^{12}C(1^+$ T = 0 and 1) reactions for $122 \leq T_p \leq 400$ MeV. They do not indicate, however, what differences exist between results using the G-matrix or the t-matrix for the effective interaction.)

At these energies, there is considerable evidence that the central pieces of the isovector effective interaction, namely the spin-dependent and spin-independent terms, $V_{\sigma\tau}$ and V_τ, respectively, are reasonably well described by the impulse approximation. For example, Glashausser et al.[8] have shown that the cross section and analyzing powers of the $^{12}C(p,p')^{12}C(15.11$ MeV 1^+ T = 1) transition are reasonably well-described by the DWIA over the energy range to 800 MeV. This transition, at low q, is excited predominantly via $V_{\sigma\tau}$. The accuracy of the impulse approximation description of V_τ is indicated in Fig. 1 where the impulse approximation values of the ratio $R = |V_{\sigma\tau}/V_\tau|$ at q = 0 as obtained from the Arndt phase shifts[9] is plotted versus bombarding energy. The data points shown are those extracted from various (p,n)[10] and (^{3}He,t)[11] measurements. The comparison in Fig. 1 shows qualitative agreement but the (p,n) results, where the reaction mechanism is presumably better understood than for (^{3}He,t), systematically yield a value of R greater than that given by the impulse approximation. This in turn suggests that $|V_\tau^{eff}|$ is smaller than the impulse approximation value in the 100 to 200 MeV region. The importance of measuring the value of R for $0°$(p,n) reactions at energies above 200 MeV has been stressed by Guichon and Kroll[12] who also suggest that the present data for energies below 200 MeV is not consistent with the impulse approximation. They point out that it is essential to understand the energy dependence of the effective value of R if one is to use higher energy $0°$(p,n) measurement to examine the Gamow Teller (governed by $V_{\sigma\tau}$) and Fermi (governed by V_τ) strength in nuclei. In this regard, it has been pointed out by Auerbach et al.[13] that T_p = 800 MeV is an especially interesting region because of the resurgance of the relative Fermi strength at this energy (R $\simeq$ 2) predicted by the impulse approximation. It is clearly of interst to test this speculation.

With this and other objectives in mind, an experimental program has been undertaken on Beam Line D (WNR) at LAMPF to make high resolution (p,n) measurements at energies between 318 and 800 MeV.[14] An overall resolution of roughly 2 to 2.5 MeV (fwhm) has been achieved using time-of-flight techniques and recently angular distributions out to 9° in the laboratory have been measured to complement

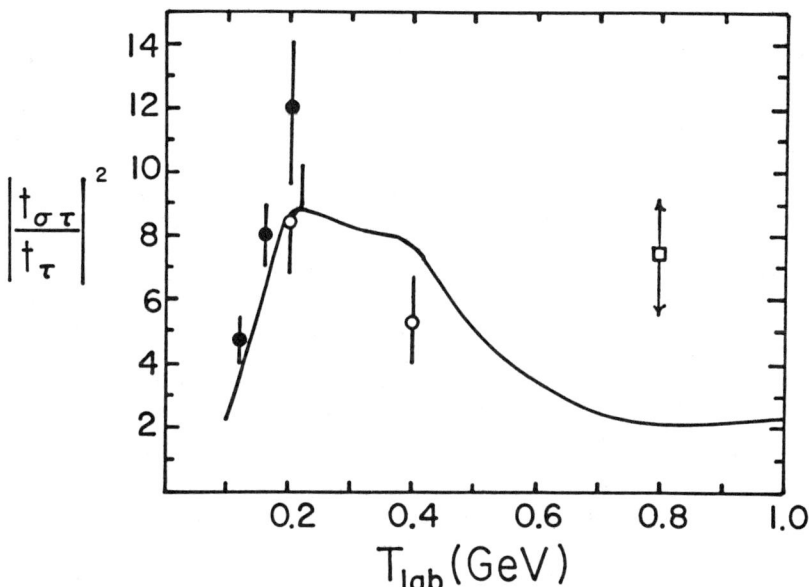

Fig. 1. The ratio $R \simeq |t_{\sigma\tau}/t_{\tau}|^2$ is plotted as a function of bombarding energy. The closed circles are values extracted from (p,n) measurements while the open circles are from (^{3}He,t) results. The solid curve is extracted from the free NN amplitude of Arndt. The tentative value extracted from the WNR data is represented by the square.

earlier 0° results. Spectra for the $T_p = 800$ MeV ^{6}Li and ^{12}C(p,n) reactions at angles of 0° through 9° are displayed in Figures 2 and 3, respectively. On the left side of these figures is a full spectrum covering an outgoing neutron energy range of ≈ 120 to 800 MeV. At high neutron energy is a relatively sharp peak to be discussed below. The other prominent feature, at T_n ≈ 500 MeV, corresponds to the creation of a Δ resonance in the scattering process. The very interesting question of how much of the observed strength is due to a simple quasi-free mechanism and how much can be attributed to more exotic processes such nuclear Δ-hole excitations will be addressed in the course of future analysis of these data. At the right in Figs. 2 and 3 are expanded spectra which show details of the high energy end ($700 \le T_n \le 800$ MeV) of the spectrum. Narrow peaks corresponding to discrete nucler levels appear at the highest neutron energies. Careful energy calibrations of the 0° spectra have conclusively shown that these peaks correspond to the Gamow Teller (+ Fermi?) states strongly excited in $T_p = 100$ to 200 MeV 0° measurements. In particular, the sharp peak in the ^{6}Li spectrum corresponds to the unresolved excitation of the 0^+ g.s. and 1.67 MeV 2^+ level in ^{6}Be. Likewise, there is a strong peak in the ^{12}C spectrum arising from the excitation of the 1^+ g.s. and 0.96 MeV 2^+ level in ^{12}N. At somewhat lower neutron energy corresponding to an excitation energy of ≈ 5 MeV in ^{12}N is another peak which has its maximum cross section at 3° or 4° suggesting that it arises from a $\Delta \ell = 1$ process. The dominance of the spin dependent piece of the isovector interaction (to be discussed below) suggests that a spin transfer is also involved in the transition. This peak can then be assumed to arise from the excitation of 0^-, 1^- and 2^- levels in ^{12}N.

In the largest angle spectra for both ^{6}Li and ^{12}C, another peak is apparent at p ≈ 1400 to 1420 MeV/c or T_n = 750 to 770 MeV. This peak appears to shift to higher neutron energies and become smaller at more forward angles. The preliminary interpretation of this feature is that it arises from quasi-elastic pn knock on which is strongly suppressed by Pauli blocking at forward angles where the momentum transfer is small. Preliminary analysis indicates that both the position an magnitude of this peak in the 9° ^{12}C spectrum are consistent with such an interpretation.

The angular distribution for the excitations of the low-lying discrete levels in the ^{6}Li and ^{12}C(p,n) reactions allow the impulse approximation description of $V_{\sigma\tau}^{eff}$ to be tested since it is predominantly this piece of the effective interaction which governs the cross sections at low q. In the language of the 100 to 200 MeV (p,n) studies, these are "Gamow-Teller" ($\Delta \ell = 0$ $\Delta s = 1$) transitions. Since distortions are relatively strong at 800 MeV, it was decided to treat them as carefully as possible which dictated a full distorted wave impulse approximation (DWIA) analysis. Both the distorted waves[15] and the effective interaction[16] were generated in the relativistic impulse approximation using the NN amplitudes of Arndt.[9] The nuclear densities used were chosen to be consistent with elastic electron scattering. Furthermore the nuclear structure information used in generating the (p,n) transition densities was checked for consistency with the inelastic electron scattering[17] for the analogue

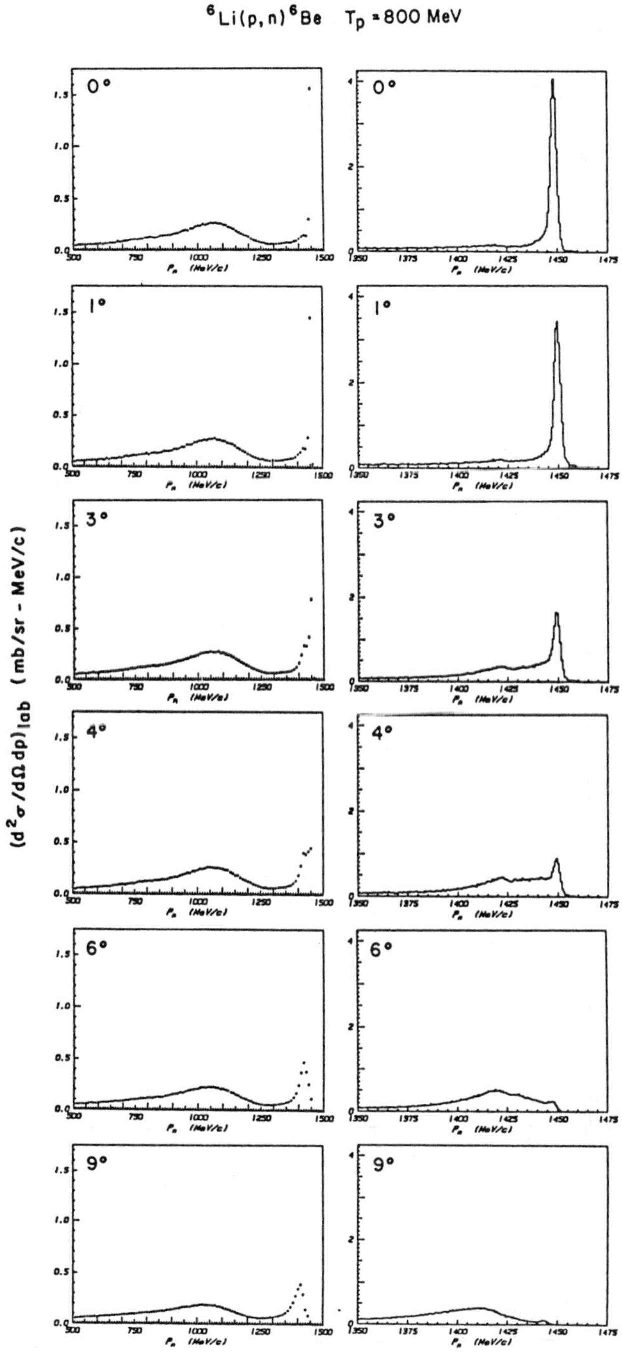

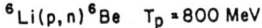

^{6}Li(p,n)^{6}Be T_p = 800 MeV

Fig. 2. Spectra for ^{6}Li(p,n) at 300 MeV from Θ_{lab} = 0° to 9°. The right hand spectra are blow ups of the high energy end of the full spectra on the left.

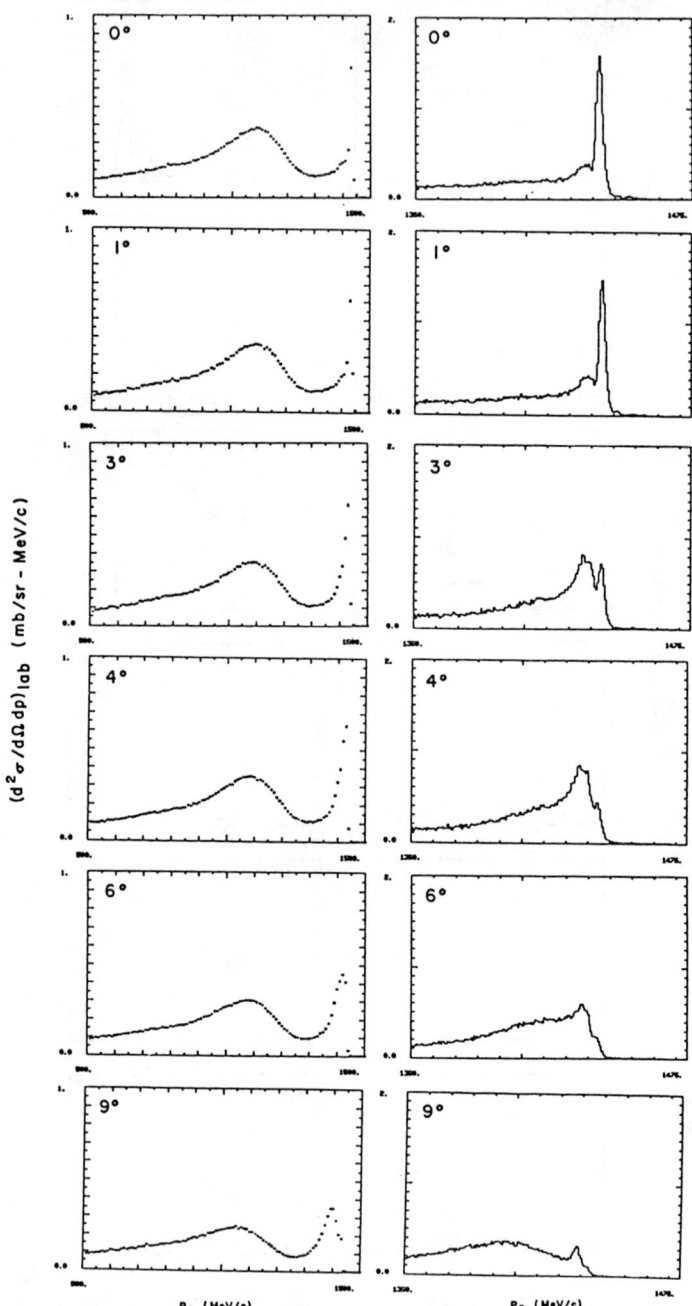

Fig. 3. Spectra for ^{12}C(p,n) of 800 MeV from θ_{lab} = 0° to 9°. The right hand spectra are blowups of the high energy end of the full spectra on the left.

transitions.[17,18] The (p,n) calculations were performed using the code DRIA[19] and the unrenormalized results are compared with data in Figs. 4 and 5. As these figures demonstrate, the agreement between the impulse approximation results and the data is excellent at all angles, even at 0° where the experimental uncertainties are quite small. This is taken as strong confirmation of the assertion made above that the impulse approximation gives an excellent quantitative description of $V_{\sigma\tau}^{eff}$ at T_p = 800 MeV.

By assuming that 0° ΔJ = 1 (p,n) transitions are dominated by the ΔL = 0 amplitude, we can relate the 0° cross section to the Gamow-Teller β-decay matrix element as has been done at lower energies.[10] The quantitative agreement between the impulse approximation and experiment noted above for Δs = 1 transitions suggests that the scale factor relating theoretical and experimental 0° cross sections can be used to relate theoretical and experimental Gamow-Teller transition rates. Of particular interest in this regard is the B_{GT} for the transition to the 3.511 MeV $3/2^-$ T = 1/2 level in ^{13}N. The energetics of β-decay forbid the excitation of the analogue of this level via ^{13}N β-decay. Consequently the only experimental determinations of its B_{GT} come from the lower energy (p,n) studies[10] where a value of B_{GT} = 0.75 to 0.83 is found. This represents less than 35% from the Cohen and Kurath value. This degree of quenching is about twice as great as what is typically encountered in the p-shell. In the present 800 MeV case, we calculate a 0° cross section for the transition to this state of 8.2 mb/sr using the Cohen and Kurath wave functions while the measured value is 5.1 mb/sr. Using the quenching factor given by the ratio of these cross sections along with the Cohen and Kurath value of B_{GT} = 2.376 we find an actual B_{GT} of 1.48 or about 62% of the Cohen-Kurath value. The origin of the discrepancy between the present result and that of the lower energy studies is not known.

Having confidence in the DRIA calculations of 0° (p,n) spin flip strength allows the estimation of the non-spin flip strength observed in the 800 MeV $^{13}C(p,n)^{13}N(g.s.)$ transition. In β-decay this transition (or more properly, its inverse) is an incoherent mixture of Gamow-Teller (Δs = 1) and Fermi (Δs = 0) processes. The Cohen-Kurath wavefunctions overestimate the Gamow-Teller decay rate by ≈ 70%; the elastic M1 from factor for ^{13}C at q = 0.5 fm^{-1} is over estimated by ≈ 55%. On this basis, the "quenching" of the Cohen-Kurath spin flip strength is estimated to be ≈ 60%. This factor is then applied to the ΔJ = 1 0° cross section calculated with DRIA using the Cohen-Kurath nuclear structure information yielding a "quenched" cross section of 1.5 mb/sr. The experimental cross section is 2.1 mb/sr suggesting that the non-spin flip cross section is 0.6 mb/sr. Since the calculated non-spin-flip (ΔJ = 0) cross section is 1.55 mb/sr, $|V_\tau^{eff}|$ must be reduced form the impulse approximation value by $\sqrt{0.6/1.55}$ = 0.62. The present analysis then suggests that

$$|V_\tau^{eff}| = 0.62 \; |V_\tau^{I.A.}| = 50 \text{ MeV fm}^3.$$

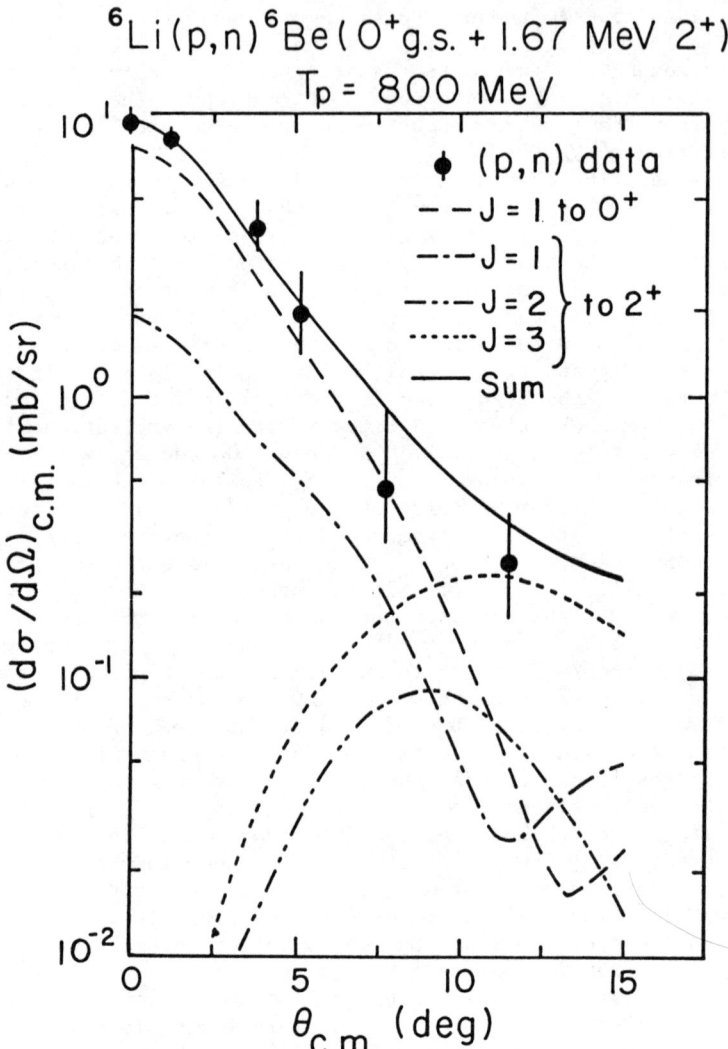

Fig. 4. Impulse approximation calculations (described in text) are compared with the ^{6}Li(p,n)^{6}Be angular distribution.

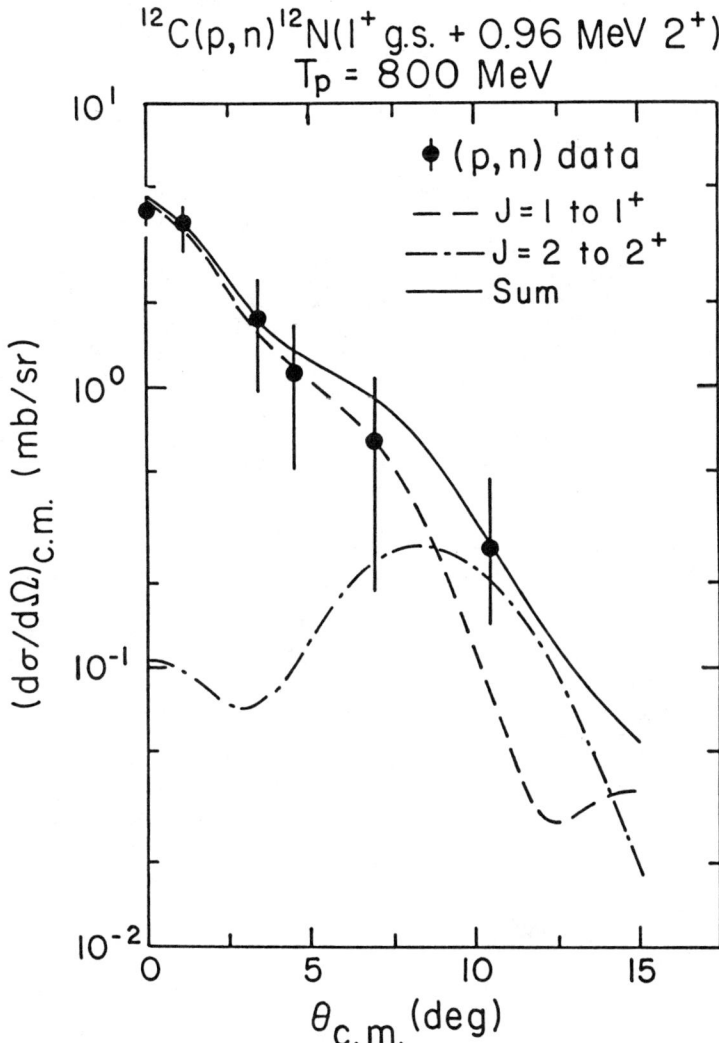

Fig. 5. Impulse approximation calculations (described in text) are compared with the $^{12}C(p,n)^{12}N$ angular distribution.

This result gives much larger value of R than the impulse approximation, namely $R = (134.8/50)^2 = 7.3$. This value of R is displayed in Figure 1. The origin of this apparent discrepancy with the impulse approximation -- whether it resides in uncertainties in the free NN interaction, the nuclear structure amplitudes, or the reaction mechanism -- is not understood.

The value of spin transfer observables in examining the properties of effective interactions is illustrated in some recent (p,n) and (p,p') studies. Goodman et al.,[20] for example, have measured 0° spin-flip probabilities for low-lying levels excited in the ^{13}C and ^{15}N$(\vec{p},\vec{n})$ reactions at T_p = 150 MeV. They "calibrate" the spin-flip effective interaction empirically by measuring the spin-flip probability for "pure" spin-flip transitions such as ^{12}C$(\vec{p},\vec{n})^{12}$N$(1^+$ g.s.). The fact that the spin-flip probability is zero for the $\Delta J = 0$ transitions allow Goodman et al. to extract the ratio of spin-flip cross sections at 0° which they in turn relate to the ratio B_{GT}/B_F. Though not addressed in the analysis of Goodman et al., comparison of such $(\vec{p},\vec{n})$ spin-flip probability with impulse approximation results will very likely be useful in testing models of both the spin-dependent and spin-independent components of the isovector effective interaction.

Even more extensive measurements of this type have been reported by McClelland et al.[21] Specifically, they measured all spin transfer observables but P and A_y for the ^{12}C$(p,p')^{12}$C*$(1^+$ T = 0 and 1) at T_p = 500 MeV. Such a complete set of spin observables test all components of the spin-dependent NN effective interaction. The analysis of McClelland et al. shows that the standard impulse approximation gives a reasonable description of their data. This conclusion is tempered to some degree by the fact that there are relatively few data points and that the experimental uncertainties are quite large. It is also true that their measurements do not include certain quantities, such as P-A_y, which are particularly sensitve to subtle details of the NN amplitude and the reaction mechanism and therefore pose an especially strong challenge to our inelastic scattering models.

In fact, recent theoretical developments[22,23] suggest that the standard formulation of the impulse approximation[24] which is typically used to analyze inelastic scattering data of the type discussed above is incomplete and, in certain instances, might give rise to misleading interpretations of experimental results. This is particularly true of a certain class of spin observables, including the quantity P-A_y. The impulse approximation of Ref. 22 differs from the standard formulation by beginnng with a representation of the NN amplitude in terms of relativistic invariants[25] rather than the usual Wolfenstein representation. The two different forms of the NN amplitude, M_{NN}, are given by:

$$M_{NN}(1,2) = A + \vec{\sigma}_1 \cdot \vec{\sigma}_2 \, B + i(\vec{\sigma}_1 + \vec{\sigma}) \cdot \hat{n} \, qC +$$

$$\vec{\sigma}_1 \cdot \vec{q} \, \vec{\sigma}_2 \cdot \vec{q} D + \vec{\sigma}_1 \cdot \hat{p} \, \vec{\sigma}_2 \cdot \hat{p} E \tag{1a}$$

$$= \bar{u}_1' \bar{u}_2' F_{NN} u_1 u_2 \qquad (1b)$$

where:

$$\vec{q} \equiv \vec{k}_f - \vec{k}_i, \quad \vec{p} = \frac{1}{2}(\vec{k}_i + \vec{k}_f), \quad \hat{n} = \frac{\vec{k}_i \times \vec{k}_f}{|\vec{k}_i \times \vec{k}_f|} = \hat{p} \times \hat{q}, \quad (2a)$$

$$u = u(\vec{k}, s) = \sqrt{\frac{E+m}{2m}} \begin{Bmatrix} 1 \\ \frac{\vec{\sigma} \cdot \vec{k}}{E+m} \end{Bmatrix} \chi_s \text{ is the free Dirac} \quad (2b)$$

spinor and χ_s is the Pauli spinor, and

$$F_{NN} = F_s + \gamma(1) \cdot \gamma(2) F_v + \gamma^5(1) \gamma^5(2) F_p$$
$$+ \gamma^5(1) \gamma^5(2) \gamma(1) \cdot \gamma(2) F_A + \sigma^{\mu\nu}(1) \sigma_{\mu\nu}(2) F_T, \quad (2c)$$

where the γ's are the usual Dirac γ-matrices,[26] which defines the relativistic invariants F_i.[25] The plane-wave impulse approximation inelastic scattering amplitude for $J_i \rightarrow J_f$ is

$$T_{fi} = \langle J_f M_f ; m_f | M_{NN} e^{-i\bar{q}\cdot\bar{r}} | J_i M_i ; m_i \rangle \quad (3)$$

where the M's and m's refer to the magnetic quantum numbers of the target and projectile, respectively. In the standard impulse using the form of M_{NN} given in Eq. 1a, the influence of the nuclear dynamics binding the target nucleon on its lower components are implicitly ignored. In consequence, nuclear convection currents are ignored. In the so-called Free Relativistic Impulse Approximation (FRIA) developed in Ref. 22, this approximation is lifted and convection currents arise in a natural way just as they do in standard microscopic treatments of inelastic electron scattering.[27] In the FRIA a new class of nuclear structure terms arises, all of which contain the probability current operator,[28]

$$\vec{j} = \frac{i}{2m}(\overleftarrow{\nabla} - \overrightarrow{\nabla}). \quad (4)$$

Specifically, these matrix elements are

$$\langle J_f M_f | \vec{j} e^{-i\bar{q}\cdot\bar{r}} | J_i M_i \rangle \quad (5a)$$

$$\langle J_f \, M_f \, | \, \vec{\sigma} \cdot \vec{j} \, e^{-i\vec{q}\cdot\vec{r}} \, | \, J_i \, M_i \rangle \qquad (5b)$$

$$\langle J_f \, M_f \, | \, \vec{\sigma} \times \vec{j} \, e^{-i\vec{q}\cdot\vec{r}} \, | \, J_i \, M_i \rangle. \qquad (5c)$$

without these terms (an without an explicit treatment of knock-on exchange) P-A_y = 0 is required; with them, non-zero values of P-A_y emerge even in the plane wave and zero-Q-value limit. Numerical impulse approximation calculations of P-A_y[23] which include distortion effects[19] are compared with data for the ^{12}C(p,p')^{12}C*(12.71 MeV 1^+ T = 0 and 15.11 MeV 1^+ T = 1) of Carey et al.[29] in Figures 6 and 7. The standard impulse approximation results are indicated by the dashed curve; they were calculated using the code DW81 and a recent panametrization[30] of Arndt's[31] SP84 NN phase-shift solution. These calculations give non-zero P-A_y because they treat knock-on exchange explicitly. Also shown in Figs. 6 and 7 are FRIA and Dynamic Relativistic Impulse Approximation (or DRIA) calculations, the latter differing from the former in that the effects of the large scalar and time-like vector potentials which characterize current relativistic models[12] are included. These calculations use an NN interaction derived from Arndt's SM84 solution; they do not include an explicit treatment of knock-on exchange. The FRIA and DRIA results describe the T = 0 data as well as or better than the standard calculation. The T = 1 data are best described by the DRIA result when the pseudovector form of the pseudoscalar invariant is used.[23] While these results can not be interpreted as unambiguously supporting the approach of Refs. 22 and 23, they do suggest that the convection currents which appear naturally in these approaches -- but are omitted in the standard treatments -- can have important effects on certain observables and should be included whether or not knock-on exchange is treated explicitly. As is discussed in some detail in Refs. 17, 22 and 23, the magnitudes of the convection current terms (Eq. 5) and the observables they affect are strongly influenced by the presence of the strong scalar and timelike vector potentials alluded to above. These effects appear in Figs. 6 and 7 as the differences between the FRIA and DRIA results. Thus, analysis of nucleon-nucleon inelastic scattering spin transfer observables may eventually tell us whether such strong potentials are actually present in nuclei.

In summary, it has been shown that a new generation of difficult measurements has provided unique information to test our reaction models and the effective interactions they employ. Many of these experimental programs involve the measurement of spin transfer observables for either the (p,p') or (p,n) reactions. Some of the lower energy (T < 100 MeV) measurements were shown to indicate specific shortcomings in the isovector portion of the M3Y G-matrix. At higher energies (T > 100 MeV) recent (p,n) measurements in particular show that the spin dependent piece of the isovector effective interaction is well-explained by the impulse approximation while the spin-independent strength is apparently over estimated.

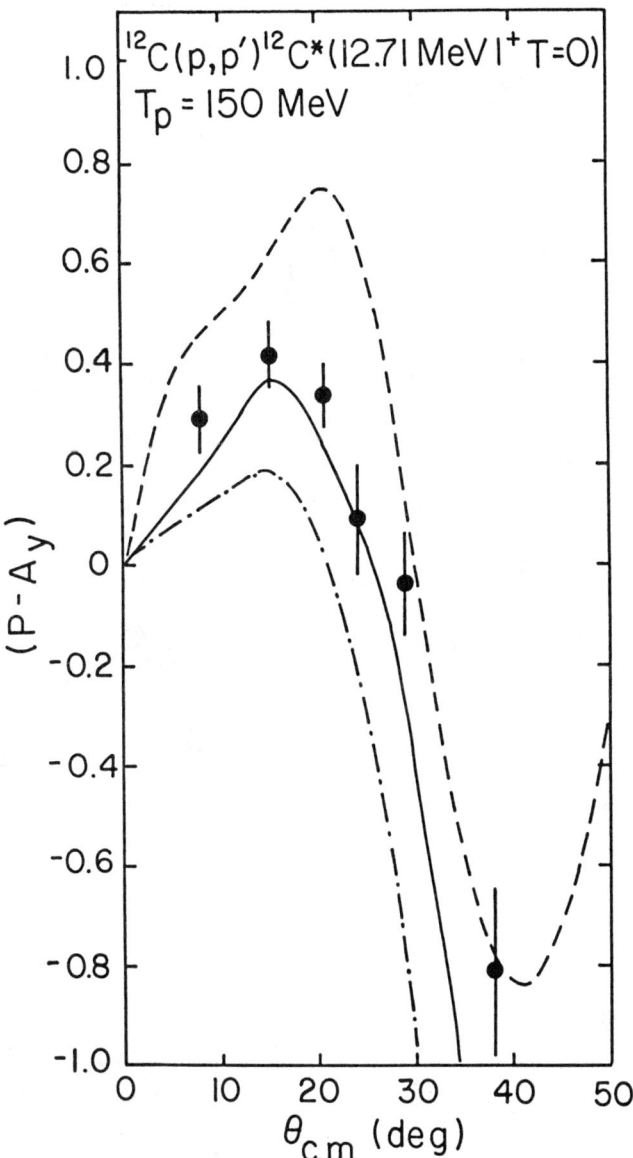

Fig. 6. Impulse approximation calculations are compared with the P-A_y data of Ref. 29. The dashed curve is the Standard I. A. result while the solid and dashed curves are F.R.I.A. and D.R.I.A. results, respectively. (See text).

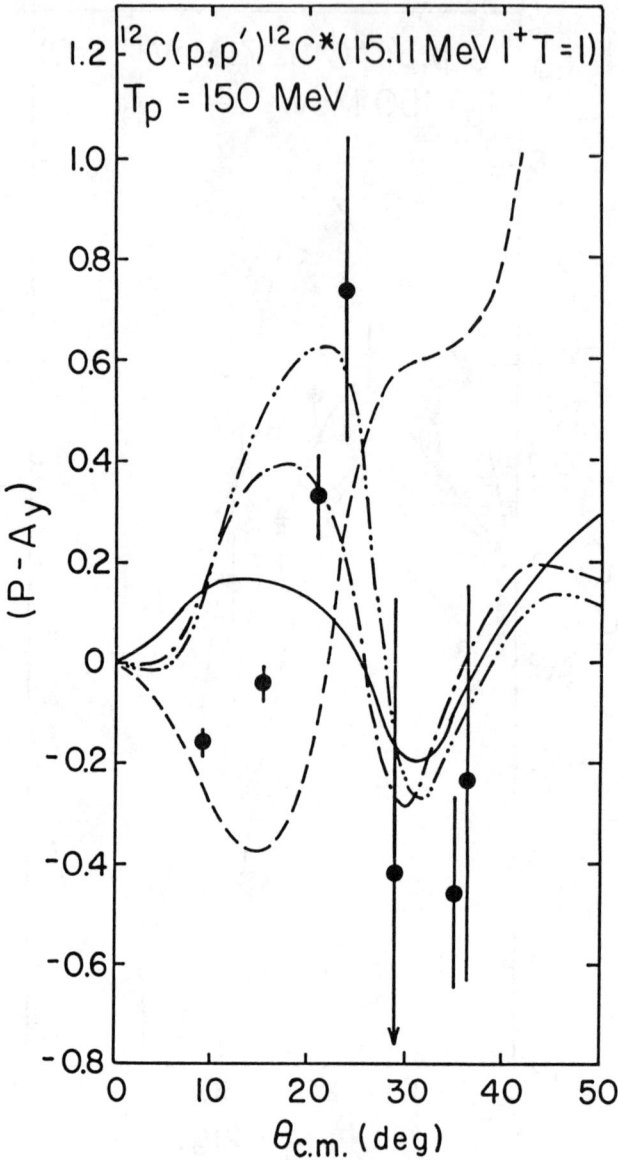

Fig. 7. Impulse approximation calculations are compared with the $P-A_y$ data of Ref. 29. The dashed curve is the Standard I.A. result while the solid curve is the D.R.I.A. calculation using the pseudoscalar parametrization of the pseudoscalar NN invariant. The dashed-dot and dashed-double dot curves employ the pseudovector parametrization with the slightly different nuclear structure inputs. (See Ref. 23).

Although the standard impulse approximation appears to work reasonably well at the higher energies, it omits contributions from convection currents which can be important for certain spin observables. Recent formulations of the impulse approximation which naturally include the current contributions have been able to reproduce P-A_y data without explicit treatment of knock-on exchange which is required in the standard treatments for P-$A_y \neq 0$. These current terms are sensitive to the presence of strong potentials such as those which occur in current relativistic models of nuclear physics.

It can be anticipated that much future experimental effort will be directed toward making the difficult type of measurements discussed above. These will include better spin transfer measurements for ($\vec{p}$,p'), more high-quality (p,n) data, especially in the $T_p > 200$ MeV range, more and better (p,n) spin transfer data, and (n,p) measurements to complement the (p,n) results. On the theoretical side, work on the construction of a non-relativistic G-matrix to use in describing lower energy inelastic processes is continuing as is the application of both G-matrix and t-matrix techniques in the higher energy realm. At the same time, further study of the FRIA and DRIA can be anticipated. At one level, a fundamental basis for the approach must be sought; on another, experience with and understanding of the present formulation must occur through analysis of existing and future data. The explicit treatment of knock-on exchange must be included. Finally, the approach must be extended to lower energies where the impulse approximation is no longer reliable. The essential ingredient in this extension -- the relativistic G-matrix- has already been studied in some detail.[32]

REFERENCES

1. G. Bertsch, J. Borysowicz, H. McManus and W. G. Love, Nucl. Phys. A$\underline{284}$ 399 (1977).
2. H. Orihara et al., Proc. 1983 RCNP Int. Symposium on Light Ion Reaction Mechanism, Osaka, May 1983, p. 176.
3. N. Anantaraman, H. Toki and G. F. Bertsch, Nucl. Phys. A$\underline{398}$, 269 (1983).
4. H. Sakai et al., Proc. Int. Symp. on Electromag. Properties of Atomic Nuclei, Tokyo, Nov. 1983, p. 278.
5. W. G. Love, The (p,n) Reaction and the Nucleon-Nucleon Force, Goodman et al., eds, Plenum, N.Y., 1980, p. 23.
6. F. Petrovich, ibid., p. 115.
7. L. Rikus, K. Nakano and M. V. von Geramb, Nucl. Phys. A$\underline{414}$, 413 (1984).
8. C. Glashausser et al., preprint.
9. R. Arndt, phase shift solution WI82, private communication.
10. T. W. Taddeucci, et al., Phys. Rev. C$\underline{25}$ 1094 (1982).
11. C. Ellegaard et al., Phys. Rev. Lett $\underline{50}$, 1745 (1983).
12. P. A. M. Guichon and P. Kroll, Phys. Lett. $\underline{132B}$, 265 (1983).
13. N. Auerbach et al., Phys. Rev. C$\underline{28}$, 280 (1983).
14. N. S. P. King, P. Lisowski, G. Morgan, C. Goulding, R. Jeppessen, J. Ullmann, D. A. Lind, C. Zafiratos, J. R. Shepard and C. D. Goodman, to be published.
15. J. A. McNeil, J. R. Shepard and S. J. Wallace, Phys. Rev. Lett. $\underline{50}$, 1439 (1983).
16. J. R. Shepard, E. Rost and J. Piekarewicz, Phys. Rev. C (in press).
17. J. R. Shepard, E. Rost, E. R. Siciliano and J. A. McNeil, Phys. Rev. C$\underline{29}$, 2243 (1984).
18. Bergstrom et al., Nucl. Phys. A327, 439 (1979).
19. E. Rost and J. R. Shepard, unpublished.
20. C. D. Goodman et al., preprint.
21. J. McClelland et al., Phys. Rev. Lett. $\underline{52}$, 98 (1984).
22. J. R. Shepard, E. Rost and J. A. McNeil, to be published.
23. D. A. Sparrow, J. Piekarewicz, J. R. Shepard, E. Rost, J. A. McNeil, T. A. Carey, and J. McClelland, to be published.
24. e.g., A. K. Kerman, H. McManus and R. M. Thaler, Ann. Phys. (N.Y.) $\underline{8}$, 551 (1959).
25. J. A. McNeil, L. Ray and S. J. Wallace, Phys. Rev. C$\underline{27}$, 2133 (1983).
26. J. D. Bjorken and S. Drell, Relativistic Quantum Mechanics (McGraw-Hill, New York, 1964).
27. e.g., T. W. Donnelly and J. D. Walecka, Ann. Rev. Nucl. Sci. $\underline{25}$, 329 (1975).
28. L. I. Schiff, Quantum Mechanics, 3rd Ed., McGraw-Hill, N.Y., 1968.
29. T. Carey et al., Phys. Rev. Lett. $\underline{49}$, 266 (1982).
30. W. G. Love and M. A. Franey, private communication.
31. R. Arndt, private communication.
32. e.g., C. J. Horowitz, private communication and to be published.

DIRAC PHENOMENOLOGY AT LOW ENERGIES

B. C. Clark, S. Hama and S. G. Kälbermann
Department of Physics, Ohio State University, Columbus, Oh 43210

E. D. Cooper
Nuclear Theory Group, TRIUMF, Vancouver B.C., Canada V6T 2A3

R. L. Mercer
Thomas J. Watson Research Lab., IBM, Yorktown Heights, NY 10598

ABSTRACT

There is increasing evidence that relativistic effects play an important role in the description of nuclear structure and reactions. The development of a relativistic impulse approximation (RIA) and its success at intermediate energies is a case in point. Although the RIA cannot be used to describe proton scattering at low energies it successfully describes recent $\bar{p}$-^{12}C data at 46.8 MeV. We discuss a phenomenological nuclear optical model approach based on the Dirac equation containing large cancelling Lorentz scalar and Lorentz vector potentials. The model is used to describe proton and neutron scattering at low energy (T<100 MeV) from various targets. Systematic features of the phenomenology and its relationship to relativistic mean field theory (RMFT) calculations are given.

Modern treatments of the nuclear many-body problem within a relativistic framework indicate the importance of relativistic effects. Relativistic quantum field theoretical approaches, sometimes termed Quantum Hadrodynamics (QHD), have been recently reviewed by Anastasio, Celenza, Pong and Shakin[1] and Serot and Walecka.[2] The success of the RIA at intermediate energies is now well documented.[3-7] Nuclear optical model studies using the Dirac equation containing large cancelling Lorentz scalar and Lorentz four-vector potentials have shown its superiority to the standard Schrödinger equation based phenomenology.[8] In this paper we extend these studies to include neutron as well as proton scattering and consider lower energies than previously studied.

Descriptions of nuclear processes in which the Dirac equation plays a central role range from the purely phenomenological optical model studies discussed here to RIA or RMFT calculations in which very few or no parameters are introduced. At low energies medium effects make the calculation of the nuclear optical potential difficult. For example, the RIA is unreliable for nucleon-nucleus scattering much below 300 MeV.[5] In view of this is it perhaps surprising that the RIA successfully describes recent $\bar{p}$-^{12}C elastic cross sections measured at 46.8 MeV. Figure 1, taken from Ref. 9, shows such a parameter free calculation. The Paris $\bar{N}N$ amplitudes[11,12] in an invariant representation, are folded with scalar and vector neutron and proton relativistic Hartree densities[13] for the target to obtain the RIA optical potentials. The agreement between the RIA calculation

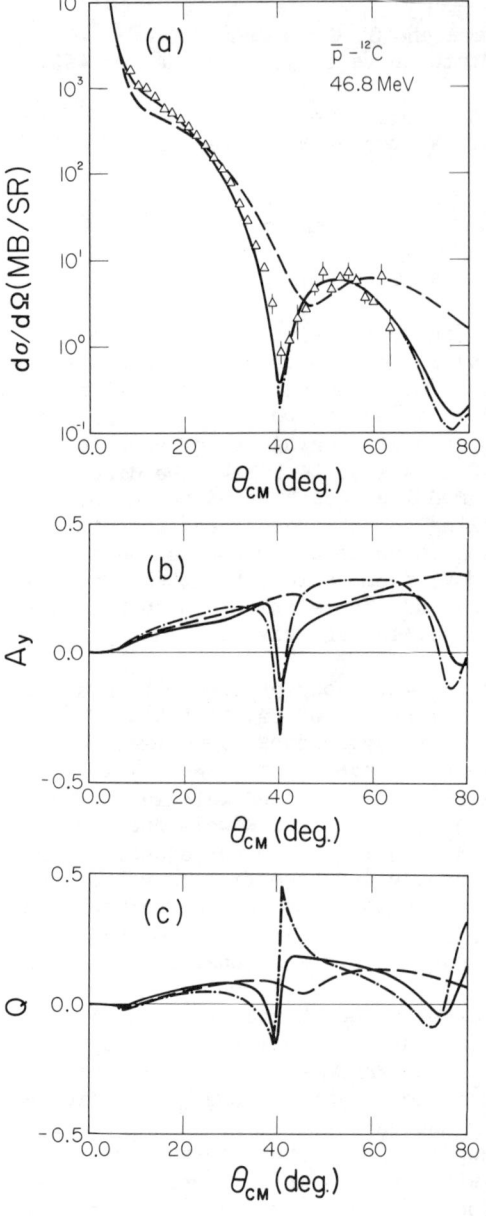

Fig. 1. Calculated differential cross sections for $\bar{p}$-^{12}C scattering at 46.8 MeV (Ref. 9). The data are from Ref. 10. The smooth curve is the RIA calculation, the dashed curve is the RIA calculation neglecting the range of the $\bar{N}N$ amplitudes, the dash-dot curve shows the non-relativistic impulse approximation.

(solid line) and the data is quite acceptable. A calculation with non-folded potentials, the "$t(o)\rho$" approximation, (dashed lines) poorly describes the cross section. The corresponding non-relativistic impulse approximation calculations (dash-dot) lines are in substantial agreement with the full RIA. The long range and magnitude of the imaginary central potential effectively prevents the antiproton from experiencing effects of the nuclear medium. Although impulse approximation may be adequate to describe $\bar{p}$-A scattering at 50 MeV this is certainly not the case for p-A or n-A elastic scattering.

One reason for comparing neutron and proton scattering from nuclei is to obtain information about the isovector part of the optical potential. This can be done by comparing optical potentials obtained from neutron and proton scattering, and we give some first results using the Dirac approach later. One could also employ a relativistic generalization of the Lane equations described in Ref. 14. In this work, observables for the direct and IAS transitions are calculated using coupled Dirac equations for the four component neutron and proton wave functions. At intermediate energies, if the data on the (p,n) reaction were available, this method would check the validity of RIA optical potentials. At lower energies, where the isovector interaction is affected by the nuclear medium, one could extract the isovector optical potential phenomenologically. Unfortunately this procedure is hampered by the lack of high quality (p,n) data, especially for spin observables. Relativistic approaches for other inelastic processes are discussed by Shepard[15] in these Proceedings.

In most work employing Dirac phenomenology, static, local, spherically symmetric potentials are used. In this case, the Dirac equation contains potentials with Lorentz scalar, S, Lorentz four-vector, V, (time-like component only) and tensor, T, components, and may be written

$$\{\vec{\alpha}\cdot\vec{p} + \beta(m + S) - (E - V) - i\beta\vec{\alpha}\cdot\hat{r}T\}\psi = 0 \quad . \tag{1}$$

The vector contains the static Coulomb potential for protons and the tensor contains a small contribution from the interaction of the anomalous moment of the proton or neutron with the charge distribution of the target. To obtain the large spin-orbit term required by experiment we need to include at least two of the three potentials in Eq.(1). This becomes obvious when one considers the second order Dirac equation[8] for the upper two component wave function given by,

$$\{\nabla^2 + (E-V)^2 - (m+S)^2 - T^2 - \frac{2}{r}T + \frac{1}{A}(\frac{\partial A}{\partial r})T - \frac{1}{r}(\vec{r}\cdot\vec{p})T$$

$$- \frac{1}{rA}\frac{\partial A}{\partial r}\vec{r}\cdot\vec{p} + (\frac{1}{rA}\frac{\partial A}{\partial r} - \frac{2T}{r})(\vec{\sigma}\cdot\vec{L})\}\psi_u = 0, \tag{2}$$

where

$$A = (m+S+E-V)/(E+m) \quad . \tag{3}$$

The standard model of Dirac phenomenology contains large cancelling scalar and vector potentials, the SV model, to produce the spin orbit enhancement. However, one could consider other combinations which rely on the tensor to obtain the large spin orbit strength. We have used scalar plus tensor (ST), vector plus tensor (VT) or all three (SVT) models to fit data at intermediate energies. In fact we have shown[16] that a wave function transformation

$$\psi(\vec{r}) = e^{i\gamma^0 F(r)} \phi(\vec{r}) \quad , \tag{4}$$

where F(r) approaches zero as r approaches infinity allows us to change from a SV model to an equivalent ST, VT or SVT model. The resulting potentials may, however, have very complicated geometries.[16] In addition, we find that we are unable to fit data in the transition energy region with ST or VT models using geometries which resemble nuclear densities. For this reason, among others, we feel the SV model to be a preferable phenomenology.

The optical potentials used in this analysis of low energy (T<50 MeV) proton and neutron elastic scattering have the form

$$S = V_s f_s(r) + iW_s g_s(r) \quad , \tag{5}$$

$$V = V_v f_v(r) + iW_v g_v(r) \quad , \tag{6}$$

where the form factors are two parameter Fermi shapes. The model contains twelve parameters, the same as the standard nonrelativistic phenomenology. We have applied this model to some of the available data on proton and neutron scattering from the calcium isotopes ^{40}Ca, ^{42}Ca, ^{44}Ca and ^{48}Ca.

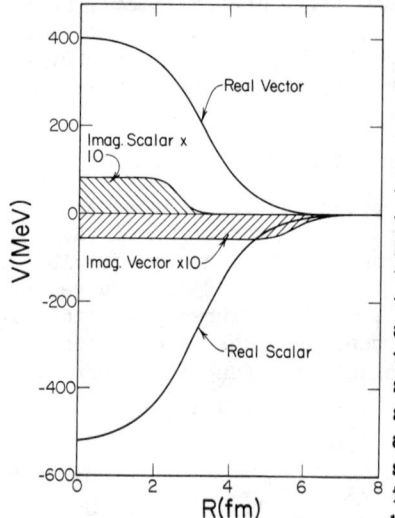

Fig. 2. Real and imaginary scalar and vector potentials determined from p-^{40}Ca at 35 MeV. The plotted imaginary potentials have been multiplied by a factor of ten.

The imaginary central potential at low energies ordinarily has a surface peaked geometry. We accomplished this while using volume form for the imaginary scalar and vector potentials through the cancellation shown in Fig. 2. In this figure the large real vector and scalar potentials for the best fit to p-^{40}Ca at 35 MeV (see Fig. 5) are shown along with the positive, short range imaginary scalar potential and the negative long range imaginary vector potential. These potentials produce the effective central and spin orbit terms shown in Fig. 3. The surface peaked character of the central absorption is evident. The imaginary spin orbit is very small but opposite in sign to the real part, just as at higher energies.[8]

Figure 4 shows a typical fit to p-^{40}Ca data in this energy region. The analyzing power is not as well represented as the cross section. The need for higher quality spin observables can not be overemphasized, and some of our

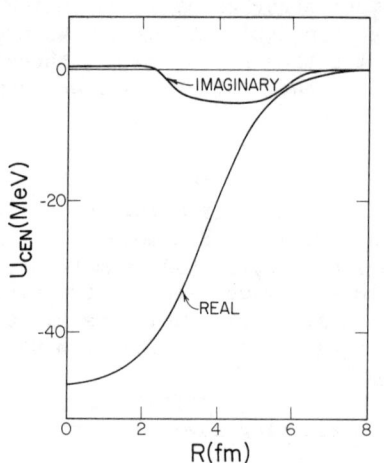 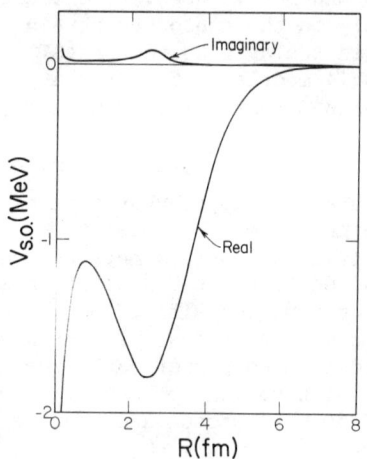

Fig. 3. Effective and central spin orbit potentials calculated from the potentials of Fig. 2.

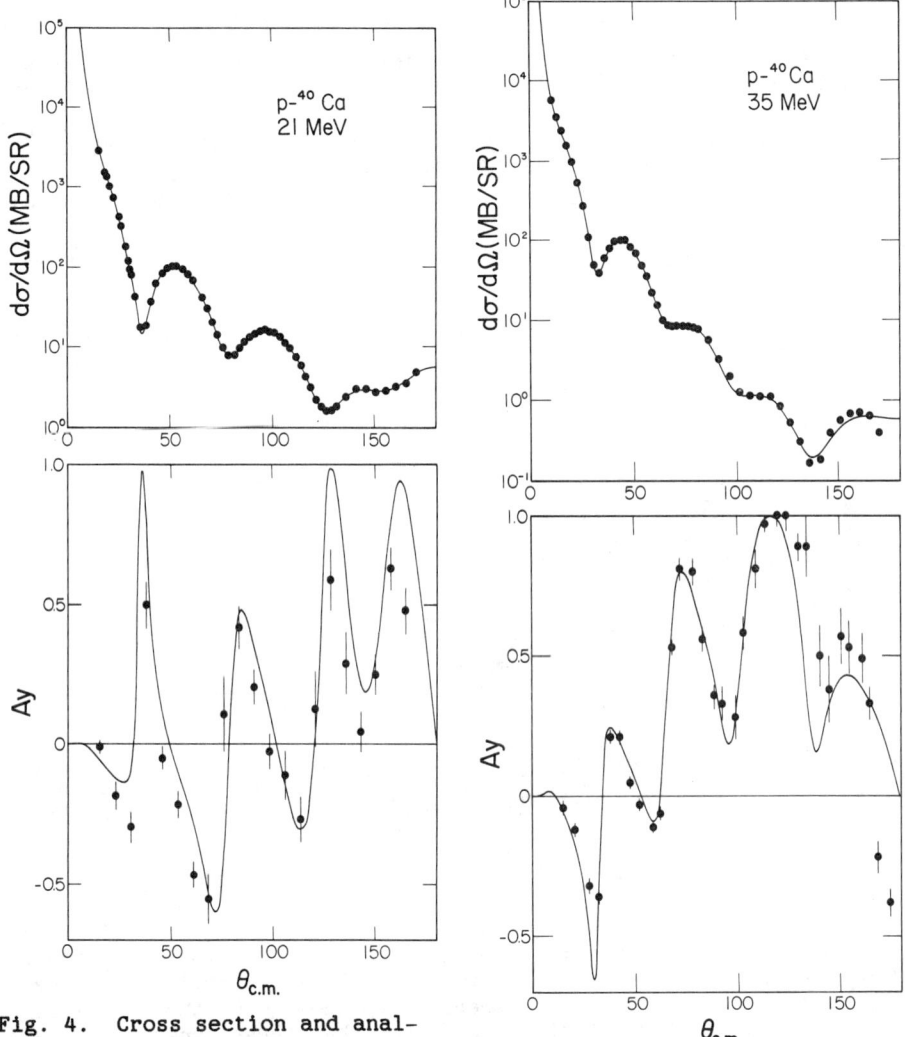

Fig. 4. Cross section and analyzing powers for the best fit to p-^{40}Ca at 21 MeV. The cross section data are from Ref. 17; the A_y data are from Ref. 18.

Fig. 5. Same as Fig. 4 for p-^{40}Ca at 35 MeV. The cross section data are from Ref. 17; the experimental analyzing power from Ref. 18.

lack of systematics is traceable to this deficiency. The agreement at large angles, a feature which persists throughout this energy region, cannot be obtained in the Schrödinger phenomenology without introducing ℓ-dependence, and additional parameters in the optical model. Figures 5 and 6 show our results for ^{40}Ca, ^{42}Ca, ^{44}Ca and ^{48}Ca at 35 MeV, again the large angle data is well represented. At lower energies,[19] T<20 MeV, we have found that a change from volume absorption to a surface form, given by

$$g(r) = -\exp((r-c)/z)/(1 + \exp((r-c)/z)) \quad ,$$

produces better agreement with experiment. We are currently investigating this point and redoing our analysis of the higher energy data using surface absorption.

Turning to neutron scattering we compare first the 40 MeV n-^{40}Ca data with a calculation using the 48 MeV p-^{40}Ca parameters, see Fig. 7. The 8 MeV energy difference is close to the average energy

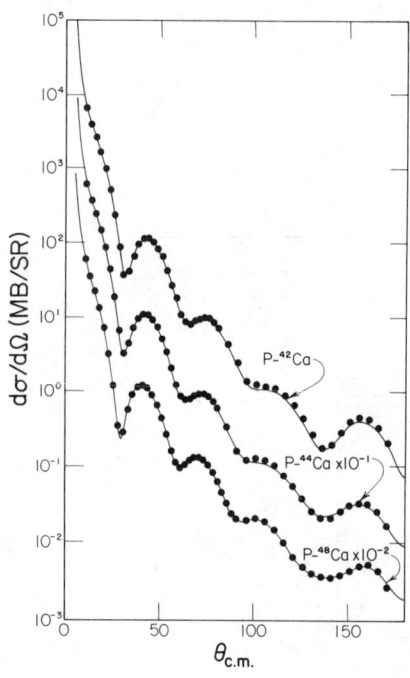

Fig. 6. Best fit cross sections for p-^{42}Ca, p-^{44}Ca and p-^{48}Ca at 35 MeV. The cross section data is from Ref. 17.

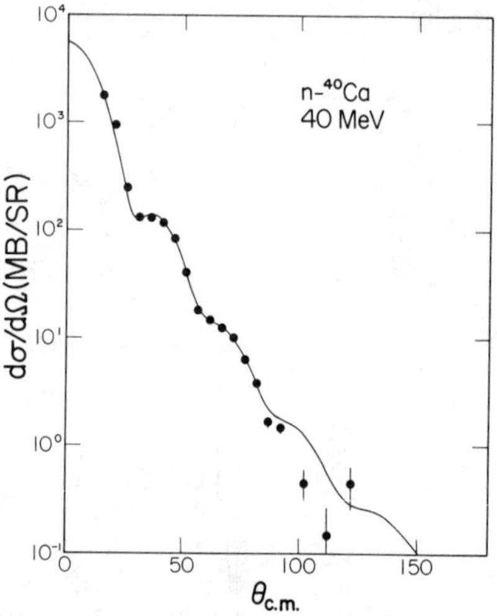

Fig. 7. Cross sections for n-^{40}Ca at 40 MeV calculated using the best fit parameters determined from 48 MeV p-^{40}Ca data of Ref. 17. The n-^{40}Ca data are from Ref. 20.

of the proton in the Coulomb potential of ^{40}Ca (7.7 MeV). While the agreement is not complete, the calculation indicates that these "energy shifted" parameters provide a good starting point for the phenomenological analysis. In Fig. 8 we show our results for n-^{40}Ca at 21.7 MeV along with the data from the Ohio University group[21] The agreement at large angles is good. At lower energies the n-^{40}Ca analyzing powers have been measured as well as the elastic cross sections. The results for n-^{40}Ca at 11.9 MeV (using volume absorption) are shown in Fig. 9 along with the data from TUNL.[22] As for proton scattering, we find that surface peaked absorption produces improved agreement with experiment and the all of the neutron data is being reanalyzed with this form for the imaginary optical potentials.

The lack of high quality spin observables for low energy

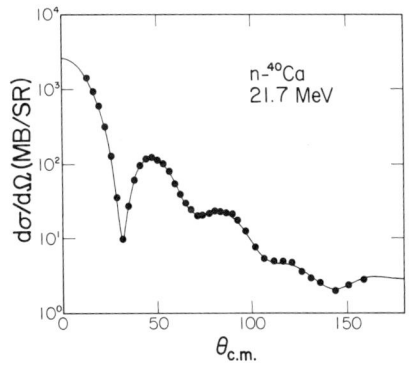

Fig. 8. Best fit cross sections for n-^{40}Ca at 21.7 MeV. The data are from Ref. 21.

scattering produces some lack of systematic behavior of the scalar and vector potentials determined at each energy. Integrated quantities, such as the rms radii of these potentials or the volume integrals of the central potential are much more systematic. In Fig. 10 we show the real and imaginary central potential volume integrals per nucleon for p-^{40}Ca (open circles) and n-^{40}Ca (open squares). The real J/A for p-^{40}Ca varies linearly with energy for 10≤T≤80 MeV. The least squares fit, shown by the dashed line in Fig. 10, is given by J/A=-447+2.2T and it is quite similar to the relativistic Hartree results[8,23] shown by the solid line. The closed circles in Fig. 10 are the results of a 16-parameter Schrödinger equation based analysis[17] of the p-^{40}Ca data of

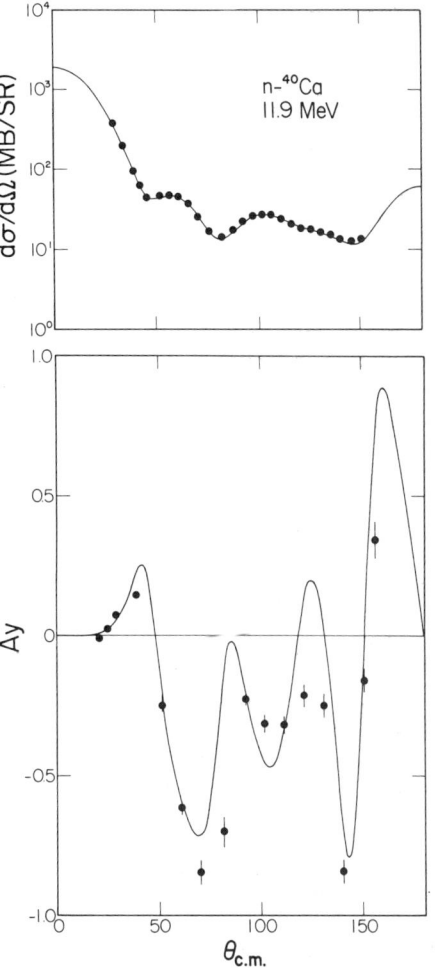

Fig. 9. Best fit cross sections and analyzing powers for n-^{40}Ca at 11.9 MeV. The data are from Ref. 22.

Ref. 17. The nonrelativistic results show a nonlinear increase in J/A as the energy decreases. The same type of nonlinearity also occurs for neutron scattering.[24] We may be observing a similar behavior starting at a somewhat lower energy for both proton and neutron volume integrals but additional analyses at low energies are required. Such analyses are currently underway. The J/A values for ^{42}Ca, ^{44}Ca and ^{48}Ca for 20≤T≤50 are in good agreement with the ^{40}Ca results shown in Fig. 10.

The rms radii of the real central potentials for p-^{40}Ca are almost independent of energy for 20≤T≤50 MeV. The value for ^{40}Ca,

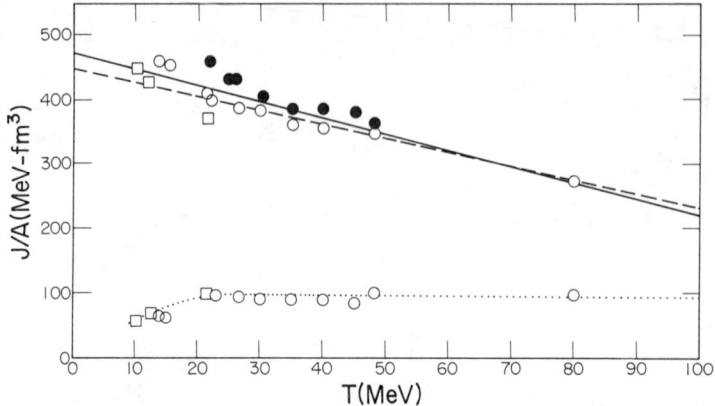

Fig. 10. Volume integrals per nucleon of the real and imaginary central potentials from the analysis of p-^{40}Ca (open circles) and n-^{40}Ca (open squares). The closed circles are the results of a 16-parameter Schrödinger based analysis of Ref. 17. The smooth line is a relativistic Hartree calculation of the real central potential; the dashed line is a least squares fit to the results of the present analysis. The dotted line is a guide to the eye.

^{48}3.998±0.061 fm, is somewhat better determined than the corresponding ^{40}Ca result 4.093±0.093 fm, due to the availability of spin observables for p-^{40}Ca. The energy dependence of the real central rms radius changes dramatically as the energy increases, becoming discontinuous where the real central volume integral passes through zero around 500 MeV. On the other hand, the individual scalar and vector rms radii are rather energy independent, having average values of 3.605±0.162 fm and 3.582±0.191 fm respectively. The individual values from analyses from 20 to 1000 MeV all agree to within 10% of these average values. This feature illustrates one of the advantages of the Dirac approach in that the geometries of the vector and scalar potentials resemble a nuclear density throughout the entire range. This feature is absent in the nonrelativistic approaches where non-standard (wine-bottle-bottom) geometries are required in the transition energy region (100-400 MeV).[25]

The rather poorly known spin observables for many of the cases produces nonsystematic behavior of the volume integral of the real spin orbit potentials shown in Fig. 11. The results of the current analyses are shown by the solid circles, and previous results are shown by open circles. The imaginary volume integral is small and essentially constant for $20 \leq T \leq 50$ MeV. The average value is Im(K/A$^{1/3}$)=3.16±0.87 MeV fm^3; while the average for Re(K/A$^{1/3}$)=-136.16 ±14.49 MeV fm^3. In every case, we obtain opposite signs for the real and imaginary spin orbit volume integrals, in contrast to the work of Walter and Delaroche.[26] The dashed line is the relativistic Hartree result[23] which exhibits energy dependence similar to the phenomenology. The dotted lines at higher energies are the results of a relativistic impulse approximation calculation.[5]

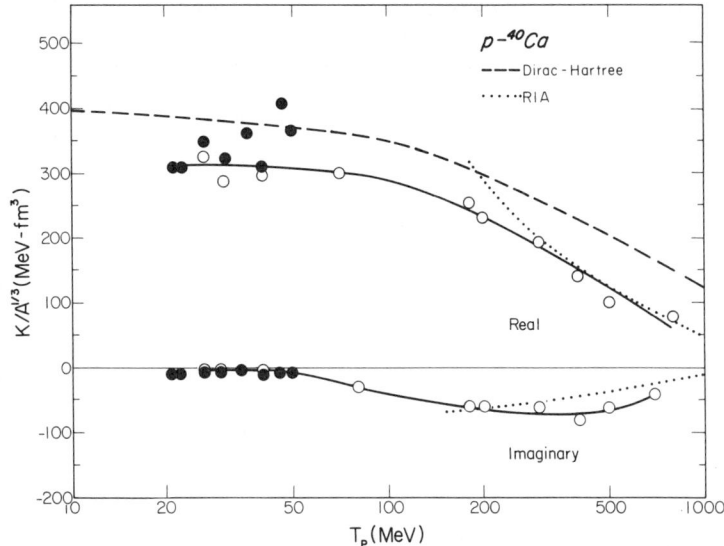

Fig. 11. Volume integrals of the real and imaginary spin orbit potential determined from fitting the p-^{40}Ca data. Open circles are from an earlier analysis, closed circles from this work.

Interesting characteristics of the second order Dirac equation in the SV model discussed in Ref. 8 are apparent from Eq.(2). They include nonlinearities in the central and spin orbit potentials, the occurrence of a nonlocal term, the Darwin term (sometimes referred to as a relativistic Perey effect), discussed by Rawitscher in these Proceedings,[26] the natural occurrence of a spin orbit term and a cross term, VV_c, between the nuclear (vector) and Coulomb potentials. This last term has been called a relativistic Coulomb correction[28] and it is complex. For the p-^{40}Ca analysis we find that the real and imaginary parts of the volume integral,

$$J_{cc} = \frac{\int V(r)V_c(r)d^3r}{2E}, \quad (7)$$

are almost energy independent. The average values are Re(J_{CC}/A)=-16.4±2.5 MeV fm^3 and Im(J_{CC}/A)=+0.8±0.1 MeV fm^3 for 20≤T≤50 MeV. Similar results are obtained for the other calcium isotopes. The values at each energy for p-^{40}Ca are shown in Fig. 12, the dashed line is the result using a relativistic Hartree calculation.[8,23] Our preliminary calculations with surface absorption indicate a somewhat larger imaginary volume integral; however, the sign is still positive and the real part is essentially unchanged. Microscopic nonrelativistic calculations[29-31] of the usual Coulomb correction term also give a small positive imaginary volume integral and a negative real volume integral. If we calculate the Coulomb correction from the volume integrals calculated from the best fits to n-^{40}Ca and p-^{40}Ca at the same energy of 21.7 MeV, we obtain $J_{CC}/A=-31.7+i6.1$ MeV fm^3, or using the energy shifted proton volume integrals[19] at 30 MeV we obtain $J_{CC}/A=-13.4+i15.3$ MeV fm^3.

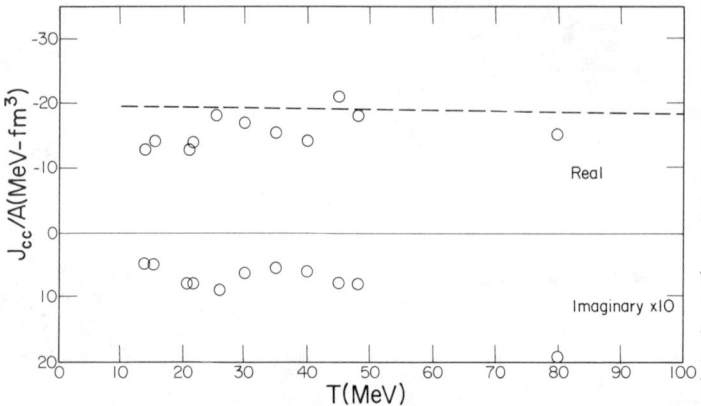

Fig. 12. Volume integrals of the real and imaginary relativistic Coulomb correction term from the analysis of p-^{40}Ca data.

In this paper we have given some results of our analysis of a substantial body of low energy proton and neutron data. The analysis is still in progress, but our preliminary results indicate that the Dirac equation based approach may prove a viable alternative to the Schrödinger based phenomenology. Fewer parameters are needed to represent the data in the Dirac case; especially noticeable is the agreement at larger angles. Integrated quantities, rms radii, volume integrals are at least as systematic as in the nonrelativistic case. Moreover, the Dirac phenomenology behaves smoothly with energy over the energy region $10 \leq T \leq 1000$ MeV. The behavior of the real and imaginary central and spin orbit potentials are shown in Figs. 13-15.

A comparison of proton and neutron analyses is underway along with additional investigations for other targets. When completed we should have additional information about the behavior of the Dirac potentials at low energies. We feel such work will

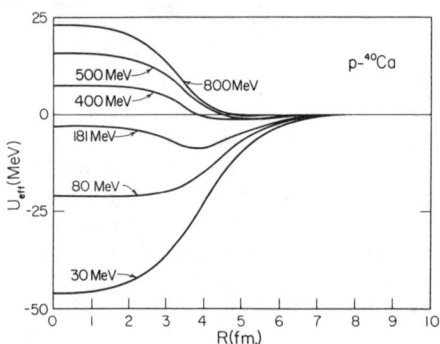

Fig. 13. Effective real central potential from the analysis of p-^{40}Ca elastic scattering data.

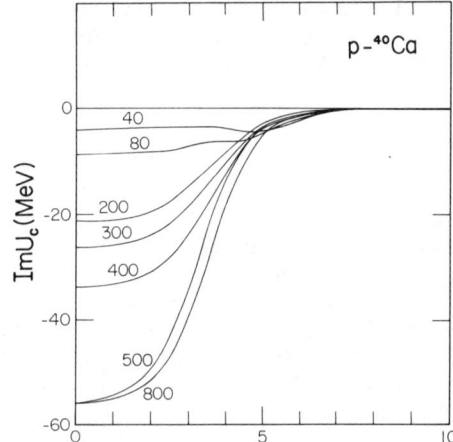

Fig. 14. Same as Fig. 13 for the imaginary spin orbit potentials.

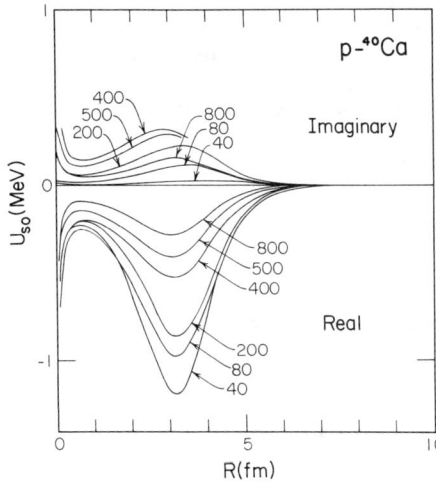

Fig. 15. Same as Fig. 13 for the real and imaginary spin orbit potentials.

provide a testing ground for various RMFT calculations. We stress that definitive tests of Schrödinger vs. Dirac phenomenology at low energies are hampered at present by the lack of high quality spin data.

This work was supported by the National Science Foundation under Grant PHY-8306268. We thank R. Alarcon, S. A. Austin, R. P. DeVito, J. C. Lombardi, R. H. McCamis, W. T. H. van Oers, and R. L. Walter for providing tables of the data used in this work.

REFERENCES

1. M. R. Anastasio, L. S. Celenza, W. S. Pong and C. M. Shakin, Physics Reports 100, 327 (1983).
2. B. D. Serot and J. D. Walecka, to be published in Advances in Nuclear Physics.
3. J. A. McNeil, J. Shepard and S. J. Wallace, Phys. Rev. Lett. 50, 1439 (1983).
4. J. Shepard, J. A. McNeil and S. J. Wallace, Phys. Rev. Lett. 50, 1443 (1983).
5. B. C. Clark, S. Hama, R. L. Mercer, L. Ray and B. D. Serot, Phys. Rev. Lett. 50, 1644 (1983) and B. C. Clark, S. Hama, R. L. Mercer, L. Ray, G. W. Hoffmann and B. D. Serot, Phys. Rev. C28 Rapid Communications, 1421 (1983).
6. M. V. Hynes, A. Pickelsimer, P. C. Tandy and R. M. Thaler, Phys. Rev. Lett. 52, 978 (1984).
7. R. D. Amado, J. Piekarewicz, D. A. Sparrow and J. A. McNeil, Phys. Rev. C28, 2180 (1983).
8. B. C. Clark, S. Hama and R. L. Mercer, AIP Conf. Proc. No. 97, The Interaction Between Medium Energy Nucleons in Nuclei - 1982, edited by H. O. Meyer (A.I.P. Press, New York, 1983), p. 260.
9. B. C. Clark, S. Hama, J. A. McNeil, R. L. Mercer, L. Ray, B. D. Serot, D. A. Sparrow and K. Stricker-Bauer, Phys. Rev. Lett., in press.
10. D. Garreta, P. Birien, G. Bruge, A. Chaumeaux, D. M. Drake, S. Janovin, D. Legrand, M. C. Mallet-Lemaire, B. Mayer, J. Pain, J. C. Peng, M. Berrada, J. P. Bocquet, E. Monnand, J. Mougey, P. Perrin, A. Erell, J. Lichtenstadt and A. I. Yavin, Phys. Lett. 135B 266 (1984).
11. J. Côte, M. Lacombe, B. Loiseau, B. Moussallam and R. Vinh Mau, Phys. Rev. Lett. 48, 1319 (1982).
12. C. B. Dover, M. E. Sainos and G. E. Walker, Phys. Rev. C28, 2368 (1983).

13. C. J. Horowitz and B. D. Serot, Nucl. Phys. A368, 503 (1981).
14. B. C. Clark, S. Hama, E. Sugarbaker, M. A. Franey, R. L. Mercer, L. Ray, G. W. Hoffmann and B. D. Serot, Phys. Rev. C30, 314 (1984).
15. J. Shepard, Proceedings of the Conference on Neutron-Nucleus Collisions, to be published.
16. B. C. Clark, S. Hama, S. G. Kälbermann, E. D. Cooper and R. L. Mercer, OSU Preprint (1984).
17. R. H. McCamis, T. N. Nasr, J. Birchall, N. E. Davison, W. T. H. van Oers, P. J. T. Verheijen, R. F. Carlson, A. J. Cox, B. C. Clark, S. Hama, E. D. Cooper and R. L. Mercer, Manitoba-OSU Preprint (1984).
18. K. H. Bray, K. S. Jayaraman, G. A. Moss, W. T. H. van Oers, D. O. Wells and Y. I. Wu, Nucl. Phys. A167, 57 (1971).
19. J. C. Lombardi, R. N. Boyd, R. Arking and A. B. Robbins, Nucl. Phys. A188, 103 (1972).
20. R. P. DeVito, S. A. Austin, W. Sterrenburg and V. E. P. Berg, Phys. Rev. Lett. 47, 628 (1981).
21. R. Alarcon and J. Rapaport, private communication (1984).
22. W. Tornow, E. Woye, G. Mack, C. E. Floyd, K. Murphy, P. P. Guss, S. A. Wender, R. C. Byrd, R. L. Walter, T. B. Clegg and H. Leeb, Nucl. Phys. A385, 373 (1982).
23. B. C. Clark, S. Hama, R. L. Mercer and B. D. Serot, OSU Preprint (1984); see also Ref. 8.
24. R. W. Finlay, J. R. M. Annand, J. S. Petler and F. S. Dietrich, Proceedings of the Conference on Neutron-Nucleus Collisions, to be published (1984).
25. L. G. Arnold, B. C. Clark, R. L. Mercer and P. Schwandt, Phys. Rev. C23, 1949 (1981).
26. R. L. Walter and J. P. Delaroche, Proceedings of the Conference on Neutron-Nucleus Collisions, to be published (1984).
27. G. H. Rawitscher, Proceedings of the Conference on Neutron-Nucleus Collisions, to be published (1984).
28. L. G. Arnold, B. C. Clark and R. L. Mercer, Phys. Rev. C23, 15 (1981).
29. F. Osterfeld, Proceedings of the Conference on Neutron-Nucleus Collisions, to be published (1984).
30. F. A. Brieva and J. R. Rook, Nucl. Phys. A307, 493 (1978).
31. J.-P. Juekenne, A. Lejeune and C. Mahaux, Phys. Rev. C15, 10 (1977); C16, 80 (1977).

INTERPRETATION OF THE PEREY-BUCK NONLOCALITY IN TERMS OF THE RELATIVISTIC OPTICAL MODEL FORMALISM.

George H. Rawitscher
Physics Department, Univ. of Conn., Storrs, CT 06268

It is suggested that the effect of a gaussian non-locality of the Perey-Buck type[1] in the non-relativistic optical potential can be represented in terms of a local relativistic optical potential[2] which operates on the upper components of a Dirac four-spinor. This result is obtained by combining the observation recently made by Fiedeldey and Sofianos[3] that a nonlocal Schroedinger equation can be transformed into a local one if a gradient term (or velocity term) is added, together with the well known fact that the upper component of the solution of a Dirac equation also obeys a second order equation which has a gradient term, the well known Darwin term.[4]

The Darwin term is given by

$$(\hbar^2/2m)\,[(dA/dr)/A]\,d/dr \qquad (1)$$

and $A(r)$ is obtained in terms of the relativistic scalar and fourth component vector potentials, U_s and U_4, respectively, as

$$A(r) = (E + mc^2 + U_s - U_4)/(E + mc^2) \qquad (2)$$

For the case of a non-local Schroedinger Eq. the solution can be shown[3] to also obey exactly a second order local equation with a Darwin-like gradient term which is of the same form as Eq.(1), where $A(r)$ is replaced by the normalized Wronskian W_ℓ

$$A(r) \longrightarrow W_\ell(r)/W_\ell(\infty) \qquad (3)$$

Here W is the wronskian of two linearly independent solutions of the non-local Schroedinger Eq. This function can be strongly dependent on the angular momentum L of the projectile relative to the target[5], while $A(r)$ is L-independent. Hence an identification between W and A is not possible in general. However, for the Perey-Buck type of non-locality the function W is only weakly dependent[3] on L, and hence this type of non-locality can be closely related to the one which arises in the second order representation of a local first order relativistic

0094-243X/85/1240135-02 $3.00 Copyright 1985 American Institute of Physics

equation.

An application is made to the case of nucleon-^{40}Ca scattering. For the older data at 24 MeV the Perey-Buck potential is available[3], and for the recently obtained data at 21.7 MeV a fit with the relativistic formalism has been obtained by the OSU group.[7] Preliminary results show that the normalized Wronskian is comparable to the relativistic function A(r), the former being closer to unity at small distances by approximately 25%. The phase equivalent local (no gradient term explicitly present) central potentials in the two formulations are also comparable, the non-relativistic one being deeper by about 25% also. This approximate agreement tentatively supports the surmise that a Perey-Buck non-locality attempts to simulate the relativistic optical model description of nucleon-nucleus scattering, however the differences are large enough to warrant further study of this interpretation.

One implication of this result is that the relativistic optical potential which is equivalent to a non-relativistic one may be less non-local than the latter; another is that a Perey-Buck damping factor used in a local non-relativistic description, is not needed in the relativistic description.

Support from The National Science Foundation, grant PHY 8214571 is gratefully acknowledged.

REFERENCES

1. F. G. Perey and B. Buck, Nucl. Phys. A **32** (1962) 353; W. E. Frahn and R. H. Lemmer, Nuovo Cimento **5** (1957) 523, 1564.

2. B.C. Clark, S. Hama and R.L. Mercer, Proceedings of the Workshop on the Interaction Between Medium Energy Nucleons in Nuclei (Indiana University Cyclotron Facility) AIP Conf. Proc. No 97, edited by H.O. Meyer (AIP, New York, 1983) p. 260.

3. H. Fiedeldey and S.A. Sofianos, Z. Phys A **311**, 339 (1983).

4. C. G. Darwin, Proc. Roy. Soc. (London), A **118** (1928) 654.

5. G. H. Rawitscher, Bull. Am. Phys. Soc. **29**, 700 (1984).

6. R. Alarcon, D. Wang, J. R. Annand and J. Rapaport, Bull. Am. Phys. Soc., **29**, 636(1984).

7. The author is indebted to B. C. Clark for communicating to him the parameters of the relativistic potential prior to publication.

FEW-NUCLEON EXPERIMENTS WITH FAST POLARIZED NEUTRONS

H.O.Klages, R.Aures, F.P.Brady[*], P.Doll, E.Finckh[**], J.Hansmeyer,
W.Heeringa, J.C.Hiebert[***], K.Hofmann, H.Krupp, Ch.Maier, W.Nitz,
P.Plischke, J.Wilczynski, H.Zankel[****] and B.Zeitnitz

Kernforschungszentrum Karlsruhe, Institut für Kernphysik 1,
D-7500 Karlsruhe, W.-Germany

ABSTRACT

At the Karlsruhe cyclotron few-nucleon systems are studied in scattering experiments of polarized fast neutrons on very light nuclei. The continuous energy distribution of the neutron beam from POLKA enables one to measure spin-dependent observables in the energy range from 15 to 50 MeV simultaneously.

In a first series of experiments analyzing power distributions $A_y(E_n,\Theta)$ have been measured for the elastic scattering of neutrons on ^1H, ^2H, ^3He and ^4He. The current work is concentrated on detailed studies of the n-p interaction. Preliminary data have been taken for the n-p spin correlation parameter A_{yy} using a "brute-force" polarized proton target.

The facility POLKA[1] at the Karlsruhe cyclotron is used to produce a well collimated beam of fast polarized neutrons by means of the $D(\vec{d},\vec{n})X$ reactions. This beam exhibits a continuous energy distribution up to 52 MeV. The polarization of the neutrons is a smooth function of energy and rises from ~ 0.2 at 15 MeV to ~ 0.5 at 30 MeV and above. These properties enable one to measure simultaneously in the energy range from 15 to 50 MeV, with an off-line determination of the incident neutron energy by the time-of-flight method.

In the first experiments with POLKA, scintillating scattering samples were used. The scattered neutrons were detected by up to 20 scintillation detectors. Multiparameter data aquisition and a careful off-line analysis provided data with low backgrounds. Extensive Monte-Carlo calculations have been performed to correct the raw data for multiple scattering and finite geometry effects. The new precise data increase considerably the quality of the data base for the few-nucleon systems in the energy range from 20 to 50 MeV.

---

[*] permanent address: UC Davis, CA 95616
[**] permanent address: Universität Erlangen, W.-Germany
[***] permanent address: Texas A&M Univ., College St. 77843
[****] Universität Graz, Austria

Final results have been published of the analyzing power A_y of the n-p scattering[2] and of the n-^{4}He scattering[3]. Phase shift analyses have been performed for both systems.

The data of the elastic and inelastic $\vec{n}$-d scattering are partially analyzed. Faddeev calculations[4] are carried out using different types of nucleon-nucleon potentials. The results for the $\vec{n}$-^{3}He scattering can be compared to other data in the A=4 isospin multiplet[5,6].

Fig.2 shows the data at 22 MeV. A phase shift analysis has been performed to reproduce the n-^{3}He scattering observables by smoothly varying phase parameters.

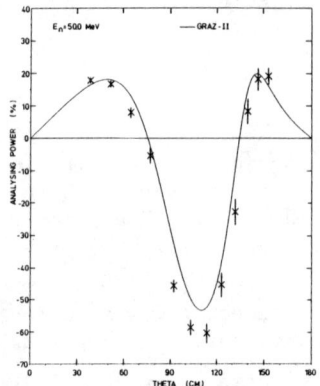

Fig.1. Analyzing power of the elastic $\vec{n}$-d scattering at 50 MeV. The solid line shows the result of a Faddeev calculation using the code of Y.Koike (RCNP Osaka) and the Graz II potential.

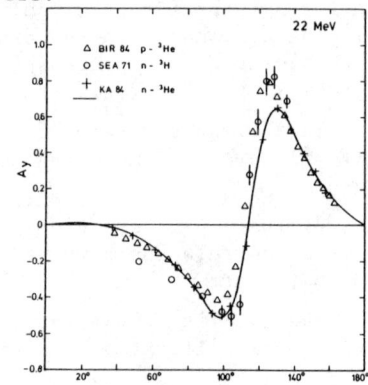

Fig.2. Analyzing power distributions in the A=4 systems.
Bir 84: p-^{3}He data from ref.5,
Sea 71: n-T data from ref.6,
Ka 84: n-^{3}He data and phase shift analysis, this work.

The first results of the $\vec{n}$-$\vec{p}$ spin correlation parameter A_{yy} are in agreement with previous data but are still lacking statistical accuracy.

REFERENCES

1. H.O.Klages et al., Nucl. Instr. Meth. 219 (1984) 269
2. J.Wilczynski et al., Nucl. Phys. A425 (1984) 458
3. H.Krupp et al., Phys. Rev. C (in print)
4. Y.Koike (RCNP Osaka), private communication
5. J.Birchall et al., Phys. Rev. C29 (1984) 2009
6. J.Seagrave, Pol. Phen. in Nucl. Phys., Madison (1971) 477

ISOVECTOR "$\Delta J^\pi = 0^-$" TRANSITIONS OBSERVED IN THE CHARGE-EXCHANGE
(p,n) REACTIONS ON $^{13,14}C$ AND ^{16}O

Hikonojo ORIHARA
Cyclotron and Radioisotope Center, Tohoku University,
Sendai 980, Japan

ABSTRACT

Isovector $0^+ \to 0^-$ ($1/2^- \to 1/2^+$) transitions have been studied by the charge-exchange (p,n) reactions on $^{13,14}C$ and ^{16}O at $E_p = 35$ MeV by means of the high-resolution neutron time-of-flight technique. Angular distributions of emitted neutrons were measured covering a wide range of the transfer momentum up to $q \sim 2.2$ fm^{-1}. DWBA analyses with the M3Y interaction give a reasonable account of the observed cross sections at small angles ($\theta_{C.M.} \lesssim 60°$), where the tensor force plays a significant role. The enhancement of the $\Delta J^\pi = 0^-$ (p,n) cross section was observed at large momentum transfer. This evidence may suggest an effect of the pionic correlation in the nuclear medium.

An isovector excitation of a particle-hole state by a $\Delta J^\pi = 0^-$ transition is of particular interest since it is only sensitive to the longitudinal part ($\vec{\sigma}\cdot\vec{q}$) of the N-N coupling. The pion field in nuclei, for which it is one of the important questions in nuclear physics whether a nucleon in a nucleus interact differently from a free nucleon, is known to affect the longitudinal response significantly. The advantage of $\Delta J^\pi = 0^-$ transition, compared with $0^+ \to 1^+$ and other unnatural transitions, is that the transverse part ($\vec{\sigma}\times\vec{q}$) of the N-N coupling vanishes[1], and hence the ρ-mesic correlations do not act to these transitions due to the vector character of the ρ-meson. We have reported the angular distribution measurement of the isovector $0^+ \to 0^-$ transition in the $^{16}O(p,n)^{16}F$ at $E_p = 35$ MeV[2], where we pointed out that the tensor force played an important role as a consequence of the one-pion exchange character of this transition and that an enhancement of the differential cross section at large momentum transfer might be due to the effects of the pion field in nuclei. Very recently, Yabe, Kubo and Toki[3] have reported an analysis for this enhancement in terms of the effect of the dimesic polarization due to the pionic correlation among nucleons.

In this report we discuss extended experimental results for the $\Delta J^\pi = 0^-$ transitions observed in the (p,n) reactions on ^{14}C and ^{13}C, where the $1/2^- \to 1/2^+$ transition contains $\Delta J^\pi = 0^-$ as well as $\Delta J^- = 1^-$. The experiment was performed with use of a 35-MeV proton from the AVF cyclotron and the time-of-flight facilities at the Cyclotron and Radioisotope Center, Tohoku University. The overall time resolution for a γ-flash was 0.9 nsec, which correspond to 90 keV for 30 MeV-neutrons measured with a flight path of 44.3 m. Figure 1 shows the angular distribution for the neutrons to the 2.39 MeV-$1/2^+$ state in ^{13}N. Curves in the figure are DWBA prediction by the code DWBA-70, where the M3Y interaction[4] is employed as the nucleon-nucleon interaction. One body transition density matrix elements were calculated by Brown.[5] A dashed curve is calculated (p,n) cross sections for $\Delta J^\pi = 1^-$, while dash-dot one is for $\Delta J^\pi = 0^-$, then the solid curve in the figure means the sum of these two components. In order to examine the tensor contribution, the dotted line is calculated for the $\Delta J^\pi = 0^-$ part without tensor forces in M3Y. It is clear that tensor part plays an important role to reproduce the second peak in the angular distribution for the $1/2^- \rightarrow 1/2^+$ transition on ^{13}C as well as the cases on ^{16}O and ^{14}C. Thus, the present analysis is quite reasonable so long as the (p,n) cross sections in $\theta_{C.M.} \lesssim 100°$ are concerned. However, the third peak appearing in $1.0 < q \leq 2.0$ fm^{-1} on the angular distribution cannot be explained by the conventional DWBA analysis, where the angular distribution of the (p,n) cross section for the $\Delta J^\pi = 0^-$ transition shows a deep valley in $\theta_{C.M.} \gtrsim 90°$ presumably due to the interference between the central and tensor contributions, while $\Delta J^\pi = 1^-$ one does monotonous decrease in this region. Though, in the $1/2^- \rightarrow 1/2^+$ transition, the main part of the (p,n) cross section comes from $\Delta J^\pi = 1^-$, we have found the second evidence for the enhancement of the (p,n) cross section peculiar to the $\Delta J^\pi = 0^-$ transition at large momenta. This work was supported by a Grant-in-Aid for Scientific Research, No. 59540147, from the Ministry of Education and Culture in Japan.

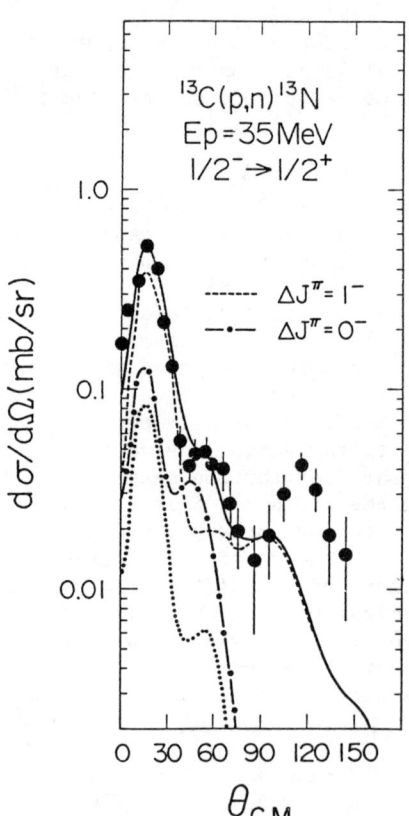

Fig. 1. (p,n) cross sections and DWBA comparison for the $\Delta J^\pi = 0^-$ transition in the ^{13}C(p,n)^{13}N reaction leading to the 2.39 MeV, $1/2^+$ state.

References

1. G. Love et al., in Proc. of the International Conf. on Spin Excitations in Nuclei, Telluride,, 1982, p. 205.
2. H. Orihara et al., Phys. Rev. Lett. 49, 1318 (1982).
3. Yabe et al., reprint.
4. G. Bertsch et al., Nucl. Phys. A284, 399 (1977).
5. A. Brown, private communication.

THE (n,p) REACTION AT ZERO DEGREES FROM ^{12}C, ^{13}C, ^{90}Zr, AND ^{209}Bi AT 65 MEV*

T.D. Ford, F.P. Brady, C.M. Castaneda, J.L. Romero
J.R. Drummond, B. McEachern, M.L. Webb
Department of Physics and Crocker Nuclear Laboratory
University of California, Davis, CA 95616

N.S.P. King
Los Alamos National Laboratory
Los Alamos, NM 87545

J. Martoff, K. Wang
Stanford University
Palo Alto, CA 94305

ABSTRACT

The first spectra from the UC Davis cyclotron's zero degree facility for (n,p) scattering have been obtained from ^{12}C, ^{13}C, ^{90}Zr, and ^{209}Bi at an incident neutron energy of 65 MeV. Cross sections for the ground state, the ∿4.4 MeV and the GDR are extracted for the carbon isotopes. Two broad resonances in ^{209}Bi at ∿7.5 MeV and ∿15 MeV and a broad resonance in ^{90}Zr at ∿22 MeV are compared with previous work done at Davis away from zero degrees.

INTRODUCTION

Giant resonances have been of interest for some time. Recent, sophisticated calculations on the angular distributions of these resonances, such as those of V. Brown,[1] has renewed interest, especially at very forward angles. Zero degree scattering ℓ(transfer) = 0 provides a tool to help separate excitation of the giant isovector monopole resonance (IVMR), expected to peak at forward angles, from higher multipoles. Further, the (n,p) reaction has the feature that only T_o+1 components of transitions are excited. Recent (π^-, π^o) data (which share the T_o+1 selection rule) from LANL[2,3] give good evidence of the IVMR.

EXPERIMENT

Measurements were made at the Davis 76 inch isochronous cyclotron using the ^{7}Li(n,p)^{7}Be neutron beam facility. Zero degree scattered particles were swept into the multiwire chamber detector system[4] via an 18.5 kG dipole magnet. Data were collected event by event into CAMAC interfaced to a PDP 15/40 and stored on 7-track tape.

---

* Supported by NSF grant PHY 81-21003

RESULTS

Cross sections have been obtained for ^{12}C and ^{13}C(n,p) to the ground states, ~4.4 MeV, and the giant dipole resonances (GDR) (fig. 1). Cross sections have also been extracted from ^{90}Zr for a broad resonance at ~22 MeV in excitation and for two broad resonances from ^{209}Bi at ~7.5 and ~15 MeV.

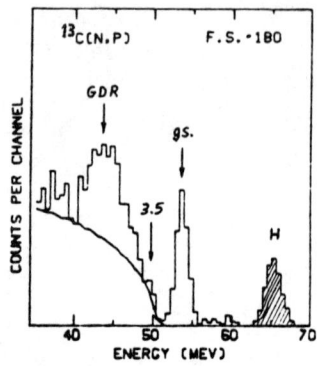

Figure 1

The bump at ~22 MeV in the ^{90}Zr (fig. 2) data was not seen in an earlier measurement made at Davis at 16 degrees.[5] The increasing cross section toward zero degrees and Auerbach et al.[6] predicting ~5 mb/sr for the IVMR in ^{90}Zr at this excitation suggest that this is the IVMR. The ~15 MeV excitation in ^{209}Bi is believed to be IVMR as well, and the ~7.5 MeV peak is the giant quadrupole resonance. Figure 3 shows the angular distribution for this resonance from previous data taken at Davis, along with theoretical predictions.[1] Included is the point from this zero degree experiment.

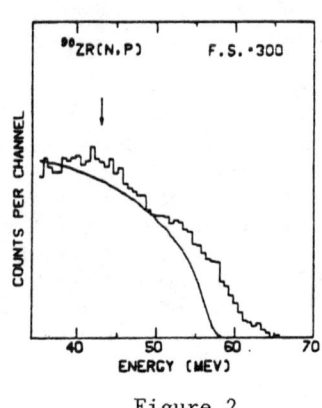

Figure 2

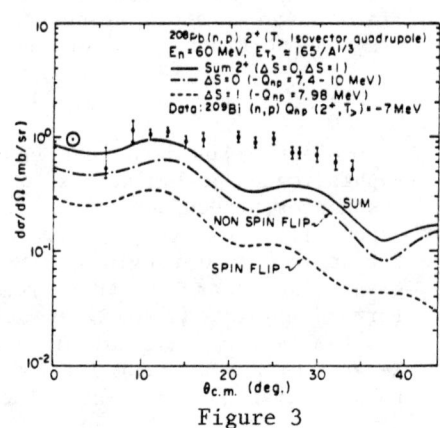

Figure 3

REFERENCES

1. F.P. Brady, V.R. Brown, C.M. Castaneda, C.H. Poppe, J.L. Romero, Proc. Int. Conf. on Spin Excitations in Nuclei, Telluride, editors F. Petrovich, G. Brown, C. Goodman, G. Love.
2. J.D. Bowman et al., Phys. Rev. Lett. 50,1195 (1983).
3. A. Erell et al., Phys. Rev. Lett. 52,2134 (1984).
4. T.D. Ford, G.A. Needham, F.P. Brady, J.L. Romero, C.M. Castaneda, NIM in press.
5. N.S.P. King, J.L. Ullmann, The (n,p) Reaction and the Nucleon-Nucleon Force, Edited by C.D. Goodman et al., Plenum Pub. Corp., p. 373 (1980).
6. N. Auerbach et al., Phys. Rev. C (1984).

THE ANALOG OF THE ^{40}Ca G.D.R. THROUGH THE (n,p) REACTION AT 65.5 MeV

C.M. Castaneda, J.L. Romero, F.P. Brady
J.R. Drummond, T.D. Ford, B. McEachern
Department of Physics and Crocker Nuclear Laboratory
University of California, Davis, CA 95616

ABSTRACT

The charge exchange reaction ^{40}Ca(n,p)^{40}K has been measured at 65.5 MeV. A strong enhancement is observed at energies corresponding to the analog of the ^{40}Ca G.D.R. The angular distribution is partially consistent with a $\Delta L=1$ transfer, but at forward angles deviates from it. Comparison with the (^{3}He,t) and (p,n) reaction is presented.

There has been recent interest and effort to use different kinds of probes and energies to separate the $\Delta S = 0$ and $\Delta S = 1$ components of the charge exchange analogs of the $\Delta L = 1$, $\Delta T = 1$ target resonance.

We have performed the ^{40}Ca(n,p) reaction to complement the information already obtained in charge exchange reaction with other probes on the same target.[1,2,3] The measurements were made at the Crocker Nuclear Laboratory 76" Cyclotron using unpolarized neutrons of 65.5 MeV. The target was a natural Calcium target with a small Hydrogen contamination which was substracted from the results. The energy resolution was measured to be 1.2 MeV.

Figure 1 shows the background subtracted spectrum at 5° from the ^{40}Ca(n,p) reaction as well as the results from the RPA calculation from ref. 3. The large enhancement observed is between 6 MeV and 16.5 MeV in excitation energy. The maximum of the peak is at E=12 MeV and one observes shoulders at E=9 MeV and at E=15 MeV. We have compared the 5° spectrum to the 4.5° (p,n) at 200 MeV[2] and the 15° to the 8° (^{3}He,t) at 197 MeV[3] which have comparable momentum transfer in both reactions. The shapes of the spectra are very similar in both cases. Figure 1 shows the background subtracted spectra together with RPA calculations from Gaarde et al. for

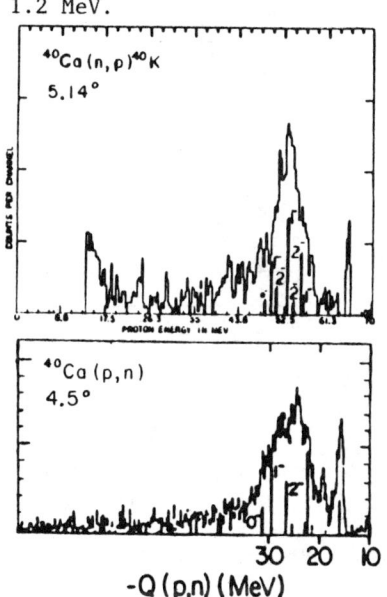

Figure 1

(p,n) at 4.5°.[2] Figure 2 shows the angular distribution for the whole region, which includes all the ΔS=1 states. The solid line is a DWBA calculation using the code DWUCK[4] and a Goldhaber-Teller macroscopic form factor.[5] There is good agreement at the back angles, but the experimental distribution does not dip at the forward angles as the calculation does. This feature is similar for the (^{3}He,t) reaction at 197 MeV.[3] If we extract how much of the energy weighted sum rule is exhausted, the strength observed is 112% of the EWSR. When the RPA calculations from (^{3}He,t) are used for placing the ΔS=0 and ΔS=1 components of the GDR and to unfold the shoulders shown in the experimental distribution as shown in figure 3, the strength of the enhancement is almost equally divided between both components.

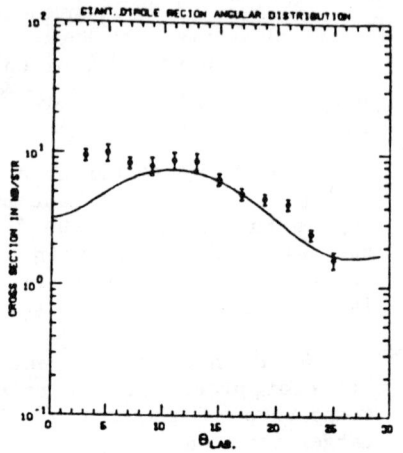

Figure 2

Our preliminary conclusion is that as expected from theory at 65 MeV, the contributions from the non-spin flip and spin flip parts of the effective interaction are roughly equal. A comparison with microscopic calculations is in progress.

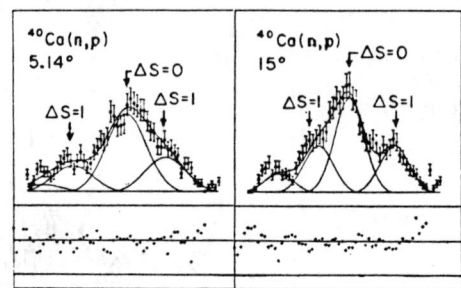

Figure 3

REFERENCES

1. H.W. Baer et al., P.R.L. 49 #19:1376-1379 (1982).
2. C. Gaarde et al., Nuclear Physics A369:281-288 (1981).
3. S.L. Tabor et al., Nuclear Physics A422:12-44 (1984).
4. Program DWUCK4, P.D. Kunz (unpublished; extended version of J.R. Comfort).
5. G.A. Needham et al., Nuclear Physics A385:349-372 (1982).

SESSION C

NUCLEAR STRUCTURE

WITH NEUTRONS

ISOVECTOR VIBRATIONAL MODES IN HEAVY NUCLEI

J. Wambach
Department of Physics, University of Illinois
at Urbana-Champaign
1110 W. Green Street
Urbana, IL   61801

ABSTRACT

The dynamics of isovector modes in nuclei reveal the role of isospin- as well as spin-isospin degrees of freedom in nuclear structure. Implications of the giant dipole- and isovector monopole oscillations for the equation of state of asymmetric nuclear matter will be discussed. Spin-isospin vibrations are preferentially excited in nucleon-nucleus collisions at intermediate energy. A recent theoretical analysis including two particle-two hole as well as isobar-hole excitations in an extended RPA-framework will be presented. The relevance of low energy β_+-strength experimentally accessible for instance in (n,p)-reactions will be mentioned in connection with neutrino cooling during collapse of type II supernovae.

1.   INTRODUCTION

Isovector vibrational motion is an old subject in nuclear physics. Soon after the discovery of the dipole resonance it was realized that the systematics of the vibrational frequency provides important information of the role of proton and neutron degrees of freedom in the nucleus. Renewed interest in isovector collective motion was stimulated by the rapid developments in the last ten years:

- There is substantial experimental progress in exciting isovector modes other than the dipole, especially in hadronic charge exchange reactions such as ($\pi^{\pm}$, π°) or (p,n). Among the vibrations identified the isovector monopole (IVM) is of special interest.

- Inelastic scattering of medium energy nucleons turned out to be an excellent spectroscopic tool for studying spin-isospin vibrations in nuclei. The role of subnuclear degrees of freedom and tensor correlations in the low energy response is still a matter of intense discussion.

- One of the astrophysical implications of isovector modes in N ≠ Z nuclei concerns the equation of state of asymmetric nuclear- and pure neutron matter. The understanding of the dynamics of the giant dipole resonance, especially the role of the surface is crucial for the empirical determination of the compression modulus as a function of neutron excess.

- The vastly different environment in which nuclei are produced in large stars especially during gravitational collapse demands

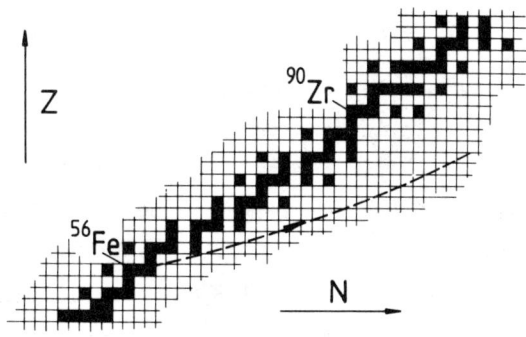

Fig. 1: Mean abundance of nuclei (dashed line) in the core of a ~10M$_\odot$ star during collapse calculated with the equation of state of Bethe et al.[1]

extrapolation of the line of β-stability. On the collapse time scale weak interactions are not equilibrated and electron capture neutronizes the nuclei. For illustration Fig. 1 shows the evolution of the mean abundance of nuclei in the core.

At the onset of the collapse the abundance peaks near ^{56}Fe but quickly moves away from the valley of stability. Near mass A = 100 corresponding to a matter density of $\sim 10^{12}$ g/cm^3 neutrinos produced in the e^--capture reactions are trapped.

The present talk deals with nuclear structure aspects of the above fields of research. In the first part I shall describe microscopic analyses of isovector vibrational modes of natural parity and their implication for the nuclear matter equation of state. Part two deals with unnatural parity excitations especially spin-isospin vibrations. Model calculations are presented which include isobar-hole (1Δ1h) as well as two-particle two-hole (2p2h) excitations. To describe e^--capture rates in supernova collapse the spin-isospin response in neutron rich nuclei at finite temperature is needed. Theoretical extrapolations and the feedback on the collapse dynamics especially from neutrino cooling will be discussed at the end.

II. NATURAL PARITY ISOVECTOR VIBRATIONS

One of the problems in extrapolating nuclear properties off the β-stability line as well as to the infinite matter limit is the uncertainty introduced by the surface properties of finite nuclei, in the present context the magnitude of surface symmetry coefficient a_δ^s. The liquid drop model parameter a_δ^s is not well constrained by the ground state properties of nuclei in particular the static dipole polarizability and information has to be obtained from excited states, in this case from the giant dipole resonance. The compression modulus K_∞ of nuclear matter as a function of the asymmetry parameter $\delta = \frac{N-Z}{A}$ is influenced by the volume symmetry energy a_δ^v and the isospin dependence of the nuclear effective mass

$m*^2$. The nuclear mode which most strongly constraints $K_\infty(\delta)$ is the IVM-vibration, in which neutrons "breathe" against protons. The systematics of this mode allow for a determination of a_δ^V and a_δ^S which is independent from that of the dipole.

II.1 THE GIANT DIPOLE RESONANCE: VOLUME VS SURFACE SYMMETRY ENERGY

In the equation of state of a compressed nucleus with $N \neq Z$ the density dependence of the symmetry energy coefficient $a_\delta(n)$ provides a pressure. In a simplified liquid drop model with constant density n_o inside a sphere of radius R_o and zero density outside one obtains

$$P_\delta(n_o) = -\frac{n_o}{3} L \delta^2 \tag{1}$$

where δ is the excess parameter and L is given by the first derivative of a_δ with respect to n_o

$$L = 3 n_o \frac{da_\delta}{dn_o} . \tag{2}$$

The second derivative corresponds to a symmetry term in the compression modulus of the nucleus

$$K_\delta = 9 n_o \frac{d^2 a_\delta}{dn_o^2} . \tag{3}$$

Both L and K_δ are known to some extent empirically,[3] but are not well constrained by microscopic theories of the nuclear ground-state. Thus a theoretical extrapolation to the infinite matter limit is rather uncertain. For illustration the nuclear matter values of L and K_δ predicted in non-relativistic Hartree-Fock theory (HF) with phenomenological two-body interactions as compiled by Blaizot[4] are given in Table 1.

$$\sigma_{-2} = 4\pi^2 \frac{\hbar c}{e^2} \int \frac{\sigma(E)}{E^2} dE \tag{4}$$

where $\sigma(E)$ is total photoabsorption cross section. In linear

	B1	D1	SKa	S1V	SIII
L	163.	18.40	75.07	63.58	10.13
K_δ	-24.80	-278.	-79.12	-140.07	-392.35

Table 1: HF nuclear matter values of L and K_δ for various effective forces[4].

Additional information on the density dependence of a_δ can be obtained from the static dipole polarizability defined as

response theory $\sigma(E)$ is obtained from the dipole response function

$$R_D(E) = \langle|D^+(E-H+E_o+i\eta)D|\rangle - |\langle|D|\rangle|^2/(E+i\eta) \tag{5a}$$

$$D = \sum_{i=1}^{A} \frac{e}{2}(1+\tau_3^i)r^i \tag{5b}$$

as

$$\sigma(E) = -\frac{\text{Im}}{\pi} R_D(E)E. \tag{6}$$

A connection between the response R_D and a_δ is provided in Fermi liquid theory where the density dependence of a_δ is related to the density dependence of the Landau parameter F_o' as

$$a_\delta(n) = k_F^2/(2m^*)(1+F_o'(n)). \tag{7}$$

In RPA we have

$$R_D(E) = \sum_{\substack{ph \\ p'h'}} (Q_{ph}^+, Q_{hp}^+) \begin{pmatrix} E-A_{php'h'}+i\eta & -B_{php'h'} \\ -B^*_{php'h'} & -E-A^*_{php'h'}-i\eta \end{pmatrix}^{-1} \begin{pmatrix} Q_{p'h'} \\ Q_{h'p'} \end{pmatrix} \tag{8}$$

with the density dependent Landau-Migdal amplitude[5]

$$F(n) = C_0 \delta(\vec{r}-\vec{r}\,') [F_0(n) + F_0'(n)\vec{\tau}\vec{\tau}\,' + G_0(n)\underset{\sim}{\sigma}\underset{\sim}{\sigma}\,' + G_0'(n)\underset{\sim}{\sigma}\underset{\sim}{\sigma}\,'\vec{\tau}\vec{\tau}\,']$$

$$C_0 = \frac{\pi^2 \hbar^2}{2 k_F m^*} = 150 \text{ MeV fm}^3 \qquad (k_F = 1.36 \text{ fm}^{-1}, \; m^* = m) \tag{9}$$

providing the residual ph-interaction, i.e.

$$A_{php'h'} = (\varepsilon_p - \varepsilon_h)\delta_{pp'}\delta_{hh'} + \langle ph|F|p'h'\rangle \tag{10a}$$

$$B_{php'h'} = \langle ph|F|h'p'\rangle. \tag{10b}$$

As usual the density dependence of the Landau parameter is chosen linear[5], i.e. for $F_0'(n)$ one has

$$F_0'(n) = F_0'^{ex} + (F_0'^{in} - F_0'^{ex})n \tag{11a}$$

$$n(r) = (1 + \exp(R-r)/\alpha)^{-1}. \tag{11b}$$

The volume symmetry coefficient $a_\delta^V = a_\delta(n_0)$ is known with some precision from semi-empirical mass formula fits[3]. The values range between 28 MeV and 37 MeV. Using eq. (7) together with eq. (11) this implies $1.2 \leq F_0'^{in} \leq 1.9$. Treating $F_0'^{ex}$ as a variable the results for σ_{-2} in ^{208}Pb are displayed in Fig. 2 for values of $F_0'^{in}$ corresponding to a_δ^V = 27 MeV, 33 MeV, and 38 MeV respectively. One concludes from this calculation, that experiment (indicated by the hatched area) does not constrain $F_0'^{ex}$ but rather determines a combination of $F_0'^{in}$ and $F_0'^{ex}$. It is clear however, that fixing $F_0'^{in}$ from the empirical a_δ^V the data require large density dependence of $F_0'(n)$ ($F_0'^{ex}/F_0'^{ex}$ = 2-4).

The consequences of strong density dependence of the isovector ph-interaction on the giant dipole strength function

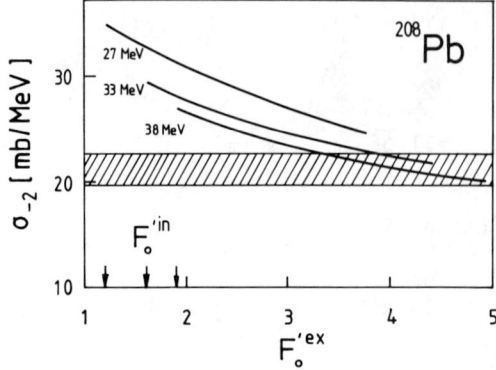

Fig. 2: Static dipole polarizability σ_{-2} for various combinations of $F_0'^{in}$ and $F_0'^{ex}$ (the value for σ_{-2} in the independent particle model is 76.8 mb/MeV).

$$S_D(E) = -\frac{Im}{\pi} R_D(E) \qquad (12)$$

have been pointed out by Treiner recently[6]. He finds that it leads to an unphysical spreading of the RPA strength distribution. Using a self-consistent scheme with the modified Skyrme interaction SKM he finds essentially two fragments around 12.3 MeV and 16 MeV. The present calculation yields similar results as shown in the upper part of Fig. 3.

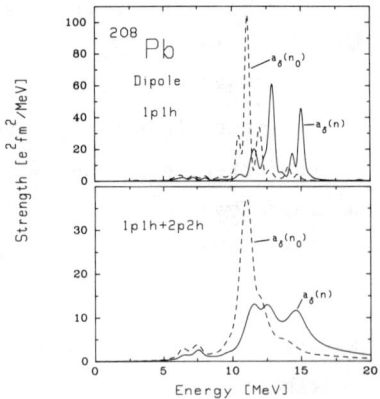

Fig. 3: Giant dipole strength function in ^{208}Pb.
Upper part: 1p1h RPA result for $F_0'^{in} = F_0'^{ex} = 1.6$ (dashed line) and $F_0'^{in} = 1.6$ $F_0'^{ex} = 4$ (full line) Lower part: 1p1h + 2p2h RPA result with the same choices for the density dependence of the residual interaction.

Since the main dipole strength in ^{208}Pb is at excitation energies where the density of 2p2h-states is already appreciable (of the order of 200 states/MeV) one may expect that the 1p1h fragments interact with each other via 2p2h excitations. If this interaction is attractive it has a focussing effect reducing the unphysically large spreading on the 1p1h level.

To study this expectation, one has to extend the RPA formalism to include 2p2h excitations. In such an extension[7,8] a frequency dependent mass operator Σ is added to the A-matrices in eq. (8) and the expression for the response function R_D becomes

$$R_D(E) = \sum_{\substack{ph \\ p',h'}} (Q^+_{ph}, Q^+_{hp}) \begin{pmatrix} E-A_{php'h'}(E)+i\eta & -B_{php'h'} \\ -B^*_{php'h'} & -E-A^*_{php'h'}(E)-i\eta \end{pmatrix}^{-1} \begin{pmatrix} Q_{p'h'} \\ Q_{h'p'} \end{pmatrix} \quad (13)$$

With

$$A_{php'h'}(E) = (\varepsilon_p - \varepsilon_h)\delta_{pp'}\delta_{hh'} + V_{php'h'} + \Sigma_{php'h'}(E) \quad (14a)$$

$$\Sigma_{php'h'}(E) = \sum_{2p2h} \frac{\langle ph|V|2p2h\rangle\langle 2p2h|V|p'h'\rangle}{E-\varepsilon_{2p2h} + i\eta} \quad (14b)$$

It should be noted, that the Landau F is now obtained from terms which are first and second order in the residual interaction in V. For finite η the mass operator Σ becomes complex adding a spreading width to the 1p1h-response. The results of the 1p1h + 2p2h RPA calculation are depicted in the lower part of Fig. 3. The agreement with experiment is improved compared to 1p1h alone. The centroid energy is reasonable, but the width of 5.5 MeV is still somewhat too large. It is expected, however, that collective effects in the 2p2h excitations[9] may reduce the width further.

To close this analysis I wish to conclude, that the energy of the giant dipole resonance in ^{208}Pb and σ_{-2} seem to demand a strong density dependent isovector interaction which is about a factor 2.5

larger in the surface than in the interior. Such density dependence is supported by Brückner calculations in nuclear matter[10].

II.2 ISOVECTOR MONOPOLE VIBRATIONS: COMPRESSION MODULUS IN ASYMMETRIC NUCLEAR AND NEUTRON MATTER

The frequency of the isovector monopole vibration, in which protons and neutrons 'breathe' in phase has been determined in many nuclei[4] and the systematics have lead to a stringent constraint on the compression modulus $K_\infty(\delta)$ in symmetric nuclear matter ($\delta=0$). The extracted value of $K_\infty(0) = 210 \pm 30$ MeV is very close to the free Fermi gas value with nucleons of $m^*=m$. This can be easily seen from the relation between $K_\infty(0)$ and the Landau parameter F_0.

$$K_\infty(0) = 6\varepsilon_F(1 + F_0(n_0)) \qquad (15)$$

where n_0 denotes the equilibrium density and ε_F the Fermi energy

$$\varepsilon_F = \hbar^2 k_F^2/2m^* \qquad (16)$$

Inserting $k_F = 1.36$ fm^{-1} and $m^*=m$ on has in the non-interacting gas ($F_0=0$) $K_\infty(0) = 230$ MeV which lies within the experimental limits.

Recently an extension of eq. (15) for the asymmetric case $\delta \neq 0$ has been given by Haensel[2]. Neglecting effective mass corrections quadratic in δ one obtains

$$\begin{aligned} K_\infty(\delta) &= K_\infty(0) + 6\varepsilon_F[5/9 + F_0'(n_0)]\delta^2 \\ &= K_\infty(0) + K_\infty^e \cdot \delta^2 \end{aligned} \qquad (17)$$

i.e. the compression modulus is changed by a term proportional to the bulk symmetry energy coefficient a_δ^v. Note that n_0 is the equilibrium density for $\delta \neq 0$ which needs not necessarily to be the same as for $\delta = 0$. Therefore the density dependence of F_0' is crucial. If one chooses nuclear matter equilibrium density one obtains with $a_\delta^v(n_0) =$

33 MeV, i.e. $F'_o(n_o) = 1.6$ a correction of $K^e_\infty = 496$ MeV. For pure neutron matter ($\delta = 1$) at density n_o of symmetric nuclear matter this implies a bulk modulus of ~700 MeV. The nuclear dipole frequency indicates that at lower density K^e_∞ is even larger.

To improve the empirical constrains on the equation of state of asymmetric matter the isovector breathing modes have to be analyzed, i.e. it has to be seen if the Fermi liquid parameters describing the dipole also reproduce the IVM. Since these modes are built on $\Delta N=2$ ph-transitions and the residual interaction is repulsive at all densities the excitation energy is very high. The hydrodynamical model predicts for the $T_<$-component $E_{T_<} = 170\ A^{-1/3}$ MeV[11]. In Fig. 4 the results of a microscopic 1p1h and 1p1h + 2p2h RPA-calculation are shown using the same interaction as for the dipole. The pure 1p1h-calculation has centroid energies somewhat lower than the hydrodynamical predictions (indicated by the arrows) and it is in good agreement with self-consistent RPA calculations by Auerbach and Klein[12] (since the present calculation is performed in a discretized single particle basis the escape width $\Gamma\uparrow$ has been added by energy average using finite η in eq. (8) and eq. (13)). As can be seen from the figure most of the damping of the resonance comes from mixing with 2p2h-states. Their level density is very large at those high frequencies. One therefore concludes that experimental detection is very difficult in the τ_o channel. In addition to the large width hadronic reactions also excite higher multipoles at forward angles due to distortion effects on the projectile wave. Thus the isolation of a single broad resonance is complicated. The situation becomes better in the charge exchange branch which involves τ_+-isospin transitions where protons in the nucleus are converted into neutrons (Fig. 5).

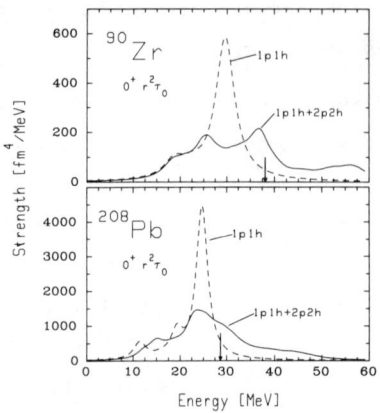

Fig. 4: IVM strength distribution in ^{90}Zr and ^{208}Pb in the long wavelength limit ($r^2\tau_0$). The arrows denote the hydrodynamical estimate $E_{T_<} = 170\ A^{-1/3}$.

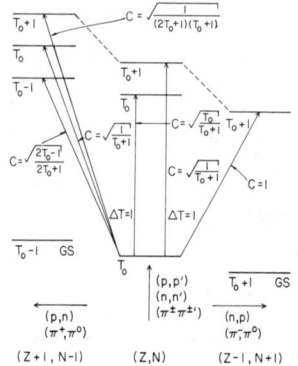

Fig. 5: Schematic representation of various isovector excitations in $N \neq Z$ nuclei. C denotes the isospin Clebsch-Gordon coefficient regulating the relative strength of the excitations. ($T_0 = (N-Z)/2$.)

In τ_+-transitions such as (π^-,π^0) or (n,p) the $T_0 + 1$-component which can only be excited is lowered by the Coulomb displacement energy Δ_c. This implies much lower level density and hence smaller width.
In (π^-,π^0) -charge exchange in fact the IVM in heavy nuclei has been seen.[13] Also in (n,p) one has some evidence for IVM-strength in ^{208}Pb.[14] The predictions from a microscopic 1p1h + 2p2h RPA calculation in all three isospin branches are shown in Fig. 6. The

width decrease in going from τ_- to τ_+ is clearly seen in ^{90}Zr and more dramatically in ^{208}Pb where in the τ_+-channel the strength is concentrated in two narrow peaks around 10 MeV and 17 MeV excitation energy relative to the parent nucleus.[15] In ^{90}Zr experimental information from (π^-,π^0) is available[13] which gives a centroid energy of 24.1 ± 1.3 MeV and a width of 13.2 ± 2.3 MeV. This should be

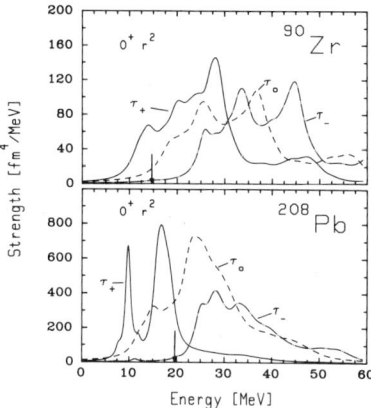

Fig. 6: IVM-response in the long wavelength limit in the three isospin branches τ_+, τ_0 and τ_-. For comparison the isobaric analog states (IAS) with energies denoted by the arrows are projected out.

compared to 26.4 MeV centroid energy and ~14 MeV width from the RPA. The calculated energy location of the IAS of 14.8 MeV in ^{90}Zr and 19.6 MeV in ^{208}Pb which is also in very reasonable agreement with experiment.

To close this section, I wish to conclude:
- from the microscopic analysis of the giant resonance energetics in heavy nuclei it seems clear, that in order to be consistent with a bulk symmetry energy coefficient a_δ^v of 28 MeV-37 MeV the effective isovector interaction has to be strongly density dependent. In the nuclear surface the force is about a factor 2.5 larger than in the interior. The unphysical width of the 1p1h RPA response is reduced by 2p2h-effects giving reasonable agreement with experiment.
- density dependent isovector interactions are compatible with the energetics of IVM-vibrations in ^{90}Zr where this mode has been

identified.[13]. In ^{208}Pb the 1p1h + 2p2h RPA calculation predicts two localized resonances around 10 MeV and 17 MeV with narrow widths. The calculated IAS energies are also in good agreement with experiment.

- a strong density dependence of the isovector interaction has important consequences for the empirical determination of the bulk modulus of neutron matter making it very sensitive to the equilibrium density n_o and small proton impurities.

III. UNNATURAL PARITY ISOVECTOR EXCITATIONS

Unnatural parity isovector states when excited in hadronic reactions dominantly involve spin-flip. Spin vibrations are of interest for many reasons. They reveal the role of the virtual pion field in nuclei and subnuclear degrees of freedom as far as they are coupled to nucleons via the pion. In the long wavelength limit they furthermore exhibit short-range correlations among the nucleons produced by the Pauli principle and the hard core of the two nucleon interaction. The role of short-range correlations in the spin response has been most beautifully demonstrated experimentally by the existence of highly collective Gamow-Teller resonances (GTR) over the whole periodic table.

In the following, I discuss a model in which the interaction among nucleons is based on one-boson exchange and the Hilbert space is spanned by 1p1h, 2p2h and 1Δ1h states. Applications to GTR and the giant spin dipole resonance (GSDR) are presented. Both the GTR and the GSDR are important in retarding low energy allowed and first forbidden β_+ and β_--decay matrix elements. The amount of quenching and its feedback in the neutrino cooling process during collapse of the supernova core will be discussed at the end.

III.1 GIANT GAMOW-TELLER RESONANCES IN HEAVY NUCLEI

Only very recently (p,n)-experiments at intermediate energies have systematically mapped out the spin-isospin response in nuclei[16]. Scattered nucleons with energies E_p between 100 MeV-400 MeV are especially suited since the central spin-isospin part $t_{\sigma\tau}^c$ of the interaction with the target nucleons at low momentum transfer is enhanced by a factor 2-4 over the non spin-flip part t_τ^c (Fig. 7).

Particularly the forward angle spectra are dominated by the GTR. A measure of the total observed strength is provided by the model independent Ikeda sum rule[18]

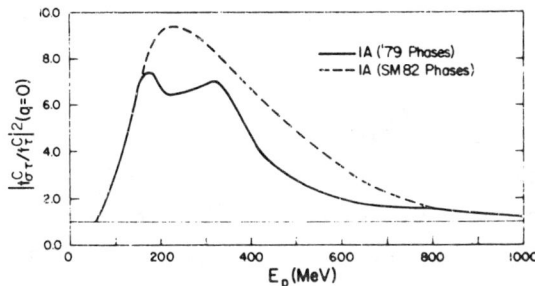

Fig. 7: Energy dependence of the ratio $|t_{\sigma\tau}^c/t_\tau^c|^2$ at zero momentum transfer. (The figure was taken from ref. 17)

$$S_{\beta_-} - S_{\beta_+} = 1/2 \sum_f \sum_\mu |\langle f| \sum_i \sigma_\mu^i (\tau_-^i - \tau_+^i)|\rangle|^2 = 3(N-Z) \qquad (18)$$

The sum over intermediate states $|f\rangle$ includes states in the (Z+1, N-1)- and (Z-1, N+1) daughter nuclei. In (p,n)-charge exchange only S_{β_-} is accessible, but it is argued that in nuclei with large neutron excess S_{β_+} is largely Pauli blocked such that S_{β_-} is essentially 3(N-Z). If S_{β_-} is evaluated by integrating the (p,n)-cross sections to high excitation energies this assumption may not be justified and S_{β_+} could be appreciable especially if tensor correlations are large[19]. Bertsch and Hamamoto[21] calculated S_{β_+} to be ~20% of 3(N-Z). More theoretical work needs to be done here[21]. On the experimental side it is vital for this question, that S_{β_+} is measured directly in (n,p)-reactions. In the following discussion I ignore the enhancement of S_{β_-} by unblocking such that 3(N-Z) becomes a lower limit of expected $S_{\beta-}$-strength. Integrating the (p,n)-cross sections up to ~40 MeV only about 65% of 3(N-Z) are found[16]. The "missing strength" has been attributed to $1\Delta 1h$-admixtures and mixing with multiparticle-multipole excitations above 40 MeV. To quantitatively account for both effects we have recently performed a calculation[22] including $1\Delta 1h$ states as well as 2p2h states interacting via a correlated $\pi+\rho$-meson exchange potential $\hat{V}_{\pi+\rho}$.

In ^{90}Zr the independent particle model (IPM) strength function has a simple structure. Since the 1g9/2-neutron shell is filled and both the 1g9/2 → and 1g7/2 proton shells are empty a n1g9/2 → p1g9/2 and a n1g9/2→h1g7/2 transition occur (left part of Fig. 8) with relative strengths

$$B(GT; \ell j_> \to \ell j_>)/B(GT; \ell j_> \to \ell j_<) = (2\ell+3)/4\ell \qquad (19)$$

(in the classical limit ($\ell \to \infty$) this ratio goes to 1/2). The energy separation is determined by the proton spin-orbit potential. In ^{208}Pb because of the larger neutron excess more 1p1h-transitions are possible with the strongest carrying about 20% of the total strength (left lower part of Fig. 8). As $\hat{V}_{\pi+\rho}$ is allowed to act among the IPM transitions the strength is pushed to higher energies but compared to experiment the theoretical energies are too low (the Fermi-liquid parameter G_o' from $\hat{V}_{\pi+\rho}$ is 1.4). Additional repulsion is provided by the 2p2h-admixtures (right part of Fig. 8). This extra repulsive contribution to the ph-interaction is identified as the perturbative equivalent of the "induced interaction" of Babu and Brown[23] (Fig. 9). Also the damping of the 1p1h response is described correctly. The FWHM in the high energy resonance peaks are about 5 MeV as compared to 4.4 MeV in ^{90}Zr[24] and 4.2 MeV in ^{208}Pb[25] respectively. The strength distributions exhibit long tails extending up to 60 MeV excitation energy. The amount of $S_{\beta-}$ which resides between 0-25 MeV and 0-40 MeV relative to 3(N-Z) is given in Table 2.

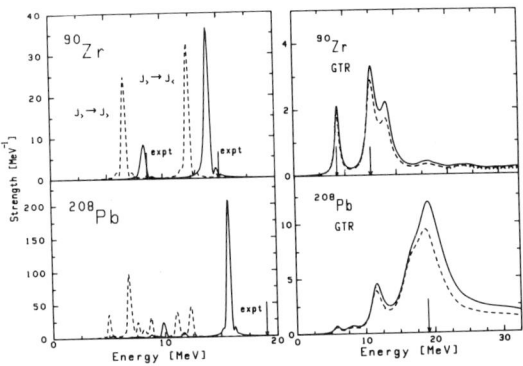

Fig. 8:
left part: IPM GT-response (dashed lines) as compared to inclusion of $\hat{V}_{\pi+\rho}$ on the 1p1h level (full lines)
right part: inclusion of 2p2h-admixtures (full lines) and 1Δ1h-excitations (dashed lines). The arrows indicate empirical centroid energies.

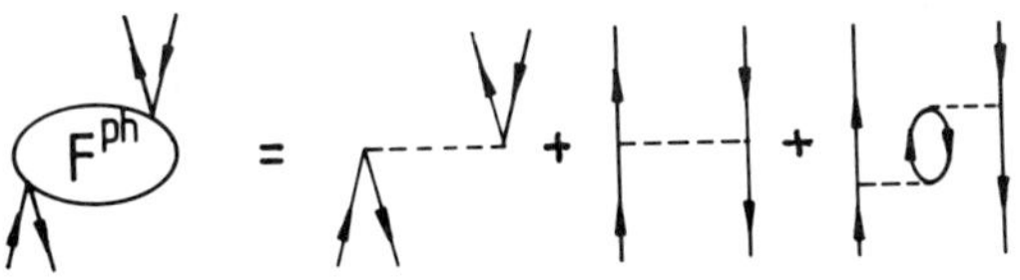

Fig. 9: Effective ph-interaction in second order RPA. The dashed lines indicate $\hat{V}_{\pi+\rho}$. The "bubble term" essentially cancels the exchange term of the bare $\hat{V}_{\pi+\rho}$, thus providing extra net repulsion.

	^{90}Zr	^{208}Pb	^{90}Zr	^{208}Pb
1p1h	100%	100%	100%	100%
1p1h + 2p2h	63%	68%	77%	84%
1p1h + 2p2h + 1Δ1h	53%	54%	63%	66%

Table 2: Amount of S_{β_-}-strength in ^{90}Zr and ^{208}Pb as calculated in Ref. 22.

These values have to be compared to 60% ± 10% found in experiment[16].

III.2 GIANT SPIN-DIPOLE RESONANCES

Shortly after the discovery of the GTR the existence of giant spin-dipole resonances in nuclei has also been established experimentally[25]. Within the 1p1h framework calculations of the (p,n) cross-section have been performed by the Jülich group[27]. However, the width remained undetermined theoretically. To improve on this point and to see whether the correlated π+ρ-model with 2p2h excitations also works in this case the cross sections have been reinvestigated[28]. To be consistent the projectile target-nucleon interaction was also chosen as $\hat{V}_{\pi+\rho}$, which is a very good approximation for the tensor part and reasonable for the central part when compared to realistic t-matrices[29]. The 160 MeV (p,n)-cross section was evaluated using the plane wave Born approximation (PWBA) expressions of Petrovich and Love[30] including optical distortions of the projectile waves via a distortion factor. The results summarized in Fig. 10 are in reasonable agreement with experiment. The integrated strength between 0 and 40 MeV is 26.2 mb/sr as compared to 32 mb/sr found experimentally. The centroid energy of 24 MeV relative to ground state of ^{208}Pb is also well reproduced. The FWHM

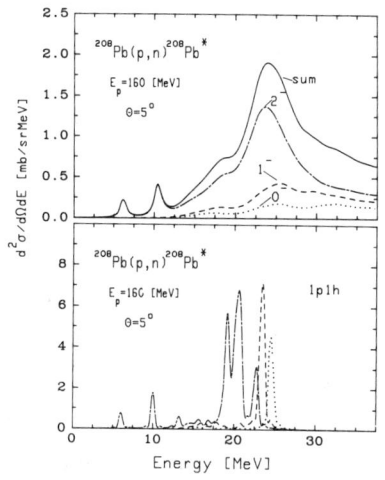

Fig. 10: $\frac{d^2\sigma}{d\Omega dE}$ for 160 MeV (p,n)-charge exchange. Upper part: 1p1h + 2p2h calculation with correlated $\pi + \rho$-exchange. Lower part: 1p1h-calculation only.

is 7.5 MeV as compared to 10 MeV in experiment. In contrast to the earlier calculation by Osterfeld et al.[27] the low multipoles 0^- an 1^- have much larger damping widths. As in the case of the GTR the coupling to 2p2h provides extra repulsion and hence an upward shift in the resonance energies of the GSDR.

To summarize the last two sections the following main results should be pointed out:

- a higher order RPA-model which includes 1p1h as well as 2p2h excitations seems to describe the main features of the nuclear spin-isospin response very well, if "realistic" $\pi + \rho$ meson exchange potentials are used.
- $\hat{V}_{\pi+\rho}$ fixes the transition potential via PCAC when $1\Delta 1h$-excitations are included in the model space. The predicted quenching of the nucleonic response by isobar degrees of freedom is of the order of 20%-25% in heavy nuclei as can be seen from Table 2.

III.3 THE ROLE OF β_+-STRENGTH IN STELLAR COLLAPSE

As pointed out in the pioneering work by Bethe et al.[30] gravitational collapse of Type II supernovae proceeds at low entropy S. Crucial in maintaining the initial S of roughly unity is the amount of e^--capture on free protons and heavy nuclei in the collapsing core. The entropy change is most conveniently related to the electron fraction per baryon Y_e. One has[31]

$$\frac{dS}{dY_e} = (\mu_e - \hat{\mu} - \langle\varepsilon_\nu\rangle^{(p)} - \langle\varepsilon_\nu\rangle^{(h)})/T \tag{20}$$

where μ_e is the electron chemical potential, $\hat{\mu}$ the neutron-proton chemical potential difference and $\langle\varepsilon_\nu\rangle^{(p)}$ and $\langle\varepsilon_\nu\rangle^{(h)}$ the average energies of neutrinos emitted after e^--capture on free protons and on heavy nuclei respectively. T denotes the matter temperature, typically of the order of 1-2 MeV. The average neutrino energies are specified by total capture cross sections σ (ε_e,T) for electrons of energy ε_e. Neglecting the electron rest mass m_e σ_{if} is given by[32]

$$\sigma_{if}(\varepsilon_e,T) = G_W^2 \frac{(\varepsilon_e - E_{if})^2}{2\pi} F(\varepsilon_e, Z) \, S \, (E_{if}, T) \tag{21}$$

where G_W^2 is the weak interaction coupling strength, E_{if} the Q-value of the reaction and F the relativistic Coulomb penetration factor. The information about the capturing species is contained in the strength function S (E,T). For e^--capture on free protons the Q-value is equal to the neutron-proton mass difference M_{np} = 1.29 MeV and F is equal to one. Therefore σ_{if} reduces to a simple form namely

$$\sigma_{if}^{(p)}(\varepsilon_e,T) = G_W^2 \frac{(\varepsilon_e - M_{np})^2}{2\pi} (g_V^2 + 3g_A^2) \tag{22}$$

and in particular does not depend on T. In expression (22) g_V and g_A denote the vector-and axial vector coupling constants respectively.

III.3a ALLOWED e^--CAPTURE ON HEAVY NEUTRON RICH NUCLEI

In analogy to free protons allowed e^--capture on nuclei is described by the β_+-Fermi and Gamow-Teller operator

$$\mathcal{O} = (g_V + g_A \underline{\sigma})\tau_+ \qquad (23)$$

It has been pointed out by Fuller et al.[33] that as the abundance of nuclei evolves from the Fe peak to heavier neutron rich nuclei allowed capture is largely prohibited because of Pauli blocking as the 1f5/2, 2p1/2 and 1g9/2 shells gradually fill (Fig. 11).

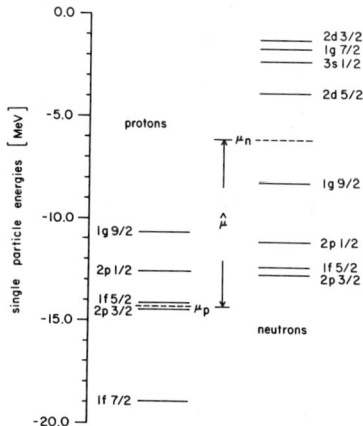

Fig. 11: Proton and neutron single particle levels in the medium heavy mass region. $\hat{\mu}$ denotes the difference in the neutron and proton chemical potentials $\mu_n - \mu_p$.

Various unblocking mechanisms have to be considered. One is configuration mixing and pairing and the other is thermal unblocking. At temperatures between 1 and 2 MeV the pairing field is mostly destroyed[34]. In the following, I describe an estimate of the amount of thermal unblocking within the finite temperature RPA framework. In this theory the response function is given by[35]

$$R(E,T) = \sum_{\substack{\mu\mu' \\ \nu,\nu'}} \mathcal{O}^*_{\mu\mu'} G_{\mu\mu'\nu\nu'}(E,T) \mathcal{O}_{\nu\nu'} \qquad (24)$$

where $\mathcal{O}_{\mu\mu'}$ are matrix elements of the one-body operator in the single particle particle basis $|\mu\rangle$ and the propagator is the finite T extension of eq. (8)

$$G_{\mu\mu'\nu\nu'}(E,T) = \{2\mathcal{N}(T)(\mathcal{N}(T)E - A(T))^{-1}\mathcal{N}(T)\}_{\mu\mu'\nu\nu'} \qquad (25a)$$

with

$$\mathcal{N}_{\mu\mu'\nu\nu'}(T) = (f_{\mu'}(T) - f_\mu(T))\delta_{\mu\nu}\delta_{\mu'\nu'} \qquad (25b)$$

and

$$A_{\mu\mu'\nu\nu'}(T) = (f_{\mu'}(T) - f_{\mu}(T))(\epsilon_{\mu} - \epsilon_{\mu'})[\delta_{\mu\nu}\delta_{\mu'\nu'} - \delta_{\mu\nu'}\delta_{\mu'\nu}]$$
$$+ 1/2 \, (f_{\mu'}(T) - f_{\mu}(T)) V_{\mu\mu'\nu\nu'}(f_{\nu'}(T) - f_{\nu}(T)). \quad (25c)$$

$f_\mu(T)$ denote the temperature dependent distribution functions

$$f_\nu(T) = 1 + e^{\beta(\epsilon_\nu - \mu)}; \quad \beta = 1/T \quad (26)$$

where μ is the chemical potential. The strength function $S(E,T)$ which is directly proportional to the finite temperature cross-section is given by

$$S_{\mathcal{O}}(E,T) = (1 - e^{-\beta E})^{-1} \frac{\text{Im}}{\pi} R_{\mathcal{O}}(E,T). \quad (27)$$

Figure 12 displays the results of a finite temperature calculation for the nucleus ^{82}Ge a mean nucleus close to the point of neutrino trapping. This calculation uses an approximation to eq. (24) as described in Ref. 31.

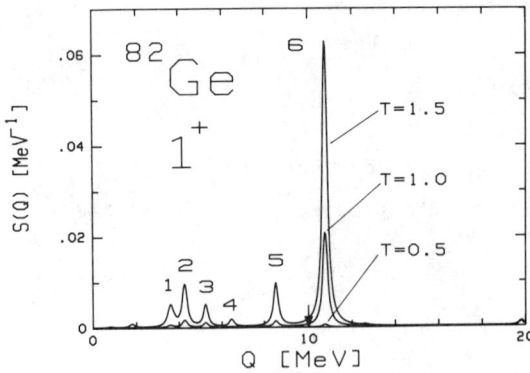

Fig. 12: Thermal unblocking of GT-strength in ^{82}Ge for various T as a function of the Q-value of the reaction. The numbers label individual single particle transitions[31]. The arrow indicates the zero temperature threshold $Q = \hat{\mu} + M_{np}$.

As a main result one finds that during the collapse the thermal unblocking probability is typically of the order of 1%, very similar to earlier estimates by Fuller et al.[33] The unblocking of the GT-strength in ^{90}Zr plays a major role. The entropy change dS/dYe remains quite small and positive, i.e., allowed capture from hot nuclei leads to cooling.

III.3b PARITY FORBIDDEN e^--CAPTURE

With allowed transitions being unblocked by ~1% also first forbidden capture has to be considered. The $\Delta \ell = 1$ parity operators that mediate the nuclear transitions have been given by Konopinski[32] as

$$\mathcal{O} = \begin{cases} g_A [\dfrac{\underline{\sigma} \cdot \underline{p}}{m} + \dfrac{\alpha Z}{2R} i\underline{\sigma} \cdot \underline{r}]\tau_+ & \Delta J^\pi = 0^- & (28.a) \\[4pt] [g_V \dfrac{\underline{p}}{m} + \dfrac{\alpha Z}{2R}(g_A \underline{\sigma} \times \underline{r} + ig_V \underline{r})]\tau_+ & \Delta J^\pi = 1^- & (28.b) \\[4pt] ig_A/\sqrt{3}\,[\underline{\sigma} \cdot \underline{r}]_m \sqrt{p_e^2 + q_\nu^2}\,\tau_+ & \Delta J^\pi = 2^- & (28.c) \end{cases}$$

where R is the nuclear radius, Z the charge and p_e and q_ν the electron and neutrino 3-momenta respectively. In general the matrix elements of these operators are about a factor of 10^2 smaller than the allowed matrix elements, but with ~99% blocking of the allowed transitions the capture rates are of the same order of magnitude. As Fuller et al.[33] pointed out unique first forbidden transitions with $\Delta J^\pi = 2^-$ are strongly favored by phase space weighting, because of the dependence of the operator on the lepton momenta. At e^--energies around 20 MeV, i.e., at matter densities around 10^{11}g/cm^2 they dominate the capture process completely. In contrast to the allowed capture they lead to a heating of the matter as can be seen by the increase of the entropy above $\rho = 10^{11}$ g/cm^3 (Fig. 13). However, the heating is sufficiently small to still ensure a net decrease of the entropy from its initial value.

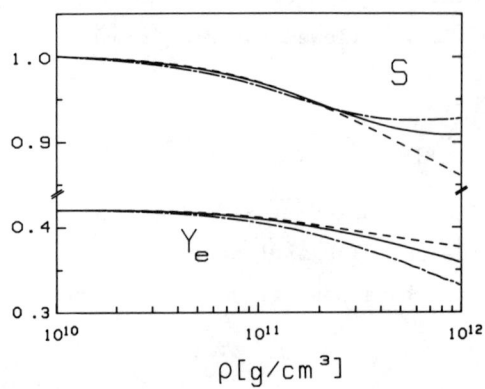

Fig. 13: Entropy S and number of e^-/baryon Ye as a function of matter density. The dashed line includes capture on free protons only. The full and dashed dotted lines include capture from heavy nuclei as well with two different assumptions for the strength of RPA-groundstate correlations[31].

Acknowledgement

This work was supported by the Physics Department of the University of Illinois at Urbana-Champaign.

References

1. H. A. Bethe, G. E. Brown, J. Cooperstein and J. R. Wilson, Nucl. Phys. A403 (1982) 625.
2. P. Haensel, Phys. Lett. 81B (1979) 276.
3. W. D. Myers and W. J. Swiatecki, Ann. Phys. 84 (1974) 211.
4. J. P. Blaizot, Phys. Rep. 64 (1980) 175.
5. A. B. Migdal, "Theory of Finite Fermi Systems and Applications to Atomic Nuclei", (Wiley, New York, 1967).
6. J. Treiner, Proceedings to the Int. Conf. on "Highly Excited States in Nuclei, HESANS", Orsay (Journal de Physique, 1984) 281.
7. C. Yannouleas, M. Dworzecka and J. J. Griffin, Nucl. Phys. A397 (1983) 239.
8. B. Schwesinger and J. Wambach, Phys. Lett. 134B (1984) 29.
9. J. Wambach, V. K. Mishra and Li Chu-hsia, Nucl. Phys. A380 (1982).
10. S. O. Bäckman, O. Sjöberg and A. D. Jackson, Nucl. Phys. A321 (1979) 10.
11. A. Bohr and B. Mottelson, "Nuclear Structure" (Benjamin, Reading, Mass., 1975), Vol. II.
12. N. Auerbach and A. Klein, Nucl. Phys. A395 (1983) 77.
13. J. D. Bowman et al. Phys. Rev. Lett. 50 (1983) 1195.
14. F. P. Brady et al. Proceedings to the Int. Conf. on "Spin Excitations, in Nuclei", Telluride (Plenum, 1984).
15. B. Schwesinger and J. Wambach, Phys. Rev. C29 (1984) 663.
16. For a recent review: J. Rapaport, IUCF Workshop "The Interaction between Medium Energy Nucleons in Nuclei", (1982).
17. W. G. Love and M. Franey, Proceedings to the Int. Conf. on "Highly Excited States in Nuclei, HESANS", Orsay (Journal de Physique, 1984).
18. K. I. Ikeda, S. Fujii and J. I. Fujita, Phys. Lett. 3 (1963) 271.
19. A. Arima and H. Hyuga, Mesons in Nuclei, (1979) 683.
20. G. F. Bertsch and I. Hamamoto, Phys. Rev. 26C (1982) 1323.

21. S. Drozdz, V. Klemt and J. Wambach, in progress.
22. D. Cha, B. Schwesinger, J. Wambach and J. Speth, Nucl. Phys. A, in press.
23. S. Babu and G. E. Brown, Ann. of Phys. $\underline{78}$ (1973) 1.
24. C. Gaarde et al., Proceedings to the Int. Conf. on "Spin Excitations in Nuclei", Telluride, (Plenum, 1984).
25. D. J. Horen et al., Phys. Lett. $\underline{95B}$ (1980) 27.
26. C. Gaarde et al., Nucl. Phys. $\underline{A369}$ (1981) 258.
27. F. Osterfeld et al., Phys. Lett. $\underline{105B}$ (1981) 257.
28. H. P. Morsch, D. Cha and J. Wambach, in progress.
29. F. Petrovich and W. G. Love, Nucl. Phys. $\underline{A354}$ (1981) 499.
30. H. A. Bethe, G. E. Brown, J. Applegate and J. M. Lattimer, Nucl. Phys. $\underline{A324}$ (1979) 487.
31. J. Cooperstein and J. Wambach, Nucl. Phys. $\underline{A420}$ (1984) 591.
32. E. J. Konopinski, "The Theory of Beta Radioactivity", Oxford University Press (1966).
33. G. M. Fuller, Astrophys. J. $\underline{252}$ (1982) 741; G. M. Fuller, W. A. Fowler and N. J. Newman, Astrophys. J. $\underline{252}$ (1982) 715.
34. A. L. Goodman, Nucl. Phys. $\underline{A352}$ (1981) 30; Nucl. Phys. $\underline{A352}$ (1981) 45.
35. M. Sommermann, Ann. of Phys. $\underline{151}$ (1983) 163; P. Ring, unpublished.

ISOSPIN EFFECTS IN NUCLEAR VIBRATIONS*

V. A. Madsen
Oregon State University, Corvallis, OR 97331
and
Lawrence Livermore National Laboratory, Livermore, CA 94550

V. R. Brown
Lawrence Livermore National Laboratory, Livermore, CA 94550

ABSTRACT

A review of the evidence that the ratio of neutron and proton multipole matrix elements for collective vibrations in single-closed-shell nuclei differ systematically from N/Z is presented. A theoretical framework is given for understanding the data on the basis of the ideas of core polarization. It follows that nuclear deformation parameters are probe dependent and that analysis of excitations by two different probes such as (p, p') and (n, n') can, in principle, give the ratio M_n/M_p. Application is made to first 2^+ states of open shell nuclei. Trends of M_n/M_p for higher 2^+ states are presented. Expected systematics of M_n/M_p ratios for giant isoscalar quadrupole transistions are presented.

INTRODUCTION

Historically, in the 1960's there were many measurements of the excitation of low-lying collective states by various external probes, electromagnetic, (p,p'), (α,α'), and (n,n'). It was noticed and expected on the basis of the collective model that for the excitation of the 2^+ first excited state of any given even nucleus, the deformation parameter β was independent of excitation mechanism. The absolute errors in the measurements would not have excluded, however, small systematic differences.

Whereas the collective model described the low-lying collective states in terms of vibrations or rotations of a homogeneous neutron-proton fluid with a single deformation parameter, the microscopic model described them in terms of linear combinations of neutron and proton single-particle transitions. Because of the shell structure and the Pauli principle, differences in the freedom of neutrons and protons to vibrate could be expected. This is particularly true in single-closed-shell (SCS) nuclei, where one type of nucleon is "frozen in" by the shell closure. The collective model and the shell model would therefore disagree on the degree of participation of the neutrons and protons in nuclear vibrations. Furthermore, since different external fields or probes such as electromagnetic, (p,p'), (n,n'), (α,α'), etc., interact differently with the nuclear neutrons and protons, one would expect on the basis of the shell-model picture that deformation parameters β determined by a collective-model analysis of excitation of vibrational states would vary greatly from one probe to another.

---

*Work performed under the auspices of the U. S. Department of Energy by the Lawrence Livermore National Laboratory under contract No. W-7405-ENG-48, and at Oregon State University under contract No. DE-AT06-79ER-10405.

The resolution of this contradiction comes from core polarization and the greater strength of the nuclear force between nucleons of the opposite type. In a proton SCS nucleus, for example, the valence neutrons will polarize "core" protons more strongly than core neutrons, thereby tending to restore the propotionality of neutron to proton amplitudes toward N/Z, in agreement with the collective model. However, the restoration need not be complete. A schematic model[1] which treats core polarization as a virtual excitation of giant resonances and includes unequal like and unlike particle forces predicts[2] that for SCS nuclei there should be systematic differences between deformation parameters obtained by analysis of data from different external fields.

In this talk, following a brief statement of the schematic model of core polarization, we review the evidence for systematic differences in the deformation parameters β_n and β_p of nuclear neutrons and protons and the consequent probe dependence of measured β values. Next an application is made to the systematics of first 2^+ states in open-shell nuclei, and a comparison is made with available data. Evidence is presented that for excitation of other low-lying 2^+ states there is a reversal in the trend of neutron and proton deformation parameters ratios from those of the first 2^+ state. Last, some expectations and information are presented concerning neutron and proton differences in excitation of giant resonances.

THEORETICAL FORMULATION OF CORE POLARIZATION

The neutron and proton multipole matrix elements are defined as

$$M_{\{{}^n_p\}} = \langle f \| \sum_{i=1}^{\{{}^N_Z\}} r_i^\lambda Y_\lambda(\hat{r}_i) \| i \rangle \tag{1}$$

The formulation of the collective model[3] relates this quantity to the nuclear deformation parameters, β such that M_p is proportional to $Z\beta$:

$$\frac{M_n}{M_p} = \frac{N\beta_n}{Z\beta_p} \tag{2}$$

We have allowed in Eq. (2) for the possibility that the nuclear neutrons and protons have different deformation parameters β_n and β_p for a particular excited vibrational state, although, as stated in the introduction, the conventional collective model assumes a homogeneous nuclear fluid, for which $\beta_n = \beta_p$.

In the microscopic picture we must deal with a limited model space, where M_n and M_p may vary rapidly from nucleus to nucleus but in which one has the capability of calculating the multipole matrix elements accurately, such as the $0\hbar\omega$ space in the shell model. It is expected that additional contributions vary slowly and systematically and can therefore be treated as core polarization.

We have formulated core polarization[1,4] within the context of the TDA or RPA approximations using separable multipole forces and breaking the Hamiltonian matrix into two pieces,

$$H = \begin{pmatrix} H_{aa} & 0 \\ 0 & H_{bb} \end{pmatrix} + \begin{pmatrix} 0 & H_{ab} \\ H_{ba} & 0 \end{pmatrix}, \quad (3)$$

where a represents the $0\hbar\omega$ model space and b, the $2\hbar\omega$ core-polarization space. The first term in Eq. (3) is to be treated exactly, giving a description of low-lying 2+ states in a and giant resonance states in b, and the second term, which connects the spaces, is to be treated in perturbation theory.

The perturbation mixes the high and low-lying states, giving the following results[4] for M_n and M_p of Eq. (1) for a state s in the lower space,

$$M_s \approx M_s^{(a)} + 2 \sum_t M_t^{(b)} \frac{[m_t^{(b)\dagger} V m_s^{(a)}] E_t^{(b)}}{[E_s^{(a)}]^2 - [E_t^{(b)}]^2}, \quad (4)$$

where V is a 2 x 2 matrix of like (diagonal) and unlike (non-diagonal) nucleon-nucleon force constants. The matrix M_s is defined as

$$M_s = \begin{pmatrix} M_{ns} \\ M_{ps} \end{pmatrix} \quad (5)$$

and m is defined similarly but with r^λ and Eq. (1) replaced by $w_\lambda(r)$, the radial function in the separable multipole force.[5] If, in Eq. (1), r^λ is replaced by $\delta(r-r')/rr'$, Eq. (4) becomes an equation for the transition densities ρ_s with their contribution from the model space $M_s^{(a)} \to \rho_s^{(a)}$ and from the $2\hbar\omega$ core-polarization space $M_t^{(b)} \to \rho_t^{(b)}$. The rest of the factor in the second term of Eq. (4) is a numerical coefficient, common for matrix elements of all operators of multipolarity λ. The presence of the $[E_s^{(a)}]^2$ and $[E_t^{(b)}]^2$, coming from the energy-squared eigenvalues of the RPA equations,[5] takes into account ground-state correlations.

If we take $w_\lambda(r) = r^\lambda$, then $m = M$, and Eq. (4) may be factored to give the result,

$$M_s = \left\{ 1 + 2 \sum_t \frac{E_t^{(b)} M_t^{(b)} M_t^{(b)\dagger} V}{[E_s^{(a)}]^2 - [E_t^{(b)}]^2} \right\} M_s^{(a)} \equiv (1 + \delta) M_s^{(a)}, \quad (6)$$

where δ is defined to be the core-polarization matrix.

$$\delta = \begin{pmatrix} \delta^{nn} & \delta^{np} \\ \delta^{pn} & \delta^{pp} \end{pmatrix} \quad (7)$$

In Eq. (7), δ^{np} is the parameter describing polarization of core neutrons by valence protons, etc. The schematic model[1,4] consists of treatment of the b-space configurations as degenerate.

Equation (6) may be transformed to the isoscalar notation, in which the core-polarization matrix δ becomes nearly diagonal. The unequal off-diagonal matrix elements cause $\delta^{np} \neq \delta^{pn}$, $\delta^{nn} \neq \delta^{pp}$, which is important in the systematics of core polarization.

REVIEW OF RESULTS FOR FIRST 2^+ STATES OF SINGLE-CLOSED-SHELL NUCLEI

In Fig. 1 is shown a collection[6] of data for the ratios M_n/M_p for the first excited 2^+ states of SCS nuclei for various pairs of probes. The symbol α/EM means, for example, that measurements of (α,α') and electromagnetic excitation of the nucleus were used together for the determination of the multipole matrix element-ratio. The mirror method[7] consists of determination of M_n in a given nucleus with $N > Z$ by measuring M_p in an isobaric analog of the 2^+ state transition in question and making corrections, where necessary, for isospin impurities. In view of the extreme ratios M_n/M_p which would result from a calculation in a limited shell-model basis (∞ or 0 for neutron or proton valence nuclei, respectively), the results are remarkably close to the homogeneous collective model value of N/Z.

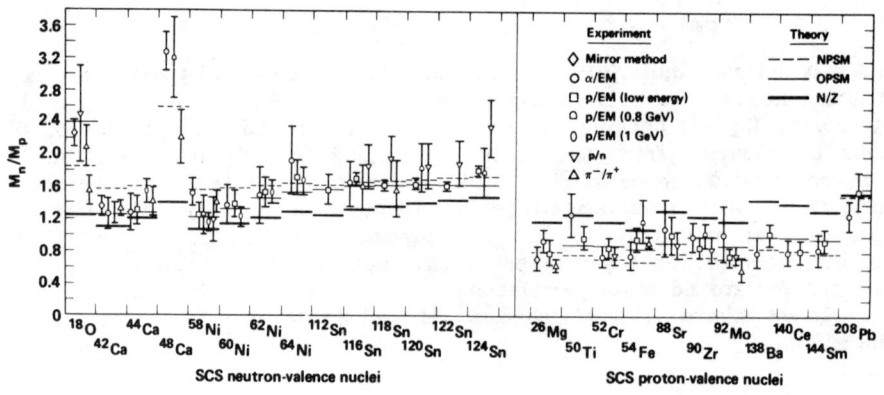

FIG. 1. Ratios of neutron to proton multipole matrix elements of single-closed-shell nuclei.

For the purposes of the neutron conference it is important to note that n/p was used in establishing M_n/M_p ratios in ^{18}O, isotopes of Ca, ^{58}Ni, ^{88}Sr, ^{90}Zr, and the isotopes of Sn.

Also shown in Fig. 1 are the results of the schematic model described in the previous section. The dashed and solid lines are the no (free) parameter and one-parameter versions of the model.[1] In the simple case of SCS nuclei, in which either $M_n^{(a)}$ or $M_p^{(a)}$ is zero, the ratio is a function only of the core-polarization parameters:

$$\frac{M_n}{M_p} = \frac{1 + \delta^{nn}}{\delta^{pn}} \quad \text{(neutron valence)}, \quad (8a)$$

$$\frac{M_n}{M_p} = \frac{\delta^{np}}{1 + \delta^{pp}} \quad \text{(proton valence)}. \quad (8b)$$

It is clear from Fig. 1 that the schematic model describes correctly the trend of the data. The data imply that $(1 + \delta^{nn})/\delta^{pn}$ is a little greater than N/Z and that $\delta^{np}/(1 + \delta^{pp})$ is a little less than N/Z. However, the closeness of the ratios M_n/M_p to N/Z tell us also that $\delta^{nn} < \delta^{pn}$, $\delta^{pp} < \delta^{np}$, and that $[(1 + \delta^{nn})/(1 + \delta^{pp})][\delta^{np}/\delta^{pn}] \approx N^2/Z^2$. In words the first two of these relationships means that nucleons polarize unlike nucleons more than like nucleons, and the third means that neutrons give larger core polarization contributions than protons. Both of these results are easy to understand. The first follows from the greater interaction between unlike than like nucleons, and the second is due to the fact of a neutron excess.

Figure 2 shows the trend of probe dependences[8] of the nuclear deformation parameters resulting from the departure of M_n/M_p from N/Z for three examples of SCS nuclei as a function of the ratio of strengths of interaction of the probe F with nuclear neutrons to that of nuclear protons. There is evidently a rising trend in $\delta = \beta R$ with increasing ratio b_n^F/b_p^F for the two neutron valence nuclei ^{116}Sn and ^{58}Ni and a falling trend for the proton valence nucleus ^{90}Zr. The schematic model calculations have been made using the relationship

$$\beta_F = \frac{N b_n^F \beta_n + Z b_p^F \beta_p}{N b_n^F + Z b_p^F}. \quad (9)$$

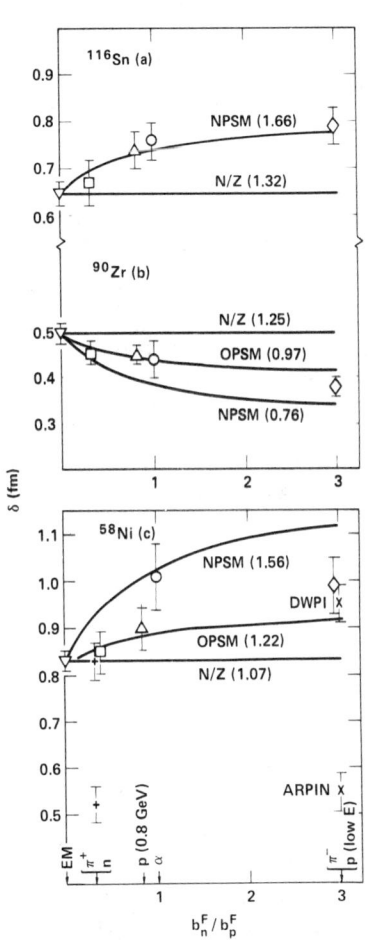

FIG. 2. Dependence of deformation parameter $\delta \equiv \beta R$ on external field (probe).

The most sensitive pairs of probes for determining M_n/M_p are those with the most extreme differences in neutron and proton strengths. From the abscissa of Fig. 2 we see that these are p/EM, n/p, and π^-/π^+.

FIRST 2+ STATES IN OPEN-SHELL NUCLEI

Equation (6) applies also to open-shell nuclei, but one requires then a model for the valence neutron and proton matrix elements $M_n^{(a)}$ and $M_p^{(a)}$. One would normally use shell-model calculations for this purpose. However, we have taken advantage of the empirical fact that the major shell closures dominate the systematics of nuclear collectivity. At or near magic numbers deformation parameters are small, and away from them they are large. We have used[9] schematic model in the $0\hbar\omega$ model space for calculating $M_n^{(a)}/M_p^{(a)}$. The use of a pairing force and the resulting pairing gap makes the rough assumption of degeneracy of the single-particle states reasonable though not accurate. In the degenerate model, the pairing equations can be solved in closed form, as can the RPA equations for degenerate two quasi-particle states. Parameters from our earlier work[1] were used in Eq. (6) for the core polarization matrix δ.

Figure 3 shows the results of these simple calculations for the trends of β_n/β_p for several sets of isotopes and isotones along with values determined from 50 MeV P/EM by Matoba[10] and a few other measurements.[9] Several features should be noted. First, as in Fig. 1, SCS nuclei such as the Sn isotopes, the N=50 isotones, and the N=82 isotones have ratios β_n/β_p rather different from 1. Second, there is a sudden jump in the direction of unity as the number of neutrons or protons is changed from a magic number by even a single pair of nucleons. This effect is seen in the isotopes of Mo which include in ^{92}Mo a SCS member and in the comparison of isotopes of $_{48}$Cd and $_{52}$Te with magic number nuclei $_{50}$Sn. Third, the crude schematic model, which ignores all details of shell structure within a major shell, but does take into account the magic numbers within a major shell, agrees with the trends of the data. We note also that away from closed shells ratios of β_n/β_p are near 1, so therefore hard to establish that $\beta_n \neq \beta_p$.

OTHER LOW-LYING 2+ STATES

Besides the first 2+ states of nuclei discussed so far, there are other low-lying states with some collective strength. It is interesting to see whether there are also trends of β differences for these states. In the SCS nucleus ^{90}Zr with N closed at magic number 50, the first 2+ state at 2.19 MeV has[11-15] β_n/β_p = .82 ± .18, so $\beta_n < \beta_p$ as anticipated from the shell structure. At 3.31 MeV there is another much weaker collective state with β_n/β_p = 1.49 ± .40, reversed from the direction of the ratio of β_n and β_p from that of the first 2+ state. This reversal behavior has been seen[11] in several nuclei, ^{18}O, ^{26}Mg, ^{30}Si, ^{42}Ca, ^{54}Fe, and ^{58}Ni. Core polarization, which treats all low-lying nuclear states nearly alike [See Eq. (6)], cannot be solely responsible when this phenomenon occurs in SCS nuclei. The reversal states must be due to some other configurations from those dominating the first 2+ state. In an SCS nucleus, such as ^{18}O, proton configurations in the low-energy range must come from more complicated configurations than $(2d)^2$, $(2d,2s)$ such as two valence particles plus two-particle two-hole configurations, known to be important for the phenomenon of nuclear coexistence.

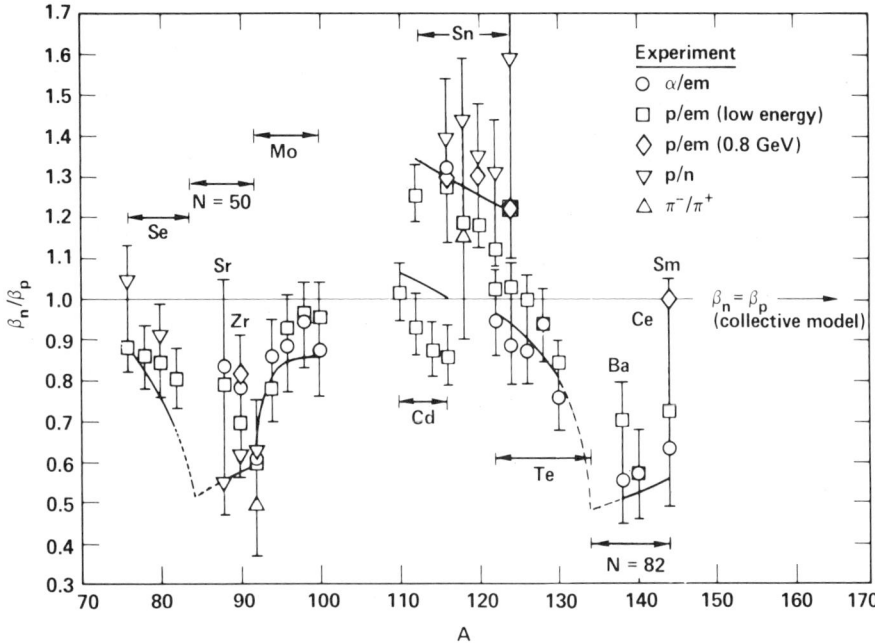

FIG. 3. Ratios of neutron and proton deformation parameters for open-shell nuclei. Lines are results of a $0\hbar\omega$ schematic model plus core polarization. Dashed lines reach to unmeasured isotopes.

The question of the sign of M_n/M_p has been discussed by Miskeman et al.[16] If $|M_n| \approx |M_p|$ the state is nearly isoscalar or nearly isovector in character. In a nucleus with only one neutron configuration and one proton configuration, the interaction would mix the two, giving one state of $M_n/M_p > 0$ and one with $M_n/M_p < 0$. The former would be pushed down in energy by the interaction and the latter would be pushed up. Furthermore, the sense of the inequality of $|M_n/M_p|$ would be reversed. If there are several neutron and several proton configurations, the interaction will likewise mix them, producing some isoscalar-like states, which will be pushed down in energy, and some isovector like states, which will be pushed up. Under these circumstances there will be a tendency for the lower lying of these states to be isoscalar like, therefore having $M_n/M_p > 0$.

For determination of the relative sign, measurements with three different probes are needed. Some interesting work from Ohio university on 2^+ states in ^{34}S will be discussed in a contributed paper in this session by Alarcon.[17] There is a strong indication from this[17,18] and other work that the measurements of M_n/M_p, including the sign, will be of considerable importance in discovering the structure of the low-lying 2^+ states in even nuclei.

GIANT RESONANCE STATES

As we have seen the systematics of M_n and M_p of first 2^+ states, the primary determiner of M_n/M_p is the nature of the shell structure in the N and Z region of the nucleus in question. The results of the previous section showed that SCS nuclei are special in the extent of the departure of M_n/M_p from the collective model ratio of N/Z. On the other hand, shell structure should be nearly irrelevant for giant states because they are inherently shell-breaking transitions.

As an example consider the isotopes of Sn. The important low-lying transitions come from neutrons within the primarily positive parity states of the N=4 harmonic oscillator, which implies that for the first 2^+ state $M_n/M_p >$ N/Z. For the giant resonances the dominant configurations involve protons in the N=2 and 3 major shells excited by $2\hbar\omega$ into the N=4 and 5 shells and neutrons in the N=3, 4, and 5 shells also excited by $2\hbar\omega$ into the N=5,6, and 7 shells. The shell structure should be practically irrelevant for the question of neutron and proton multipole matrix elements. What effect remains from shell structure is, in fact, in the opposite direction from that of the low-lying states. In Sn greater neutron strength in the low-lying states implies simply on the basis of the sum rule that the neutron strength in the giant resonance will be decreased. This is a weak effect (typically ~10%) however, because of the large disparity in the energies of the $0\hbar\omega$ and $2\hbar\omega$ transitions. Another way of saying the same physics is that the core polarization of giant 2^+ states due to low-lying 2^+ states is small and in the opposite direction from the core polarization of the latter by the former. This effect can be seen in Eq. (6) with the reversed identification of b with the $0\hbar\omega$ space and a with the $2\hbar\omega$ space.

Although the schematic model[1,4] is normally used for calculating M_n and M_p for low-lying states, its results are also relevant for giant resonances because M^b used by Eq. (6) is the 2 x 1 matrix of neutron and proton multipole matrix elements for the giant resonance states (without the small coupling to the low-lying states). In a neutron-excess nucleus the neutrons have a larger number of $2\hbar\omega$ particle-hole pairs than do the protons. However the isoscalar interaction mixes the neutron and proton configurations and tends to give the protons a larger weighting, reducing M_n/M_p and producing a nearly isoscalar vibration.

The schematic model[1] with pure harmonic-oscillator wave functions, a separable quadrupole interaction, degenerate $2\hbar\omega$ particle-hole states, and the constraint of equal RMS radii $r_n = r_p$ gives the ratio

$$M_n/M_p \approx (1 + \frac{2}{3}\frac{V_0}{V_0 - V_\tau}\frac{N-Z}{A}) / (1 - \frac{2}{3}\frac{V_0}{V_0 - V_\tau}\frac{N-Z}{A}) \qquad (10)$$

for an assumed relation of isoscalar to $\tau\bullet\tau$ interaction strength $V_0 = -2V_\tau$. This ratio is $M_n/M_p = 1.15$ for ^{118}Sn compared to N/Z = 1.36. If we allow the $r_n = r_p + .2$ fm, the calculated value becomes $M_n/M_p = 1.26$, still somewhat lower than N/Z. These results are well corroborated by our full RPA calculations in a harmonic-oscillator basis, also using separable quadrupole interactions. The latter calculations for ^{208}Pb with equal neutron and proton rms radii, experimental single-particle energies, and a single choice of quadrupole interaction strength give a satisfactory description of the energy, the B(E2) value, and M_n/M_p of the first 2^+ state and of the isoscalar giant resonance energy (although slightly low). They are also in good agreement with an RPA calculation[19] using the Migdal force and a discritized continuum, except that M_p in our calculations is about 10% larger than Ref. 19. This may indicate a suppression of proton strength when Coulomb forces are included.

Experimentally, M_n/M_p for giant resonances may be measured with two different probes just as for low-lying states. Excitation by inelastic scattering of both π^+ and π^- projectiles gives, in principle, a fairly sensitive measurement of M_n/M_p because of 3:1 ratio of the interaction of $V_{\pi^-,n}/V_{\pi^-,p} = V_{\pi^+,p}/V_{\pi^+,n}$. Recent experiments[20,21] have given surprisingly large ratios of M_n/M_p. Ignoring differences in the scattering between π^- and π^+ and neglecting terms of the order $(M_n/M_p)^2$, we have

$$\frac{\sigma_{\pi^-}}{\sigma_{\pi^+}} \approx \left| \frac{3M_n + M_p}{M_n + 3M_p} \right|^2 \approx \frac{M_n}{M_p} . \qquad (11)$$

For ^{118}Sn the measured ratio[20] is 1.9 and for ^{208}Pb it is[21] 2.8, for which Eq. (11) would give $M_n/M_p \approx 1.9$ and 2.8 respectively, much larger than the schematic-model predictions.

A recent theoretical calculation[22] does self consistent RPA using Skyrme III forces, based on Hartree-Fock neutron and proton single-particle orbitals including the continuum. Calculated and experimental results are shown in Table I. Several important points are to be noted in the table: First M_n/M_p is somewhat larger than the schematic-model results for Sn and generally is not proportional to N/Z. The ratio of cross sections $R = \sigma_{\pi^-}/\sigma_{\pi^+}$ is not proportional to M_n/M_p, showing that distortion effects make a difference for π^- and π^+ projectiles. The large ratios R_{exp} do agree within errors with the RPA calcuations for Ca and Sn but disagree for Pb. The RPA calculations treated ^{120}Sn as a double-closed-shell nucleus and so did not find a low-lying 2^+ state. Its inclusion and the correction for the fact that the measurement was done in ^{118}Sn would reduce the calculated value by 8-10%, reducing R_{calc} to ~ 1.5, barely on the limits of error compared with the experimental ratio. A ratio M_n/M_p would also be close to a schematic value result given above.

It should also be mentioned that the pion scattering experiment, while indicating quite a large ratio M_n/M_p for the giant resonance in Sn, gives a small value $\leq$ N/Z $\approx$ 1.36 for the lowest-lying 2^+ state. The systematics[6] of multipole moment ratios for Sn give M_n/M_p ~ 1.6 to 1.8, larger than N/Z because of the shell closure on protons.

All of these comparisons make it difficult to understand why the pion scattering experiments are giving such large values of M_n/M_p for giant resonances. It would be good to have corroborating experiments. At low energy (p,p') and (n,n') used together are about equally as sensitive as pion scattering. If we consider energies of 135 to 300 MeV, at which giant resonances could readily be excited, the V_τ/V_0 ratio has dropped to so low a value that n/p is rather insensitive to neutron-proton differences. At 60-70 MeV, however, there is still considerable sensitivity, so (n,n') and (p,p') scattering experiments on giant resonances at this energy would be valuable in establishing systematics of M_n/M_p values for giant resonances. Interpretation of such experiments would not be trivial. Background subtraction problems, which have sometimes given misleading information for giant resonances, would have somehow to be solved.

TABLE I. Results from Ref. 22: Calculated neutron and proton multipole matrix elements and comparison of calculated cross section ratios with experiment.

Nucleus	M_n/M_p	N/Z	R^a_{calc}	R_{exp}
^{40}Ca	.91	1.00	1.19	1.05±.35
^{48}Ca	1.40	1.40	1.29	- - -
^{90}Zr	1.10	1.25	1.22	- - -
^{120}Sn[b]	1.45	1.40	1.61	1.9±.4
^{208}Pb	1.44	1.54	1.70	2.8±.8
^{208}Pb[c]	1.55			

[a] $R = \sigma_{\pi^-}/\sigma_{\pi^+}$
[b] measurement for ^{118}Sn
[c] Ref. 19.

SUMMARY

We have seen that for first 2⁺ states of even nuclei there are systematic departures of multipole matrix element ratios M_n/M_p from N/Z in SCS nuclei. The nearness of M_n/M_p to N/Z is explained by the greater polarization of core nucleons of the opposite type than of the same type by valence nucleons. For open-shell nuclei M_n/M_p may also depart from N/Z but the differences tend to be much smaller than for SCS nuclei. All of these trends are explainable in terms of simple schematic models.

For higher 2⁺ states there is also collective strength found, and it typically has a reversal of the sense of ratios of M_n/M_p from those of the ground states. This tendency and the establishment of the sign of M_n/M_p for higher 2⁺ states promises to become an important experimental tool for investigating the nuclear structure of low-lying states.

For giant states, shell effects should play a very minor role. Schematic model values for M_n/M_p are lower than N/Z, because of the strong coupling between neutron and proton configurations. Self-consistent RPA calculations give somewhat higher ratios, indicating possible importance of the differences between the neutron and proton orbits in the continuum. Inelastic pion experiments on ^{118}Sb and ^{208}Pb give ratios which are even larger and difficult to understand on the basis of theory. Corroborating experiments are needed.

REFERENCES

1. V. R. Brown and V. A. Madsen, **Phys. Rev.** C, 11, 1298 (1975).

2. V. A. Madsen, V. R. Brown, and J. D. Anderson, **Phys. Rev.** C, 12, 1205 (1975).

3. A. Bohr and B. R. Mottelson, K. Dan. Vidensk Selsk., Mat.-Fys. Medd. 27, No. 16 (1953).

4. V. R. Brown and V. A. Madsen, to be published.

5. S. Yoshida, Nucl. Phys. 38, 380 (1962).

6. A. M. Bernstein, V. R. Brown, and V. A. Madsen, Comments on Nuclear and Particle Physics 11 No. 5, 203 (1983); Phys. Rev. Lett, B103, 255 (1981), and references therein.

7. A. M. Bernstein, V. R. Brown, and V. A. Madsen, Phys. Rev. Lett. 42, 425 (1979).

8. A. M. Bernstein, V. R. Brown, and V. A. Madsen, Phys. Lett. B103, 255 (1981).

9. V. A. Madsen and V. R. Brown, Phys. Rev. Lett. 52, 176 (1984) and references therein.

10. M. Matoba, Phys. Lett. 88B, 249 (1979).

11. A. M. Bernstein, V. R. Brown, and V. A. Madsen, and references therein, to be published.

12. L. T. Van der Bijl, H. P. Blok, I. F. A. Van Hienen, and J. Blok, Nucl. Phys. A393, 173 (1983).

13. F. Todd Baker, et al., Nucl. Phys. A393, 283 (1983).

14. M. M. Gazzaly, N. M. Hintz, G. S. Kyle, R. K. Owen, G. W. Hoffmann, M. Barlett, and G. Blanpied, Phys. Rev. C. 25, 408 (1982).

15. J. Heisenberg, private communication, December 1983.

16. R. A. Miskeman, A. M. Bernstein, B. Quinn, S. A. Wood, M. V. Hynes, G. S. Blanpied, B. G. Ritchie, and V. R. Brown, Phys. Lett. 131B, 26 (1983).

17. R. Alarcon, Neutron Nucleus Collisions - - Probe of Nuclear Structure, Burr Oak, Glouster, Ohio, September 1984 (this conference), to be published.

18. A. Brown, Neutron Nucleus Collisions - - Probe of Nuclear Structure, Burr Oak, Glouster, Ohio, September 1984 (this conference), to be published.

19. J. Wambach, F. Osterfeld, J. Speth, and V. A. Madsen, Nucl. Phys. A324, 77 (1979).

20. J. L. Ullman, J. J. Kraushaar, T. G. Masterson, R. J. Peterson, R. S. Raymond, R. A. Ristinen, N.S.P.King, R. L. Boudrie, C. L. Morris, W. W. True, R. E. Anderson, and E. R. Siciliano, Phys. Rev. Lett, 51, 1038 (1983), and to be published.

21. C. L. Morris, S. J. Seestrom-Morris, and L. C. Bland, Proc. **HESANS 1983 Symp.** (Orsay, France, 1983), to be published.

22. N. Auerbach, Amir Klein, and E. R. Siciliano, submitted to **Phys. Rev. Lett.**.

NEW ASPECTS OF NUCLEAR STRUCTURE FROM THE VIEWPOINT OF NEUTRON INDUCED REACTIONS

B. Alex Brown
Cyclotron Laboratory,
Michigan State University,
E. Lansing MI 48824

ABSTRACT

Elastic and inelastic scattering of neutrons from nuclei can be used to study unique aspects about the nuclear structure of the ground states and transitions to excited states. The strengths of (n,p) Gamow-Teller type transitions in nuclei with N>Z are particularly sensitive to correlations in the ground-state wave functions. The relationship between (n,p) and (p,n) Gamow-Teller reactions and their relevance to the problem of double beta decay will be discussed. Combined analysis of (n,n') and (p,p') reactions to the low-lying 2^+ states can be used to separate the isoscalar and isovector components of the transitions. Recent experimental results for nuclei in the sd shell will be compared to the latest theoretical calculations.

I Introduction

The primary goal in our work is to test theoretical models of nuclear structure which can be used to quantitatively understand the experimental results related to the low-lying nuclear states. The experimental information is often difficult to obtain with the needed precision and sometimes the interpretation relies on some other assumptions such as the reaction theory in the case of hadronic-induced reactions. For this reason, experiments related to the electromagnetic and weak interaction with the nucleus often provide the most sensitive test of the nuclear structure. However, the electromagnetic and weak interaction often provide an incomplete picture, and hadronic-induced reactions are important for providing more complete information. Some examples: The coulomb form factors in electron scattering and the B(EL) electromagnetic transition rates only provide information on the proton wave functions; elastic and inelastic hadron scattering is needed to provide the complementary information on the neutron components. The number of states populated in Gamow-Teller beta decay are limited by the Q-value of the decay, but the entire spectrum of final states can be studied in (p,n) and (n,p) reactions at medium energies (200 MeV) where the nucleon-nucleus potential turns out to be dominated by the $\sigma\tau$ term.

The secondary goal is to use the best available theoretical models to calculate the nuclear matrix elements which are needed for the interpretation of nuclear experiments directed at determining some "fundamental" properties of the elementary particles and their interactions. This goal is secondary only in the sense that its success depends to some extent upon the successful completion of the first goal. Experiments in this category include those that look for effects due to charge asymmetry and parity violation in the nuclear interaction and those such as double-beta decay which depend upon the fundamental properties of the neutrino.

In this paper we will examine these goals with respect to some new aspects of nuclear structure which can be studied with neutron induced reations. We do not attempt a comprehensive comparison with experiment, indeed, very little data exists. Rather, we will concentrate on the theoretical problems with the goal of exciting some interest in doing new experments. It is clear that the (p,n) reaction is a good probe for studying particle-hole states in general and a unique probe at medium energy (200 MeV) and forward angles for studying the Gamow-Teller "giant" resonance. In Sec. II we will discuss the nuclear structure sum rules associated with the (n,p) Gamow-Teller transitions. In Sec. III we will discuss the nuclear matrix elements for double-beta decay in general and their relationship to sums over products of the form (p,n)x(n,p) to intermediate Gamow-Teller type states. Finally, in Sec. IV we consider some theoretical calculations for the proton and neutron components of the transitions to low-lying 2^+ states in some sd shell nuclei and how they compare with experiment including recent (n,n') data (Ref 1, Ref 2).

II. (p,n) and (n,p) Gamow-Teller Sum Rules

In this section we will discuss the results of shell-model calculations for the summed Gamow-Teller strength which should be seen in (p,n) and (n,p) reactions. We start by considering the triply reduced (reduced in both spin and isospin space) matrix element of the $\sigma\tau$ operator

$$M(GT) = \langle J_f, T_f ||| O(GT) ||| J_i, T_i \rangle$$

where $O(GT) = \sum_k \sigma_k \tau_k$ \hfill (1)

{Our reduced matrix element convention is that used in the books of Edmonds (Ref 3) and de-Shalit and Talmi (Ref 4). In our equations the notation |J,T> is a shorthand for all of the quantum numbers associated with the many-particle states and |j> is a shorthand for all of the quantum numbers associated with the spherical single-particle states.}

The reduced transition probabilities B(GT) used in the analysis of Gamow-Teller beta decay and (p,n) reactions is obtained from the matrix element M(GT) by

$$B(GT) = [\ 3J(T_f,1,T_i,-T_{zf},T_{zf}-T_{zi},T_{zi})\ M(GT)\ ]^2 / 2(2J_i+1) \quad (2)$$

where $3J(a,b,c,d,e,f)$ is the three-j symbol with arguments a-b-c in the first row and d-e-f in the second row.

Let us denote the Gamow-Teller strength observed in β^- decay or (p,n) reactions by B(GT-) and the strength observed in β^+ decay or (n,p) reactions by B(GT+). It is well known that these strengths obey the sum rule (Ref 5)

$$S(GT) = B(GT-) - B(GT+) = 3(N-Z) \quad (3)$$

For GT- reactions on nuclei with N>Z the isospin of the final states can have $T_f = T_i-1$, T_i and T_i+1 whose strengths will be denoted by B(GT-<), B(GT-0) and B(GT->), respectively. The GT+ reactions of course can only go to the $T_f=T_i+1$ states with strength B(GT+>). From Eq. (2) it is easy to obtain the result

$$B(GT+>) = (T_i+1)(2T_i+1)\ B(GT->) \quad (4)$$

and the sum rule can be expressed entirely in terms of B(GT-)

$$S(GT) = B(GT-<) + B(GT-0) - T_i(2T_i+3)\ B(GT->) \quad (5)$$

In (p,n) experiments the $T_>$ state is usually difficult to see because it is weak and lies high in excitation energy upon a large background of T_0 and $T_<$ states. It has been possibly observed in a few experiments such as ^{26}Mg(p,n) (Ref 6) but the large background prevents an accurate extraction of the strength.

This $T_>$ state may be important in the sum rule because of the large weight is has in Eq. (5). (For example, for ^{208}Pb, $T_i(2T_i+3)=1034$.) In addition, as we will discuss below, the value of B(GT>) itself provides important information about the ground-state correlations in the initial wave function. Presumably B(GT+>) can be more easily obtained in (n,p) reactions where the strength is stronger due to the isospin factors in Eq. (4) and where the background from T_0 and $T_<$ states does not exist.

Here we will discuss the results of calculations for the isotones with 14 neutrons in the sd shell (i.e. ^{22}O - ^{28}Si). We do not propose that experiments actually be carried out on these nuclei. But rather the calculations are meant to illustrate the differences between truncated shell-model calculations and the full 0hw space calculations which are possible for these nuclei. The results may be applicable by analogy to heavier nuclei, in particular to the isotones with 28 neutrons in the fp shell.

In Table 1 calculations carried out in the full 0hw sd model space are compared to those in a restricted space. We use the "USD" residual interaction of Wildenthal (Ref 7). The restrictions are to limit the model space to $(d_{5/2})^n$ for the initial state and then allow both $(d_{5/2})^n$ and $(d_{5/2})^{n-1}(d_{3/2},s_{1/2})^1$ for the final states. For example, in this resticted space the ^{28}Si ground state has a cloed-shell configuration and the excited states are just one-particle one-hole. These restrictions are similar to those that are required in order to carry out shell-model calculations for heavy nuclei (A>40) and are the minimal required in order to fulfill the sum rule of Eq. (3).

The important point is that the quenching factor Q(GT) is strongly dependent upon the isospin of the final state with the $T_>$ states being most quenched, by as much as a factor of six (for the T=3 state in ^{24}Ne). The study of the $T_>$ strength by (n,p) reactions can definitely provide a sensitive test for any 0hw shell-model calculation and may even shed some light on the role of higher-order configuration mixing (beyond 0hw) (Ref 8, Ref 9) which is responsible for at least some of the observed quenching of the (p,n) Gamow-Teller strength (Ref 10, Ref 11).

Table 1: full space and restricted space calculations for the summed B(GT) strength for the N=14 isotones in the sd shell. Q(GT) is the ratio of B(GT) for the full over the restricted calculation.

Nucleus	Z	Initial 2T	$2T_z$	Final 2T	$2T_z$		B(GT) full	restricted	Q(GT)
22O	8	6	-6	4	-4	B(GT-<)	17.13	16.40	1.04
				6	-4	B(GT-0)	0.87	1.60	0.54
				8	-4	B(GT->)	0	0	
				Sum		B(GT-)	18.00	18.00	1.00
24Ne	10	4	-4	2	-2	B(GT-<)	11.02	12.32	0.90
				4	-2	B(GT-0)	1.46	2.67	0.55
				6	-2	B(GT-<)	0.034	0.213	0.16
				Sum		B(GT-)	12.51	15.20	0.82
26Mg	12	2	-2	0	0	B(GT-<)	4.89	6.53	0.75
				2	0	B(GT-0)	2.60	4.80	0.54
				4	0	B(GT->)	0.30	1.07	0.28
				Sum		B(GT-)	7.78	12.40	0.63
28Si	14	0	0	2	2	B(GT->)	3.89	9.60	0.40

III.A Relationship Between Double-Beta Decay and Gamow-Teller Matrix Elements

The study of double-beta decay in nuclei is important for setting limits on the Majorana mass of the neutrino (Ref 12). In the closure approximation the double-beta decay operator can be considered simply as a product of two Fermi decay operators or two Gamow-Teller decay operators (Ref 13). The Fermi-type operator can only connect the initial state to its double analogue, for example, the T=4 state of ^{48}Ti in the case of the ^{48}Ca double-beta decay, and this transition is energetically forbidden. The Fermi-type operator can connect to the ground state of ^{48}Ti only by isospin mixing and is presumably relatively small compared to the isospin allowed Gamow-Teller mode (however, this remains to be quantitatively checked). Here we will concentrate on the the Gamow-Teller type operator and see how the nuclear matrix elements can be related to sums over intermediate states.

We want to evaluate the triply-reduced matrix element for double-beta decay (DBD)

$$M(DBD) = \langle J_i,T_i ||| O(DBD) |||J_f,T_f\rangle \quad (6)$$

where the operator is given by a cross product of two Gamow-Teller operators coupled to total spin P=0-2 and total isospin Q=2

$$O(DBD) = (1/2) [O(GT) \times O(GT)]^{(P,Q)}$$

and $O(GT)$ is the Gamow-Teller operator given by Eq. (1) which has spin and isospin rank $(p,q) = (1,1)$. {For P=0 the matrix element M(DBD) is related to the matrix element M_{GT} used in Ref 13 and Ref 14 by M_{GT} = $-M(DBD) \times [3/(2J_i+1)]^{1/2} \times 3J(T_f,2,T_i,-T_{zf},T_{zf}-T_{zi},T_{zi})/2$.} The examples given below will be for $J^\pi = 0^+$ to 0^+ ground state to ground state transitions with P=0, however, the formulae can also be applied to 0^+ to 2^+ transitions with P=2.

The connection between M(DBD) and the individual Gamow-Teller transitions is straightforwardly made by inserting a complete set of intermediate states between the two Gamow-Teller operators. (In fact this is just the reverse of the procedure used in (Ref 13) and leads back to the original expression which contains an energy denominator which depends on the energy of the intermediate states. In the closure approximation one replaces this energy denominator with an average value in order to arrive at Eq. (6). In this work we focus on the nuclear structure aspects of the matrix element and for the moment ignore the dependence on the energy denominator.)

The result in terms of a sum over the intermediate states with spin J_c and isospin T_c can be obtained using Eq. A.3.d13 of Ref 15

$$M(DBD) = (1/2) (-1)^{J_f+P+J_i+T_f+Q+T_i} [(2P+1)(2Q+1)]^{1/2}$$

$$\times \sum_{J_c,T_c} SJ(J_i,J_f,P,p,p,J_c) \; SJ(T_i,T_f,Q,q,q,T_c)$$

$$\times M(f,c) M(i,c) \quad (7)$$

where $M(f,c) = \langle J_f,T_f ||| O(GT) |||J_c,T_c\rangle$

and $M(i,c) = \langle J_i,T_i ||| O(GT) |||J_c,T_c\rangle$

$SJ(a,b,c,d,e,f)$ is the six-j symbol with arguments a-b-c in the first row and d-e-f in the second row. Of course, for 0^+ to 0^+ DBD in nuclei with a neutron excess, the intermediate states must have $J_c = 1$ and $T_c = T_i-1$. As a simple example we will calculate M(DBD) for the ^{48}Ca to ^{48}Ti decay using pure the $f_{7/2}$ shell wave functions obtained with the "42SC" interaction of Ref 16. In this model space there is only one intermediate 1^+ state in the nucleus ^{48}Sc to consider. The first two lines of Eq. (7) are just 0.1091 and the calculated values of the two reduced matrix elements in the last line are -0.718 and -13.61, respectively. Thus, we obtain M(DBD) = 1.07.

{With the "MBZ" interaction of Ref 17 we obtain M(DBD) = 0.62 in agreement with the results of Zamick and Auerbach (Ref 14) and Khodel (Ref 18) - see the comment in the {} bracket above concerning the relation between M(DBD) and M_{GT}. The difference between 1.07 and 0.62 is some indication of the sensitivity of the result to the $f_{7/2}$ shell interaction. Below we go beyond the work of Zamick and Auerbach by expanding the model space to include all fp shell orbits.}

Is this a small or a large number? To set the scale we can consider the sum to all final states in ^{48}Ti. In the $f_{7/2}$ shell model there are only three T=2 states and one T=4 state to consider. The reduced matrix elements from these four 0^+ states to the 1^+ state in ^{48}Sc in order of excitation energy are -0.718, 1.87, 7.29 and -13.61. The sum of M^2(DBD) to all states in ^{48}Ti is 534. Thus, M^2(DBD) to the ^{48}Ti ground state is relatively very small and exhausts only 0.2% of the sum.

In the remainder of this section we will discuss how to obtain some deeper understanding of the smallness of this DBD matrix element (Sec III.B) and will investigate the effects of using a larger model space (Sec III.C).

III.B Relationship Between Double-Beta Decay and
     Two-Particle Parentage Amplitudes

M(DBD) can be evaluated with shell-model codes such as OXBASH (Ref 19) by expanding the matrix element in terms of a sum of products of two-body transition densities (TBTD) times two-body matrix elements (TBME) (Ref 15 and Ref 20)

$$M(DBD) = \sum_{\substack{j_1,j_2,j_3,j_4 \\ J_{12},J_{34},T_{12},T_{34}}} TBTD \times TBME \qquad (8)$$

where $TBTD = [(2P+1)(2Q+1)]^{-1/2}$

$\times \langle J_f, T_f ||| [A^+(J_{12},T_{34}) \times \tilde{A}(J_{34},T_{34})]^{P,Q} ||| J_i, T_i \rangle$

and TBME = $\langle j_1,j_2,J_{12},T_{12} ||| O(DBD) ||| j_3,j_4,J_{34},T_{34}\rangle$.

A^+ and $A^\sim$ are the two particle creation and annihilation operators

$$A^+(J_{12},T_{12}) = -[a^+(j_1) \times a^+(j_2)]^{J_{12},T_{12}}/[1+\delta_{j_1,j_2}]^{1/2}$$

$$A^\sim(J_{34},T_{34}) = +[a^\sim(j_3) \times a^\sim(j_4)]^{J_{34},T_{34}}/[1+\delta_{j_3,j_4}]^{1/2}.$$

The (antisymmetric) two-body matrix element can be easily evaluated in terms of nine-j symbols and the single-particle matrix elements of O(GT) using expression (A.3.d4) in Ref 15. Since $Q=2$, $T_{12}=T_{34}=1$.

As a numerical example we consider again the ^{48}Ca to ^{48}Ti decay within the $f_{7/2}$ shell-model space. The TBTD calculated with the "42SC" wave functions of Ref 16 are 1.246, -0.497, -0.033 and -0.010 for $J_{12}=J_{34}=0$, 2, 4 and 6, respectively, and the TBME for $j_1=j_2=j_3=j_4=f_{7/2}$ turn out to be 3.32, 6.01, 3.64 and -3.99, respectively. The sum of products gives M(DBD) = 1.07 which agrees with the result obtained in Sec. III.A. Several things are apparent from this exercise. First, we note that the products for $J_{12}=0$ and 2 dominate. Secondly, these two terms tend to cancel due to the change in sign of the TBTD.

Some insight into this change in sign can be obtained by expanding the TBTD in terms of sums over the spins (J_o) and isospins (T_o) of the parent states with (A-2) nucleons. Using again Eq. A.3.d13 of Ref 15 we obtain

$$TBTD = \sum_{J_o,T_o} (-1)^{J_f+P+J_o+J_{12}+T_f+Q+T_o+T_{12}}$$

$$\times SJ(J_i,J_f,P,J_{12},J_{34},J_o) \; SJ(T_i,T_f,Q,T_{12},T_{34},T_o)$$

$$\times [(2J_i+1)(2J_f+1)(2T_i+1)(2T_f+1)]^{1/2}$$

$$\times TNSA(J_f,T_f) \; TNSA(J_i,T_i) \quad (9)$$

The TNSA are the two-nucleon spectroscopic amplitudes defined by

$$TNSA(J_i,T_i) = [(2J_i+1)(2T_i+1)]^{-1/2}$$
$$\times \langle J_i,T_i |||A^+(J_{12},T_{12})|||J_o,T_o\rangle$$

$$TNSA(J_f,T_f) = [(2J_f+1)(2T_f+1)]^{-1/2}$$
$$\times \langle J_f,T_f |||A^+(J_{34},T_{34})|||J_o,T_o\rangle$$

The (A-2) core nucleus in our example is ^{46}Ca and in the $f_{7/2}$ model space there are just three states to consider, those with $(J_O,T_O) = (0,3)$, $(2,3)$, $(4,3)$ and $(6,3)$. We must have $J_O=J_{12}=J_{34}$ (for $J_i=J_f=0$). The constants in the first three lines of Eq. (9) give 1.134 $(2J_O+1)^{-1/2}$. The ^{48}Ca->^{46}Ca TNSA for these four parent states are calculated to be -1, -2.236, -3, and -3.606, respectively, while the ^{48}Ti->^{46}Ca TNSA are -1.098, 0.438, 0.029 and 0.008, respectively. It is now easy to obtain the TBTD($J_{12}=J_O$) given above.

The opposite signs for the TBTD with $J_O=0$ and $J_O=2$ are thus related to the fact that the relative TNSA products leading to the 0^+ and 2^+ states in ^{46}Ca have the opposite sign. The ^{48}Ca TNSA are trivial to understand since, in the $f_{7/2}$ shell model, ^{48}Ca has a closed shell of neutrons, and the two-nucleon spectroscopic factors (TNSA^2) are just equal to the $(2J_O+1)$ sum rule for two-nucleon pickup. The ^{48}Ti TNSA are less easy to "guess" but are related to the fact that the ^{48}Ti wave function has a two-proton particle two-neutron hole configuration. The point is that for this example, the double-beta decay can be directly related to sums over two-nucleon spectroscopic amplitudes whose magnitude can in principle be studied experimentally in two-nucleon transfer reactions.

In the more general case, of course, we must consider the effect of many orbits in the shell-model space. One can still carry out the double-beta decay calculation by summing over the (A-2) core states. But the connection with two-nucleon transfer experiments is not so direct since the "weights" in the sum appropriate for double-beta decay will in general be different than those needed for two-nucleon transfer. Nevertheless, this may be an interesting method to pursue.

As an example of a many-orbit calculation, we will present the results for ^{48}Ca using a model space which includes the $(f_{7/2})^7(f_{5/2},p_{3/2},p_{1/2})^1_8$ configurations as well as the $(f_{7/2})^8$ configurations for all of the states (this will be refered to as the "8-71" model space). The wave functions in this model space were obtained with the interaction of van Hees and Glaudemans (HG) (Ref 21). The non-zero contributions to Eq. (8) are given in Table 2. Note that there are many cancellations and that the total (-0.32) has the opposite sign and is smaller than the value (1.07) obtained in the simple $f_{7/2}$ model.

Table 2: Results of the TBTD expansion [Eq. (8)] for the ^{48}Ca -> ^{48}Ti double-beta decay matrix element calculated in the "8-71" model space with the HG interaction.

2j1	2j2	2j3	2j4	J12	TBTD	TBME	M(DBD)
7	7	7	7	0	1.075	3.32	3.57
7	7	7	7	2	-0.562	6.01	-3.38
7	7	7	7	4	0.096	3.64	0.35
7	7	7	7	6	-0.030	-3.99	0.12
7	7	5	7	2	0.059	-5.77	-0.34
7	7	5	7	4	0.033	-12.00	-0.40
7	7	5	7	6	0.019	-13.03	-0.25
Sum							-0.32

III.C Sensitivity of Double-Beta Decay to the Model Space and the Importance of the (n,p) Reaction

In terms of the sums over the products of Gamow-Teller matrix elements, it is particularly clear that the simple $f_{7/2}$ model is inadequate for the ^{48}Ca double-beta decay calculation. In this simple model there is only one low-lying 1^+ state in ^{48}Sc. We need to allow at least one particle into the $f_{5/2}$ orbit in order to describe the entire Gamow-Teller giant resonance which has been observed in the ^{48}Ca(p,n) reaction (Ref 22). The crucial question is how these states in ^{48}Sc are connected to the ground state of ^{48}Ti. Here we will theoretically investigate this question.

First we will consider the calculation in the 8-71 model space with the HG interaction (see Sec. III.B). In this model-space there are 19 1^+ T=3 states in ^{48}Sc. In Table 3 we give the calculated excitation energy of these states in ^{48}Ti (Ex), the reduced matrix elements needed for the sum in Eq. (7) [M(f) for the ^{48}Ti -> ^{48}Sc transitions and M(i) for the ^{48}Ca -> ^{48}Sc transitions] and the partial contribution to the DBD matrix element M(DBD)

Table 3: Results of the Gamow-Teller intermediate state expansion [Eq. (7)] for the ^{48}Ca -> ^{48}Ti double-beta decay matrix element calculated in the "8-71" model space.

	--- HG interaction ----				- modified HG interaction -			
n	Ex	M(f)	M(i)	M(DBD)	Ex	M(f)	M(i)	M(DBD)
1	11.99	1.28	4.43	0.62	12.25	0.92	5.92	0.59
2	12.70	0.80	-1.04	-0.09	12.61	-0.54	1.89	-0.11
3	13.37	1.64	1.73	0.31	13.37	-0.63	-2.78	0.19
4	13.92	-4.48	-2.61	1.28	14.23	0.39	0.32	0.01
5	14.27	0.59	-1.23	-0.08	14.77	-0.90	-0.94	0.09
6	14.83	0.08	1.50	0.01	15.07	0.72	-0.71	-0.06
7	15.15	0.28	-2.03	-0.06	15.40	0.54	1.65	0.10
8	15.45	-0.04	0.09	0.00	15.77	4.78	3.22	1.68
9	15.64	0.27	-2.25	-0.07	16.72	0.34	-1.65	-0.06
10	16.77	-0.47	5.78	-0.30	16.90	1.00	-3.88	-0.42
11	16.95	0.69	-7.57	-0.57	17.41	-0.66	5.18	-0.37
12	17.17	0.82	-8.20	-0.74	17.77	-0.56	3.84	-0.24
13	17.79	0.29	-4.76	-0.15	18.98	0.92	-9.30	-0.93
14	18.22	-0.44	6.56	-0.32	19.78	-0.59	9.82	-0.63
15	19.12	-0.08	8.98	-0.08	20.87	-0.02	-7.59	0.01
16	19.73	0.09	-4.54	-0.05	21.56	-0.08	3.58	-0.03
17	21.08	-0.02	0.62	0.00	22.94	-0.03	0.96	0.00
18	21.80	-0.05	-3.83	0.02	23.67	-0.05	-3.54	0.02
19	23.59	-0.10	4.22	-0.05	25.45	-0.08	3.75	-0.03
Sum				-0.32				-0.19

The total M(DBD) is -0.32 (in agreement with the result obtained in Sec. III.B). We note that this results from a cancellation between the postive partial products for the lowest four states and the negative partial products for the higher states in the ^{48}Ca(p,n) giant resonance region. To get some idea about the sensitivity of this result to the interaction, we have repeated the calculation with a modified HG interaction obtained by adding an additional 2 MeV onto the single-particle energy of the $f_{5/2}$ orbit. This modified interaction, in fact, improves the agreement between the calculated and experimental B(GT) strengths obtained in the ^{48}Ca(p,n) reaction (Ref 23). The GT matrix elements are given on the right hand side of Table 3 and the B(GT) values for the ^{48}Ca spectrum are compared to experiment in Fig. 1.

Figure 1. The experimental spectrum for the ^{48}Ca reaction (Ref 23) compared to the theoretical results (solid bars) obtained with the modified HG interaction (see the right-hand side of Table 3).

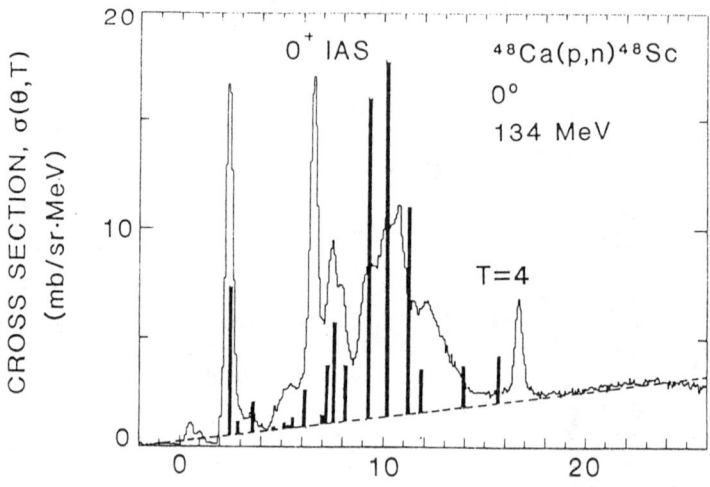

The sum of the DBD products does not change much (-0.19 compared to -0.32), but the strong ^{48}Ti M(f) matrix element has moved from the 4'th state at 13.92 MeV to the 8'th state at 15.77 MeV. It is interesting to note in the calculation that the DBD products are predominantly positive for the 1^+ states which lie below the isobaric analogue of the T=4 0^+ ^{48}Ca ground state and negative for the 1^+ states which lie above. (The calculated energy of the T=4 state in ^{48}Ti is 16.29 MeV with the modified HG interaction, compared to the experimental value of 17.38 MeV (Ref 24). At present we do not have an explanation for this.

The important point is that there is much uncertainty in the double-beta decay calculation related to the cancellations noted above. Experimental information on the ^{48}Ti -> ^{48}Sc Gamow-Teller transitions is very important in order to check the DBD calculations. These could perhaps most easily by studied by the (n,p) reaction.

IV. (n,n') Reactions to Low Lying 2^+ States

Properties of the proton component of the longitudinal transition density between the ground state and excited states are determined by the gamma decay transition probabilities (or coulomb excitation) and inelastic electron scattering between the states. However, in comparing various theoretical models it is also important to have experimental information on the neutron component. The gross relationships between the proton and neutron components are characterized by collective motion in which the protons and neutrons move with equal amplitude in phase (the isoscalar mode) and out of phase (the isovector mode) and in the single-particle models with pure proton or neutron configurations. With the interacting shell model our goal is to describe the collective and single-particle states simultaneously and a great variety of "in between" modes are possible. We will discuss in this section some calculations for E2 transitions in the sd shell nuclei (16<A<40) and how they compare with experiments including recent (n,n') experiments (Ref 1, Ref 2).

We start by calculating the proton and neutron matrix elements A_p and A_n for the sd shell wave functions

$$A_{p/n} = <sd,f|| O(E2)_{p/n} ||sd,i>$$

where $O(E2)_{p/n} = \sum_{\text{protons/neutrons}} r_i^2 Y^{(2)}(r_i)$

For the total proton matrix element it well known that the core-polarization of the core protons by the valence protons and neutrons is important for the E2 operator (Ref 25). In the schematic model (Ref 26 Ref 27) the contribution due to the valence protons is proportional to A_p with a proportionality constant δ_{pp} and the contribution due to the valence neutrons is proportional to A_n with a proportionality constant δ_{pn} and thus M_p is given by

$$M_p = A_p + A_p \delta_{pp} + A_n \delta_{pn}$$

By proton/neutron symmetry the total neutron matrix element M_n is given by

$$M_n = A_n + A_n \delta_{nn} + A_p \delta_{np}$$

Since the nearest closed shells of ^{16}O and ^{40}Ca have equal numbers of core protons and neutrons, it is a good approximation to assume $\delta_{nn} = \delta_{pp}$ and $\delta_{np} = \delta_{pn}$ and write

$$M_p = A_p e_p + A_n e_n$$

$$M_n = A_n e_p + A_p e_n$$

where $e_p = 1 + \delta e_{pp}$ and $e_n = \delta e_{pn}$ are the conventional proton and neutron effective charges.

Most E2 transitions turn out to be between collective states with the protons and neutrons moving in phase and have $A_p \simeq A_n$. These transitions obviously depend only on the isoscalar combination of effective charges $e_p + e_n$. This combination is in fact well determined and quite constant through the sd shell with a value of $e_p + e_n = 1.70$. The isoscalar effective charge can be related to properties of the giant quadrupole resonance and the strength with which it couples to the single-particle states (Ref 25, Ref 27, Ref 26, Ref 28).

The smaller isovector component is important since it reveals the noncollective nature of the wave functions and gives information about the coupling to the giant isovector E2 resonance. For a given transition we want to look at the difference between M_p and M_n in order to isolate the isovector part

$$\Delta M = [M_p - M_n] / |\Delta T_z| = \Delta A (e_p - e_n)$$

where

$$\Delta A = [A_p - A_n] / |\Delta T_z|$$

The experimental value can be obtained from existing gamma decay data if we assume mirror symmetry

$$M_n(\text{neutron rich}) = M_p(\text{proton rich}) \tag{10}$$

and thus

$$\Delta M(\exp) = [|M_p(\text{proton rich})| +/- |M_p(\text{neutron rich})|] / |\Delta T_z| \tag{11}$$

We use $|M_p|$ in equation Eq. (11) since only its magnitude is measured in gamma decay experiments $[B(E2) = M_p^2/(2J_i+1)]$. The +/- sign in Eq. (11) must be inferred from theoretical or other experimental considerations (see below); usually the - sign is appropriate.

In Fig. 2, we show a plot taken from Ref 26 of the experimental values $\Delta M(\exp)$ compared with the theoretical values ΔA calculated with the

Chung-Wildenthal "CW" sd shell wave functions together with harmonic-oscillator radial wave-functions. $\Delta M(exp)$ was obtained using the $-$ sign in Eq. (11) except for the 0-2' transition in ^{34}S which is shown in the figure by two crosses. The upper cross corresponds to the $-$ sign and the lower cross to the $+$ sign. Clustering of the the points around the $45°$ line indicates agreement between experiment and theory and also that the isovector effective charge has the free nucleon value $e_p-e_n=1$. In general, points in the $(+,+)$ quadrant are those which are single-particle in character and those in the $(-,-)$ quadrant are those which are single-hole in character.

Since our work reported in (Ref 26) a new sd shell interaction, the universal sd interaction "USD", which provides a better and more "universal" fit to binding energies and excitation energies for the sd shell nuclei has been obtained by Wildenthal (Ref 7).

The agreement between experiment and theory is somewhat better with the "USD" interaction (Fig. 3) than with the "CW" interaction (Fig. 2) especially in the $(+,+)$ quadrant. In addition, we note that $e_p=e_n = 1$ (the $45°$ line) gives about the best agreement. However, in Ref 26 we have noted that the slope of the best fit line decreases when the "coulomb" corrections to the theory are estimated by using Woods-Saxon radial wave functions. The best fit isovector effective comes out to be $e_p-e_n = 0.6-0.7$ which is reduced from the free-nucleon value in agreement with calculations which take into account coupling with the isovector giant resonance (Ref 25, Ref 27, Ref 26, Ref 28).

In both Fig. 2 and Fig. 3 the agreement between experiment and theory is rather poor for several transitions in the $(-,-)$ quadrant. (The "coulomb" corrections are small (<5%) for these cases, since the states are in the upper sd shell and have well-bound single-particle components.) In the remainder of this section we will discuss some of the problems associated with the 0^+ to 2^+ transitions in ^{26}Mg, ^{30}Si and ^{34}S.

First we consider the problem associated with the sign ambiquity in Eq. (11). In general, M_p is a linear function of T_z (Ref 29)

$$M_p = M_0 + T_z \Delta M$$

Figure 2: Comparison of experimental and theoretical isovector E2 matrix elements for sd shell nuclei. The theoretical values (ΔA) were obtained with the "CW" sd shell wave functions together with harmonic-oscillator radial wave functions. The theoretical value is given by the x coordinate and the experimental value by the y coordinate. The second state of a given spin is indicated by J'.

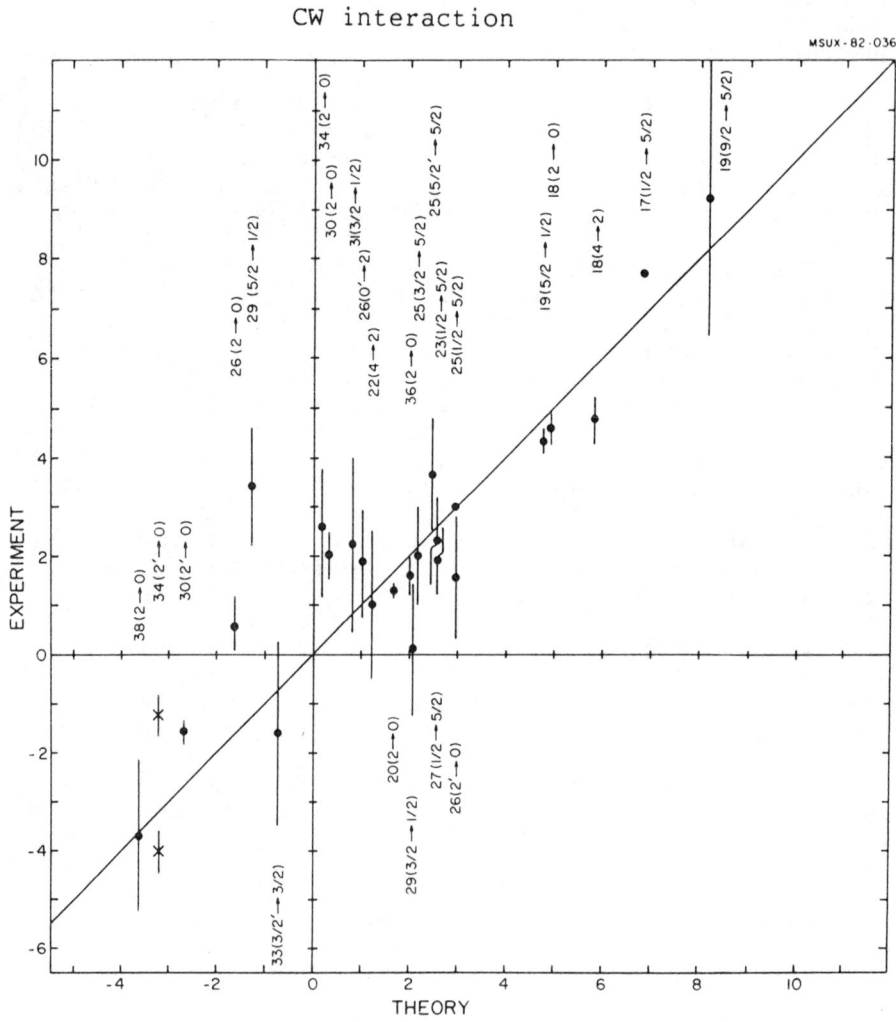

Figure 3: Same as Fig. 2, but with the theoretical matrix elements calculated with the "USD" sd shell wave functions. (The J,T values associated with the transitions can be inferred from the labels in Fig. 2.)

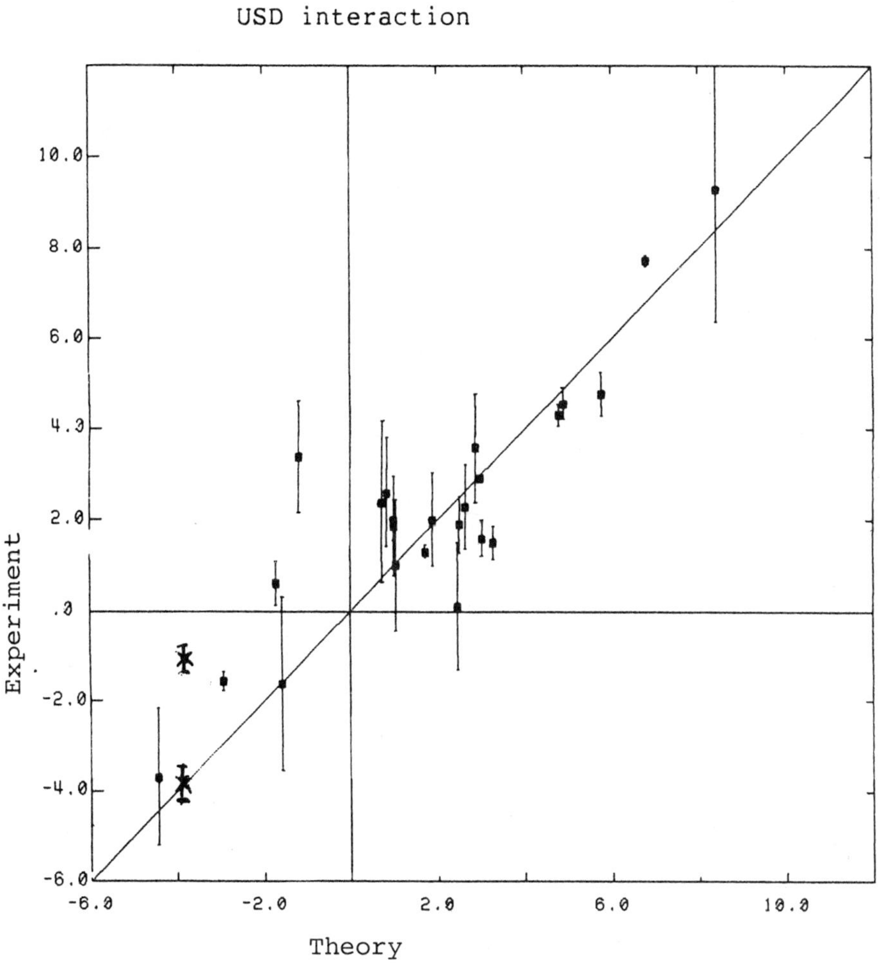

USD interaction

If we plot $|M_p|$ vs T_z for three values of T_z (e.g. $T_z=-1$, 0 and 1), the line should be straight in the usual case where the M_0 term dominates or the line should have a "V" shape in the unusal case where the ΔM term dominates. In fact, for all cases where $|M_p|$ has been measured for three values of T_z the results are consistent with a straight line (see Table I in Ref 26).

With the "CW" wave functions, two of the transitions considered in Ref 26 were predicted to be of the "V" type, namely, the 2'->0 transitions A=30 and A=34. But with the new "USD" wave functions only the A=34 transition is predicted to have a "V" shape. Experiment and theory are compared in Table 4.

Table 4: Comparison of experimental and theoretical M_p matrix elements (in units of e fm^2) for the A=26, 30 and 34 2->0 transitions. The theoretical matrix elements were calculated with $e_p = 1.35$ and $e_n = 0.35$. The notation 2' denotes the second excited state.

T_z	$2^+ \to 0^+$				$2'^+ \to 0^+$			
	$\|M_p(exp)\|$		M_p CW	M_p USD	$\|M_p(exp)\|$		M_p CW	M_p USD
^{26}Mg	17.7(0.4)	a	16.90	17.31	3.0(0.2)	a	4.05	3.00
^{26}Al	17.7(2.1)	a	15.22	15.58	3.8(0.5)	c	6.98	6.28
^{26}Si	18.8(0.9)	a	13.53	13.86	6.0(0.7)	a	9.92	9.55
^{30}Si	14.2(0.4)	b	16.44	14.05	6.5(0.3)	b	4.57	7.89
^{30}P	15.5(0.9)	b	16.74	15.04	5.7(1.1)	b	1.76	4.96
^{30}S	18.2(0.8)	b	17.03	16.02	3.4(0.3)	b	-1.04	2.03
^{34}S	14.0(0.3)	b	14.11	12.78	4.9(0.2)	b	2.96	4.76
^{34}Cl	17.3(0.8)	d	14.25	13.62	5.1(1.0)	d	-0.23	0.94
^{34}Ar	21.1(3.2)	b	14.38	14.45	2.8(0.5)	b	-3.42	-2.88

a) (Ref 30)
b) See references in Table I of Ref 26
c) (Ref 31)
d) (Ref 32)

The new USD calculations are better than CW for A=30 and also for the ^{34}S and ^{34}Ar transitions. But the result of the recent measurement from Ref 32 for ^{34}Cl is a surprise, indicating an isoscalar dominance rather than the predicted isovector dominance. What could be wrong? From the theoretical side, we might question the assumed validity of charge symmetry for the mirror states, or the sd shell wave functions might simply be wrong. From the experimental side, we might question the systematic errors in the analysis measurements carried out by the Doppler-shift attenuation method due to uncertainties in the stopping power of the recoiling reaction products. Inelastic hadron scattering experiments are important for providing an independent measurement of both M_p and M_n for the stable neutron rich nuclei which can be used as a targets (Ref 33, Ref 34).

There are many hadronic reactions which can be used and many arguments as to why one is superior to the other in determining M_p and M_n. In addition to the low-energy (n,n') and (p,p') reactions discussed here, we can mention recent work for sd shell nuclei with pions at energies near the (3,3) resonance (Ref 35, Ref 36), low-energy pions (Ref 37), alphas (Ref 38, Ref 39) and high-energy protons (Ref 40, Ref 41, Ref 42).

We can qualitatively understand the magnitude of the direct (single step) part of the hadronic cross sections by the relation (Ref 33)

$$\sigma(F) = [\ b_p(F)\ M_p + b_n(F)\ M_n\ ]^2 \qquad (12)$$

where $b_{p/n}(F)$ represent the "coupling constants" between the hadronic "external field" (F) and the protons/neutrons in the nucleus. These coupling constants can either be calculated in some fundamental way or calibrated by looking at transitions where M_p and M_n are known by other means, such as those in nuclei with N=Z where $M_p \approx M_n$. In addition to the summary of values for b(F) given in Ref 33 we note the very large value of $b_p/b_n = -20$ obtained for 50 MeV π^- scattering in Ref 37. (More quantitatively, the cross sections of course depend in a more complex way upon the shape of the transition density, but empirically Eq. (12) provides a good approximation when the transition density has a collective shape.)

Since the 0->2' transitions are rather weak, the experimental data is often difficult to obtain and one must be careful to include the important two-step processes in the reaction analysis. Thus, the analyses so far have been primarily concentrated on just determining the relative sign of M_p and M_n and taking the values from the gamma decay experiments (Table 4) by making use of Eq. (10). The experiments published thus far are in agreement with USD theory that the relative sign is positive for ^{26}Mg and ^{30}S (Ref 39, Ref 40). However, for ^{34}S the published results are contradictory with a high-energy proton scattering experiment indicating a negative sign (Ref 40) and an alpha scattering experiment indicating a positive sign (Ref 38) (in agreement with the gamma decay data).

New information about the ^{34}S transition has been obtained from an analysis of recent (n,n') and (p,p') experiments reported at this meeting by Alarcon et al. (Ref 2). One advantage of the low-energy (n,n') and (p,p') reactions are that the coupling constants can be based on the extensive literature which exists for the elastic and inelastic scattering optical-model parameters. In addition, it is useful to look at the ratio between the (n,n') and (p,p') cross sections since some of the two-step contributions cancel out in the ratio. The coupled channel analysis is consistent with a relative positive sign between the M_p and M_n matrix elements for the ^{34}S 0->2' transition (Ref 2). The relative sensitivity to protons and neutrons in Eq. (12) can be inferred from the value of $b_n/b_p \approx 0.5(2.0)$ for neutron (proton) scattering deduced from the optical-model potentials needed to fit the elastic and inelastic scattering to the lowest 2^+ state in the coupled channel analysis. Additional comparisons with theory concerning the first and third 2^+ states in ^{34}S are discussed in Ref 2.

Finally, we discuss the transition to the lowest 2^+ state ^{26}Mg. Other than the A=38 2 -> 0 transition, this is the only matrix element between the ground state and lowest 2^+ state which is predicted to have $M_p > M_n$ for the neutron rich nucleus (^{26}Mg in this case). The reason for this is that in the simplest shell model the lowest few levels in ^{26}Mg (0^+, 2^+ and 4^+) are described by two-proton $d_{5/2}$ hole states in ^{28}Si and in this extreme model has $A_n=0$ and $A_p=6.9$ ($M_n=2.4$ and $M_p=9.3$ with $e_p=1.35$ and $e_n=0.35$). In the full sd shell model it retains a little of this simple character in both the CW and USD wave functions. Thus, it is somewhat surprising that the gamma-decay matrix elements disagree with this expectation (see Table 4).

This is a case which has again been studied by a combined analysis of the (n,n') and (p,p') reactions (Ref 1) as well as a number of other reactions. In Table 5 we compare the ratio M_n/M_p from different analyses. The differences between different experiments tend to be outside the error bars which indicates some systematic problems in the reaction theory analysis which must be understood. It is gratifying that the (n,n')+(p,p') analysis agrees with the mirror gamma-decay data, but unfortunately they disagree with our theory which gives for both the CW and USD wave functions M_n/M_p

= 0.80 with harmonic-oscillator wave functions (see Table 4) and 0.83 with Woods-Saxon wave functions (see Table V in Ref 26).

Table 5: M_n/M_p for the $0^+ \rightarrow 2^+$ transition in ^{26}Mg obtained from the mirror gamma-decay data and analyses of hadron scattering experiments.

M_n/M_p	Year	Method
0.62(14)	1980	160 MeV pions (Ref 35)
0.74(12)	1982	800 MeV pions (Ref 42)
1.00(5)	1982	$[M_p(^{26}\text{Si})/M_p(^{26}\text{Mg})]/1.06$ where M_p are from Table 4 and 1.06 is the coulomb correction based on Table V of Ref 26.
1.02(5)	1983	Low energy (n,n') and (p,p') (Ref 1)
0.83(6)	1984	50 MeV pions (Ref 37)

Our conclusion in this section is that there are still some problems to be solved both theoretically and experimentally concerning the E2 transitions in the sd shell. We have discussed specific problems in ^{34}S and ^{26}Mg. The experimental consensus is that the proton and neutron matrix element have the same relative sign for the E2 transition to the second 2^+ state in ^{34}S in contradiction to the present theory. (In Figs. 2 and 3 we should use the upper cross.) For the transition to the lowest 2^+ state in ^{26}Mg the low-energy (n,n')+(p,p') and gamma ray data give $M_n/M_p \approx 1$ which is about 20% larger than theory, however, smaller values for the ratio have been obtained from pion experiments. Perhaps at this level one must put into the hadron scattering analysis more detail (beyond the usual collective model assumptions) about the shape of the transition density (potential) in order to obtain better agreement between various probes. In any case, these will be some of the interesting cases with which to evaluate the success or failure of other nuclear structure calculations and future improvements in the present ones.

Acknowledgements: All of the sd-shell calculations were carried out in collaboration with Hobson Wildenthal at Drexel University. This work was supported in part by the National Science Foundation under grant no. PHY-83-12245.

References:

1. R. C. Tailor, J. Rapaport, R. W. Finlay and G. Randers-Pehrson, Nucl. Phys. A401, 237 (1983).

2. R. Alarcon, D. Wang, J. Rapaport, R. T. Kouzes, W. H. Moore and B. A. Brown, abstract submitted to this meeting.

3. A. R. Edmonds, Angular Momentum in Quantum Mechanics, (Princeton Univ. Press, Princeton, New Jersey) 1960.

4. A. de-Shalit and I. Talmi, Nuclear Shell Theory, (Academic Press, New York and London) 1963.

5. C. Gaarde, J. S. Larson, M. N. Harakeh, S. Y. van der Werf, M. Igarashi and A. Muller-Arnke, Nucl. Phys. A334, 248 (1980).

6. R. Madey, B. D. Anderson, J. W. Watson, A. R. Baldwin, B. S. Flanders, C. Lebo, C. C. Foster, S. M. Austin, A. Galonsky and B. H. Wildenthal, Bull. Am. Phys. Soc. 27, 731 (1982).

7. B. H. Wildenthal, Bull. Am. Phys. Soc. 27 (1982), and lectures presented at the International School on Nuclear Physics; 7th course: Mesons, Quarks and Nuclear Excitations, Erice-Trapany-Sicily, 6-18 April 1983 (to be published in Progress in Particle and Nuclear Physics Vol 11).

8. G. F. Bertsch and I. Hamamoto, Phys. Rev. C26, 1323 (1982).

9. I. S. Towner and F. C. Khanna, Nucl. Phys. A399, 334 (1983).

10. C. Gaarde, J. S. Larsen and J. Rapaport, Spin Excitations in Nuclei, Telluride Conference, March (1982), Plenun Press.

11. C. D. Goodman, Proceedings of the International Conference on Nuclear Physics (Tipografia Compositori, Bologna) edited by B. Blasi and R. A. Ricci, Vol. 2 (1983), p. 165.

12. H. Primakoff and S. P. Rosen, Ann. Rev. Nucl. Part. Sci. 31, 145 (1981).

13. W. C. Haxton, G. J. Stephenson and D. Strottman, Phys. Rev. Lett. 47, 153 (1981).

14. L. Zamick and N. Auerbach, Phys. Rev. C26, 2185 (1982).

15. P. J. Brussard and P. W. M. Glaudemans, Shell Model Applications in Nuclear Spectroscopy (North Holland, Amsterdam, 1977).

16. W. Kutschera, B. A. Brown and K. Ogawa, Rivista del Nuovo Cimento, Serie 3, Vol. 1, N. 12, 1978.

17. J. D. McCullen, B. F. Bayman and L. Zamick, Phys. Rev. 134B, 515 (1964).

18. V. A. Khodel, Phys. Lett. 32B, 583 (1970).

19. W. M. D. Rae, A. Etchegoyen, N. S. Godwin and B. A. Brown, The Oxford-Buenos Aires Shell Model Code, unpublished.

20. B. A. Brown, W. A. Richter and N. S. Godwin, Phys. Rev. Lett. 45, 1681 (1980).

21. A. G. M. van Hees and P. W. M. Glaudemans, Z. Phys. A303, 267 (1981).

22. B. D. Anderson, J. N. Knudson, P. C. Tandy, J. W. Watson, R. Madey and C. C. Foster, Phys. Rev. Lett. 45, 699 (1980).

23. B. D. Anderson, T. Chittrakarn, A. R. Baldwin, C. Lebo, R. Madey, J. W. Watson, B. A. Brown and C. C. Foster, submitted to Phys. Rev. C.

24. R. T. Kouzes, P. Kutt, D. Mueller and R. Sherr, Nucl. Phys. A309, 329 (1978).

25. B. A. Brown, A. Arima and J. B. McGrory, Nucl. Phys. A277, 77 (1977) an references therein.

26. B. A. Brown, B. H. Wildenthal, W. Chung, S. E. Massen, M. Bernas, A. M. Bernstein, R. Mickimen, V. R. Brown and V. A. Madsen, Phys. Rev. C26, 2247 (1982).

27. V. R. Brown and V. A. Madsen, Phys. Rev. C11, 1298 (1975); 17, 1943 (1978).

28. H. Sagawa and B. A. Brown, to be published in Nucl. Phys. A.

29. E. K. Warburton and J. Weneser, Isospin in Nuclear Physics, edited by D. H. Wilkinson (North-Holland), Amsterdam, 1969), p. 173.

30. T. K. Alexander, G. C. Gill, J. S. Foster, W. G. Davies, I. V. Mitchell and H.-B. Mak, Bull. Am. Phys. Soc. 26, 1127 (1981), Phys. Lett. 113B, 132 (1982).

31. P. de Wit, E. L. Bakkum, C. van der Leun and P. M. Endt, Phys. Lett. 113B, 137 (1982).

32. J. Keinonen, A. Luukkainen and M. Bister, Nucl. Phys. A412, 101 (1984).

33. A. M. Bernstein, V. R. Brown and V. A. Madsen, Comments in Nuclear and Particle Physics, XI, No. 5 (1983) and references therein.

34. B. A. Brown and B. H. Wildenthal, Phys. Rev. C21, 2107 (1980).

35. C. A. Wiedner, K. R. Cordell, W. Saathoff, S. T. Thornton, J. Bolger, E. Boschitz, G. Proebstle and J. Zichy, Phys. Lett. 97B, 37 (1980).

36. S. Iversen, H. Nann, A. Obst, K. K. Seth, N. Tanaka, C. L. Morris, H. A. Thiessen, K. Boyer, W. Cottingame, C. F. Moore, R. L. Boudrie and D. Dehnhard, Phys. Lett. 82B, 51 (1979).

37. R. Tacik, K. L. Erdman, R. R. Johnson, H. W. Roser, D. R. Gill, E. W. Blackmore, R. J. Sobie, T. E. Drake, S. Martin and C. A. Wiedner, Phys. Rev. Lett. 52, 1276 (1984).

38. A. Saha, K. K. Seth, M. Artuso, B. Harris, R. Seth, H. Nann and W. W. Jacobs, Phys. Rev. Lett. 52, 1876 (1984).

39. A. Saha, K. K. Seth, D. Barlow, H. Nann and W. W. Jacobs, Phys. Lett. 144B, 419 (1982).

40. A. M. Bernstein, R. A. Miskimen, B. Quinn, S. A. Wood, M. V. Hynes, G. S. Blanpied, B. G. Ritchie and V. R. Brown, Phys. Rev. Lett. 49, 451 (1982).

41. R. A. Miskimen, A. M. Bernstein, B. Quinn, S. A. Wood, M. V. Hynes, G. S. Blanpied, B. G. Ritchie and V. R. Brown, Phys. Lett. 131B, 26 (1983).

42. G. S. Blanpied, N. M. Hintz, G. S. Kyle, M. A. Franey, S. J. Seestrom-Morris, R. K. Owen, J. W. Palm, D. Dehnhard, M. L. Barlett, C. J. Harvey, G. W. Hoffmann, J. A. McGill, R. P. Liljestrand and L. Ray, Phys. Rev. C25, 422 (1982).

SENSITIVITY OF NEUTRON SCATTERING
TO LOW ENERGY COLLECTIVE EXCITATIONS.*

M. T. McEllistrem

Department of Physics and Astronomy
University of Kentucky, Lexington, Kentucky 40506

ABSTRACT

The strong coupling between scattering to ground and first excited 2^+ levels of even A nuclei makes low energy neutron scattering a powerful tool for assessing their collective excitation strengths. Higher even parity excited levels are much less strongly coupled to ground state scattering, and the corresponding transition amplitudes are extracted from neutron scattering with less confidence. The sensitivity with which coupling can be assessed for several levels of nuclei, and the sensitivity to details of nuclear structure, will be examined. At incident energies well above low lying collective excitations, elastic scattering is not sensitive to nuclear structure details. The enhanced sensitivity of elastic scattering to structure properties at lower incident energies will be illustrated for the shape transitional nuclei near A = 190. Recent results of microscopic model calculations will also be presented, to illustrate the structure insights which can be obtained with those methods.
*Supported in part by the National Science Foundation

INTRODUCTION

The strong coupling between scattering to ground and excited levels of even A nuclei has made neutron scattering an effective tool for the examination of collective nuclear dynamics of some excited levels, yielding information about collective properties not available from scattering with other probes. For example, several years ago the combined analyses of neutron and proton scattering[1] from deformed nuclei, using coupled channels methods, showed that the quadrupole moment per target nucleon was, within about 5%, the same for target protons and neutrons. Similar scattering studies, emphasized in other presentations to this conference, show markedly different behavior for target protons and neutrons for spherical nuclei near single closed shells. Thus we see the importance of accurate cross sections for different hadronic probes for contrasting the different types of collective behavior in nuclei of differing character.

Even parity excited levels other than the first 2^+ level are less strongly coupled to scattering from the ground state, and the degree to which neutron scattering can be used to probe the dynamics associated with these levels is not as clear as is the case for the first excited levels. The spirit of this review is to focus on the questions associated with extracting collective properties for states other than those of the ground state band in deformed nuclei, or other than those of the first 2^+ level. It is also important to continue to confirm the confidence with which properties of the first 2^+ levels, or of ground state bands, are delineated with different hadronic probes.

Structure properties can be extracted only from relatively unambiguous analyses of scattering data, for example combined analyses of elastic and inelastic scattering in coupled channels models employing well determined complex scattering potentials. The potentials employed reflect the models of the target nuclei used; and ultimately one would have a model from which target dynamics and scattering potential would both be consistently derived. Present phenomenological analyses assume a specific structural character for the target, and within that context develop a scattering potential and structure parameters to represent all scattering observables. One principal advantage of neutron scattering is that many observables are available to fix potentials and structure.

Neutron scattering observables include total cross sections and very low energy scattering properties as well as differential scattering cross sections, and sometimes analyzing powers and polarizations. The usual and best method of procedure begins with careful fits to total cross sections, s-and p-wave strength functions calculated below 10 keV neutron energy, and potential scattering lengths. The entire procedure has been systematized as the SPRT method by the theory and analysis group at Bruyeres-le-Chatel (BRC).[2] After these properties have been fitted, structure parameter adjustments are made to accommodate the elastic and inelastic scattering cross sections and other observables. Usually a combined fit to low energy scattering properties and total cross sections will result in a rather well determined potential and set of structure parameters. Only modest adjustments of parameters are then needed to accomodate differential scattering cross sections and polarizations or analyzing powers. Also potential parameters are little effected by changes in choices of target structure model, as long as large changes in model space are not involved.

A scattering analysis may be thought of, then, as consisting of two tasks. One is fixing the properties of the potential and the other is the determination of the specific structure properties of the target which are the goal of the study.

One may think of a collective model as characterized by a set of transition and reorientation matrix elements which the scattering analysis determines; a typical set of such matrix elements, which fixes the coupling between elastic and inelastic scattering channels, is illustrated in Fig. 1. An original set of matrix elements is provided by a model, and then the model is altered to provide altered matrix elements consistent both with the spirit of the model and with the scattering data. Presented

Level Schemes and Transitions

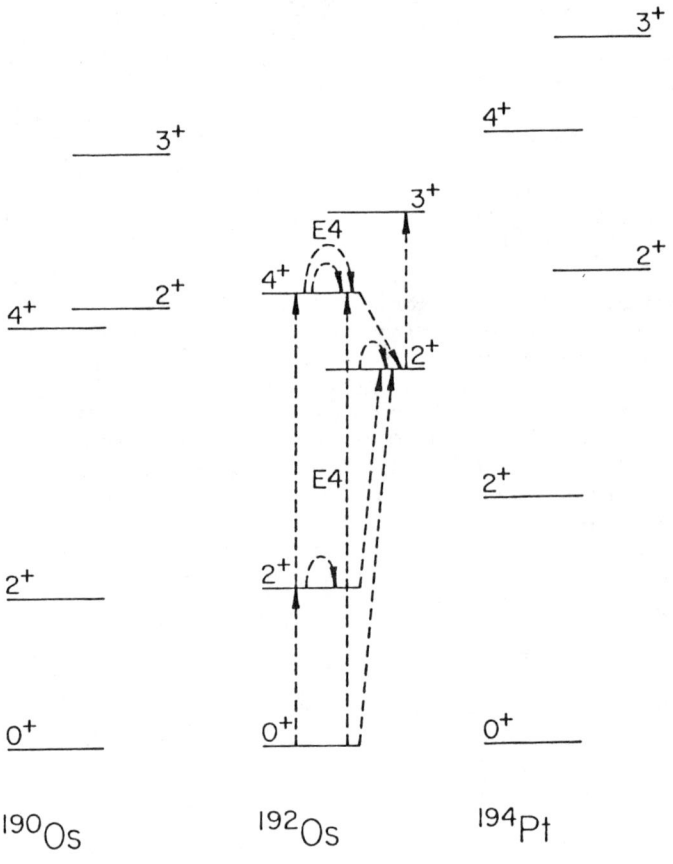

Fig. 1. Lowest levels of three transitional nuclei. Dashed curves are transition matrix elements. Labelled ones are E4; others are E2.

here are the relevant low-lying collective levels of the shape
transitional nuclei near A = 190. The set of E2 and E4 matrix
elements which, experiences teaches us, seem to be most important
for describing neutron scattering below 10 MeV are shown for
^{192}Os. The scattering potential yields wave functions for the
various coupled channels which, together with the matrix
elements, determine the scattering amplitudes and observables
calculated from them.

The spirit of this review reflects the fact that scattering
potentials are altered little when different structure models
with the same model space are employed. Thus it is proposed that
the actual structure properties, as represented by the matrix
elements found from fitting scattering observables, can be
determined by these analysis methods, much as they are determined
in Coulomb excitation experiments, for example. In fact, as
noted above for stably deformed nuclei and at least for the
ground state band, the same E2 matrix elements are obtained from
Coulomb excitation and from neutron scattering. The range of
nuclei for which the dynamics excited by neutrons is the same as
that excited by other hadrons is a particularly interesting
question in these studies.

REVIEW OF COLLECTIVE EFFECTS ON SCATTERING

To review some of the recent efforts to determine collective
properties of excited levels, I have rather arbitrarily selected
studies which seemed to me to address the question of which
collective properties would be best determined in neutron
scattering. A particular limitation of neutron studies comes
from the fact that usually experiments must be done at modest
incident energies to have adequate energy resolution to resolve
the closely spaced collective levels of heavy nuclei. Thus it is
fortunate that collective effects manifest themselves strongly at
low energies.

To illustrate the clear appearance of these effects at quite
low energies, I present a few results from work of two Russian
groups, those of Y. Popov at Moscow, and that of Sitko and
colleagues at Kiev, as reviewed by Konobeevskii recently.[3]
Figure 2 shows the neutron inelastic scattering cross sections
only 300 keV above the threshold for exciting the first 2^+ level
of nuclei, presented as a function of A for the upper half of the
periodic table. Where the cross section is large, one sees the
direct influence of collective effects. Illustrations of studies
in a couple of regions where shape transitions in nuclei seem to
be important will be given below. The first set will be for the
Se nuclei, and the other locus of much current interest is

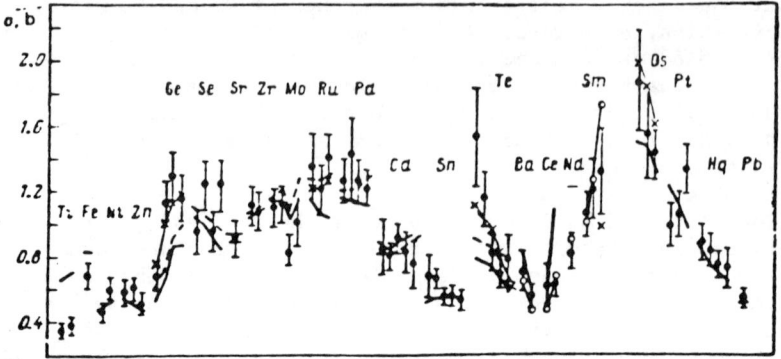

Fig. 2. Neutron inelastic scattering cross sections for scattering to first 2^+ levels at incident energies 300 keV above the excitation threshold.

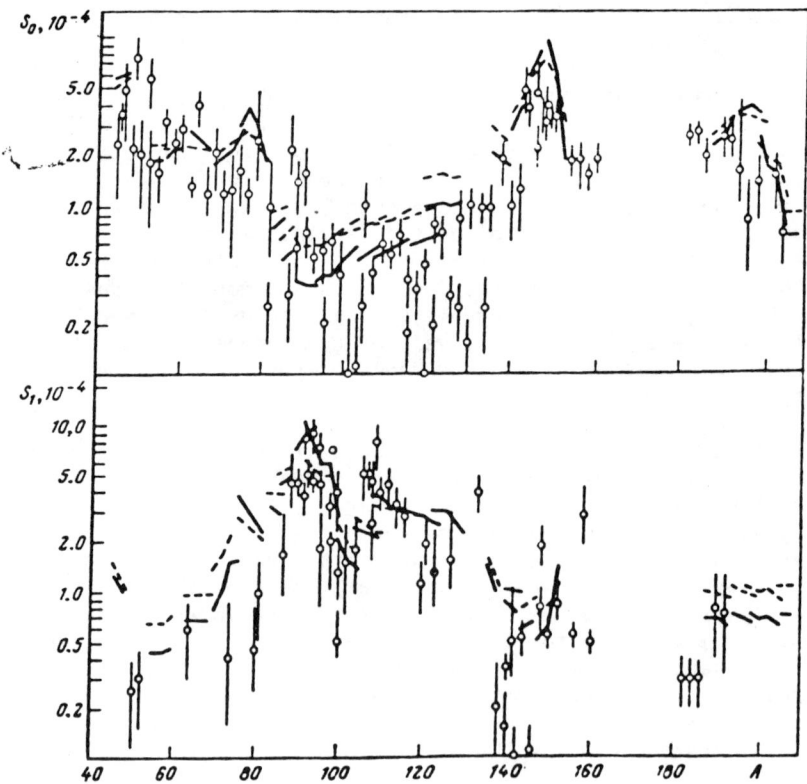

Fig. 3. The s- and p-wave strength functions vs. atomic weight A.

near the shape transitional region, the Os and Pt isotopes.

That these collective effects also manifest themselves directly in other scattering observables is apparent when one examines Fig. 3, which shows the low energy s-and p-wave strength functions through the same mass region. The solid curves in these two panels represent coupled channels calculations in which scattering to the first excited 2^+ level is coupled to ground state scattering. Particularly in the upper panel, for s-waves, one sees the same peaks as in the previous figure showing the inelastic scattering cross sections. The strong coupling which makes the inelastic scattering cross sections large also enhances the s-wave strength functions.

SCATTERING STUDIES FOR A FROM 70 TO 80

Turning now to particular scattering studies, I start by reviewing the culmination of several neutron scattering experiments for the Se isotopes. An experiment completed 10 years ago[4] reported cross sections for elastic and inelastic scattering to the first 2^+ level for the four even A Se isotopes with A between 76 and 82. The analysis, which included only the coupling between scattering to the ground state and the first excited levels, indicated two problems. The first was that the coupling strengths found were not consistent with those extracted from Coulomb excitation or proton scattering; they were too small. The other deficiency was a reported neutron excess for the imaginary part of the scattering potential which was much stronger than that usually found for such potentials, and certainly much stronger than neutron excess dependencies expected on the basis of i-spin dependence of the scattering potential.[4,5]

A recent experiment of Kurup et al.[5] remeasured the differential cross sections for ^{76}Se and ^{80}Se at 8 and 10 MeV, but included data for four additional excited levels of these two nuclei. These new measurements for elastic and inelastic scattering to the first excited levels are in reasonably good agreement with the earlier measurements, but now the data are interpreted in terms of nuclear structure models which include several levels of each nucleus; the models provide more realistic approaches to the structure of these nuclei. Some model calculations are shown in Fig. 4. In the two panels of this figure the dashed curves are second order (harmonic) vibrational model calculations, an extension of the model used in the older study.[4] In both nuclei the two phonon states are not well fit; this is good, since it is now well known that these nuclei have rather large quadrupole moments in their first excited levels.[6] A spherical model should not work if the neutron scattering data are sufficiently sensitive to the quadrupole moment, for example. The solid curves of Fig. 4 are interesting, because they result from inserting the E2 transition matrix elements measured in Coulomb excitation measurements.[6] The agreement for both nuclei is excellent, showing that the coupling strength problem of an earlier[4] scattering analysis can be resolved through including more of the collective strength of these nuclei than that of the first excited levels. The schematic model,[7] discussed elsewhere in this conference, would suggest that as soon as a nucleus is a few nucleons away from a single closed shell, then the quadrupole coupling strengths measured by protons, neutrons, and Coulomb excitation should all be the same; the solid curves confirm this nicely for neutron scattering in the Se isotopes.

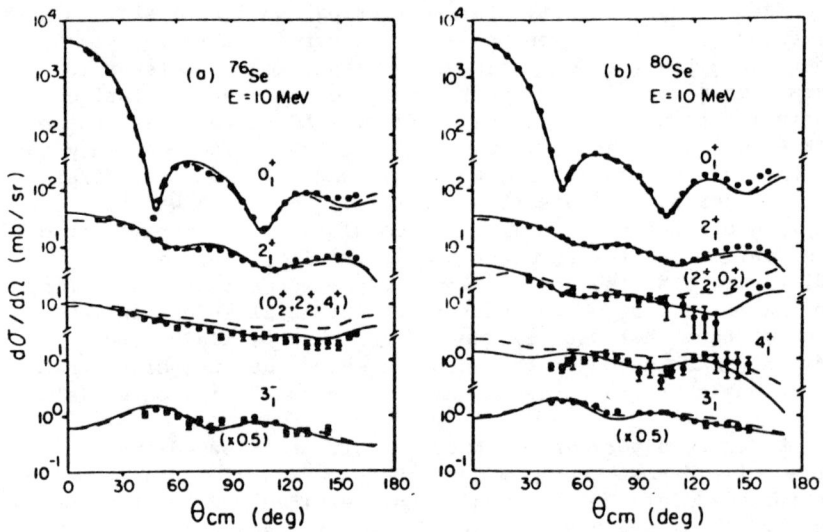

Fig. 4. Scattering cross sections for two Se isotopes. Dashed curves are for spherical vibrator calculations. Solid curves result from use of matrix elements deduced from Coulomb excitation measurements.

Two models more realistic than the harmonic vibrator are tested in Fig. 5. The solid curves here are results of a rotation-vibration model,[8] in which one assumes the nucleus is intrinsically deformed but vibrates around a deformed intrinsic shape. The results look rather good, especially for ^{76}Se, which has the largest measured quadrupole moment[6] of these isotopes.

The dashed curves are from an asymmetric rigid rotor model (ARM), with the excitation energies of the two lowest 2+ levels used to fix the departure from axial symmetry according to the model of Davydov and Filippov.[9] Here we see sufficient sensitivity in the measured cross sections for the second 2+ level to be able to reject that as a model for these nuclei. That these nuclei are soft, rather than rigidly deformed, seems also most consistent with the E2 matrix elements extracted in the Coulomb excitation experiment.[6]

From the Se studies we learn several things. First, the information extracted even for the most strongly coupled first 2+ excited level can be in error unless many of the levels which can couple directly to that state are also included in the analyses. Second, reasonably precise and accurate inelastic scattering cross sections are sufficiently sensitive to distinguish between different assumptions about the collective character of the second 2+ excited level. But one older problem remains. Coupling in more of the even parity collective excited levels decreased substantially the residual neutron excess dependence of the absorptive potential, closer to the dependence expected from i-spin arguments than that first reported.[4] But the strength of that dependence in the Se nuclei may still be a factor of two larger than expected.

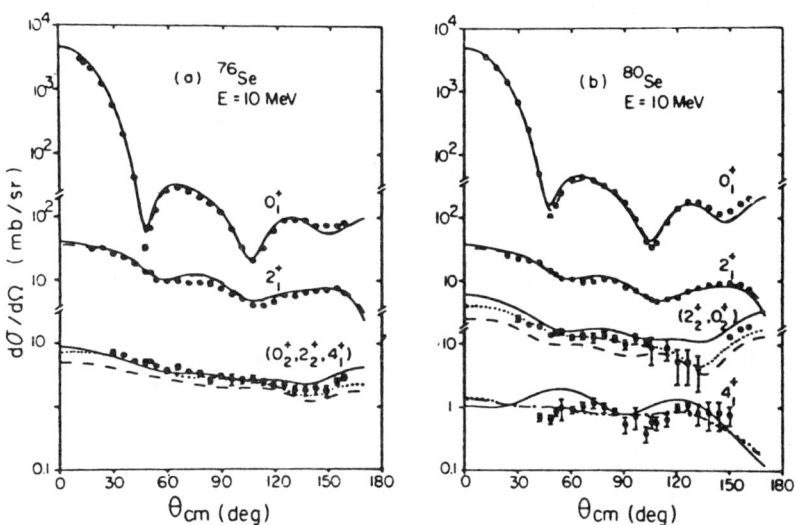

Fig. 5. Measured points are those of Fig. 4. The solid curves are from the rotation-vibration model. The dashed curves are from the ARM model.

SCATTERING STUDIES FOR A FROM 180 TO 200

The other mass region from which I draw examples of the sensitivity of hadron scattering to collective excitations not of the ground state band is the Os, Pt region with A from 190 to 200. This shape transitional region is marked as one in which stable, oblate deformations are found, and is a region in which non-axially symmetric quadrupole excitations are known to be quite important.

The scattering studies of the Os and Pt nuclei are ongoing ones, and to date only the low energy scattering work for ^{194}Pt is reasonably complete; even for that nuclide, more analytical work needs to be done. This study of these three nuclei is a project involving experiments and analyses at two laboratories, Kentucky and Bruyeres-le-Chatel; and as we shall see, it is being approached from several different perspectives. One reason for the choice of these nuclei is illustrated in Fig. 1, which shows their low-lying collective levels. Note that in contrast to well deformed nuclei, such as the light W isotopes for example, the second 2^+ level is the second excited level; the coupling between ground and γ-bands is very strong in these nuclei, which is reflected in the fact that excitation energies in the two bands are comparable. Sensitivity to γ-band excitations in neutron scattering might be large for these nuclei.

The first experiments were scattering studies done at low energies by Mirzaa et al.,[10] with the analysis efforts led by Delaroche and others.[11] A potential and set of models were developed first to fit low energy scattering properties and total cross sections measured[12,13] in the period from 1960 to 1971. After developing the model and associated potential parameters, new and more accurate total cross section data were made available from an experiment of Poenitz et al.[14] The fit to the new data provided by our model is excellent, within experimental uncertainties for neutron energies from 2 MeV to 20 MeV. There is some risk in accepting this, because the data are for natural abundance Pt. Separated isotope data do not exist at this time. The structure models we used are illustrated in Fig. 6, which reports data and fits for inelastic neutron scattering to several interesting collective levels. Since we are at low energy, we must account carefully for the statistical model, or compound system cross sections. This is not the place to discuss in detail the elaborate methods we use to do that; but let me point out that at these energies the direct coupling cross sections to the unnatural parity state, 3^+, are very small; they are of the order of 1% of the statistical model values. Thus the fit shown for that level in Fig. 6 is confirmation that our model correctly provides the compound system cross sections.

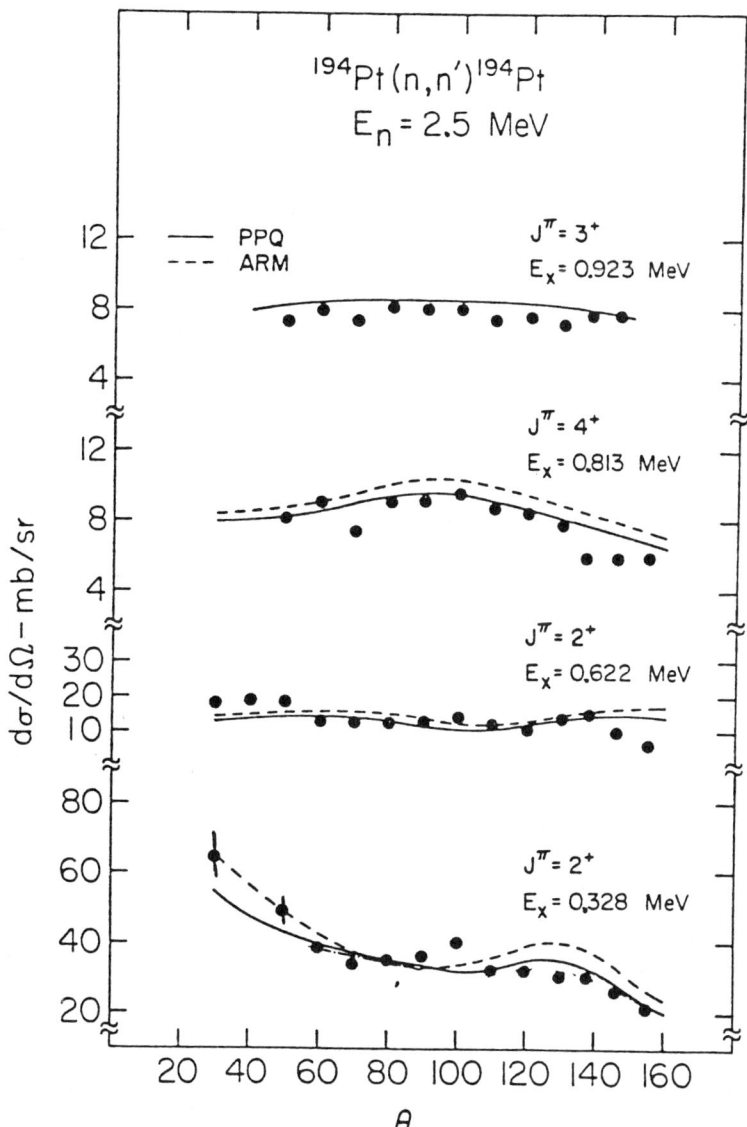

Fig. 6. Measured and calculated inelastic scattering cross sections for the first four excited levels. Solid curves are for the PPQ or IBA models. The dashed curves are for the ARM model.

The two sets of curves in Fig. 6 are for two different methods of treating the second 2^+ level. The solid curves employ the dynamic deformation theory (PPQ) of K. Kumar,[15] which provides excellent fits to level energies and electromagnetic transition rates throughout this mass region.[15] This model is roughly equivalent to a γ-soft character for ^{194}Pt, or one which does not imply a definite deformation and shape for the nucleus. The other model of Fig. 6 is the asymmetric rotor (ARM) model of Davydov and Filippov.[9] As one can see, the best data for the first 2^+ level are those for 60° and beyond. For these and for the 4^+ level data, a goodness of fit criterion, Q^2, is constructed proportional to the usual χ^2 of a least squares fit. This parameter is a factor of 2.5 larger for the ARM than for the PPQ model, or for an Interacting Boson Model (IBA) also developed for this nucleus.[16] Further model tests are contained in the comparisons for the three models with the elastic scattering differential cross sections. Once again, the parameter Q^2 is 2.5 times larger (worse) for the ARM than for the other two models. The elastic scattering data itself are sufficiently sensitive to the different treatments of the second 2^+ level to provide some model discrimination. Measurements for this nucleus have been completed also near 5 MeV, but that study is incomplete. Tentatively, the conclusions are the same.

After developing the scattering potential and models for the neutron scattering experiment, the usual i-spin dependencies were used to transform it into a proton scattering model, and the energy dependence of the potential was used to extrapolate it to 35 MeV, where tests could be made through comparisons with the measured cross sections of Deason et al.[17] They provided scattering sections for several even A isotopes of Pt. The fits of the two models to inelastic scattering cross sections are shown in Fig. 7. Here the ARM model provides an unacceptable description of scattering to the second 2^+ level. In contrast to neutron scattering at low energy, the calculations for elastic proton scattering at 35 MeV are quite satisfactory for all three models; that is, the elastic scattering cross sections provide no discrimination between different collective models. Thus it is in the inelastic scattering data of Fig. 7 that we have direct confirmation of the conclusions of the neutron scattering study, namely that the ARM model is not appropriate, but the γ-soft or γ-unstable limit of the IBA model provides an excellent description of the measurements. The E2 matrix elements needed for the IBA model were kindly provided by R. Bijker from the model of Bijker et al.[16] Their model provides excellent fits to level energies and γ-ray transition rates for ^{194}Pt, and for several Os isotopes. It is gratifying that a structure model which does an excellent job for the bound state structure of the nucleus also provides the information necessary to describe both proton and neutron scattering, and that the scattering experiments enable us to discriminate between alternate models which might be considered as realistic for nuclei in this mass region.

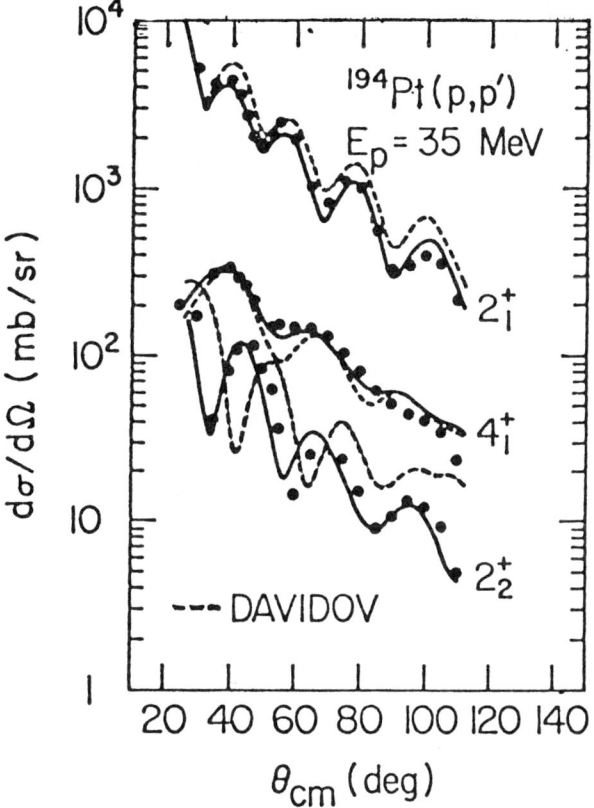

Fig. 7 Inelastic scattering cross sections for lowest three excited levels. The solid curves are for the IBA model.

COLLECTIVE EFFECTS IN LOW ENERGY ELASTIC SCATTERING

A special interest of present studies in the Os and Pt isotopes is the examination of the strength and visibility of collective effects at low incident energies. Earlier work in the Sm isotopes[18] suggested that elastic scattering cross sections were especially sensitive to these effects at incident energies below 3 MeV. It is also well known[19] that low energy total cross sections are quite sensitive to details of the models used in coupling scattering channels. On the other hand elastic scattering cross sections at higher energies, 7 MeV for neutrons,[1] or 35 MeV for proton scattering, show no sensitivity to differences in coupling models. At these energies and above, all the sensitivity is in the inelastic scattering cross sections and analyzing powers.

As an illustration of enhanced sensitivity of low energy elastic scattering cross sections, some preliminary results are shown for scattering from ^{192}Os. Total cross sections have been measured with an isotopically enriched sample,[20] and are shown together with complex potential model fits in Fig. 8. The curve shown there represents both one channel model and two coupled channels model fits. The second model couples ground state and first excited 2^+ level scattering, using first order vibrational coupling. The scattering potential is different for the two models, principally in the strength and energy dependence of the absorptive part,

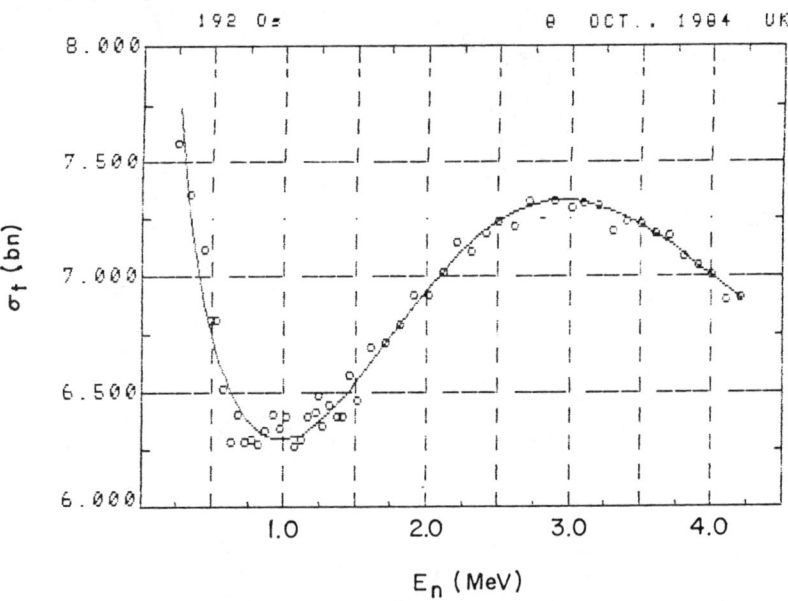

Fig. 8. Total cross sections for neutron scattering from ^{192}Os. The curve is calculated with either spherical or coupled-channels model.

but the same fit is achieved with either model. With these two scattering models constrained to fit the data of Fig. 8, the differential scattering cross sections measured at two incident energies, 1.6 MeV and 2.5 MeV, were calculated. The measurements and calculations are shown in Figs. 9 and 10, where one sees that neither model can provide a successful fit to the measured cross sections. The discrepancies are most serious for the single channel model, shown as dashed curves; but they

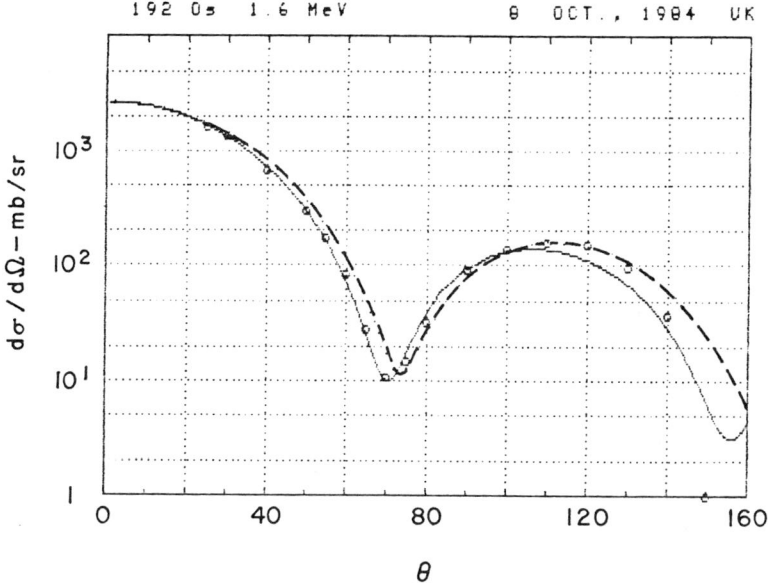

Fig. 9. Shape elastic scattering cross sections for ^{192}Os at 1.6 MeV. Curves are for spherical (dashed) and coupled-channels models fits to total cross sections.

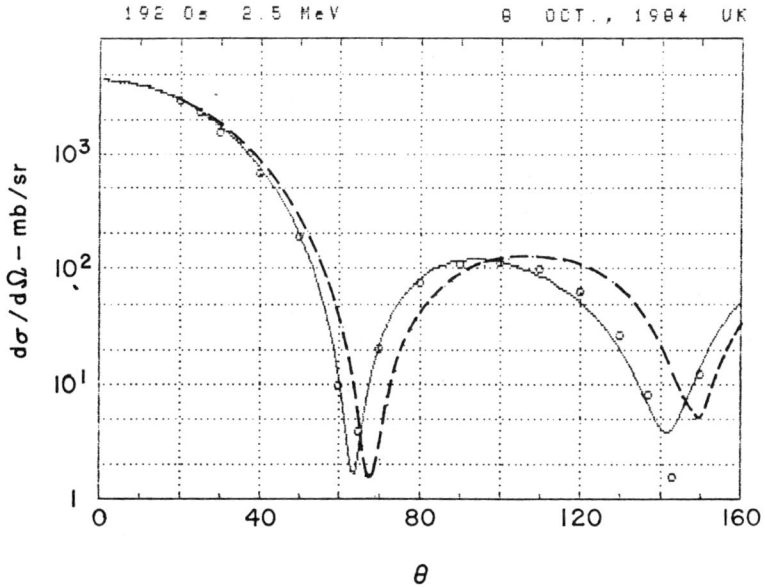

Fig. 10 Shape elastic scattering cross sections at 2.5 MeV. Curves as in Fig. 12.

are serious for both descriptions at both incident energies. To aid in visualizing the potential scattering, the compound elastic or statistical model cross sections have been subtracted from the measured values, again for both energies, so that both the curves and the points represent potential scattering only. The best representation is that of the two channels model for 2.5 MeV energy, but even that fit is unacceptably poor for the data beyond 100°.

The problem at either energy observed in Figs. 9 and 10 can be solved by adjusting potential parameters separately at each energy and for each model, but then the total cross sections shown in Fig. 8 cannot be fitted with a consistent, energy dependent potential. It is the combination of total cross sections and differential elastic scattering cross sections which provides the sensitivity to discriminate against these models, both of which ignore the second 2^+ level. Also the results for the two channels model illustrate the fact that the discrimination sensitivity of the elastic scattering decreases as the incident neutron energy is increased. The calculations are worst at 1.6 MeV.

The use of such measurements does necessitate the confident extraction of the compound system cross sections. The very thorough and careful work of P. A. Moldauer[21] and the Heidelberg theory group[22] leaves the calculations of these compound system, or compound nucleus, cross sections quite well determined, if the scattering potential is well fixed through examination of low energy scattering properties and total cross sections.

MICROSCOPIC MODEL CALCULATIONS

As noted in the Introduction, an ultimate goal in providing a comprehensive description of nuclear structure and scattering would be to provide a microscopic calculation of the nuclear structure from an assumed model Hamiltonian, which would then also be used to provide the scattering wave functions to be inserted into a coupled channels model for scattering. At present, with the scattering models employed above, one must parameterize a scattering potential, assume a form for the potentials which couple the different scattering channels, and specify typically two to four coupling strengths for the different bands of the target nucleus, and also for different multipole orders. Enough data exists to fix all of the parameters; but an approach which reduced the number of

parameters needed would provide a more powerful test of structure
models, and a more direct test of the consistency between bound
levels and scattering information.

A strong step in the direction of a microscopically based
scattering model has recently been taken by two teams associated
with analysis and theory at BRC, Lawrence Livermore Laboratory
(LLL), and the University of Middle Tennessee at Cookeville
(MTC). J. P. Delaroche of BRC and F. S. Dietrich of LLL are
collaborating on one microscopic model for the Pt isotopes, using
the Wilets and Jean Hamiltonian[23] for o-unstable nuclei. They
have used this Hamiltonian in a first test of their newly
developed methods for handling microscopic calculations partly
because much of the work of solving it can be done
analytically,[24] with less reliance on extensive numerical
calculations than would be needed for other, current models for
these nuclei. The model as adapted to ^{196}Pt provides a nucleus
whose structure is formally very similar to that of the O(6)
subgroup limit of the full IBA Hamiltonian. That nucleus had
been characterized[25] as one at the O(6) limit of the IBA
symmetries. Delaroche and Dietrich insert the Wilets and Jean
wave functions into a proton scattering calculation for
comparison with the 35 MeV scattering experiment of Deason et
al.[17] The scattering potential for the protons is taken from
Ref. 17. The resulting calculated cross sections are shown in
Fig. 11, together with the measured inelastic scattering cross
sections. Since the elastic scattering cross sections are not
very model dependent at high proton energies, they are not shown.
The agreement for the first 2^+ level is gratifying, but the phase
and magnitude of the calculations for the second 2^+ level, shown
as the heavy solid curve, are not so good. Delaroche and Dietrich
point out that the Wilets and Jean Hamiltonian leaves the first
4^+ and second 2^+ levels degenerate. That symmetry can be broken,
thereby lowering the second 2^+ level below the first 4^+ level,
by adding a term to the Hamiltonian which will mainly mix the two
low-lying 2^+ levels. This they have done, providing the results
shown as the dashed curves in Fig. 11 for the two 2^+ levels.
This marked improvement in fit, without altering noticeably
results for other levels, comes from a very small admixture of
the first 2^+ level wave function into that of the other level.
The admixture amplitude is only 0.1, which illustrates how
sensitive the cross sections are to small admixtures.

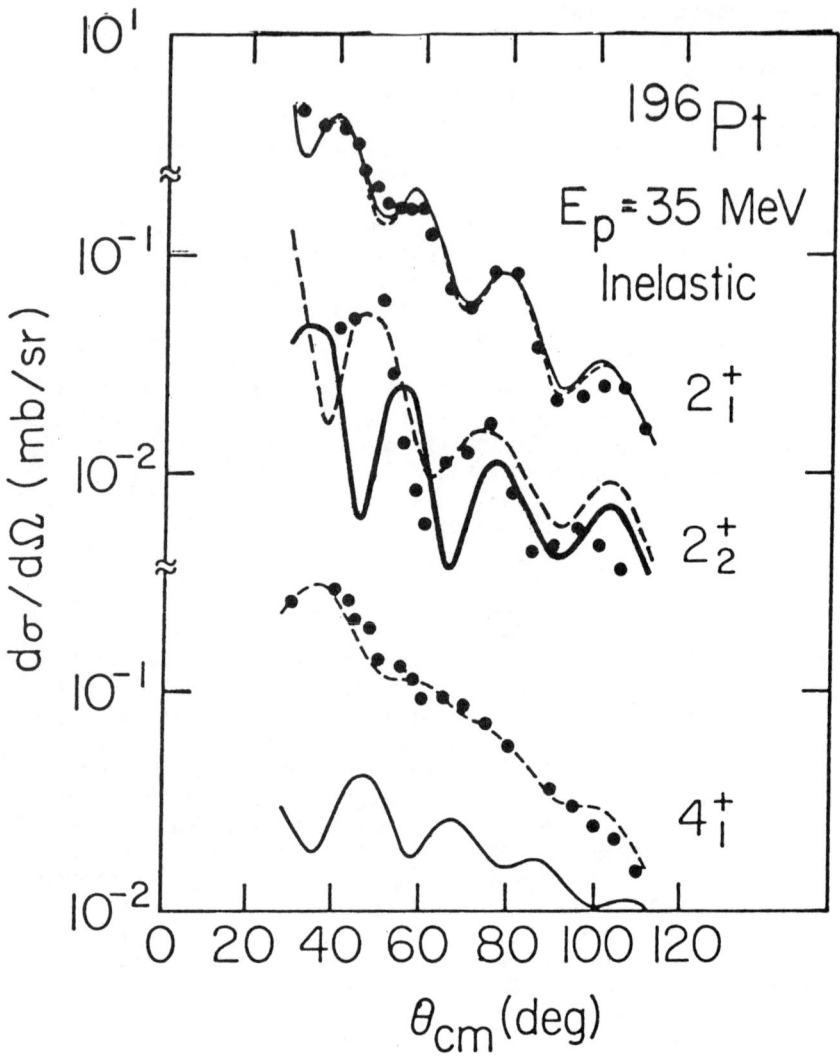

Fig. 11. Measured and calculated proton inelastic scattering cross sections. Solid curves are for the Wilets-Jean model. Dashed curves include 2^+ level mixing and one-step E4 for 4_1^+.

This last calculation illustrates an important difference between a microscopic model and the phenomenological methods used above. With the phenomenological method one might isolate a problem like this to one or a few E2 matrix elements which had

been inserted into the coupled channels calculation. But with
the presently illustrated approach, one is taken directly back to
the Hamiltonian for the problem, and the model alteration needed
to solve it. There is also the point that no ad-hoc assumptions
about the radial form of the coupling potentials are needed, nor
is one faced with the ambiguities of relating deformation lengths
or deformation parameters of coupling potentials having different
radial forms,[26] since the coupling forms and strengths are direct
results of the model.

Another microscopic calculation is available from Ch.
Lagrange, using a recently developed method for providing wave
functions for an extended DDT model of Kumar and Lagrange.[27] In
this method Lagrange finds target potential surfaces and state
wave functions as a function of two conventional deformation
parameters throughout the allowed ranges of those parameters.
The density distributions from the DDT Hamiltonian obtained in
this way weight the different deformations by the probability of
their occurrence in the different states. This model has been
solved by Kumar,[27] and the wave functions from it and a neutron
scattering potential for 8 MeV neutron scattering from ^{194}Pt were
inserted into Lagrange's microscopic coupled channels code.[26]

Since this was essentially the same model as that which was
employed to represent the neutron scattering measurements in the
older phenomenological model,[15] a comparison could be made
between the new calculation and the phenomenological model, using
the E2 matrix elements of Kumar's earlier structure calculation[15]
in the phemonenological model calculations. The results are
shown in Fig. 12 for 8 MeV neutron scattering to two 2^+ levels
of ^{194}Pt. A measurement program for 8 MeV scattering is in
progress at BRC. Since the same model is at the base of both
calculations, it is quite encouraging that the agreement between
the two calculations is so good. The calculated elastic
scattering cross sections from the two models are in excellent
agreement. There is a definite but slight difference between the
two calculations for scattering to the second 2^+ level, the lower
pair of curves in Fig. 12, which may reflect the deficiency of
the assumed radial form factor of the coupling potential in the
phenomenological calculations, or slight differences in the new[27]
and older[15] forms of the DDT model. In this example we see that
the same nuclear structure information can be extracted in two
ways. One can parameterize it in terms of E2 matrix elements
which describe scattering, and then separately seek the altered
model which provides the needed matrix elements. Or, with the
microscopic model, one can make a direct connection between bound
level structure and scattering cross sections.

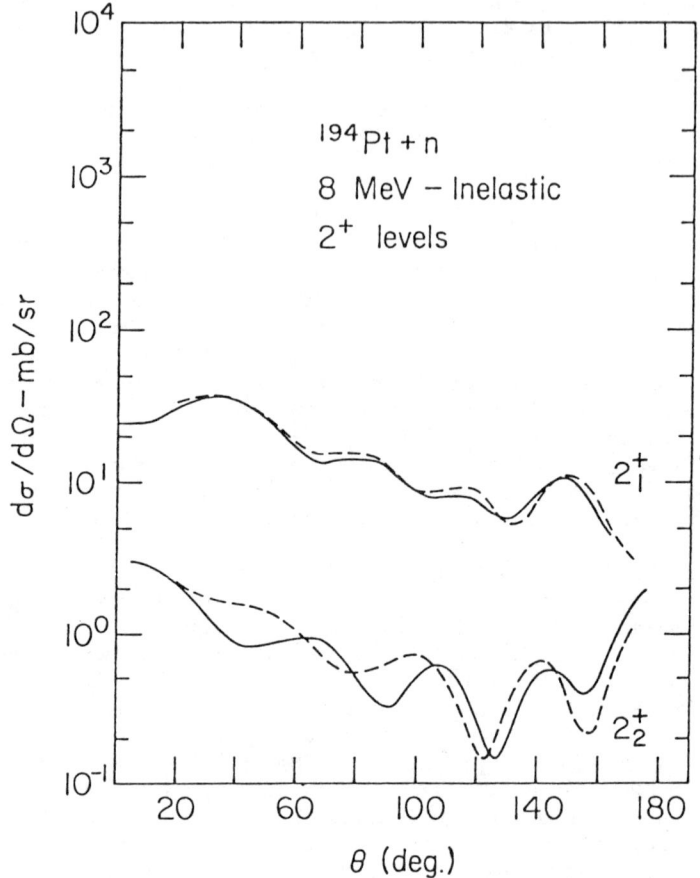

Fig. 12. Comparison of phenomenological and semi-microscopic models for inelastic scattering. Solid curves are the microscopic model results.

SUMMARY

The improved sensitivity and resolution achievable in neutron scattering experiments during the last ten years has allowed their use for the study of collective nuclear dynamics of heavy deformed and shape transitional nuclei. Several experiments show that the quadrupole excitations of the ground state band obtained from neutron scattering are the same as those inferred from Coulomb excitation, or other hadron scattering, as long as the target nuclei are more than a few nucleons removed from single closed

shell nuclei. A few experiments suggest that that is also the
case for collective excitations other than those of the ground
state band, but more detailed and complete tests of that
suggestion are needed. While neutron scattering shows sufficient
model sensitivity to discriminate between different dynamical
models for the non-axial quadrupole excitations, the sensitivity
is not large. Accurate measurements and careful data
interpretation are required.

The sensitivity to low-lying collective nuclear dynamics
seems to be especially high at low neutron energies, particularly
in the combination of total and differential elastic scattering
cross sections. This means that successful interpretation of
both low and medium energy scattering experiments should be quite
helpful in delineating the structural character of particular
nuclei. Most importantly, the low energy data are best for
proper determination of the scattering potential, an important
requirement to communicate inferred structural properties of
nuclei with confidence.

The semi-microscopic models of scattering combine target
wave functions from a structure model with a phenomenological
scattering potential, and thus provide direct and consistent
tests in which the model used to describe bound levels of the
target also provides the structure information needed directly in
the scattering description. The current success of these
descriptions of scattering is encouraging, in that it allows the
analysis of the scattering data to be done directly in terms of
parameters of a model Hamiltonian. This method also minimizes
the number of parameters to be adjusted for an interpretation.
The one comparison in which scattering descriptions used the same
nuclear structure model as the basis for both types of analyses,
phenomenological and semi-microscopic, provided good agreement
between the results of both methods for neutron scattering at 8
MeV.

ACKNOWLEDGEMENTS

The author is particularly indebted to Prof. J. L. Weil for
his cooperation and assistance with many aspects of this study of
collective effects in scattering, and indebted also to Ch.
Lagrange and to J. P. Delaroche of Bruyeres-le-Chatel with whom
he has had the pleasure of collaborating on several occasions. I
acknowledge with gratitude semi-microscopic model results
received from Ch. Lagrange. Results of the model employed by J.
P. Delaroche and F. Dietrich of Lawrence Livermore Laboratory
are also gratefully acknowledged. A collaboration at BRC with M.
Gerard Haouat of that laboratory and Prof. Tom Clegg of
North Carolina also contributed to an understanding of collective
excitation effects in neutron scattering, which is much
appreciated.

REFERENCES

1. Ch. Lagrange, J. Lachkar, G. Haouat, R. E. Shamu, and M. T. McEllistrem, Nucl. Phys. $\underline{A345}$, 193 (1980).

2. J. P. Delaroche, G. Haouat, J. Lachkar, Y. Patin, J. Sigaud, and J. Chardine, Phys. Rev. $C\underline{23}$, 136 (1982).

3. E. S. Konobeevskii, R. M. Musaelyan, V. I. Popov, and I. V. Surkova, Sov. J. Part. Nucl. $\underline{13}$, 124 (1982).

4. J. Lachkar, M. T. McEllistrem, G. Haouat, Y. Patin, J. Sigaud, and F. Cocu, Phys. Rev. $C\underline{14}$, 933 (1976).

5. R. G. Kurup, R. W. Finlay, and J. Rapaport, Nucl. Phys. $\underline{A420}$, 237 (1984).

6. R. Lecompte, P. Paradiso, J. Barette, M. Barette, G. Lamoureux, and S. Monaro, Nucl. Phys. $\underline{A284}$, 123 (1977).

7. V. R. Brown and V. A. Madsen, Phys. Rev. $C\underline{11}$, 1298 (1975) ibid, $C\underline{17}$, 1943 (1978).

8. J. P. Delaroche, R. L. Warner, T. B. Clegg, R. E. Anderson, B. L. Burks, E. J. Ludwig, W. J. Thompson, and J. I. Wilkerson, Nucl. Phys. $\underline{A414}$, 113 (1984).

9. A. S. Davydov and G. F. Filippov, Nucl. Phys. $\underline{8}$, 237 (1958) A. S. Davydov and U. S. Rostosky, ibid.,$\underline{12}$, 58 (1959).

10. M. Mirzaa, J. P. Delaroche, S. W. Yates, J. L. Weil, and M. T. McEllistrem, to be published in The Physical Review C.

11. J. P. Delaroche, M. Mirzaa, S. W. Yates, J. L. Weil, and M. T. McEllistrem, to be submitted to The Physical Review C.

12. S. F. Mughabghab, M. Divadeenam, and N. E. Holden, Neutron Cross Sections, vol. 2, Part B, (Academic Press, New York, N. Y. (1984).

13. D. I. Garber and R. R. Kinsey, Brookhaven National Laboratory Report BNL 325, Third Edition, Vol. II (1976).

14. W. P. Poenitz and J. F. Whalen, Argonne National Laboratory Report ANL/NDM-80 (1983).

15. K. Kumar, Phys. Lett. $\underline{29B}$, 25 (1969).

16. R. Bijker, A. E. L. Dieperink, O. Scholten, and R. E. Spanhoff, Nucl. Phys. $\underline{A344}$, 207 (1980).

17. P. T. Deason, C. H. King, R. M. Ronningen, T. L. Khoo, F. M. Bernthal, and J. A. Nolen, Phys. Rev. C$\underline{23}$, 1414 (1981).

18. D. F. Coope, S. N. Tripathi, M. C. Schell, J. L. Weil, and M. T. McEllistrem, Phys. Rev. C$\underline{16}$, 2223 (1977).

19. J. P. Delaroche, Phys. Rev. C$\underline{26}$, 1899 (1982).

20. S. E. Hicks, Z. Cao, J. Hanly, and M. T. McEllistrem, Bull. Am. Phy. Soc. $\underline{28}$, 984 (1983).

21. P. A. Moldauer, Proc. Int'l. Conf. Interactions of Neutrons with Nuclei, ERDA Report CONF-760715-P1 (ed. by E. Sheldon, USERDA, TIC, Oak Ridge, TN, 37831 (1976)), and references therein.

22. J. W. Tepel, H. M. Hofmann, and M. Herman, Proc. Conf. Nuclear Cross Sections for Technology, Report NBS-594, 765 (ed. by J. L. Fowler, C. H. Johnson, and C. D. Bowman, National Bureau of Standards, Washington, D. C. 20234 (1980)).

23. L. Wilets and M. Jean, Phys. Rev. $\underline{102}$, 788 (1956).

24. F. Dietrich and J. P. Delaroche, Centre d'Etudes de Bruyeres-le-Chatel, private comm.

25. R. F. Casten and J. A. Cizewski, Nucl. Phys. $\underline{A309}$, 477 (1978).

26. Ch. Lagrange, Centre d'Etudes de Bruyeres-le-Chatel, private comm.

27. K. Kumar and Ch. Lagrange, to be published.

DETERMINATION OF NUCLEAR TRANSITION DENSITIES
WITH VARIOUS PROBES

J. A. Carr and F. Petrovich
The Florida State University, Tallahassee, FL 32306

J. J. Kelly
University of Maryland, College Park, MD 20742

ABSTRACT

We propose a new method for the systematic analysis of elastic and inelastic scattering data which maximizes the information that can be deduced and quantifies the comparison with theoretical models. Our present application studies the utility of nucleons and pions in the determination of matter distributions for normal parity transitions using pseudodata generated for the first 2^+ state of ^{18}O. The intent of this work is to provide perspective on the role of neutron-nucleus scattering in these determinations. We conclude that the model dependence of analyses of this type can be reduced by having available both neutrons and protons in the same energy region. Some additional remarks on the sensitivity of neutrons to unnatural parity spin and current distributions are included.

1. ELASTIC AND INELASTIC SCATTERING

Much of the discussion at this conference concerns the use of elastic and inelastic neutron-nucleus scattering data in nuclear reaction and structure studies. Nearly all of the precision neutron data currently available is restricted to incident energies $E_n < 27$ MeV. This paper will demonstrate that neutron-nucleus scattering data for $E_n > 50$ MeV will provide important new information, complementing that available from other probes (in particular, electrons, protons, and pions) and thereby increasing our knowledge of effective interactions and nuclear densities. The illustration is made within the framework of a theoretically based, phenomenological approach to the ordinary scattering processes which is made practical through the use of linear expansion techniques.[1] We assert that this guided phenomenological approach (and variations thereof) is the most useful way to interpret experimental data on elastic and inelastic scattering. Two general issues will be addressed: the sensitivity of the probes of interest to the various components of the nuclear response and the capability of the probes to precisely determine the radial form of the relevant nuclear density. We proceed under the assumption that non-relativistic, single scattering ideas (optical potential, distorted wave approximation, nucleus independent effective interactions) are applicable in describing the scattering of the various probes of interest from nuclei at the incident energies to be considered. There has been considerable progress in the theory of both the nucleon-nucleon[2-4] and pion-nucleus[5-7] effective interactions in recent years, and many of the major features of the nucleon-nucleus and pion-nucleus scattering data are qualitatively understood within the single scattering framework.[8-13] More systematic studies, based on the theoretically guided phenomenology adopted here, can raise comparisons between theory and experiment to a more quantitative level.

The essential ideas behind the single scattering picture are most easily illustrated using the momentum space representation and assuming local, density independent interactions. Then, the transition amplitude for the elastic and inelastic scattering of an elementary probe from a nucleus contains terms of the form[14]

$$T \sim \frac{2}{\pi} \int k^2 dk\; D(k,E_p,\theta) t(k,E_p)\; \rho(k,E_x) \qquad (1)$$

where D is a distortion function depending on the continuum wave functions describing the motion of the probe with respect to the target nucleus, t is the effective two-body interaction between the probe and the target nucleons, and ρ is a nuclear density which is reaction independent. The E_p, θ, and E_x appearing as arguments of D, t, and ρ are the incident probe energy, the scattering angle, and the excitation energy of the nucleus, respectively. For elastic scattering ρ is a ground state density (ρ_{gs}) and the distortion function is constructed from a plane wave and the appropriate scattering solution to the Schrodinger equation containing the optical potential $U_{opt} = t\rho_{gs}$. For inelastic scattering ρ is a transition density (ρ_{tr}) and the distortion function is constructed from the elastic scattering wave functions as prescribed by the distorted wave approximation (DWA). The transition potential $U_{tr} = t\rho_{tr}$ is just the usual DWA form factor. The simple product relation for the scattering potentials $U = t\rho$ is destroyed when realistic, density dependent effective interactions[2-7] are encountered; however, there is still a definite relation between U, t, and ρ in this case and the basic idea conveyed by Eq. (1), that the transition amplitude is comprised of three primary factors, is preserved. Similar remarks apply to the non-localities associated with the exchange amplitudes arising from the antisymmetrization required in describing nucleon-nucleus scattering. These effects can generally be incorporated in the effective interaction.[15]

The essential point behind Eq. (1) is that elastic and inelastic scattering data contain information on t and ρ. The quality of this information is governed by D, which is inherently connected to the properties of the coupling interaction. For example, when t is strong the distortion significantly obscures the effect of t and ρ in the nuclear interior. Data can be used to gain information on ρ when t is known and, conversely, when ρ is known the data provide information on t. The same t should be used to construct the appropriate scattering potentials for both elastic and inelastic scattering, while the same ρ is used for different reactions populating the same state. Given a sufficient data base, information on t and ρ can be determined simultaneously. These are the basic ideas which underly most of the discussion of this paper.

The above discussion of the transition amplitude for elastic and inelastic scattering is schematic and somewhat oversimplified. In reality, the complete transition amplitude for a given probe scattering from a nucleus is a sum of terms like Eq. (1) containing different densities and interaction components. Corresponding to the coordinates $(\vec{r}_i, \vec{p}_i, \vec{\sigma}_i, \vec{\tau}_i)$ for a single nucleon in the target, the complete set of nuclear densities are conveniently designated[16] ρ^m, ρ^s, ρ^ℓ, and $\rho^{\ell s}$, with m, s, ℓ, and ℓs representing matter, spin, current, and spin-current, respectively. The $\vec{\tau}_i$ coordinate is dealt with by labeling the densities ρ_p, ρ_n, ρ_0, and ρ_1 for proton, neutron, isoscalar, and isovector, respectively. By making the Born approximation (where $D \sim \delta(k-q)/kq$ with q representing the asymptotic momentum transfer) and working with a cartesian representation of the transition densities and effective interactions, it is possible to write the scattering cross sections for all the reactions of interest here in a compact, common form which reveals the full complexity of the scattering problem. For $0^+ \to J^\pi$ transitions, these expressions are[17]

$$\frac{d\sigma^N}{d\Omega} = 4\pi \left(\frac{\mu}{2\pi\hbar^2}\right)^2 (2J+1) \{ |\bar{t}_m \rho_J^m|^2 + |\bar{t}_{\ell s} \rho_J^{\ell s}|^2 + I_{m,\ell s}$$

$$+ |\bar{t}_{s\perp} \rho_J^s|^2 + |\bar{t}_\ell \rho_J^{\ell\perp}|^2 + I_{s,\ell\perp} \} \quad (2)$$

and

$$\frac{d\sigma^U}{d\Omega} = 4\pi \left(\frac{\mu}{2\pi\hbar^2}\right)^2 (2J+1) \{ |\bar{t}_{s\|} \rho_J^{s\|}|^2 + |\bar{t}_{s\perp} \rho_J^{s\perp}|^2$$

$$+ |\bar{t}_\ell \rho_s^\ell|^2 + I_{s\perp,\ell} \} \quad (3)$$

where N and U refer to natural and unnatural parity, respectively, μ is the reduced relativistic mass of the incident probe, the $\bar{t}_\alpha$ are combinations of the components of t and the Born approximation D which couple solely to a given ρ^α, and the I represent interference terms. The arguments of $\bar{t}_\alpha$ and ρ^α have been suppressed for convenience. In addition, it is to be noted that each $\bar{t}_\alpha \rho^\alpha$ factor in Eq. (2) and Eq. (3) contains an implied sum over charge indices, i.e. T = 0 and 1, or p and n. The appearance of the symbols || and $\perp$ (for longitudinal and transverse) indicate directions parallel and perpendicular to the momentum transfer. Further information on Eq. (2) and Eq. (3) and the associated notation, in particular the precise definition of the ρ^α and the relationship between the $\bar{t}_\alpha$ and the more conventional central, spin-orbit, and tensor interaction components can be found in ref. 16 and 17. For the present purpose it suffices to say that the above "$\bar{t}\rho$" form for the cross sections makes it easy to keep track of and discuss the relevant physics.

Since several $\bar{t}_\alpha \rho^\alpha$ contribute to a given cross section according to Eq. (2) and Eq. (3), it should be clear that a single cross section measurement is, in general, not sufficient to allow the extraction of information on a particular $\bar{t}_\alpha$ or ρ^α. There are a number of special classes of transitions for which the properties of

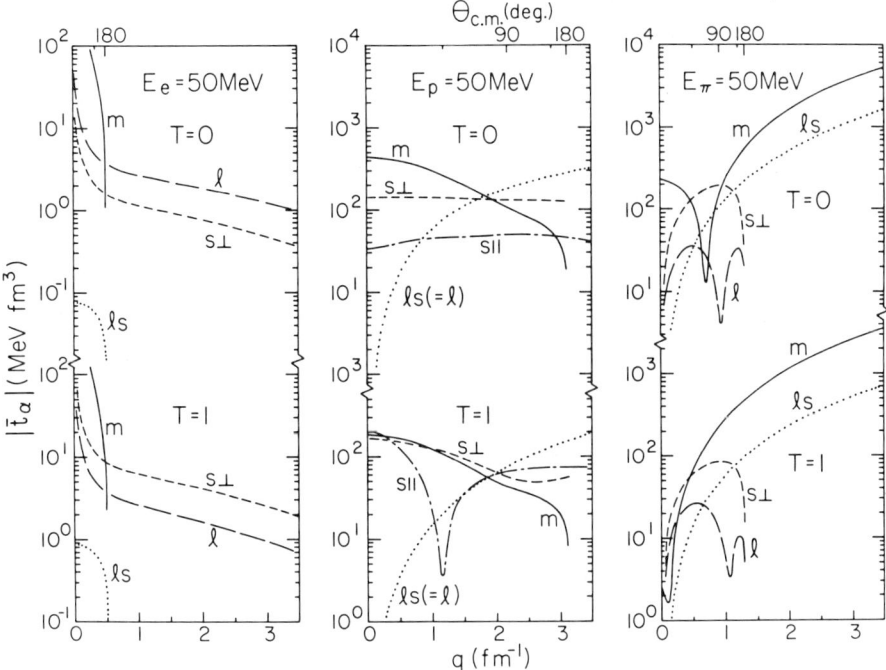

Fig. 1. Moduli of the isoscalar (top) and isovector (bottom) $\bar{t}$ "interaction" components for 50 MeV electrons, nucleons and pions as a function of momentum transfer. The scale across the top shows the correspondence between scattering angle and momentum transfer for each case.

the $\bar{t}_\alpha$ and ρ^α are such that Eq. (2) and Eq. (3) can effectively be reduced to expressions containing only one $\bar{t}_\alpha \rho^\alpha$ factor. Even when such a reduction is possible, there are still two unknown factors since each $\bar{t}_\alpha \rho^\alpha$ contains an implied sum over charge indices. An exception to this would be the case of self-conjugate nuclei for which it is reasonable to assume that the transitions are purely isoscalar or isovector in character. For natural parity transitions to low lying collective states in the target nucleus, ρ^m is much larger than either ρ^s, ρ^ℓ, or $\rho^{\ell s}$. In addition, for the probes of interest here, $\bar{t}_m$ is much larger than either $\bar{t}_s$, $\bar{t}_\ell$, or $\bar{t}_{\ell s}$, so it is sufficient to consider only the $\bar{t}_m \rho^m$ in Eq. (2) in discussing these transitions. For strong $0^+ \to 1^+$ unnatural parity transitions (Gamow-Teller and giant M1 transitions) and unnatural parity transitions to high spin states of stretched configuration, ρ^ℓ is small and definite relationships exist between $\rho^{s\parallel}$ and $\rho^{s\perp}$ which can be exploited to reduce the complexity of Eq. (3). Unnatural parity $0^+ \to 0^-$ transitions are another important class of transitions for which only the $\bar{t}_{s\parallel} \rho^{s\parallel}$ term in Eq. (3) is important. All of the above has been discussed in some detail in Ref. 1, 8, 9, 11, 13, and 17 and references cited therein. Here we will focus mainly on natural parity transitions for which it is safe to assume that only $t_m \rho^m$ contributes to the transition. Comments on unnatural parity transitions are made in Sect. 5 of this paper.

To provide some feeling for the basic properties of the complete effective interactions for electrons, nucleons, and pions, the $\bar{t}_\alpha$ of Eq. (2) and Eq. (3) for incident energies of 50 MeV are displayed as a function of q in Fig. 1. The electron-nucleon interaction is the usual electromagnetic interaction including the effect of the finite size of the nucleon and relativistic corrections to order M^{-2} in the nucleon mass.[18] The nucleon-nucleon interaction is just the G-matrix interaction of Bertsch et al.[19] (M3Y) and the pion-nucleon interaction is a local form based on the Rowe, Salomon, and Landau phase shifts.[20] A similar set of curves, for higher probe energies, is given in Ref. 17. The dominance of $\bar{t}_m$, particularly in the T = 0 interaction channel, is clear for all three probes. This persists at higher energies. Note also that only nucleons have $\bar{t}_{s\parallel} \neq 0$. Although all of the interactions displayed can be considered "realistic", the electron-nucleon interaction has the firmest theoretical base. Correspondingly, information from electron-nucleus scattering plays a very special role in any considerations of the use of elastic and inelastic scattering in the study of atomic nuclei.

2. NATURAL PARITY TRANSITIONS

As discussed above, natural parity transitions to low-lying collective states in the target nucleus depend essentially only on the matter transition densities. There are two matter densities to be considered -- the proton distribution and the neutron distribution. For this class of transitions we can rewrite Eq. (2) in the following approximate form which makes the dependence on ρ_p^m and ρ_n^m explicit.

$$\frac{d\sigma_\nu}{d\Omega} = 4\pi \left(\frac{\mu}{2\pi\hbar^2}\right)^2 (2J+1) |\bar{t}_m \rho^m|^2$$

$$\approx 4\pi \left(\frac{\mu}{2\pi\hbar^2}\right)^2 (2J+1) |\bar{t}_m^{\nu p}(0)\rho_p^m + \bar{t}_m^{\nu n}(0)\rho_n^m|^2$$

$$\approx 4\pi \left(\frac{\mu}{2\pi\hbar^2}\right)^2 (2J+1) |\bar{t}_m^{\nu p}(0)|^2 |\rho_p^m + R_{\nu n}\rho_n^m|^2 \qquad (4)$$

Here ν is a label specifying the incident probe and $\bar{t}_m^{\nu p}(0)$ and $\bar{t}_m^{\nu n}(0)$ are the volume integrals of the proton and neutron components of the interaction between the probe and the target nucleons. In the last line of Eq. (4) we have introduced the ratio $R_{\nu n} = \bar{t}_m^{\nu n}(0)/\bar{t}_m^{\nu p}(0)$ which provides a measure of the sensitivity of the probe to the neutron transition density ρ_n^m relative to the proton transition density ρ_p^m. It will prove convenient to define $R_e = R_{en}$, $R_N = R_{pn} =$

R_{nn}^{-1}, and $R_\pi = R_{\pi^-n} = R_{\pi^+n}^{-1}$ for discussions of this sensitivity factor.

The most reliable information on ρ_p^m is obtained from electron scattering because $R_e \approx 0$ and the Coulomb interaction is well established. Information on ρ_n^m can be obtained from simultaneous considerations of electron and hadron scattering since R_N and R_π do not vanish. Information on ρ_p^m and ρ_n^m can also be obtained independently of electron scattering data provided data for both charge states of the nucleon or the pion are available. It is important to note that values of R_N and R_π that differ from unity are essential for the determination of differences between ρ_p^m and ρ_n^m using the two nucleon (p and n) or pion (π^+ and π^-) charge states in this manner. That this is so is clearly seen from the relation

$$\frac{d\sigma_\nu/d\Omega}{d\sigma_{\nu'}/d\Omega} = \left| \frac{\rho_p^m + R_\gamma \rho_n^m}{R_\gamma \rho_p^m + \rho_n^m} \right|^2 \tag{5}$$

where $\gamma = N$ or π with $\nu, \nu' = p, n$ or π^-, π^+, respectively. When $R_\gamma = 1$ the cross section ratio in Eq. (5) is unity independent of ρ_p^m and ρ_n^m and the probe possesses no intrinsic capability for distinguishing between proton and neutron transition densities. Even when electron scattering data are available, a value of R_γ that is substantially different from unity has the advantage of decoupling the components of $\bar{t}_m$ and ρ^m in attempts to obtain information on $\bar{t}_m$ and ρ^m via simultaneous analysis of data from electromagnetic and hadronic probes.

These arguments are strictly valid only in the plane wave limit. The presence of distortion reduces the clarity with which these ideas can be applied and a more quantitative analysis is required to extract neutron-proton differences. Electrons are affected mainly by the mean Coulomb field of the nucleus. For high incident energies the distortion effects are relatively small except for heavier targets and high momentum transfers. For hadronic probes the distortion arises from both the Coulomb and strong hadronic mean fields. The mean hadronic field is mainly that associated with the strong central part of the optical potential which is given approximately by

$$U_{opt}^\nu \approx \bar{t}_m^{\nu p}(0)\rho_p^m + \bar{t}_m^{\nu n}(0)\rho_n^m$$

$$\approx \bar{t}_m^{\nu 0}(0)\rho_0^m \tag{6}$$

where the ρ^m denote ground state densities. In the second line of Eq. (6) we have switched from the neutron-proton to the isospin representation and made use of the fact that typically $\bar{t}_m^0 > \bar{t}_m^1$ and $\rho_0^m \gg \rho_1^m$. The essential point is that distortion effects for hadronic projectiles are driven by the isoscalar part of the probe-nucleon effective interaction $\bar{t}_m^0$. The stronger $\bar{t}_m^0$ is, the more severe the distortion effects.

The general features of the sensitivity factor R for the two hadronic probes of interest can be understood from qualitative arguments. The nucleon-nucleon effective interaction at low energies is essentially even state in character with $R_N \approx 2 - 3$ corresponding to $R_N^T = \bar{t}_m^0(0)/\bar{t}_m^1(0) \approx -(3 - 2)$ for the ratio of the isoscalar to isovector nucleon-nucleon interaction components. The deviation from the Serber limit ($R_N = 2$) is associated with the strong second order tensor contributions to the central triplet even component of the effective interaction.[2-4,21,22] The pion-nucleon interaction for $E_\pi < 300$ MeV is dominated by the coupling to intermediate isobar-hole states.[5-7] A simple estimate based on the Clebsch-Gordon coefficient for forming the J=3/2, T=3/2 Δ-resonance leads to $R_\pi = 3$. This corresponds to $R_\pi^T = 2$. Thus protons and π^- couple more strongly to ρ_n^m while neutrons and π^+ are more sensitive to ρ_p^m in the incident energy region considered above.

To be more precise about the utility of a particular probe in an actual experiment, it is necessary to examine R and $\bar{t}_m^0$ for realistic interactions. We have computed R and $\bar{t}_m^0$ for nucleons and pions for incident energies over the range 0 - 800 MeV. The nucleon results were obtained using the Love and Franey parameterization[23] of the free t-matrix for $E_N > 100$ MeV, while the results for $E_N < 50$ MeV and 50 MeV $< E_N < 100$ MeV were obtained from the M3Y interaction[19] and the free limit of the Paris G-matrix interactions,[24] respectively. The pion results were obtained from Rowe, Salomon and Landau phase shifts[20] for $E_\pi < 350$ MeV and from the phase shifts of Davies[25] for $E_\pi > 350$ MeV. The results are shown in Fig. 2. The top half of the figure shows $|R|^2$ and the lower half shows $|\bar{t}_m^0(0)|$. Values of $|R|^2 \neq 1$ indicate regions of sensitivity to proton-neutron

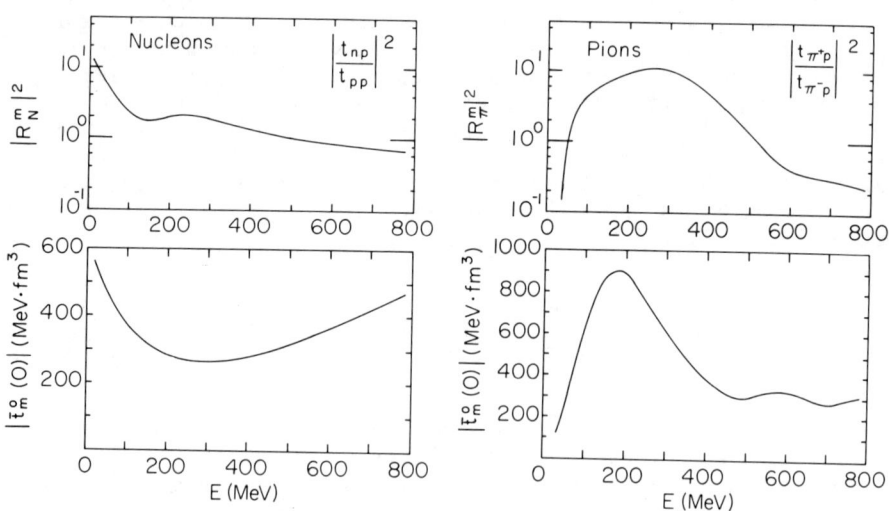

Fig. 2. Moduli squared of the ratios R_N and R_π defined in the text (top) and moduli of the T=0 matter interaction (bottom) as a function of bombarding energy for nucleons (left) and pions (right).

differences. A large value of $|\bar{t}_m^0|$ indicates large distortion effects which inhibit the determination of $\bar{t}_m$ or ρ^m from cross section data.

The nucleon results on the left indicate that the sensitivity to proton-neutron differences, measured by deviations of $|R|^2$ from unity, is large at low energies; however, distortion effects are also large in this energy region due to the strong refractive potentials produced by $\bar{t}_m^0$. Distortion effects are less significant as energy increases, with the region 150 MeV < E_N < 450 MeV exhibiting the highest transparency. Unfortunately, the sensitivity to proton-neutron differences is reduced in this region. The sensitivity increases again at high energy, but the effects of distortion also increase as the matter coupling becomes stronger and more absorptive above the energy for the opening of the pion channel. Our two criteria of $|R|^2$ differing significantly from unity and weak distortion are met for nucleons in the energy region 50 MeV < E_N < 100 MeV since the distortion effects are rapidly decreasing while the value of R^2, although decreasing, is still larger than two. This energy region would also appear to be the most easily accessible experimentally, i.e. with regard to the construction of a facility for precision (n,n') measurements.

The pion results on the right show that $|R_\pi|^2 = 9$ is obtained near the peak of the Δ-resonance at $E_\pi = 180$ MeV as expected from the simple model of the pion-nucleon interaction discussed above. However, the strong $|\bar{t}_m^0(0)|$ at this energy indicates that very strong absorption will obscure information from the nuclear interior. Note that the sensitivity decreases in the region above and below the resonance where the distortion is smaller. There appear to be two regions, $E_\pi < 50$ MeV and $E_\pi > 600$ MeV, where distortion is small and the sensitivity improves since R becomes less than one. Both of these regions have been discussed elsewhere.[26,27]

As a concrete example of the points made above, we have performed some sample calculations for inelastic scattering of nucleons and pions populating the 3^- state in ^{208}Pb. We assumed $\rho_n^m = (N/Z)\rho_p^m$, with ρ_p^m modeled with a pure $(1h_{9/2}1i_{13/2}^{-1})$ harmonic oscillator wave function. This is for purposes of illustration only, and does not reflect any fit to data. Results of a Born approximation calculation are shown in Fig. 3. It is clear that the sensitivity of nucleon and pion probes to neutron-proton density differences would be easy to exploit if there were no distortion effects. The peak cross section ratios of $\sigma_p/\sigma_n = 1.27$ ($\sigma_-/\sigma_+ = 1.45$) for nucleon (pion) scattering reflect the simple formula in Eq. (5) with $R_N = 1.8$ ($R_\pi = 2.6$) obtained from the nucleon (pion) interactions at the appropriate momentum transfer ($q = 0.6$ fm^{-1}). The absence of distortion makes the connection between the cross section ratios and density ratios very straightforward. The curves are proportional to one another because of the choice $\rho_n^m(q) \propto \rho_p^m(q)$. Differences in the shape of the neutron and proton densities would be manifest in shape differences in the cross sections.

The simple relationships evident in Fig. 3 are only valid in Born approximation, which neglects the effect of the interaction of the probe with the mean fields of the target nucleus in the entrance

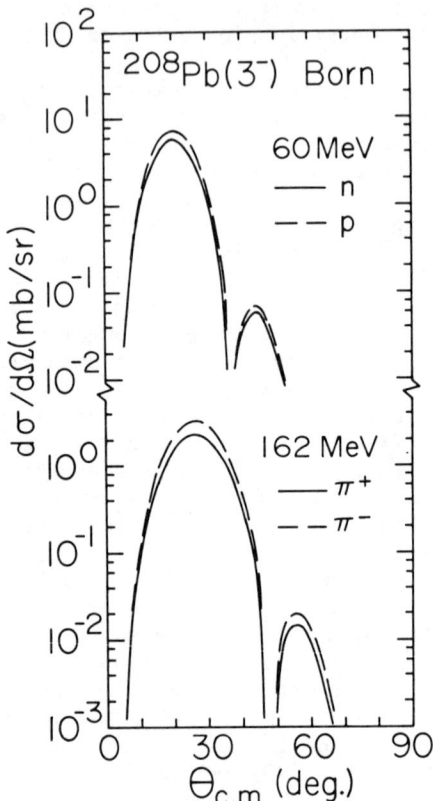

Fig. 3. Born approximation results for excitation of the 3^- state in ^{208}Pb with 60 MeV nucleons and 162 MeV pions.

and exit channels before and after the inelastic event. It is these processes which give rise to the distortion effects embodied in the distortion function D appearing in Eq. (1). The entrance and exit channel interactions cause D to deviate from its $\delta(k-q)$ Born approximation behavior. Typically D is smeared out over a range of k and the effective area it subtends is reduced relative to the Born limit. The Coulomb interaction causes D to be different for the distinct charge states of the same probe. This is illustrated in Fig. 4 where the results of DWA calculations are shown. Realistic distorting potentials have been assumed. The large difference between distortion effects for the two charge states makes a simple application of ratio techniques impossible. (Note that this example is a worst case. Simple ratio arguments are more justified for scattering from light nuclei.)

A model must be introduced to deal with the shape differences produced by the distortion, and disentangle them from effects due to possible shape differences in the scattering potentials. One procedure for dealing with this problem is to use a collective model for

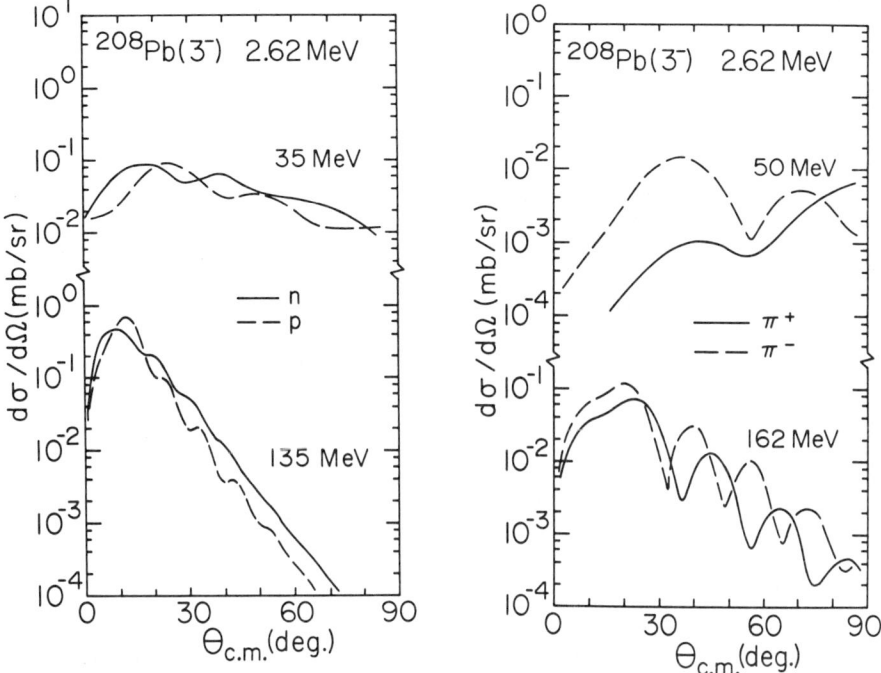

Fig. 4. DWA results for excitation of the 3^- state in ^{208}Pb with nucleons (left) and pions (right) at the indicated energies.

the combined effect of D, t, and ρ in Eq. (1). The extracted values of β_p and β_n determine the corresponding multipole matrix elements of ρ^m using the prescription of Bernstein.[28] The assumption that the shapes of the neutron and proton scattering potentials are properly represented by such a model has been discussed in detail elsewhere.[29] Another approach is to assume that ρ_n^m is proportional to ρ_p^m, make a microscopic calculation assuming a particular effective interaction and phenomenological optical potentials, and adjust the constant of proportionality between ρ_n^m and ρ_p^m to fit the experimental data.[21,30,31] In any case, these methods make limited use of the information in the data and more sophisticated methods are available as will be discussed in the next section of this paper.

To summarize, we have noted that low-lying natural parity excitations are usually dominated by the coupling to the matter transition density. An examination of the sensitivity of the coupling interactions to neutron-proton density differences indicates that nucleons below about 100 MeV are sufficiently sensitive to these differences to be useful in spectroscopic studies. Lower energies provide the strongest isovector coupling and hence the greatest sensitivity to n-p differences, but distortion effects are large. These effects decrease as energy increases and would appear to be reaching a manageable level at about 50 MeV. Neutron scattering at E_n > 50 MeV is required to provide a complete picture of nucleon-nucleus scattering in this energy region. The importance of having both charge states of hadronic probes is evident from the useful results of studies of π^+ and π^--nucleus scattering near the Δ-resonance where the pion-nucleon coupling is not even optimal.[8,32]

3. MODELING HADRONIC SCATTERING

The customary procedure for evaluating the adequacy of models for elastic and inelastic scattering has been to make a qualitative comparison between calculations and data for many individual cases. The analysis of elastic scattering data has often been restricted to parameter searches within the rather arbitrary analytic confines of the Woods-Saxon optical model. The relationship between these parameters and microscopic interactions is tenuous at best. The ability of an interaction model to reproduce inelastic scattering is rarely quantified beyond the level of global scale factors. The fact that the same matter interaction that generates the optical potential in a folding model is also responsible for the excitation of the matter part of the nuclear response for inelastic scattering is also often ignored. Similarly, the customary procedure for extracting nuclear structure information from hadronic inelastic scattering data has been to vary a scale factor that has been applied to the interesting structure component of some promising model. These methods limit the information extracted from the data to that contained in the magnitude of the first diffraction peak in the angular distribution. Moreover, the analysis of each experiment is in many cases performed in isolation, despite the fact that inelastic scattering between the same states share a common set of structure variables. Clearly, a simultaneous analysis of all data available for a given transition is preferable.

All of the above procedures are holdovers from the past when the quality of the experimental data, the status of the theories of effective interactions, and the computing facilities required for more extensive comparisons between theory and experiment were considerably behind where they are at present. Aside from the obvious lack of high energy neutron-nucleus scattering data, the quality of electron, proton, and pion-nucleus data available today is truly remarkable. The advances in the theories of effective interactions have also been significant. The fact that we have a good qualitative picture of the elastic and inelastic scattering data is clear from Ref. 2-13 and many of the papers cited therein. Two papers at this conference provide additional information on this point. One is that of von Geramb[33] who discussed the consistent treatment of high energy elastic and inelastic nucleon-nucleus scattering data. The other is that of Dietrich[34] who discussed the simultaneous study of low energy nucleon elastic, inelastic, and charge exchange data using consistent density-dependent effective interactions. The time has come to update our approach.

Here we propose a new philosophy for the analysis of elastic and inelastic scattering. We shall also briefly describe some possible applications of this philosophy to the inelastic scattering of nucleons. Although our specific remarks in the next section will be

restricted to the extraction of neutron transition densities from
inelastic scattering data, the proposed philosophy is far more general and can be easily expanded to include elastic scattering and
charge exchange reactions as well as the study of effective interactions and various nuclear structure questions. This approach will
reduce the ambiguities that occur in more limited fitting procedures,
will provide a clear statement of the precision with which structure
variables can be determined, will permit the detailed calibration of
the density dependent effective interactions for nucleon and pion
scattering, and will allow for the simultaneous analysis of all
electron, nucleon, and pion scattering data available for a given
transition.

Our proposal is to perform a linear expansion analysis within
the context of a folding model.[1] The quantity being studied is expanded as a linear series of basis functions times coefficients.
The reaction calculations are then performed for each basis function
individually and stored. The coefficients are then optimized with
respect to the largest possible data set. Folding models can differ
greatly in details and techniques, but all are characterized by the
common "$t\rho$" form representing the folding of an effective interaction
t with a nuclear transition density ρ outlined in the first section
of this paper. For any particular transition, both t and ρ have matter, spin and convection components which are coupled together as
described schematically in Eq. (2) and Eq. (3). The crucial idea is
that the control of either factor, interaction or structure, permits
the systematic evaluation of the other.

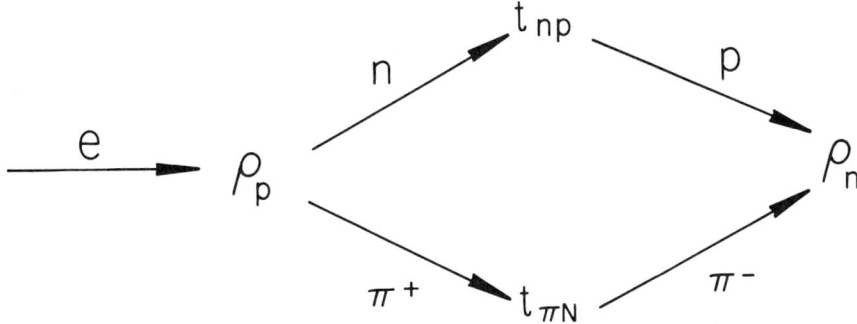

Fig. 5. Schematic picture of the procedure for determining t and ρ_n
under the condition that R_N and $R_\pi = \infty$.

A schematic picture of this concept is shown in Fig. 5, which
corresponds to the situation when the sensitivity ratios R_N and R_π
discussed above are infinity. In such a case we can use electron
scattering to define the proton matter density. With ρ_p^m known, we
can use neutron (or π^+) scattering to define t_{pn} (or $t_{\pi N} = t_{\pi^+ p} = t_{\pi^- n}$) at some bombarding energy. Then we can use proton (or π^-)
scattering to obtain ρ_n^m with this interaction. Although this picture

is oversimplified, it conveys the essential concepts that underly a sensible phenomenology. For finite R, we must first adjust the isoscalar components of the interaction to reproduce data for transitions in self-conjugate nuclei whose ρ_p^m is available from electron scattering. The remaining quantitites for $N \neq Z$ nuclei may then be determined from a simultaneous fit to all three data sets. Note that if we have electrons, nucleons <u>and</u> pions available, the fitting problem is substantially overdetermined and the reaction theory is rigorously tested by requiring a consistent description of the entire data set. Other bombarding energies and transitions can be added to further test the validity of the models used in the calculations.

This last point is crucial since a theoretical calculation is generally subject to truncation and approximations. It is often difficult or impossible to estimate the accuracy of these approximations, much less the impact of higher-order corrections. Thus we rarely expect quantitative accuracy from parameter free calculations. Nevertheless, if the results of a theory are amenable to a linearized parameterization, their physical interpretation and functional form can be used as the foundation for a guided phenomenology. The theory may be judged successful if only modest adjustments of the calculated parameters are required to fit a large body of systematic data.

If a reaction is to be a useful probe of nuclear structure, it must provide an accurate description of those transtions whose structure is known. Otherwise, the residual deficiencies of the reaction model will be artificially propagated into the deduced nuclear structure quantities with uncontrollable and unpredictable consequences. Because we cannot realistically expect quantitative accuracy from <u>ab initio</u> calculations, the reaction model must be optimized by fitting data. As we do not expect the remaining deficiencies in the models of several different reactions, or even of a single reaction at several energies, to be strongly correlated, their impact can be minimized only by a simultaneous analysis of the largest possible data set.

One important question that is under current investigation concerns the density dependent modifications of the two-nucleon interaction in the nuclear environment for proton energies below 200 MeV.[2-4,10-12,33,34] The proposed modeling methods can examine the premises of the local density approximation independent of the predictive power of nuclear matter theory, provided the relevant nuclear structure variables can be accurately determined from some independent source. This condition is satisfied for the class of natural parity isoscalar transitions in a self-conjugate nucleus for which the measured transverse electromagnetic form factors are negligible over a wide range of momentum transfer. The relevant structure variable for this class is then reduced to a single radial transition density whose proton component can be precisely determined from electron scattering measurements and whose neutron component can be equated with the proton component by invoking charge symmetry. For this class of transitions, the nuclear structure can be considered known and the effective interaction t is isolated for study. We now perform a linear expansion of t about a suitable initial value t_0 as

$$t = t_0 + \sum_{i=1}^{N} a_i t_i \qquad (7)$$

where the correction terms t_i may depend upon momentum transfer, the local density, or any state-independent degrees of freedom. A useful parameterization of the medium corrections, guided and motivated by nuclear matter theory and the local density approximation, was proposed in Ref. 35. It was subsequently demonstrated that, at 135 MeV, a single empirical interaction is capable of simultaneously fitting five normal parity multipolarities, 0^+ to 4^+, in a single nucleus, ^{16}O.[1,36] The success of this procedure depended in part upon the differential radial sensitivity of the transitions considered. The transition densities for low multipolarity states peak in the nuclear interior and are sensitive to the effective interaction at high density. High multipolarity states are surface peaked and are sensitive to the low density limit of the interaction.

It is important to note that there is no inherent restriction of t to local or nonrelativistic forms. The technique is quite general and versatile. All that is really required is that the important variables represent properties characteristic of the interaction and the nuclear medium rather than properties specific to each nuclear state.

We can now envision applying these techniques to the other components of the effective interaction. Once the contributing components of the interaction are under control, we can proceed to study the detailed structure of particular nuclear transitions. A general transition will contain contributions from the matter, spin, convection, and spin-current degrees of freedom of the nucleus as shown in Eq. (2) and Eq. (3). It usually will be impossible to disentangle these contributions in a model independent fashion. One approach is to expand these transitions in an appropriate shell model basis and determine the spectroscopic amplitudes for a fixed set of radial basis functions. For some cases, a core plus valence model may be appropriate. The availability of spin observables should do much to reduce the ambiguity in such an analysis.

Fortunately, there are a number of natural and unnatural parity transitions for which a single type of transition density (containing both proton and neutron components) is unknown. These include $0^+ \to 0^-$ transitions and the excitation of "stretched" states, referred to in Sect. 1, as well as the low-lying collective states we have primarily been discussing. These radial densities can be expanded in a linear basis as

$$\rho(r) = \sum_{i=1}^{N} a_i f_i(r) \qquad (8)$$

where the $f_i(r)$ belong to any convenient complete set of radial basis functions. The expansion parameters a_i can be fit by minimizing the composite χ^2 for all available data. The error envelope can then be constructed from the covariance matrix using standard methods.[37] This error envelope includes contributions from the statistics and the range of the data and from the penetrability of the probe. Therefore, this density and its uncertainty represents the most detailed structure information that can be deduced from the data, assuming an accurate reaction theory.

These analyses are almost model independent in the sense that rather than fitting with a restrictive, and arbitrary, analytic form, the significant coefficients of an expansion in a complete basis of radial functions are fitted. The resulting density is then not biased by the choice of model. Virtually any radial function can be represented in this manner, with only minor restrictions being introduced by the truncation of the series.

Only limited use has been made of these techniques to data. Some "model independent" studies of neutron densities from high energy proton scattering data have been made[38], but the question of consistency has not been addressed. This method and the empirical interaction described above has been used to determine the neutron radial density distributions for the three lowest 2^+ states in ^{18}O.[1,39] Many questions on the uniqueness and universality of these results remain. Definitive answers await the compilation and analysis of a more extensive and systematic data set.

Further discussion is superfluous to the present purposes -- suffice it to say that implementation of the proposed modeling philosophy is an eminently practical affair. The crucial idea is that most questions of physical interest can be posed in linearized form and as such are amenable to the present linear expansion analysis. In order to illustrate these ideas and obtain important results for the planning of future experiments, we have chosen to analyze the expected sensitivity of nucleon and pion experiments to the radial neutron density for a particular inelastic transition.

4. SENSITIVITY TO RADIAL DENSITY DISTRIBUTIONS

The ideal radial sensitivity of nucleons and pions can be compared using a pseudodata approach which circumvents the uncertainties of reaction theory. First we select a representative nuclear transition and construct realistic transition densities. Next we perform scattering amplitude calculations for each reaction of interest using the best interaction models available. Then we produce pseudodata based upon the results of the first two steps, adding random fluctuations according to a normal distribution whose width represents the customary experimental precision. The number and angular range of the pseudodata should also reflect experimental practice. We assume the proton transition density is accurately known from electron scattering. Then we fit the neutron transition density to the pseudodata and construct the associated error envelope. The reaction theory within this approach is perfect by construction. Therefore, the deduced error envelope represents a quantitative measure of the radial sensitivity of each probe. This sensitivity reflects the wavelength of the probe, the transparency or lack thereof to the probe, and the relative importance of the coupling of the probe to the target neutrons. A long-wavelength probe is incapable of revealing detailed radial variation. A strongly absorbed probe views the surface but

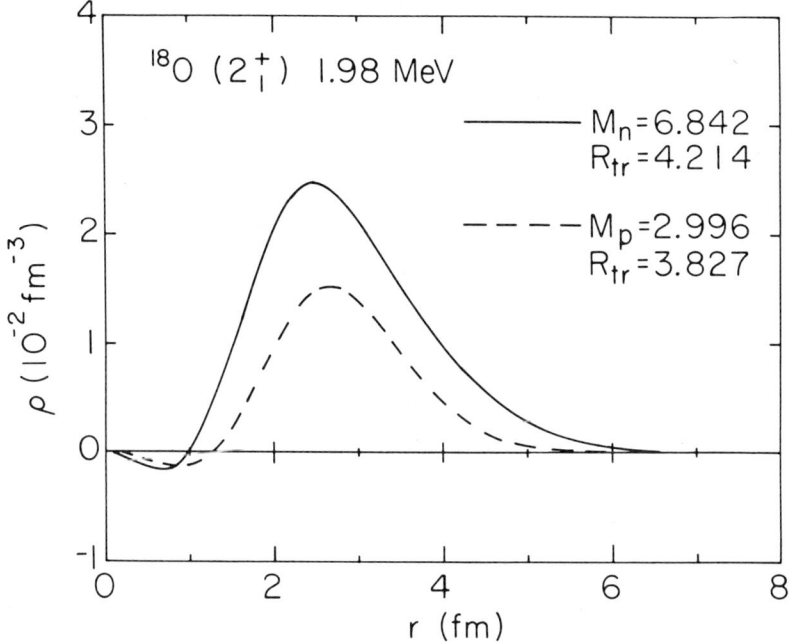

Fig. 6. Neutron and proton matter densities of the first 2^+ state in ^{18}O used for input to the pseudodata calculations.

doesn't penetrate the interior. Scattering data dominated by the probe-proton interaction are insensitive to neutron densities. All of these considerations can be assessed quantitatively with the pseudodata approach.

We illustrate this approach here using the first 2^+ state of ^{18}O as an example. The proton transition density was taken from the electron scattering measurements of Ref. 40. The neutron transition density is the result of an analysis of 135 MeV proton scattering data.[1,39] These densities are shown in Fig. 6. Cross section pseudodata were produced with 10% uncertainties and were limited in momentum transfer to a maximum 2.5 fm^{-1}, a value characteristic of electron scattering data and representative of twice the Fermi momentum. For 20 MeV nucleons, pseudodata were produced for q = 0.1 to 1.8 fm^{-1}, where the decreased upper limit represents the restrictions on the momentum transfer available at this energy. For 60 and 135 MeV nucleons, pseudodata were produced for q = 0.4 to 2.5 fm^{-1}. For 50 MeV pions, pseudo-data were produced for q = 0.3 to 1.26 fm^{-1}, a limit set by the momentum of the beam. At higher energies the data were produced over an angular range typical of existing LAMPF data. At 162 MeV we used q = 0.35 to q = 1.7 fm^{-1}, while at 230 MeV we used q = 0.37 to 2.1 fm^{-1}.

All the nucleon calculations used the local density

approximation[2-4,10-12] and a local approximation for exchange.[15] The distorted waves were produced by consistent microscopic optical potentials. At 20 MeV, the nuclear matter G-matrix based upon the Hamada-Johnston potential was used.[3] For higher energies, the effective interaction was based upon the Paris potential.[24] All pion calculations used the Kisslinger potential form with medium modifications to account for important second-order corrections to the impulse approximation.[5,6] The distorted waves were produced by consistent microscopic optical potentials. The parameters, from Ref. 5 and Ref. 20, generally describe existing data in an average way.

The choice of radial basis functions is largely a matter of convenience. Electron scattering analyses commonly employ some version of the Fourier-Bessel expansion (FBE).[37] This expansion has two principal advantages. First, each basis function displays a moderate degree of localization in momentum transfer. Thus each coefficient is clearly related to the form factor at a particular momentum transfer and is largely uncorrelated from nearby coefficients. Second, any radial dependence can be represented using sufficiently many terms. However, the expansion also suffers from two major defects. First, each basis function is oscillatory at large momentum transfer. Thus, unless care is taken to bias the fitted coefficients or limit the number of terms, unphysical short-wavelength structure can creep into the results. Second, each basis function is oscillatory at large radii. Thus, unless the asymptotic radial behavior is suitably biased, undesirable oscillations will persist at very large radii.

For the present analysis, we have chosen a polynomial times gaussion expansion (PGE)

$$\rho_L(r) = \alpha^3 (\alpha r)^L e^{-\alpha^2 r^2} \sum_{k=0}^{N} a_k (\alpha r)^{2k} \quad (9)$$

which is complete for any α. Naturally, we choose α on the basis of the harmonic oscillator model, thereby introducing a natural radial scale and minimizing the number of coefficients required to represent any radial variation likely to be encountered in practice. The advantages and disadvantages of this expansion are complementary to those of the FBE. The PGE declines exponentially for both large radii and large momentum transfer. However, each contribution is localized in radius rather than momentum transfer. Moreover, this radial localization is only modest - neighboring basis functions overlap. Therefore, the fitted coefficients tend to be highly correlated. Nevertheless, we have generally found the PGE more convenient to use.

The pion scattering results are shown first. The fitted neutron densities and their error envelopes are given in Fig. 7. The π^+ and π^- results are compared on the left hand side for E_π = 162 MeV. The greater sensitivity of π^- to neutrons leads to tighter error bands in this case. The right hand side shows the energy dependence of the pion radial sensitivity. The large error band at 50 MeV is a consequence of the limited momentum transfer and large (9.8 fm) wavelength of this probe. The error band at 230 MeV is tighter in the surface and worse in the interior. Pions at 162 and 230 MeV do not see the nuclear interior very well because of the strong absorption in the

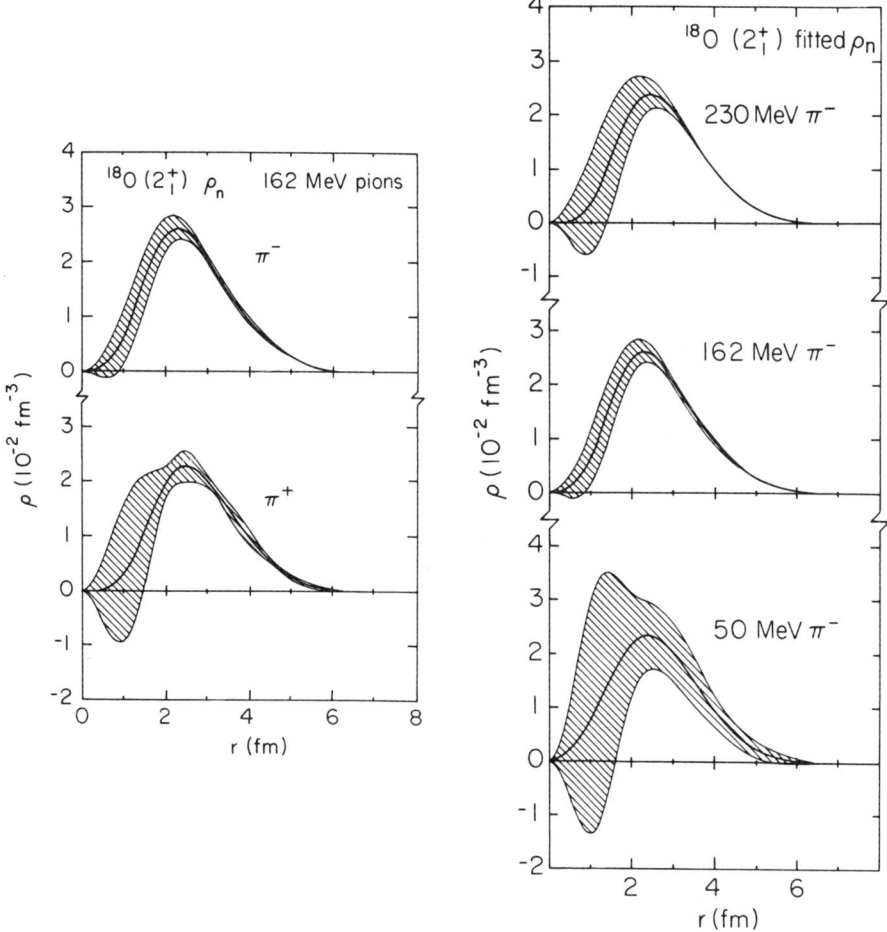

Fig. 7. Neutron densities and error envelopes determined from pion scattering pseudodata. Results for π^+ and π^- at 162 MeV are compared on the left, and the energy dependence for π^- is shown at the right.

optical potential.

The neutron densities resulting from the nucleon scattering study are shown in Fig. 8. The left hand side compares the results for 60 MeV neutrons and protons to ρ_n. The greater sensitivity of protons to the neutron density is clear from the tighter error band in the surface region. It appears that the additional distortion by the Coulomb field reduces the advantage for protons in the interior. The energy dependence of nucleon radial sensitivity is shown on the right hand side. The large error band at 20 MeV is a result of the long wavelength and strong refraction in the low energy optical

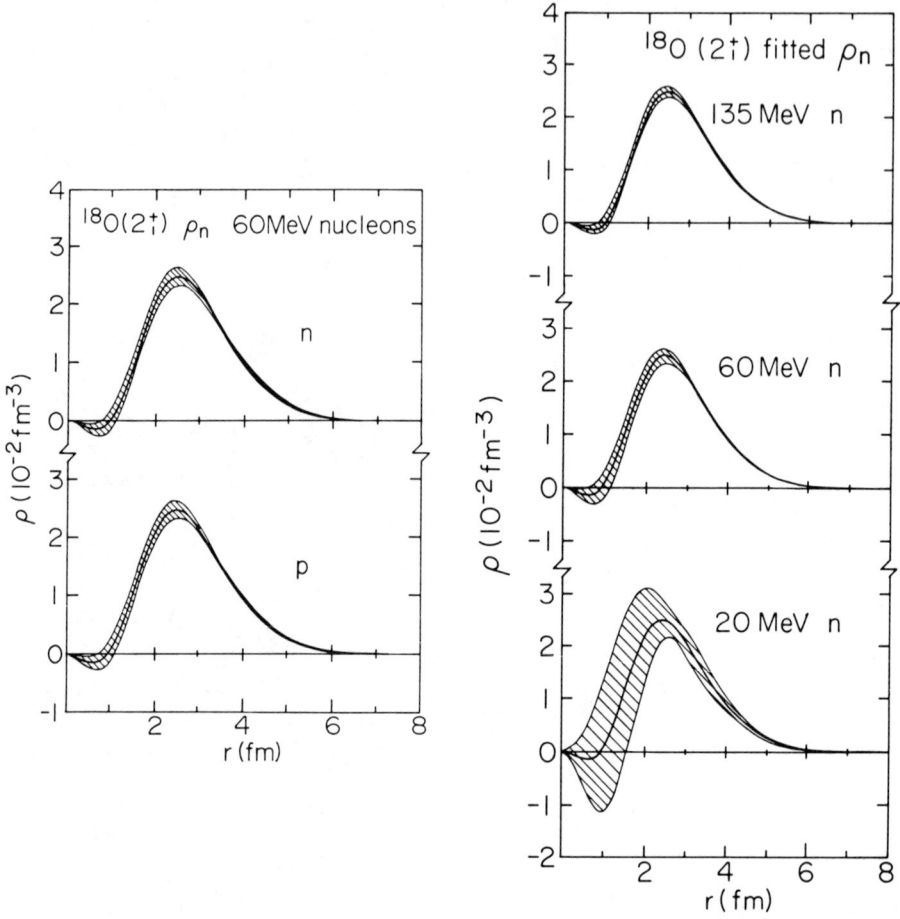

Fig. 8. Neutron densities and error envelopes determined from nucleon scattering pseudodata. Results for n and p projectiles at 60 MeV are compared on the left, and the energy dependence for n projectile is shown at the right.

potential. The error bands at 60 MeV are only slightly worse than at 135 MeV, indicating that most of the benefits of the increasing transparency of the optical potential have already been felt by 60 MeV. The dominant contribution to the error bands is the limited momentum transfer, which is (in practice) the same at these two energies. The clarity with which 60 MeV nucleons see the nuclear interior is somewhat contrary to prevailing prejudices. Nucleons in this energy region are intrinsically superior to pions as a probe of the radial distribution of nuclear transition densities.

The results for both probes are summarized in the graph of $\delta\rho$

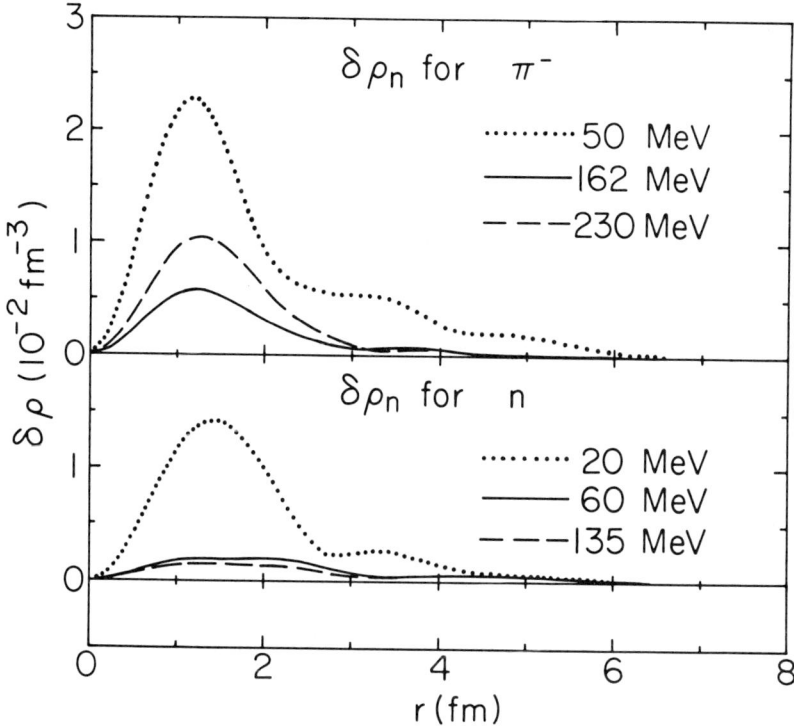

Fig. 9. Uncertainty in the fitted neutron density, $\delta\rho$, as a function of r for pions (top) and nucleons (bottom). The curves correspond to the error bands on the right hand side of Figs. 7 and 8 for pions and nucleons respectively.

shown in Fig. 9. The two dominant variables determining $\delta\rho$ are the wavelength and transparency of the probe. Large wavelengths limit us to low q, producing wide error envelopes for both low energy pions and nucleons. Once sufficient momentum transfer is available, nucleons and pions are roughly equivalent in the surface region (r > 3 fm in this case). The greater transparency of nucleons is evident in the interior region. The sensitivity of 60 MeV nucleons (with 10% uncertainties in the "data") is clearly adequate for very detailed studies on the nucleus.

The multipole matrix elements M_ν^J (ν = n or p), defined as a moment integral of the transition density

$$M_\nu^J = \int \rho_{J\nu}(r)\, r^{J+2} dr$$

$$= \frac{(2J+1)!!}{q^J} \rho_{J\nu}(q=0) \qquad (10)$$

Table I. Percent uncertainty in M_n^2, the multipole moment of the neutron density for the first 2^+ state in ^{18}O as determined from pseudodata with the indicated probe and bombarding energy.

Nucleons			Pions		
E	n	p	E	π^+	π^-
20 MeV	6.7%	2.6%	50 MeV	7.8%	21.1%
60 MeV	4.8%	2.6%	162 MeV	4.0%	2.3%
135 MeV	3.6%	2.9%	230 MeV	4.4%	2.1%

are commonly quoted in studies of low-lying collective states. The proton matrix elements are related to the γ-decay rates. Our pseudodata analysis provides a realistic estimate of the intrinsic precision with which these matrix elements can be determined from the hadronic data alone. The precision of the present results for the moments of the neutron transition density first 2^+ state in ^{18}O are summarized in Table I. Because the moments are determined primarily by the density at large radii, the effects of distortion are not of major importance. The dominant variables affecting the precision with which the moments are determined are the wavelength and the n-p sensitivity ratio of the probe. The uncertainties for pions at $E_\pi > 100$ MeV are similar to those for nucleons at all energies. There is sufficient momentum transfer available to clearly see the surface in all of these cases. Protons (π^-) are better than neutrons (π^+) in determining M_n^2 because they are more sensitive to neutron excitations in the target. The low energy pion case appears anomalous. Low energy π^+ are more sensitive to the surface neutrons because of the strong velocity dependence of the potential and the effect of the Coulomb repulsion. The net result is that the radial neutron density is better defined by low energy π^+ scattering in the model used for this study.

We have found that nucleons are sensitive to the nuclear interior for $E_N > 60$ MeV, i.e. the transparency of the nuclear medium is not an obstacle. The range of momentum transfer available for nucleous with $E_N > 60$ MeV is more than adequate and matches the range available from electron scattering data. For the idealized examples shown here, the intrinsic precision with which transition densities can be determined is, in fact, very similar to that which can be achieved with electron scattering. These remarks apply equally well in regard to determining the properties of the effective interaction over the full range of nucleon densities. As mentioned above, this information can be gained only by examining transitions covering a wide range of multipolarities.

5. UNNATURAL PARITY TRANSITIONS

The preceding discussion has focussed on the matter coupling for two reasons: (1) the matter interaction is strongest at low energy and (2) we have been examining collective natural parity states where ρ^m makes the dominant contribution to the cross section. Nucleons and pions are also sensitive to the target current and spin degrees of freedom, as has been discussed elsewhere.[8,9,11,13,17] A feature unique to nucleons is the longitudinal spin (s $\parallel$) coupling, which contributes to unnatural parity transitions as shown by Eq. (3). Therefore, it behooves us to study unnatural parity spin excitations with both proton and neutron beams since a simultaneous analysis of data for both probes will then provide information about the proton and neutron composition of the longitudinal spin interaction and of the corresponding nuclear structure. A second unique aspect is that the spin and matter couplings for nucleons are both finite at q = 0, as can be clearly seen in Fig. 1. This is the reason that nucleons are useful for examining low multipolarity spin excitations where the bulk of the transition density is at low momentum transfer.[17] For isovector transitions, the cross sections at zero momentum transfer are directly related to β-decay matrix elements.[41]

The differential sensitivity of a particular probe to the neutron and proton components of spin excitations can be illustrated by defining the ratios $R_Y^{s\perp}$ and $R_Y^{s\parallel}$ in analogy to the definition of R_Y in Eqs. (4) and (5) in Sect. 2. For electrons $R_e^{s\perp} = -0.68$ independent of energy and q. The negative ratio greatly enhances the sensitivity of electrons to the isovector part of $\rho^{s\perp}$ over the corresponding isoscalar part. Thus, electrons provide detailed information about one linear combination of the transverse spin densities. To illustrate the sensitivity of hadronic probes to the components of the spin response, we show the moduli squared of these ratios for pions and nucleons in Fig. 10.

The pion graph at the top of Fig. 10 shows that $R_\pi^{s\perp} \approx 9$ for pions with $E_\pi <$ 250 MeV. This occurs because the spin-dependent part of the pion-nucleon interaction is dominated by the Δ-resonance, which gives R = 9 for the same reasons discussed in Sect. 2 with respect to the matter interaction. This ratio indicates that pions, in contrast to electrons, are more sensitive to isoscalar than isovector transverse spin densities. The greatest sensitivity to n-p differences is near the peak of the resonance where distortion effects (c. f. Fig. 2) are the largest. The reduced effects of distortion and the large value of $(R_\pi^{s\perp})^{-1} \sim$ 100 or more near 550 MeV makes this region nearly ideal for studying $\rho^{s\perp}$, but unfortunately $t_{s\perp}$ is quite weak in that region which makes spin excitations difficult to measure.[26] The main lesson is that pion experiments at $E_\pi <$ 250 MeV, preferably in conjunction with electron scattering data, can be used to determine both components of the transverse spin density.

The nucleon graphs at the bottom of Fig. 10 indicate that the sensitivity to neutron-proton differences in $\rho^{s\perp}$ increases with decreasing energy for $E_N <$ 100 MeV. In contrast, the sensitivity to n-p differences in $\rho^{s\parallel}$ reaches a peak near 100 MeV. The latter

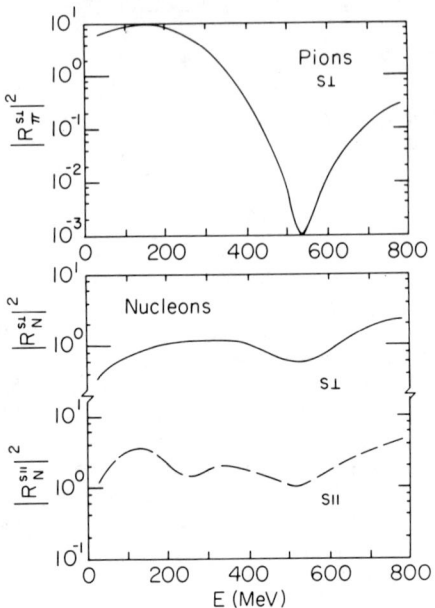

Fig. 10. Moduli squared of the force ratios for the s⊥ and s∥ interaction components for pions (top, s⊥ only) and nucleons (bottom) as a function of energy.

is quite interesting because we expect distortion to be weaker in that energy region. On this basis alone one would expect that the region 50 MeV $< E_N <$ 150 MeV would be a promising region for study, since it appears that neutrons and protons together could identify the neutron-proton components of the longitudinal spin response provided only $\rho^{s\parallel}$ contributes (as in $0^+ \to 0^-$ transitions) or $\rho^{s\perp}$ is known from electron and pion scattering experiments.

The radial dependence of the longitudinal spin response in nuclei can only be measured with nucleon scattering experiments. The determination of the isospin composition of this response requires both proton and neutron beams with energies around 100 MeV. In addition, the measurement of discrete, often weak, spin states requires optimal resolution for the neutron beams.

6. CONCLUSIONS

There is now a vast industry of intermediate energy physics based on a single scattering description. To the extent that this description is valid, all data should be analyzed in a single consistent framework. We have proposed a versatile linear expansion method based on a folding model for elastic and inelastic scattering. This permits a phenomenological analysis of the effective interaction or of nuclear structure. When the nuclear structure is known, data can be used to refine the parameters of an empirical interaction whose form has been selected on theoretical grounds. When the interaction is known, detailed nuclear structure information can be extracted from the data. In particular, the radial dependence of a single unknown transition density can be determined.

The accuracy of this type of analysis is enhanced by the availability of several complementary reactions. While interesting studies can be performed with a single probe, systematic errors in reaction theories or interaction models can not be checked. Comparing the scatter in the otherwise identical fits to different reactions provides a measure of such errors. A combined analysis will average over the inaccuracies, perhaps at the price of some precision in the final result. Consistency becomes the key test for any empirical interaction, reaction theory, or fitted density. In this regard, electrons, pions and nucleons all have a role to play.

These methods have provided perspective on the desirable characteristics of proposed neutron scattering facilities. Specifically, we have investigated the sensitivity of nucleon and pion scattering to the nuclear interior using a pseudodata analysis. We found that the nucleus is sufficiently transparent to nucleons to permit a precise measurement of neutron transition densities in the nuclear interior using bombarding energies as low as 60 MeV. The greater intrinsic sensitivity of pions to neutron-proton differences is more than offset by the greater opacity of the nucleus to pions.

Different aspects of nuclear structure are best studied with projectile energies chosen for maximum sensitivity to the component of interest. Natural parity matter excitations can be adequately studied with neutron energies as low as 60 MeV, although energies near 200 MeV might be best. The study of the longitudinal spin component of unnatural parity transitions is optimized for incident neutron energies near 100 MeV. Medium energy neutrons allow the study of the nucleus in a way that has not been exploited previously.

It is important to note that we are arguing for an in-depth study of bound states of the nucleus with the precision of a quality accelerator complemented by a sophisticated method of analysis. This is work that was left undone in the past. We note that studies of particular bound states with inelastic scattering typically require good resolution, which argues for high quality neutron beams in the 50 to 150 MeV range. (Studies of continuum spin excitations impose somewhat different criteria[42] than we have selected here.) Such a facility must be capable of analyzing power measurements. It is very important that such a neutron facility be matched to a proton scattering facility that has similar capabilities.

We wish to emphasize, in closing, that there is little to be gained from a sophisticated facility if most of the information in the data is ignored by using primitive analysis methods. As much effort must be put into developing a sensible phenomenology for correlating the experimental information at a quantitative level as goes into designing equipment to acquire precise data. Such a phenomenology must be guided and informed by a good theory. Theory provides the fitting functions and the relevant variables. Theory also provides an estimate for the coefficients. A guided phenomenological analysis of data then produces refined, or "measured", values for these coefficients and provides a quantitative measure of the success of any particular interaction model. It is this quantitative account of the strengths and weaknesses of a reaction theory that is so blatantly missing from the accumulated legacy of qualitative comparisons between fitted parameters of arbitrary functions. The quality and quantity of data now and soon-to-be available demands a quantitative confrontation between theory and experiment. Justice must be done.

This work was supported in part by the National Science Foundation.

REFERENCES

1. J. J. Kelly, W. Bertozzi, and F. Petrovich, Comments Nucl. Part. Phys., to be published.
2. J. P. Jeukenne, A. Lejeune, and C. Mahaux, Phys. Rev. C10, 1391 (1974); C15, 10 (1977); and C16, 80 (1977).
3. F. A. Brieva and J. R. Rook, Nucl. Phys. A291, 299 (1977); A291, 317 (1977); A297, 206 (1978); and A307, 493 (1978).
4. Microscopic Optical Potentials, ed. H. V. von Geramb, Lecture Notes in Physics Vol. 89 (Springer, Berlin, 1979).
5. K. Stricker, H. McManus, and J. A. Carr, Phys. Rev. C19, 929 (1979); J. A. Carr, H. McManus, and K. Stricker-Bauer, Phys. Rev. C25, 952 (1982).
6. A. W. Thomas and R. H. Landau, Phys. Rev. 58C, 121 (1980).
7. F. Lenz, M. Theis, and Y. Horikawa, Ann. Phys. (N.Y.) 140, 266 (1982).
8. F. Petrovich and W. G. Love, Nucl. Phys. A354, 499c (1981).
9. The (p,n) Reaction and the Nucleon-Nucleon Force, ed. C. D. Goodman, et al. (Plenum, New York, 1980).
10. J. J. Kelly, et al, Phys. Rev. Lett. 45, 2012 (1980).
11. The Interaction Between Medium Energy Nucleons in Nuclei, AIP Conf. Proc. No. 97, ed. H. O. Meyer, (AIP, New York, 1983).
12. F. S. Dietrich, R. W. Finlay, S. Mellema, G. Randers-Pehrson, and F. Petrovich, Phys. Rev. Lett. 51, 1629 (1983); S. Mellema, R. W. Finlay, F. S. Dietrich, and F. Petrovich, Phys. Rev. C28, 2267 (1983).
13. Spin Excitations in Nuclei, ed. F. Petrovich et al (Plenum, New York, 1984).
14. F. Petrovich, Nucl. Phys. A251, 143 (1975).
15. F. Petrovich, H. McManus, V. Madsen, and J. Atkinson, Phys. Rev. Lett. 22, 895 (1969); W. G. Love, Nucl. Phys. A312, 160 (1978).
16. F. Petrovich, R. J. Philpott, A. W. Carpenter, and J. A. Carr, Nucl. Phys. A425, 609 (1984).
17. F. Petrovich, J. A. Carr, R. J. Philpott, and H. McManus, Proc. Symp. on Electromagnetic Properties of Atomic Nuclei, ed. H. Horie and H. Ohnuma (Tokyo Institute of Technology, Tokyo, 1984) p. 312.
18. J. M. Eisenberg and W. Greiner, Nuclear Theory 2: Excitation

Mechanisms of the Nucleus (North Holland, Amsterdam, 1970)
19. G. Bertsch, J. Borysowicz, H. McManus, and W. G. Love, Nucl. Phys. A284, 399 (1977).
20. G. Rowe, M. Salomon, and R. H. Landau, Phys. Rev. C18, 584, (1978).
21. G. R. Hammerstein, R. H. Howell, and F. Petrovich, Nucl. Phys. A213, 45 (1973).
22. F. Petrovich, H. McManus, J. Borysowicz, and G. R. Hammerstein, Phys. Rev. C16, 839 (1977).
23. W. G. Love and M. A. Franey, Phys. Rev. C24, 1073 (1981).
24. H. von Geramb, Ref. 11, p. 44.
25. A. T. Davies, Nucl. Phys. B21, 359 (1970).
26. J. A. Carr, Proc. of the Second LAMPF II Workshop, LANL Rep. LA-9572-C (1982) p. 708.
27. R. Tacik, et al., Phys. Rev. Lett. 52, 1276 (1984).
28. A. Bernstein, Advances in Nuclear Physics, ed. M. Baranger and E. Vogt (Plenum, New York, 1969) Vol. 3, p. 325.
29. A. Bernstein, V. R. Brown, and V. A. Madsen, Comments Nucl. Part. Phys. 11, 203 (1983); and references therein.
30. A. Scott, N. P. Mathur, and F. Petrovich, Nucl. Phys. A285, 222 (1977).
31. R. Schaeffer, Nucl. Phys. A135, 231 (1969); see also Y. Terrien, Nucl. Phys. A199, 65 (1973); A215, 29 (1973).
32. See for example S. Seestrom-Morris, D. B. Holtkamp, and W. B. Cottingame, Ref. 13, p. 291 and references therein.
33. H. von Geramb, these proceedings.
34. F. Dietrich and F. Petrovich, these proceedings.
35. J. Kelly, Ref. 11, p. 153.
36. J. Kelly, et al, Phys. Rev. C, to be published.
37. J. Heisenberg, Adv. in Nucl. Phys. 12, 61 (1981) and references therein.
38. L. Ray, W. R. Coker, and G. W. Hoffmann, Phys. Rev. C18, 2641 (1978); G. W. Hoffmann, et al., Phys. Rev. C21, 1488 (1980); L. Ray and G. W. Hoffmann, Phys. Rev. C27, 2133 (1983).
39. J. Kelly, et al., submitted to Phys. Rev. Lett.
40. B. Norum, et al., Phys. Rev. C25, 1778 (1982).
41. C. D. Goodman, et al., Phys. Rev. Lett. 44, 1755 (1980).
42. As F. Osterfeld pointed out at the meeting, if one is only interested in continuum spin excitations then resolution is not the most important issue. Instead, the relative magnitudes of the spin and matter couplings, which govern the signal to noise ratio in the spectrum, are the important factors. From this point of view, neutron and proton induced charge exchange reactions at incident energies near 200 MeV are desirable.

NUCLEON SCATTERING FROM ^{34}S[†]

R. Alarcon and J. Rapaport
Ohio University, Athens, OH 45701

R.T. Kouzes and W.H. Moore
Princeton University, Princeton, NJ 08547

B.A. Brown
Michigan State University, East Lansing, MI 48824

In a recent paper Bernstein et al. [1] discuss the sensitivity of inelastic hadron scattering to determine relative signs of neutron, M_n, and proton, M_p, multipole matrix elements. As an example they present proton inelastic scattering data from ^{34}S at 650 MeV and conclude that the relative sign of M_n and M_p for the $0^+ \rightarrow 2^+_2$ transition in ^{34}S is negative in agreement with shell model calculations reported by Wildenthal [2]. This conclusion has been challenged by Saha et al. [3]. These authors report measurement of the alpha inelastic scattering cross sections for the first two 2^+ states in ^{34}S to show that the relative sign of M_n and M_p for both 2^+ states are positive. This is in agreement with the recently reported EM measurements in the T = 0 nucleus ^{34}Cl [4].

In both refs. [1] and [3] EM data and hadronic data are needed to evaluate values for M_n and M_p. In this paper we present hadronic data, neutron and proton scattering data in ^{34}S, which in a self-consistent analysis, are used to estimate the values and relative signs of M_n and M_p for the three 2^+ transitions in ^{34}S.

Coupled channel calculations were performed with the code ECIS [5]. The ground state and the first three 2^+ states were coupled using a vibrational model form factor for the coupling matrix elements. For a given transition, these matrix elements may be calculated from the optical potentials and the values of M_n and M_p [6]. Using the shell-model predictions for M_n and M_p the code ECIS was used to estimate the inelastic nucleon cross sections. The results are shown in figure 1 as dashed lines. It is quite clear that the shell-model predictions, with opposite sign for M_n and M_p, do not agree with the $0^+ \rightarrow 2^+_2$ transition data. It is also clear that shell-model predictions underestimate the results for the $0^+ \rightarrow 2^+_3$ transition.

The solid lines in figure 1 are the result of considering the empirical EM values of M_n and M_p for the $0^+ \rightarrow 2^+_1$, $0^+ \rightarrow 2^+_2$, and $2^+_1 \rightarrow 2^+_2$ transitions as reported in ref. [3]. For the $0^+ \rightarrow 2^+_3$

transition we deduced the M_p value from EM measurements [7] and we have estimated the M_n value from the results reported in the alpha-inelastic experiment [3]. The agreement with the measured data seems quite good.

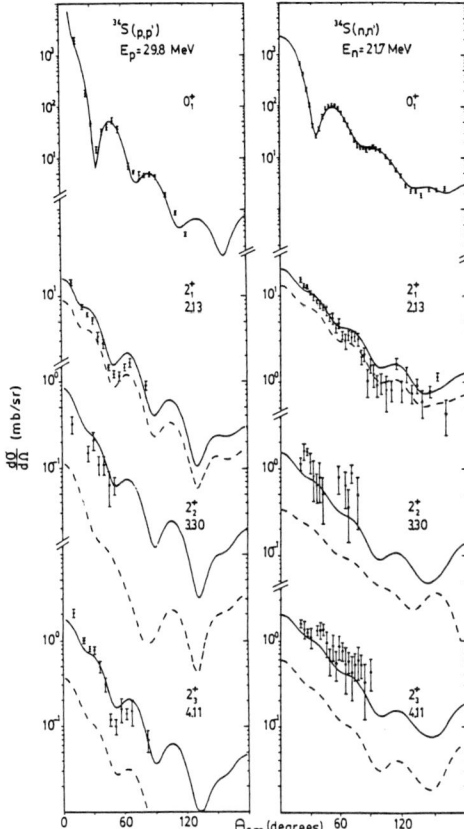

FIG. 1 Nucleon angular distribution for ^{34}S.

The present results agree with the conclusion of Saha et al. [3] that the excitation of the ^{34}S (2_2^+) state requires that M_n and M_p have the same sign. This is in disagreement with the shell model calculation results and those inferred from the 650 MeV inelastic proton data [1]. The simultaneous analysis of our nucleon data gives the a ratio $M_n/M_p = 1.25 \pm 0.35$ for the $0^+ \to 2_1^+$ transition in good agreement with the value $M_n/M_p = 1.50 \pm 0.30$ obtained by considering only EM results. For the $0^+ \to 2_2^+$ transition M_n/M_p is particularly sensitive to the interaction strengths values. Our analysis gives $M_n/M_p = 0.30 \pm 0.18$ if the interaction strengths are calculated from coupled channel optical potentials and $M_n/M_p = 0.46 \pm 0.16$ if we use the spherical optical potentials. The EM result is $M_n/M_p = 0.55 \pm 0.15$. Finally, our analysis gives the value $M_n/M_p = 1.10 \pm 0.35$ for the $0^+ \to 2_3^+$ transition.

REFERENCES

1) A.M. Bernstein et al., Phys. Rev. Lett. 49, 451 (1982).
2) B.H. Wildenthal, Bull. Amer. Phys. Soc. 27, 725 (1982).
3) A. Saha et al., Phys. Rev. Lett. 52, 1876 (1984).
4) J. Keinonen et al., Nucl. Phys. A412, 101 (1984).
5) J. Raynal, Report IAEA-SMR-918 (1972), p. 281.
6) P.W. F. Alons et al., Nucl. Phys. A367, 41 (1981).
7) P.M. Endt and C. van der Leun, Nucl. Phys. A310, 1 (1978).

SESSION DA

APPLICATIONS AND

TECHNIQUES

NUCLEAR PHYSICS IN THE 10-300 MeV ENERGY RANGE USING A PULSED WHITE NEUTRON SOURCE

C.D. Bowman, S.A. Wender, and G.F. Auchampaugh
Los Alamos National Laboratory, Los Alamos, NM 87545

ABSTRACT

A new pulsed white neutron source is under construction at the Los Alamos WNR facility. The neutrons are produced by LAMPF proton micropulses striking thick targets of various materials. Beam parameters include energy of 800 MeV, pulse rate of approximately 50,000 Hz, 0.4 nsec pulse width, average current as high as 6μa, and a useful neutron energy range from 3 to 300 MeV. The facility will receive beam approximately 80% of the time LAMPF is operational; it increases by a factor of 1000 the experimental capability over the present system at the WNR when beam intensity, angular distribution, and availability are taken into account. In addition to established white source techniques, the facility is also highly competitive with monoenergetic sources for a wide class of experiments such as neutron capture γ ray spectroscopy and neutron-induced charged particle reactions. The facility should be operational in about nine months. Arrangements are underway to make the facility readily accessible to visiting experimenters.

INTRODUCTION

A new pulsed white neutron source is presently under development as part of the Low Alamos Weapons Neutron Research (WNR) facility. The WNR facility, which is located to the south of the primary Los Alamos Meson Physics Facility (LAMPF) experimental area, receives LAMPF beam for production of neutrons or neutrinos at three target areas. Presently nuclear physics in the 3-300 MeV range is conducted using a white source at WNR Target Area 1. However this part of the facility is being modified for use as an epithermal pulsed neutron source for condensed matter physics studies using neutron scattering techniques. The MeV neutron physics capabilities are therefore being reestablished at Target Area 2 (the Blue Room). Greatly improved running conditions will be possible as a result of greater average beam current, higher repetition rate, neutron emission at forward angles, and a larger fraction of total LAMPF operating hours.

This paper describes the capabilities and operating characteristics of the new white source with particular attention to the high neutron intensity possible. The experimental program presently planned for the facility is described along with a summary of other possible research opportunities including some which are not traditionally pursued using white source methods. It will be shown that this intense white source is competitive for many experiments ordinarily pursued with monoenergetic sources. As an example the performance of a possible facility at the WNR/PSR for (n,p) studies in the 10-300 MeV neutron energy range is compared with the traditional monoenergetic source approach. These white source facilities will be available to visitors for collaborative research during most of LAMPF's operation. It might be possible to provide neutron beams for spectrometer construction and research led by outside teams of scientists. The new facility is expected to begin operation with a limited array of experimental capabilities in mid-1985. Further improvements to the facility will be made during LAMPF down periods over the next two years in accordance with basic and applied research needs and available funding.

II FACILITY DESCRIPTION

Fig. 1 shows the WNR/PSR facility. The Proton Storage Ring (PSR) shown in the upper right hand corner of the figure will become operational in the summer of 1985. A new neutrino facility, Target Area 3, located off the left hand side of the figure, was completed in 1984. This facility will become operational in 1985. The rectangular building in the center is the location of the Target Area 1 which is devoted primarily to neutron production for materials science studies. The dome shaped structure is Target Area 2 which is being prepared as an intense white source of MeV neutrons.

Each of the three target areas requires different beam conditions in terms of pulse rate and pulse width; however all use the same beam energy of 800 MeV for their primary programs, and this makes a fully multiplexed operation possible. Target area 1 accepts the PSR output of 270 nsec wide pulses with a 12-Hz rate and with an average current of 100 µa. Target Area 3 (neutrinos) now is able to accept approximately 20 µa of LAMPF beam supplied as 850-µsec long pulses at the rate of about 2 Hz; this current might grow significantly in the future. Target Area 2 of primary interest here is being prepared to accept up to 6 µa of beam at a rate of 50,000 Hz and a width of 0.4 nsec.

The means of beam delivery to Target Area 2 is somewhat complex. The H^- beam for this mode is accelerated

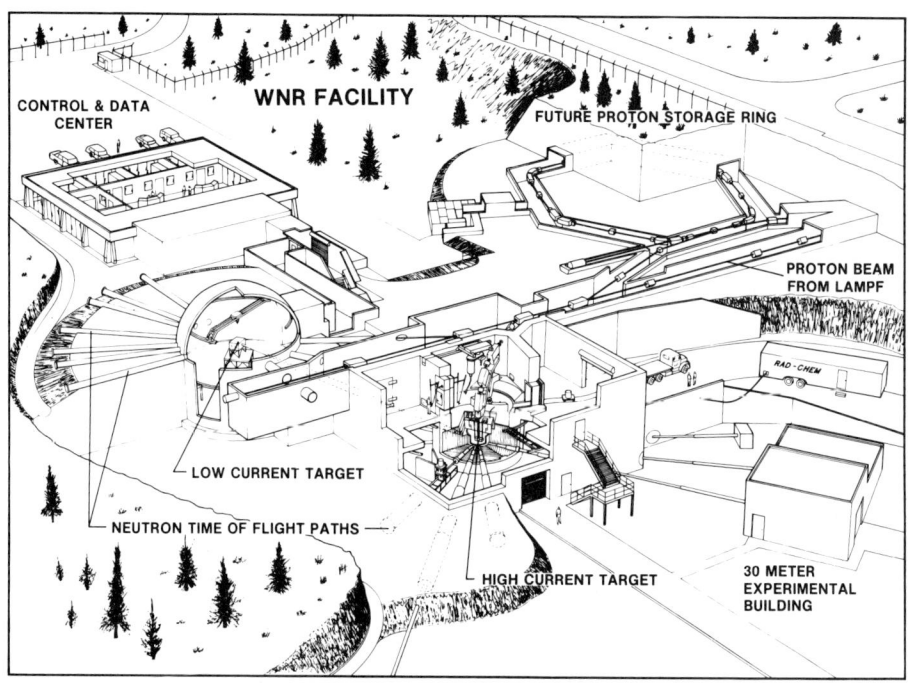

Fig. 1 - The storage ring in the upper right hand corner feeds the high current target area in the center where condensed matter physics and low energy neutron nuclear physics is conducted. The dome structure is the site of the white source discussed here.

simultaneously with the high current H^+ beam to LAMPF Target Area A. That is, two injectors operate during the same LAMPF macropulse, one producing H^- and the other H^+ beam. The H^+ beam consists of the normal LAMPF microstructure with the 5 nsec spacing between microbursts associated with the 200 MHz LAMPF frequency. The H^- beam is chopped so as to increase the spacing between pulses from the natural 5 nsec to 1 μsec. In the process a buncher in the H^- injector approximately doubles the number of H^- ions per burst. The average H^- current is therefore 1% of the H^+ beam; the H^- bursts are accelerated 180° out of phase with the H^+ beam. The two beams are magnetically separated in the LAMPF switchyard after acceleration and the H^- beam bent 90° toward the WNR/PSR complex.

Table I shows the present plan for LAMPF beam distribution for the immediate future. Note that Line X receives polarized H⁻ beam with variable energy at a rate of 40 Hz. Since the new H⁻ injector contructed as part of the PSR project can inject the same high current beam as the H⁺ injector, the LAMPF accelerator can be fully loaded with H⁻ beam. This H⁻ beam is directed to the PSR (and then to Target Area 1) and to Target Area 3 at the approximate rate of 12 and 2.5 Hz respectively. The remainder (65.5 Hz) of the time the LAMPF accelerator is in the dual H⁺ - H⁻ mode and supplies beam simultaneously to Area A and to Target Area 2. The average beam current in Target Area 2 when LAMPF is operating at full capacity is therefore:

$$1.3 \times 10^3 \text{ } \mu a \times (65.5/120) \times 0.01 = 7 \text{ } \mu a.$$

Table I - LAMPF Beam Distribution

Experimental Area	Macropulse Rep. Rate Hz	Beam* Type	Energy (MeV)
Line X	40	p⁻	Variable
Area A	65.5+	H⁺	800
Target Area 2 MeV Neutron Nuclear Science	65.5	H⁻	800
Target Area 1 PSR-Neutron Scattering	12	H⁻	800
Target Area 3 Neutrino Research	2.5	H⁻	800
	120.		

* H⁺ - protons, H⁻ - negative hydrogen ions, p⁻ - polarized negative hydrogen ions.

+ These two modes run simultaneously. The frequency actually will be somewhat different since LAMPF cannot run at half-integer rates.

A typical current of 6 μa is assumed for nominal operation here. The repetition rate is about 50,000 Hz.

Part of the development of the facility is a beam transport system which allows full multiplexing of the beam so that all these facilities can be operated simultaneously. This is shown in Fig. 2, which is not drawn to scale. Not shown at the top of the figure is a 120 Hz switching magnet in the LAMPF switchyard which will be required to allow beam

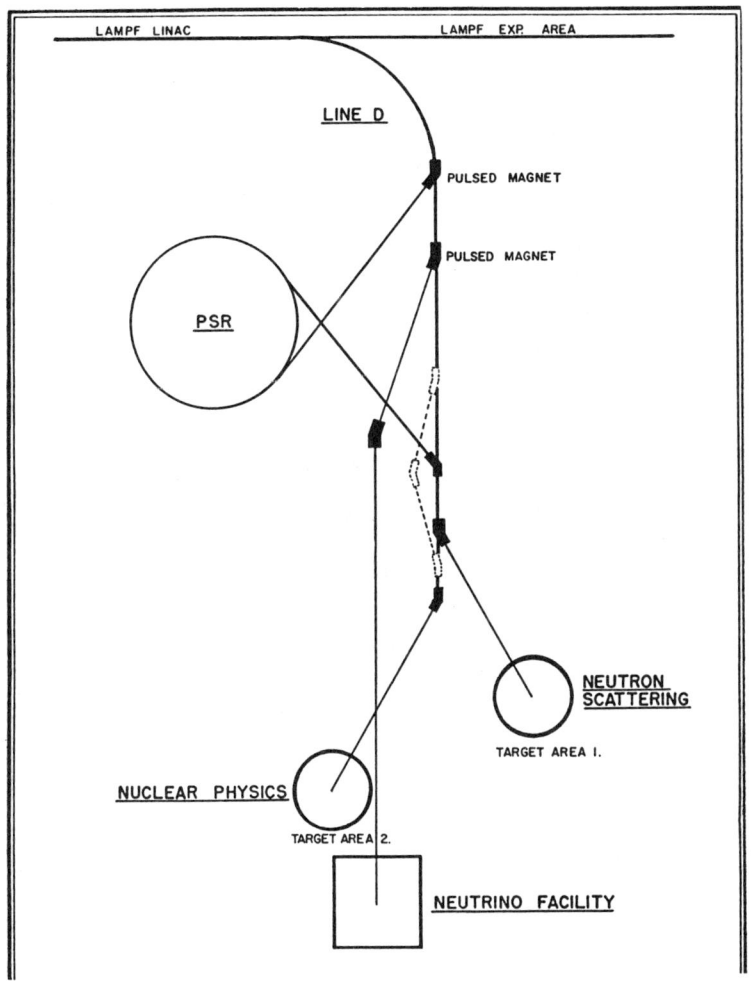

Fig. 2 - The beam distribution system for the WNR/PSR. The first pulsed magnet at the top of the page directs beam into the PSR and from there to Target Area 1. The second pulsed magnet directs some of the remaining beam to the neutrino facility. The remaining beam is directed by the by-pass line (components indicated by dashed lines) to Target Area 2 for MeV neutron nuclear physics.

from any LAMPF macropulse to be directed into the WNR/PSR complex. This magnet will replace a 24-Hz magnet presently in the same position. The line drawings show three routes for beam in the complex. All beam paths shown in solid lines are now or will be in place by spring of 1985.

1) Using the first pulsed magnet in the line at the top of the figure beam can be directed into the PSR, bunched, and then directed to Target Area 1 as shown.

2) Some of the beam, which was not removed by the first pulsed magnet, may be removed by the second pulsed magnet for direction to the neutrino facility. This line is now in place and operational at 20 μa.

3) Beam not removed by either of the two pulsed magnets can be diverted by a system of d.c. magnets into Target Area 2 for MeV neutron nuclear physics studies. The bypass line shown as dashed lines in the drawing is not in place, but installation is planned for 1985. Beam is presently transported on an exclusive use basis as indicated in Fig. 2 into Target Area 2. Note that it will be possible on occasion to transport small amounts of PSR beam ($\sim 10^{15}$ neutrons per 270 nsec burst) into Target Area 2 using the route shown by the solid line, although this mode is expected to be used only on rare occasions.

A comparison of the new capability under construction at Target Area 2 with the present capability at Target Area 1 and with a proposed short burst (1 nsec) mode of the PSR is given in Table II. The latter capability is optimized for the 0.1-3 MeV range, but there are presently no firm plans for implementing it. Note that the advantages of the new arrangement in Target Area 2 over that in Target Area 1 are:

TABLE II

WNR/PSR White Source Comparison

	Target Area 1	Target Area 2	Proposed
H^- beam energy (MeV)	800	800	800
H^- pulse spacing (μsec)	1	1	1380
H^- pulse rate (Hz)	8000	50,000	720
H^- pulse width (nsec)	0.4	0.4	1.0
H^- average current (μamps)	0.07	6	12
Target	W		
Angle of Neutron Emission (Degrees)	90°		

- approximately 100 times the average current
- enhanced intensity of higher energy neutrons using the forward angle
- approximately six times the repetition rate
- much greater facility availabiltiy

These factors combined together represent an extraordinary step beyond the capability of Target Area 1 for MeV neutron nuclear physics studies. The new source strength at Target Area 2 integrated over all angles using a ^{238}U target is 10^{15} n/s with a substantial enhancement of the intensity for neutrons above 10 MeV in the forward direction. The neutron spectrum[1] as a function of energy at 90° is given in Fig. 3. The intensity is given per unit lethargy ($U \equiv \ln E_o/E$). It is easily shown that

$$N(E) = N(U) \, (1/E).$$

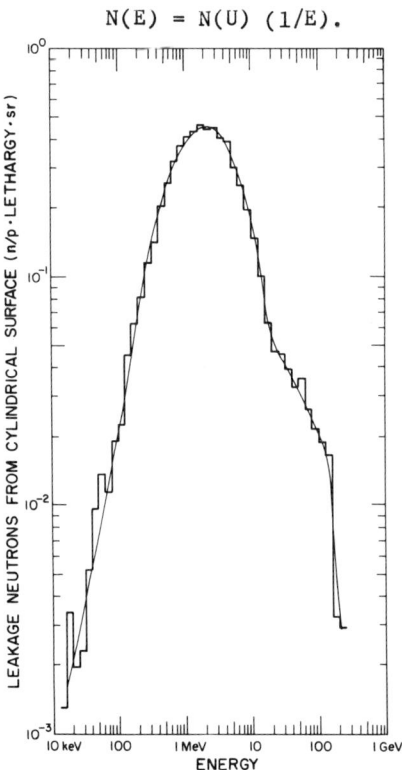

Fig. 3 - The neutron spectrum observed at 90° to the proton beam produced by protons striking a thick ^{238}U target. The definition of lethargy is given in the text.

The neutron intensity at 100 MeV in a 1-MeV wide band is found to be $(3.6 \times 10^{13}$ p/sec$) \times (0.02$ n/p-sr-lethargy$) \times (1/100$ MeV$) \times (1$MeV$) \times (2.5$ (forward angle enhancement factor)) $= 1.8 \times 10^{10}$ n/sec-sr-MeV.

A vertical section through the Target Area 2 is shown in Fig. 4. A 4-ft. thick steel-concrete enclosure around the white source is planed as shown to enclose the white source which is located at the center of the 40 ft. diameter dome shaped cell. Limited working space in a relatively high background area outside of the 4-ft. thick shield will be available with a typical path length of 5 meters. As shown in Fig. 1 several neutron drift tubes penetrate the shielding to the outside of the dome where experiments can be established at a flight path of about 22 meters or beyond. Space is available at 0° to the proton beam direction for a drift tube extending to 250 Meters. Experience with 0° (p,n) measurements in the 500-800 MeV range demonstrate[2] that the effects of air scattering on the neutron spectrum is unimportant for drift paths of that length and for neutrons in that energy range.

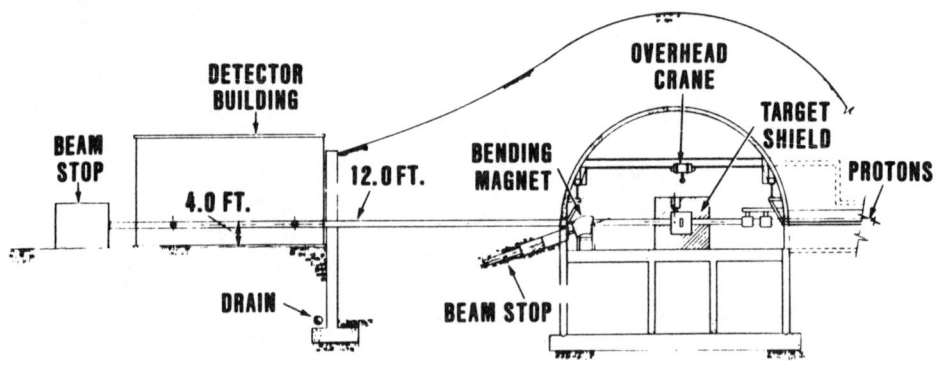

Fig. 4 - A cross sectional view of one of the neutron drift tubes associated with the MeV white source under construction.

III EXPERIMENTAL RESEARCH PROGRAM

An exploratory research program has been carried out in Target Area 1 on (n, γ_o) $(n, n' \gamma)$ and neutron total cross sections. The arrangement for the experiments involving γ-rays is shown in Fig. 5.

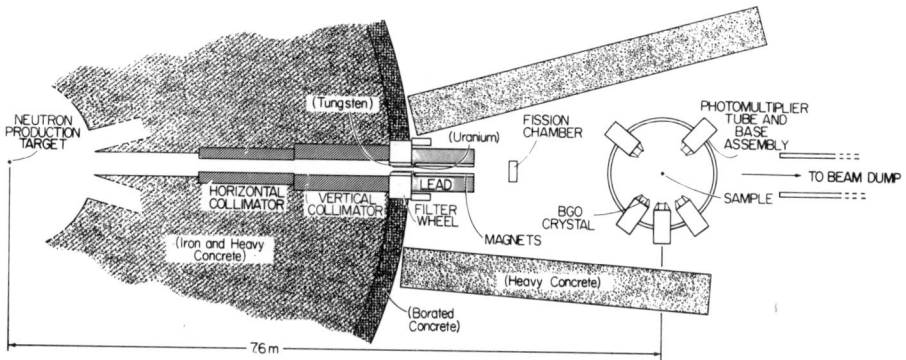

Fig. 5 - The experimental apparatus for (n,γ_o) and $(n,n'\gamma)$ spectral and angular distribution measurements as a function of neutron energy. The BGO crystals provide adequate resolution without the neutron sensitivity of NaI.

Five BGO detectors can be arranged at various angles around the sample which is located at 7.6 M from the source. The ^{235}U fission chamber is used as a flux monitor. Two examples of measurements with this system are shown in Figs. 6 and 7. Fig. 6 shows the A_0 coefficient in the angular distribution for production of the 2.12 MeV γ-ray between the first excited and ground states of ^{11}B. The upper portion of the figure shows the data between 2 and 20 MeV and the lower portion the data between 2 and 200 MeV. Ten or more peaks can be associated with known levels in the ^{12}B nucleus. Fig. 7 shows measurements of the reaction ^{40}Ca(n,γ_o). The abscissa is the neutron energy. The peak at 10.5 MeV incident neutron energy is the giant dipole resonance located at 19 MeV excitation energy in ^{41}C. The latter measurement was marginal at Target Area 1. At Target Area 2 the greatly enhanced neutron intensity, duty cycle, and resolution will allow much higher quality data. New facilities at Target Area 2 for the γ-ray measurements should be operational in the summer of 1985.

 A number of other types of neutron-induced reactions are under consideration. Aspects of the fission process include total fission cross sections, energy dependent angular distibutions, energy dependent mass distribution, and (n,2n) or (n,3n) reactions on fissile targets. The full spectrum of neutron-induced reactions with charged particle in the final state such as (n,p) and (n,α) studies is under consideration. It appears that a useful technique might be developed for (n,n') studies.[3]

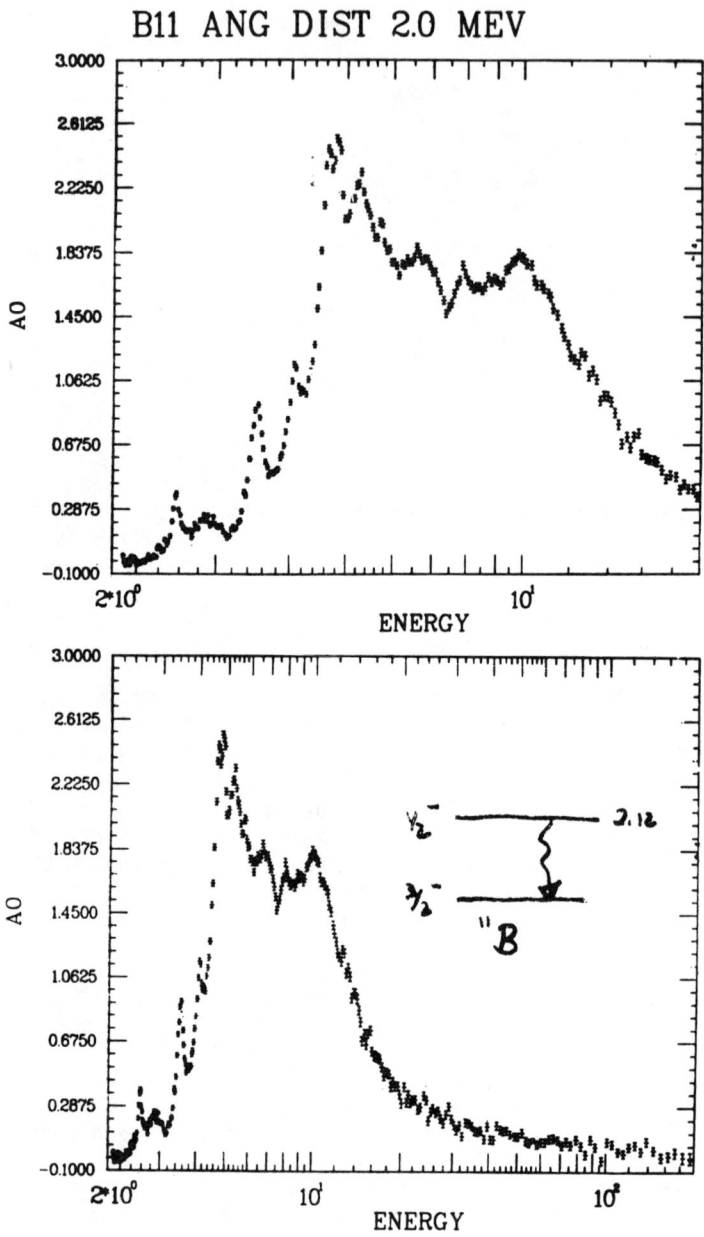

Fig. 6 - The A_0 coefficient of the angular distribution for production of the 2.12 MeV γ-ray in ^{11}B via the (n,n') reaction over the neutron energy range from 2 to 200 MeV.

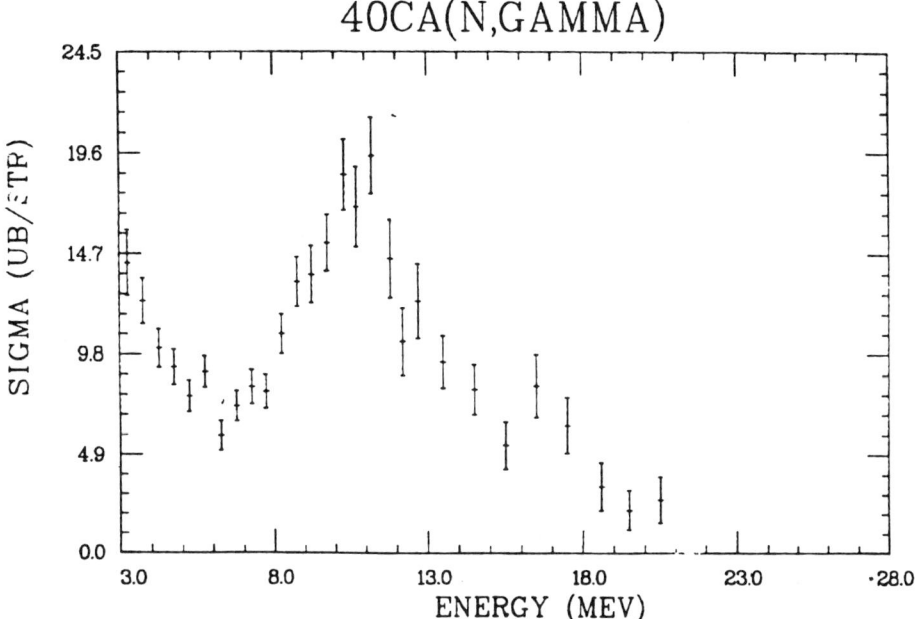

Fig. 7 - The production of the ground state gamma ray following neutron capture on ^{40}Ca. The peak is interpreted as the giant electric dipole resonance in ^{41}Ca.

Many of these experiments are traditionally done using monoenergetic variable energy neutron source techniques. A white neutron source has several advantages over monoenergetic sources especially where the energy dependence of a phenomenon is measured. Since a white source allows the measurement of all energies simultaneously, systematic errors associated with sequential measurements are reduced. The white source technique also reduces problems with time variations in neutron intensity, detector efficiency, bias levels, etc. A further advantage of the white source is that a number of different experiments can run simultaneously.

To illustrate the performance of a white source, an analysis was made for the (n,p) reaction including a comparison with the monoenergetic source technique. We assume the monoenergetic source reaction to be ^{7}Li (p,n) ^{7}Be at 0° and use the cross sections of Fig. 8 measured for proton energies up to 100 MeV. Neutron spectral measurements for this reaction indicate a fairly clean monoenergetic source with a resolution probably determined by the 0.431 MeV

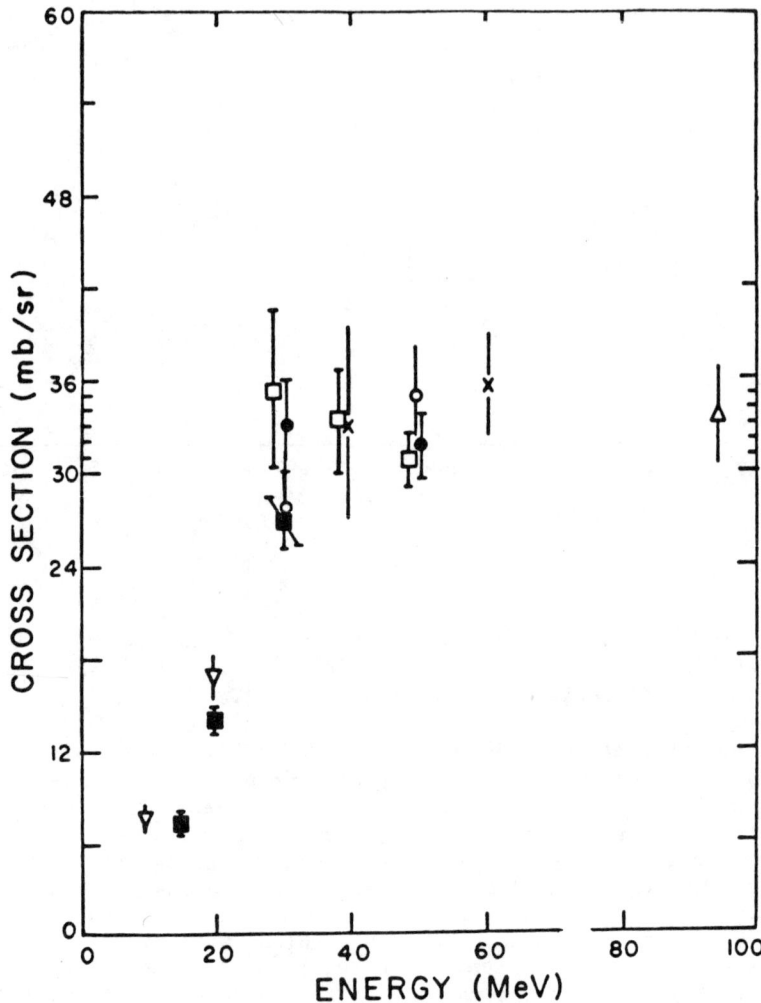

Fig. 8 - The cross section of the ^{7}Li(p,n) reaction at 0° as a function of neutron energy.

separation between the ground and first excited states of ^{7}Be. Above 100 MeV a flat cross section of 33 mb/sr-MeV is assumed. In calculating the neutron yield the lithium target thickness is varied to allow a 1% energy resolution at every energy. (This resolution is limited at lower energies by the 0.431 MeV separation referred to above). Thus for 200 MeV neutrons the target is about 2 MeV thick. A magnet would be required to sweep the transmitted proton beam away from the forward direction, and shielding would be included to reduce backgrounds from off-angle neutrons striking the magnetic spectrometer. It is assumed that the object foil for the spectrometer is located 3 meters from the lithium target to allow space for the magnet and shielding. The neutron intensity on the object foil is given as n/cm^2-µa vs neutron energy in MeV by the indicated curve in Fig. 9.

To calculate the white source effective strength, we use intensities calculated by Russell[1] for thick target emission at 90° which are shown in Fig. 3 and thick target angular distributions measured by Howe.[5] The Target Area 2 layout requires that this experiment be set up at 22 meters which determines the neutron resolution at a value better than 1%.

The band width of the magnetic spectrometer is an important element in the comparison of the two techniques and its influence on the comparison depends to some degree on the nature of the experiment. We assume here that the purpose of the experiment is to measure the excitation function for the (n,p) reaction leading to a particular final state in the product nucleus. The data collection rate for the white source then depends directly on the band width of the magnetic spectrometer, which we take to be 20%. Under these conditions the neutron intensity per cm^2-µa is given as the indicated curve in Fig. 9. The two curves cross at about 300 MeV showing that this is an approximate transition point for the comparative value of the two techniques assuming the same proton current.

Since the maximum current for both the WNR/PSR and the Indiana University Cyclotron Facility (ICUF) is about the same, Fig. 9 should be directly useful for comparison of these facilities for the particular (n,p) experiment described here. Since the ICUF is over subscribed by about 300%, perhaps some of the load could be absorbed at the WNR/PSR. Above 300 MeV, the monoenergetic source appears to be superior if a facility is available which supplies the necessary current and energy. Presently there is no such facility in the U.S. so the usefulness of the WNR/PSR might be extended to energies significantly higher than 300 MeV. A proposed variable energy P$^-$ or H$^-$ source of neutrons being planned for Area B at LAMPF could fill in the higher energy gap and would be highly complementary to the WNR/PSR facility.

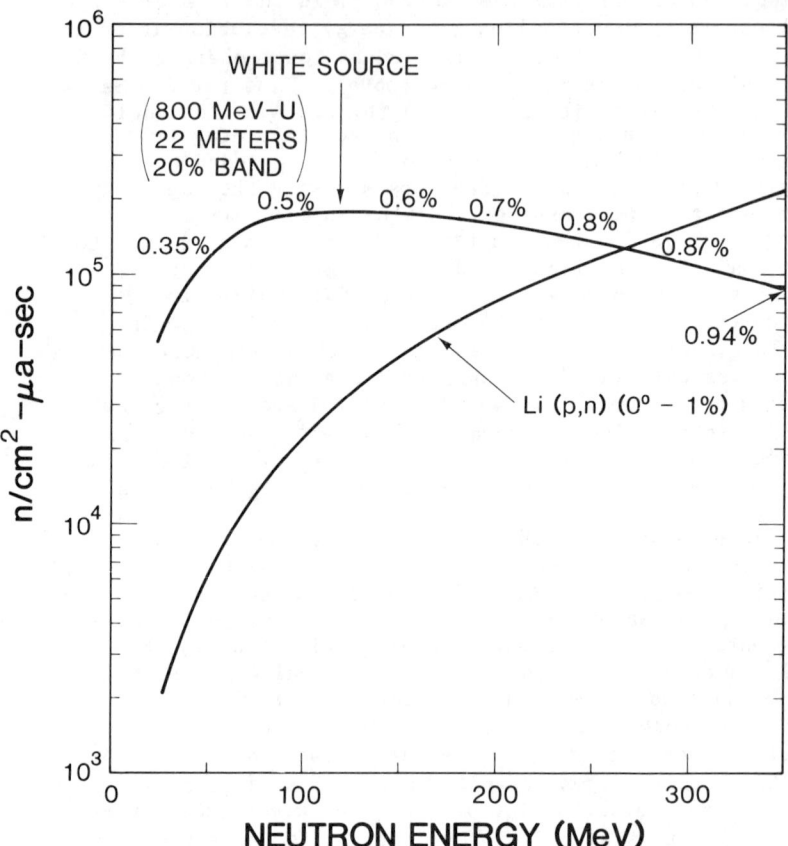

Fig. 9 - A comparison of white vs. monoenergetic neutron source effectiveness for the measurement of (n,p) reactions. The Li(p,n) monoenergetic source intensity is calculated for conditions which allow a 1% resolution in the neutron energy and therefore approximately the same resolution in the proton spectrum. It is assumed that the proton spectrometer allows a coverage of 100% of the proton energy spectrum. The curve for the white source is calculated assuming 800 MeV protons striking a thick ^{238}U target with emission angle of 15°. The proton spectrometer placed at 22 meters allows a 20% band width to be measured at one setting. The numbers above the curve indicate the % neutron resolution at various energies.

SUMMARY

Target Area 2 is being developed as a multiuse white neutron source for a variety of measurements in the 10-300 MeV range. Average current is expected to be about 6 µa at a rate of approximately 50,000 Hz and with a pulse width of about 0.4 nsec. The facility will be operational about 80% of the time that LAMPF operates and will have several flight paths as sites for independent experiments. The facility will begin operation in the summer of 1985. While Los Alamos staff will use the facility for a variety of basic and applied research, it is expected that substantial use of the facility can be made by outside investigators. An outside group might also be able to construct and operate its own spectrometer at the WNR/PSR. Outside use is encouraged and interested persons are urged to contact one of the authors.

REFERENCES

1. G. J. Russell, P. A. Seeger and R. G. Fluharty, "Parametric Studies of Target/Moderator Configurations for the Weapons Neutron Research (WNR) Facility," Los Alamos National Laboratory Report LA-6020, UC-340 (1977).

2. N. S. P. King, P. W. Lisowski, G. L. Morgan, C. Goulding, T. R. Shepard, C. D. Zafiratos, J. L. Ullmann, R. G. Jeppesen, and C. D. Goodman, "High Resolution (p,n) Reaction Measurements at $0°$ and 800 MeV," LAUR-83-2747 (1983).

3. P. Brady, University of California, Davis, Private Communication (1984).

4. M. A. Lone, " Intense Fast Neutron Sources," Symposium on Neutron Cross Sections from 10 to 40 MeV, BNL-NCS-50681, p. 79 (1977).

5. S. D. Howe, "Angular Distributions of Neutrons Produced by 800 MeV Protons on Thick Targets," Los Alamos National Laboratory, (1981).

NEUTRON SCATTERING ABOVE 25 MeV WITH MONOENERGETIC NEUTRONS

Roger W. Finlay
Ohio University, Athens, Ohio 45701 USA

ABSTRACT

Technical problems encountered in measuring elastic and inelastic neutron scattering at kinetic energies well above 25 MeV are considered by extrapolating from experience with lower energy neutrons. It is concluded that all of the techniques required for a high-quality, systematic study of neutron scattering--with the notable exception of a suitable accelerator facility--are at hand.

INTRODUCTION

Although the measurement of differential elastic and inelastic neutron scattering cross sections has a history almost as old as nuclear physics, it is only relatively recently that the facilities and programs have been established for systematic studies from a wide variety of target nuclei over a wide energy range. A survey of existing facilities is included in the proceedings of this Conference, and several contributions demonstrate the rather advanced state of our art. Briefly, several laboratories are active in the energy range $0 < E_n \lesssim 10$ MeV, a few (TUNL and Bruyères le Chatel) have neutron beams with continuously variable neutron energies up to ~ 17 MeV, Ohio University has an active program for $18 < E_n < 26$ MeV. New facilities are being developed at Studsvik [1] (22 MeV) and Lawrence Livermore National Laboratory [2] and will be welcome additions to the field.

Above ~ 26 MeV we are in trouble. Apart from a brief but important program at Michigan State University, [3] which measured neutron scattering (elastic only) from five target nuclei at 30 and 40 MeV, there exists no facility capable of measuring elastic and inelastic scattering at the higher energies of interest at this Conference. While the theoretical motivation for measuring neutron scattering at higher energies will no doubt be discussed by other participants at this Conference, it is worth mentioning that the energy dependence of effective interactions in a microscopic theory of the optical model, [4] the parameterization of the absorptive term in a phenomenological model [5], the details of the isovector potential, [6] and the imaginary part of the Coulomb correction term [7,8] are a few of the important topics that have benefitted substantially from the brief look at 40 MeV neutron scattering afforded us by the

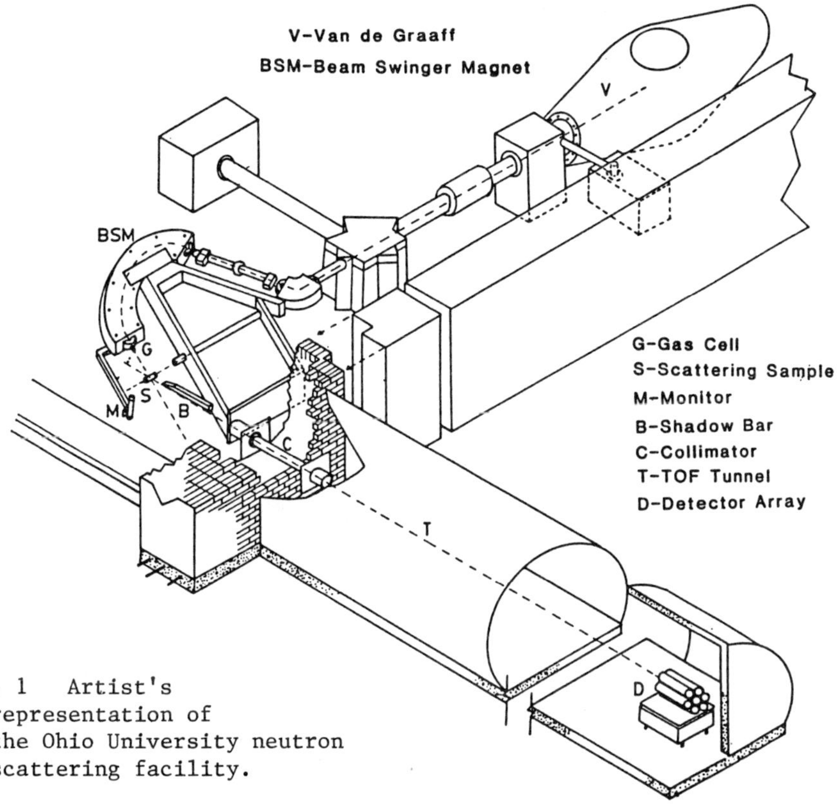

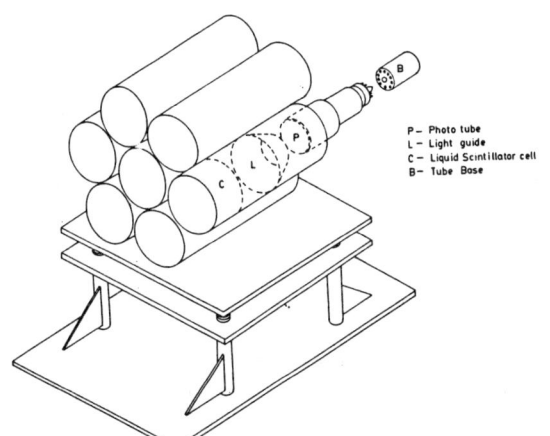

Fig. 1 Artist's representation of the Ohio University neutron scattering facility.

Fig. 2 Seven-detector array of liquid scintillator detectors.

now-terminated MSU program. The feasibility of extending this work to a) other nuclei, b) higher energy, and c) inelastic scattering is the subject of the present contribution.

PRESENT CAPABILITIES AT 24-26 MeV

We can approach the problem of neutron scattering at higher energies in a very straightforward manner, i.e., by examining what works at 25 MeV and calculating what changes at 50, 75, . . . MeV. If this doesn't work, we can start over and design a whole new approach. The emphasis will be on scattering of unpolarized neutrons. The extension to polarized neutrons is a very important topic that is discussed by several other contributors to this Conference.

The Ohio University beam-swinger time-of-flight spectrometer has been previously described in the literature. [9] The important features are illustrated in Figs. 1 and 2 and can be summarized as:
1) angular distributions are measured by rotating the incident beam instead of the neutron detector thus allowing the installation of massive, fixed shielding and collimation, and
2) the flight path may be varied over a wide range without incurring large building-modification costs.

The principal disadvantage of the present system is that it does not permit simultaneous measurements at several scattering angles, but this problem was solved at a similar installation at Colorado. [10]

The importance of massive shielding at high energy cannot be overemphasized. In 1978 we attempted to measure inelastic scattering to highly excited stated in ^{12}C with conventional techniques. The results (Fig. 3a) are compared with a recent beam swinger measurement in 1981 of the same spectrum (Fig. 3b). The 1978 experiment was abandoned but the beam swinger facility has enabled us to extract cross sections for all well-established states in ^{12}C up to 15 MeV of excitation energy. [11]

A typical set of elastic and inelastic scattering differential cross section is shown in Fig. 4 to illustrate the wide angular range and detail that are possible with the beam swinger geometry. This example was chosen to emphasize the capacity to measure weak (2^-) inelastic groups at high excitation energy with passable accuracy.

NEUTRON SOURCE REACTIONS

There is an extensive literature on the subject of mono-energetic neutron source reactions that is not reviewed here in detail. Instead we show (Fig. 5) the general trends of the 0° cross section of three popular source reactions that could

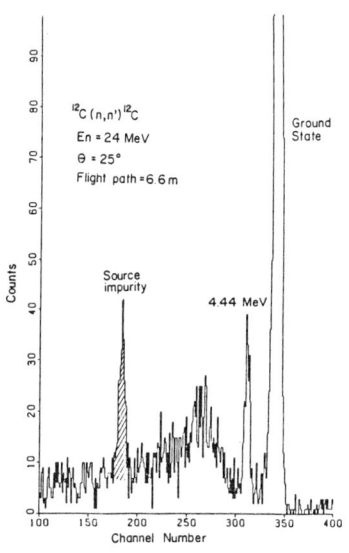
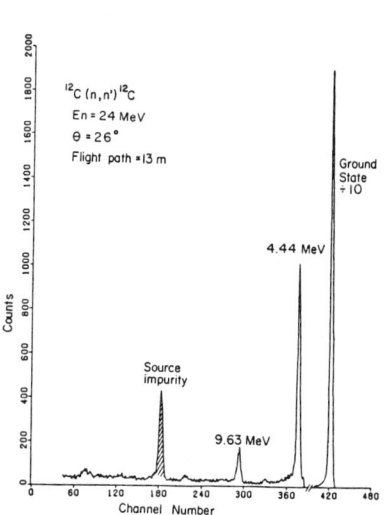

Fig. 3 Direct time-of-flight spectra (without background subtraction) for 24 MeV neutron scattering from ^{12}C. a) conventional time-of-flight spectrometer, b) beam swinger spectrometer.

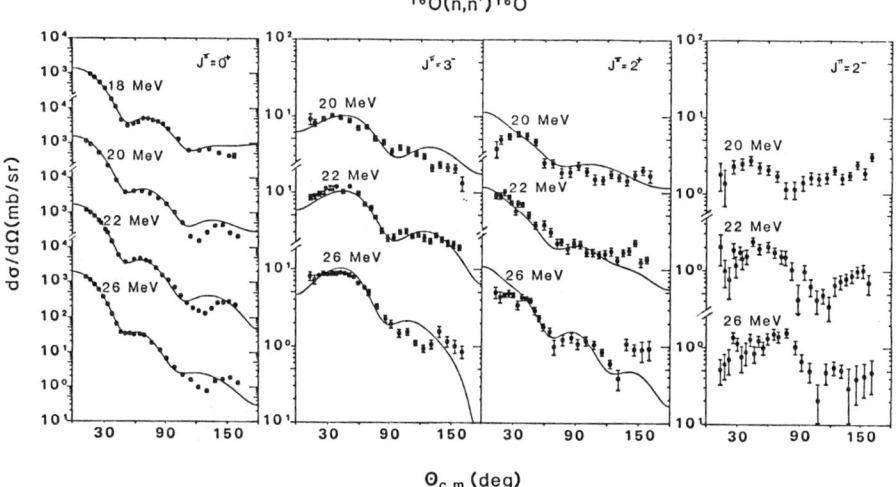

Fig. 4 Elastic and inelastic neutron scattering from ^{16}O.

produce neutrons in the 0-100 MeV region. Solid lines are based on well-measured or carefully evaluated cross sections. Dashed lines indicate less-well-established data, and the dotted line for the T(d,n)^{4}He reaction is a rough guess (by the author) of this important but unmeasured cross section.

All experimentalists understand that there are no purely monoenergetic source reactions that produce neutrons with energy greater than $\sim$ 21.3 MeV. As the energy of the charged projectile increases, additional excited states in the final nucleus are populated or (more importantly) multi-particle breakup becomes possible usually producing a huge flux of unwanted low energy neutrons. The situation is illustrated for the T(d,n) reaction in Fig. 6

ENERGY RESOLUTION

One of the earliest ideas we are taught about neutron time-of-flight spectroscopy is that the energy resolution gets rapidly worse as the neutron energy increases.

Since

$$E = \frac{1}{2} mv^2 = \frac{1}{2} m \frac{\ell^2}{t^2} \qquad (1)$$

then

$$\Delta E = 2E \frac{\Delta t}{t} = \frac{(2E)^{3/2}}{\sqrt{m_n}} \frac{\Delta t}{t} \qquad (2)$$

Thus for fixed flight path, ℓ, and for fixed timing precision, Δt, ΔE increases as $E^{3/2}$ and would surely compromise our efforts to measure neutron scattering at higher energy. For example, an energy resolution of 360 keV at 25 MeV would explode into 1.9 MeV at 75 MeV and render inelastic scattering impossible for all but ^{12}C and ^{16}O.

The analysis is *fundamentally inappropriate* for neutron scattering. Δt in eq. 2 is anything but fixed. A careful analysis of the many factors that contribute to Δt has been performed by Mellema[12] who wrote the program TOFSIM for this purpose. The program calculates time spreads due to accelerator beam burst duration, energy loss and straggling of the incident beam in the neutron source reaction target, transit time differences of the charged particle and the neutron in a gas target cell, neutron and photon transit time differences in a thick scintillator, and time and energy spreads due to the non-zero solid angles subtended by the scattering sample and the neutron detector. The predictions of TOFSIM for energy resolution have been well verified ($\sim$ 10%) in many neutron scattering experiments at Ohio University.

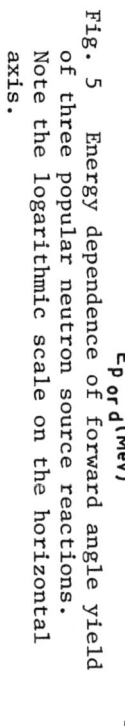
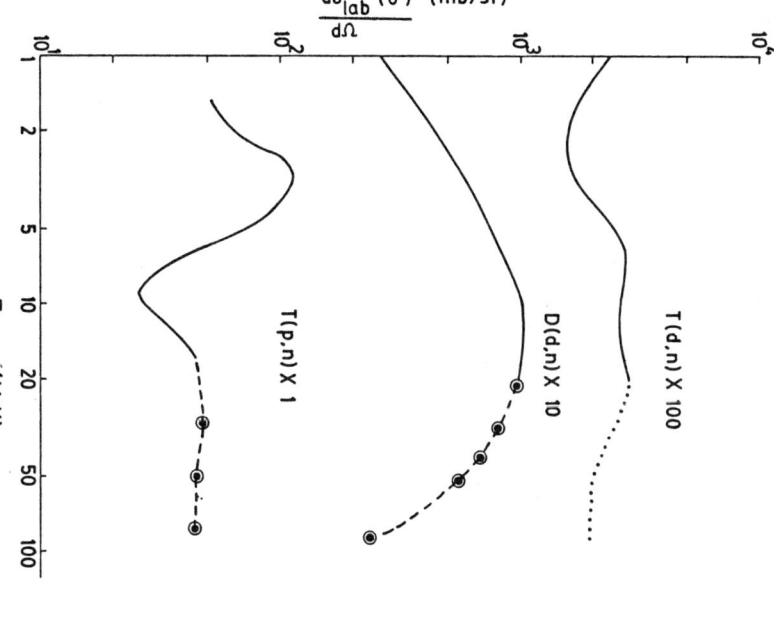

Fig. 5 Energy dependence of forward angle yield of three popular neutron source reactions. Note the logarithmic scale on the horizontal axis.

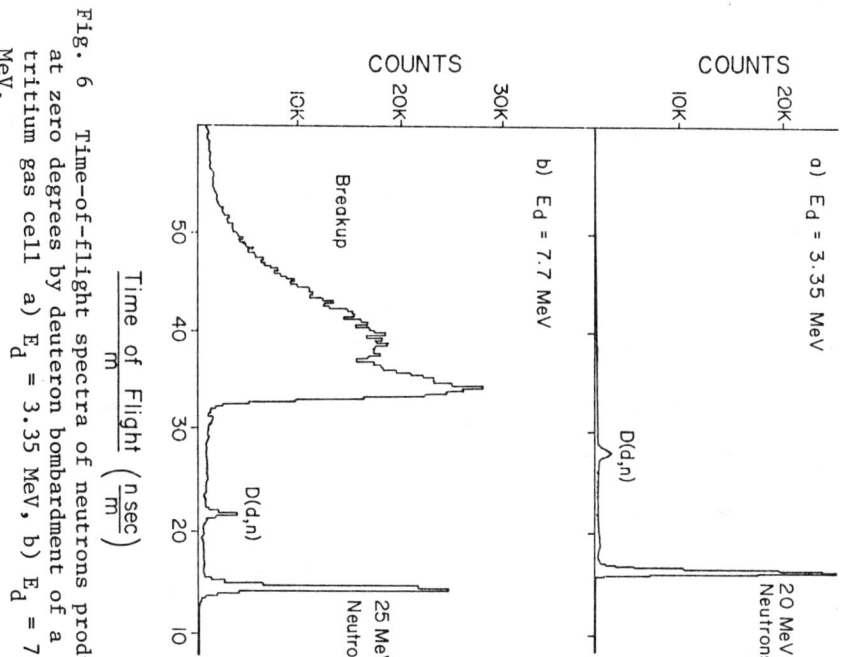

Fig. 6 Time-of-flight spectra of neutrons produced at zero degrees by deuteron bombardment of a tritium gas cell a) E_d = 3.35 MeV, b) E_d = 7.7 MeV.

Table I shows the predictions of TOFSIM for one possible approach to high energy neutron scattering, i.e., the T(d,n) source reaction, beam swinger geometry, 15 m flight path and a high energy deuteron beam. Parameters for 25 MeV neutrons are essentially those used by Mellema in his study [13] of neutron scattering from ^{54,56}Fe. As the deuteron energy increases, energy and angle straggling in the foil decrease, energy loss in the gas decreases, the source reaction energy spread remains roughly constant, transit time spreads in the detector fall by half and the energy resolution--now dominated by the initial burst duration from the accelerator--increases much more slowly than $E^{3/2}$ with increasing neutron energy, i.e., $\Delta E/E \approx 0.0135$ independent of neutron energy (Fig. 7a). This series of calculations is designated by the symbol I and plotted as the solid line in Figs. 7a-d and 8. It will be the reference case for all subsequent comparisons.

Table II shows the anticipated resolution at twice the flight path (30 m). As expected, the predicted resolution improved by a factor that approaches 2 as the energy increases and $\Delta E/E \rightarrow 0.007$.

COUNTING RATES

TOFSIM also provides a rough estimate of the anticipated counting rate in counts/hour/mb/sr. It is encouraging to see that--without any changes in experimental conditions--this rate decreases by only about 25% from 25 to 75 MeV.

Thus we conclude that the ^{54}Fe experiment could be performed at 50 MeV with the T(d,n) source reaction, a 33 MeV deuteron beam of 3 µA average current, flight path of 15 m and energy resolution of 650 keV in about 10% more time (per mb) than it actually took to perform the recently published experiment at 26 MeV. [13] This assertion rests on the hidden assumptions that the estimates of the source reaction cross section at E_d = 33 MeV are not too far wrong and that the detector efficiency remains constant as the neutron energy increases. Experience with the (p,n) program at Indiana University Cyclotron Facility indicates that this assumption is, if anything, a bit pessimistic.

In the calculations of Table I we assume that the deuteron beam burst duration on target held constant at 750 ps --a typical value for the Ohio University Tandem. When we consider the ways in which the 33 MeV deuteron beam might be produced, we see that it should be possible to do somewhat better than this. For example a small, separated sector cyclotron with a radially increasing RF voltage distribution would tend to decrease the beam burst duration. Typical values of 300 ps are commonly achieved at IUCF (see Fig. 7b).

281

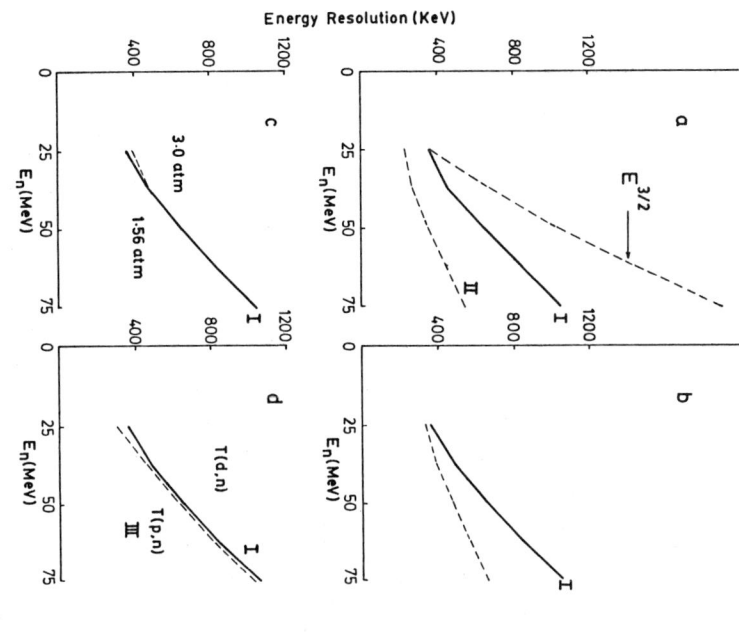
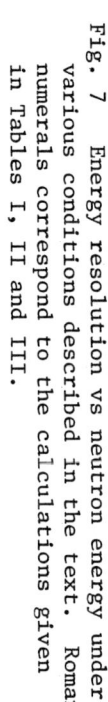

Fig. 7 Energy resolution vs neutron energy under various conditions described in the text. Roman numerals correspond to the calculations given in Tables I, II and III.

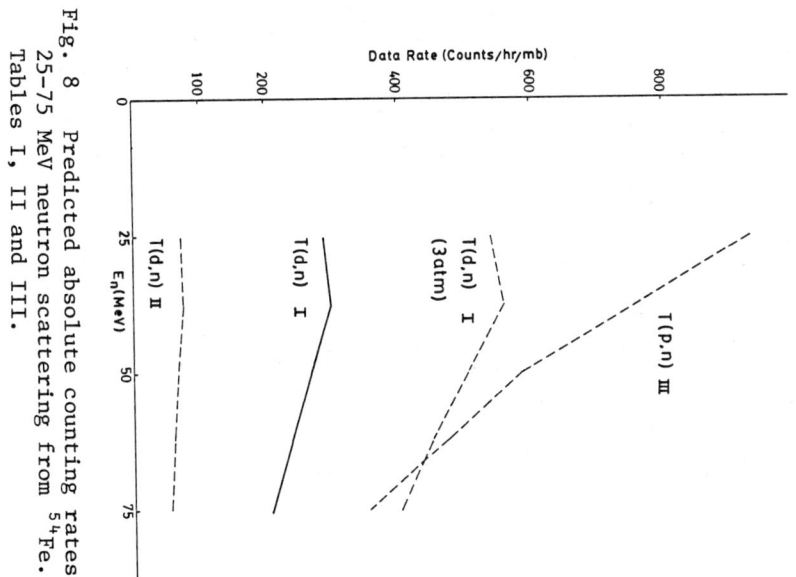

Fig. 8 Predicted absolute counting rates for 25-75 MeV neutron scattering from ^{54}Fe. See Tables I, II and III.

So far we have made no attempt to optimize the planned experiment. For example, at E_d = 33 MeV at flight path of 15 m the gas cell is clearly too "thin" considering the resolution that was finally attained. We could double the target gas pressure without any degradation in energy resolution (Fig. 7c). At Ohio we have always limited the pressure in our tritium cell to 1.7 atm in the interest of safety. In their contribution to this Conference, Olsson and Trostell [1] report that at Studsvik they have safely operated a tritium cell with 4 atm pressure.

If it becomes necessary to go to longer flight paths at 75 MeV, the inverse square loss in counting rate is a matter of obvious concern. The best hope for regaining the count rate lies in improved design of large-area detectors. Annand et al. [14] have shown that a 6.8 ℓ detector can be built with timing properties only slightly inferior to the 2.9 ℓ detector used at 26 MeV (and assumed in the calculations of Table I). A 11.6 ℓ detector was considerably worse although it might be possible to improve its performance with a different light pipe design. All of the calculations assume that an array of seven identical detectors is used in the time-of-flight spectrometer [9]

Even though our first attempt to design experimental conditions for elastic and inelastic neutron scattering at 50 and 75 MeV was successful, it is certainly worthwhile to see if there might be a better way to do the experiments. Consider, for example, the T(p,n) reaction as the neutron source. Protons of 51 and 76 MeV would be required. Since these beams have about the same magnetic rigidity as the 33 MeV beams in the previous example, it should be possible to design a cyclotron to do either job. The appropriate TOFSIM calculations are shown in Table III. We see that the energy resolution obtained with the T(p,n) reaction is essentially the same as with the T(d,n) reaction (Fig. 7d). The counting rate is more than twice as high at 25 MeV with the T(p,n) source (for the same gas pressure, flight path, etc.) and remains significantly higher than with the T(d,n) source at all energies (Fig. 8).

It is tempting to conclude that the T(p,n) reaction would be the preferred source reaction for this program, and, indeed, in many cases it would be, particularly if high pressure tritium gas tragets could be perfected. However, this reaction is also not "clean" in the sense that T(p,pn)D and T(p,p2n)H reactions produce a broad continuum of secondary neutrons with energies reaching up to about 6 MeV below the energy of interest. These neutrons will be elastically scattered from the sample producing counts in the time-of-flight spectrum that are indistinguishable from inelastic scattering of the primary neutrons. If a goal of the program is to measure inelastic scattering to states at high excitation energies (e.g., giant resonances), the T(d,n) reaction with its + 17.16 MeV Q-value and its ∿ 14 MeV of "free spectral range" would be the source reaction of choice. Of course, the T(p,n) reaction would be very well suited to a rapid survey of elastic scattering or inelastic scattering to selected low-lying collective states.

PROBLEMS

Until experiments of the type just described are actually attempted, any list of practical difficulties is bound to be incomplete, and the author would welcome comments from his colleagues concerning the important problems he has omitted. The most important concern should be the signal-to-noise ratio of the experiments. We can calculate that data rates are reasonable compared to a 25 MeV experiment, but it is difficult to estimate the background. Simple scaling in terms of the neutron attenuation length in concrete or copper is not very useful for predicting the shape of the background spectrum. We can be encouraged by the experience at 26 MeV. The "pinhole-camera" approach to collimation at the beam swinger facility lends itself well to upscaling and seems to be better suited to higher energies than the massive movable detector shield of the conventional geometry. But these generalities hardly anticipate all the sources of background that might be experienced at 75 MeV.

One very practical concern is in designing a beam stop for the charged particle beam. A 3 µA beam of 75 MeV protons travels 50 mm in water and delivers 225 W. A flowing water beam stop--perhaps preceded by a superconducting clearing magnet as proposed by the Colorado group [15]--should be adequate, but in neutron scattering studies there is no opportunity to transport the beam to a remote beam dump without losing some range of motion of the beam swinger. Neutron and charged particle activation of the material near the beam stop could pose a hazzard.

Finally, it should be acknowledged that the calculations given in this report take no account of the pulse repetition rate or duty factor of the required accelerator. For neutron time-of-flight measurements, the duty factor should be low (in contrast with the requirements for modern electron accelerators). The simultaneous requirements of long time intervals between beam bursts and moderate average current (3 µA) should not be too difficult a challenge in accelerator design, but the details have not been addressed in this work.

CONCLUSIONS

The exercise of planning neutron elastic and inelastic scattering measurements at 50-75 MeV has produced remarkably encouraging results. Techniques that are presently in use can be applied with confidence at the higher energies. Several of the factors that presently limit the energy resolution and efficiency (e.g., straggling in the neutron production target or time resolution vs. thickness of a scintillation counter) become more manageable at higher energy. Recent developments in gas target technology provide the experimenter with a wide range of choice in the planning of a specific experiment. All that is lacking is a dedicated accelerator facility with the suitable pulsed beam.

The author wishes to thank Jack Rapaport and Frank Dietrich for their continuing encouragement of this work and Steve Mellema for his original development of and diligent improvements to the code TOFSIM.

REFERENCES

1. N. Olsson and B. Trostell, contribution to this Conference.
2. L. Hansen, F.S. Dietrich, B. Pohl, C. Poppe and C. Wong, Trans. Amer. Nucl. Soc. $\underline{46}$, 627 (1984).
3. R.P. DeVito, Ph.D. dissertation, Michigan State University, 1979 (unpublished).
4. F.S. Dietrich, R.W. Finlay, S. Mellema, G. Randers-Pehrson and F. Petrovich, Phys. Rev. Lett. $\underline{51}$, 1629 (1983).
5. R.W. Finlay, J.R.M. Annand, T.S. Cheema, J. Rapaport and F.S. Dietrich, Phys. Rev. $\underline{C30}$, 796 (1984).
6. R. De Leo and S. Micheletti, Phys. Lett. $\underline{98B}$, 233 (1981).
7. J. Rapaport, Phys. Lett. $\underline{92B}$, 233 (1980).
8. R.P. DeVito, Sam M. Austin, U.E.P. Berg, R. De Leo and W.A. Sterrenburg, Phys. Rev. $\underline{C28}$, 2530 (1983).
9. R.W. Finlay, C.E. Brient, D.E. Carter, A. Marcinkowski, S. Mellema, G. Randers-Pehrson and J. Rapaport, Nucl. Instrum. Methods $\underline{198}$, 197 (1982).
10. D.A. Lind, J.D. Carlson, S.D. Schery and C. Zafiratos, Nucl. Instrum. Methods, $\underline{130}$, 93 (1975).
11. A.S. Meigooni, R.W. Finlay, J.S. Petler and J.P. Delaroche, contribution to this Conference; submitted to Nucl. Phys.
12. S. Mellema, Ph.D. dissertation, Ohio University, 1983 (unpublished).
13. S. Mellema, R.W. Finlay, F.S. Dietrich and F. Petrovich, Phys. Rev. $\underline{C28}$, 2267 (1983); $\underline{C29}$, 2385 (1984).
14. J.R.M. Annand, R.W. Finlay and M. Polster, Nucl. Instrum. Methods (in press).
15. Proposal for a National Light Ion Accelerator Facility, University of Colorado, NPL-813A, 1980 (unpublished).

TABLE I** T(d,n) reaction with 15 m flight path

Ion Energy (MeV)	Eloss Foil (MeV)	Straggling Foil (MeV)	(FWHM) (Rad)	Eloss Gas (MeV)	Source Neutrons (MeV)	(FWHM)	Resolution (FWHM) Detctr (nsec)	Gascel (nsec)	Total (nsec)	(keV)	Count Rate (*)
8.079	.319	.051	.081	.087	25.00	(.119)	0.99	0.97	1.57	361	290
20.113	.179	.047	.035	.039	37.50	(.083)	0.72	0.48	1.15	484	300
32.775	.128	.045	.022	.026	50.00	(.085)	0.57	0.34	1.00	652	271
45.630	.102	.045	.016	.020	62.50	(.092)	0.46	0.28	0.93	841	241
58.569	.085	.044	.013	.016	75.00	(.102)	0.39	0.24	0.88	1052	216

TABLE II** T(d,n) reaction with 30 m flight path

Ion Energy (MeV)	Eloss Foil (MeV)	Straggling Foil (MeV)	(FWHM) (Rad)	Eloss Gas (MeV)	Source Neutrons (MeV)	(FWHM)	Resolution (FWHM) Detctr (nsec)	Gascel (nsec)	Total (nsec)	(keV)	Count Rate (*)
8.079	.319	.051	.081	.087	25.00	(.119)	0.99	1.51	1.96	225	72
20.113	.179	.047	.035	.039	37.50	(.083)	0.72	0.73	1.28	269	75
32.775	.128	.045	.022	.026	50.00	(.085)	0.57	0.54	1.08	352	67
45.630	.102	.045	.016	.020	62.50	(.092)	0.46	0.44	0.99	448	60
58.569	.085	.044	.013	.016	75.00	(.102)	0.39	0.39	0.93	554	54

TABLE III** T(p,n) reaction with 15 m flight path

Ion Energy (MeV)	Eloss Foil (MeV)	Straggling Foil (MeV)	(FWHM) (Rad)	Eloss Gas (MeV)	Source Neutrons (MeV)	(FWHM)	Resolution (FWHM) Detctr (nsec)	Gascel (nsec)	Total (nsec)	(keV)	Count Rate (*)
25.869	.093	.044	.027	.018	25.00	(.067)	0.99	0.47	1.33	305	935
38.342	.071	.043	.019	.013	37.50	(0.74)	0.72	0.33	1.09	461	763
50.827	.058	.043	.015	.010	50.00	(.082)	0.57	0.26	0.98	634	589
63.318	.049	.043	.012	.008	62.50	(.091)	0.46	0.22	0.91	825	473
75.812	.043	.043	.010	.007	75.00	(.100)	0.39	0.19	0.87	1040	357

* Count rate given in counts/hours/mb/sr.
** In all cases: 3.0 μA beam current, 0.75 ns beam burst duration, 3.0 cm gas cell at 1.56 atm, 56 gm of ^{54}Fe scattering sample, seven detector array 2.8 ℓ each, source-sample distance 15 cm, scattering angle 20°.

NEUTRON MEASUREMENTS FOR BIOMEDICAL AND FUSION TECHNOLOGY APPLICATIONS

H.H. Barschall
University of Wisconsin, Madison, Wisconsin 53706

ABSTRACT

Measurements of reaction cross sections of neutrons of energy above 5 MeV yield important information about reaction mechanisms. The main impetus for such measurements has, however, recently come from applications. Measurements on light elements are needed for neutron dosimetry, primarily for radiotherapy. Measurements on heavier nuclides provide information for fusion technology, both for the assessment of radiation damage and for the management of radioactive wastes.

INTRODUCTION

Measurements of nuclear cross sections for neutrons from thermal energy to a few MeV have yielded much information about the nucleon-nucleus interactions; they led for example to the development of the compound nucleus model and of the optical model, and provided information about the statistical properties of nuclear energy levels near 8-MeV excitation energy. In recent years there has been an interest in extending accurate cross-section measurements to higher neutron energies. These measurements also yield new information about nuclear reaction mechanisms.

One of the early important measurements at higher neutron energies was Louis Rosen's study [1] of the secondary neutron spectra from the interaction of 14-MeV neutrons with heavy nuclei. Fig. 1 shows the angular distribution of the secondary neutrons produced in the interaction of 14-MeV neutrons with Ta. The angular distribution of the neutrons of energy below 4 MeV is isotropic, while the higher energy neutrons are forward peaked. Although the results were originally not interpreted in that way, they gave the first evidence for preequilibrium processes.

Neutron cross-section measurements at energies above about 10 MeV are much more difficult than measurements at lower energy, partly because of the lack of intense sources of monoenergetic neutrons of variable higher energy, and partly because of the many reaction channels that are important. Much of the impetus for recent work in the energy region from 10 to 30 MeV has come from the needs in applications. I want to discuss two areas in which such cross section are of importance, some of the information that is needed, and some of the results that have been obtained. The two areas are radiotherapy and fusion technology.

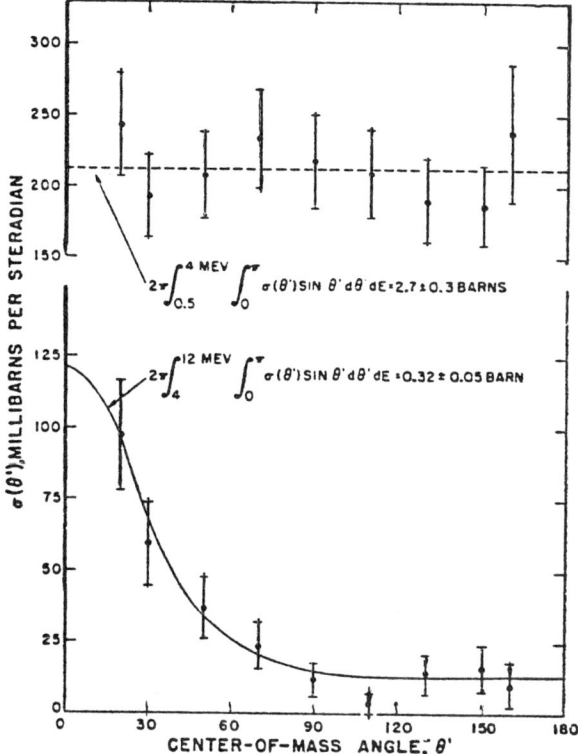

Fig. 1   Angular distribution of secondary neutrons from the interaction of 14-MeV neutrons with Ta measured in 1957. The forward peaking of the neutrons of energies between 4 and 12 MeV is the earliest evidence for preequilibrium processes. (From Reference 1).

RADIOTHERAPY

The use of high-energy neutrons in radiotherapy was proposed by E. O. Lawrence [2] in 1936. Between 1938 and 1943 200 cancer patients were treated with neutrons from the Berkeley cyclotron. Although there appeared to be some benefits from the treatment, many patients suffered severe side effects, and the physicians who had directed the neutron treatment concluded [3] that these complications made neutron treatment an undesirable procedure. A more recent evaluation of the Berkeley treatments [4] has shown, however, that, on the basis of our present knowledge of the biological effects of neutrons, the patients received too large doses of radiation. At the end of the fifties new radiobiological data became available which raised the hope that fast neutron therapy might cure some malignant diseases which do not respond to conventional x-ray therapy. The clinical use of neutrons resumed in 1966 at the Hammersmith hospital in London. The reports from Hammersmith [5] were so encouraging that many radiotherapists began to use neutrons. There are now about two dozen neutron radiotherapy facilities in Europe, Japan, and the USA, and several thousand patients have been treated. After the initial great expectations present judgment of the advantages of neutron therapy over x-ray therapy is more guarded [6], but some radiotherapists think that the advantages of neutrons have been underestimated, because most facilities that have been used do not produce optimal neutron fields. In particular, higher neutron energies would produce better penetration.

DOSIMETRY

A challenge to the physicist is the determination of the neutron dose used in the treatment. If the dose is 5% too low, the treatment may be ineffective, if it is 5% too high, complications are likely to occur. As an example, Fig. 2 shows the dose-effect relationship for tumor control and for skin and intestinal damage

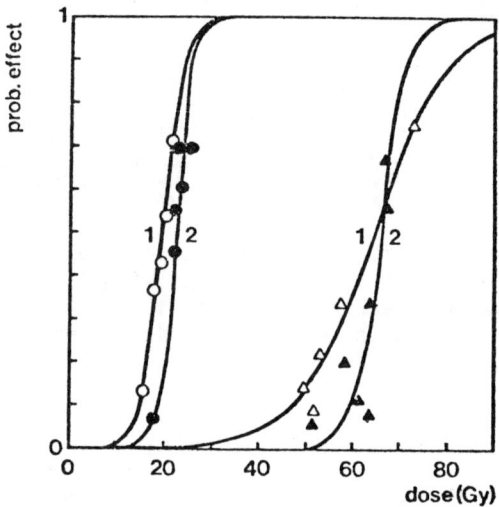

Fig. 2 Effect versus dose for tumor control (open circles) after irradiation of bladder tumors by 14-MeV neutrons. The solid circles give the corresponding results for skin and intestinal damage after neutron irradiation. Triangles show tumor control and damage, respectively, for photon irradiation. (From Reference 7).

after neutron and photon irradiation of bladder tumors [7].
Determination of the neutron dose with an accuracy of a few percent is very difficult for several reasons: Most clinical neutron beams have continuous energy spectra extending up to 30 MeV or even 60 MeV neutron energy. In addition to neutrons, there are always γ-rays present, and the neutron dose needs to be known at the position of the tumor, i.e. inside the body where measurements can usually not be performed.

Neutron dose is defined as energy deposited per unit mass. The most direct dose measurement uses a calorimeter [8], but the temperature rise is so small that this method is not practical for routine clinical use. In addition, the calorimeter does not distinguish heating by neutrons and by γ-rays, but a given neutron dose has a much larger biological effect than the same γ-ray dose. Routine dose determinations are usually carried out with tissue equivalent ionization chambers [9], although they do not distinguish neutron and γ-ray dose either. These chambers actually measure the kinetic energy of the charged particles released by the neutrons in a plastic that has properties similar to tissue. The energy of the charged particles per unit mass is called kerma, and, for practical purposes, is the same as dose. Tissue-equivalent chambers were developed for use with x-rays. In order to make the wall of the chamber a solid conductor, plastic materials have been developed in which the oxygen in tissue, especially the oxygen in H_2O, is replaced by carbon. The problem is that for neutrons the charged-particle production per unit mass of oxygen and of carbon differ substantially, and the difference depends on neutron energy. Hence the cross sections for charged-particle production in both oxygen and carbon need to be known for the neutron spectrum used in the treatment. Unfortunately the relevant cross sections are very poorly known. As an example, Fig. 3 is taken from a recent evaluation [10] of the ^{12}C cross section and shows the $^{12}C(n, n_3)$ cross section as a function of energy. This reaction makes one of the largest contributions to the kerma and the latest two evaluations (ENDF/B-V and JENDL-3) differ by more than a factor of two at some energies between 10 and 20 MeV. Efforts to improve the data have been made at Ohio University [11], UC-Davis [12], and at Lawrence Livermore National Laboratory [13]. As an example of the uncertainties in our knowledge of these cross sections, Table I taken from reference 13 shows a factor of almost three between the most recent measurements of the $^{12}C(n, 3\alpha)$ cross section which is the reaction that makes the largest contribution to the kerma of carbon, even though these measurements are performed at an energy of 14 MeV where measurements are easiest, since intense sources of monoenergetic neutrons are available.

Because of the unavailability of adequate cross-section data, integral kerma measurements have also been performed [14]. In these measurements the energy-loss distribution of the charged particles produced by neutrons is measured in a small counter, and neutron and γ-ray induced events can be distinguished. According to a

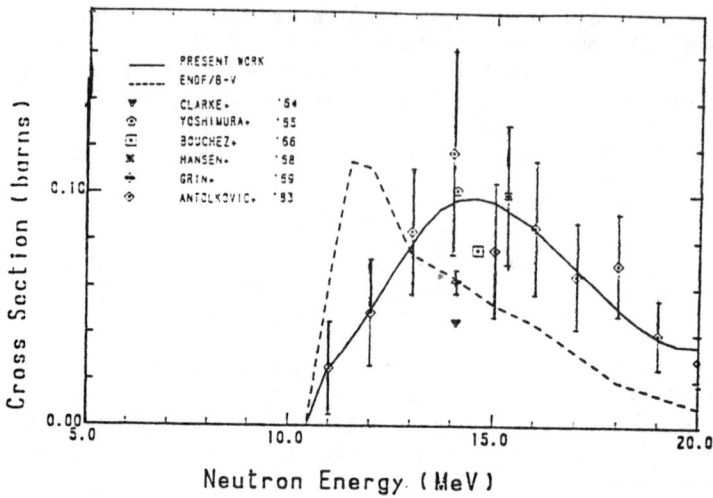

Fig. 3 Measured and evaluated cross sections for the ^{12}C(n, n$_3$) 3α reaction. (From Reference 10).

theory developed by Bragg and Gray, under carefully chosen conditions the kerma in the wall of the counter can be deduced from the energy loss of charged particles in a small cavity. Suitable counters were developed by Rossi [15]. The neutron flux density required for such a measurement can be readily achieved with low-current accelerators. These counters satisfy the Bragg-Gray condition more accurately for alpha particles than for carbon recoils. Preliminary results on carbon indicate that the alpha-production cross section in carbon varies rapidly with energy so that some of the discrepancies shown in Table I may be due to differences in the neutron spectra used in the measurements.

Table I
Cross Section for the Reaction ^{12}C(n, n′)3α

Year	Method	Neutron Energy (MeV)	Cross Section (mb)
1955	Emulsion	14.1	230 ± 50
1958	Emulsion	14.0	176 ± 82
1969	Scintillator	14.1	190 ± 20
1973	Scintillator	14.0	190 ± 20
1976	Scintillator	14.2	202 ± 30
1983	Emulsion	14.0	301 ± 89
1984	Quadrupole Spectrometer	14.1	110 ± 15

RADIATION DAMAGE

The other application of neutron cross-section measurements that I want to discuss is to fusion technology. All near-term fusion reactor designs use the D-T reaction in which most of the energy is released in the form of 14-MeV neutrons. The first wall and other components of the reactor will be exposed to a high neutron flux density, typically 10^{14} cm^{-2}s^{-1}, mostly at elevated temperatures and high mechanical stresses. Under these conditions severe deterioration of many mechanical properties is expected. Although there is much information available about the radiation damage caused by fission neutrons, 14-MeV neutrons produce additional effects which cause additional damage. Most important is the very much higher rate of production of hydrogen and helium isotopes. These gases will produce swelling and embrittlement. For estimates of the useful lifetimes of various alloys a knowledge of the gas production cross section in the constituents is necessary. Although considerable progress has been made in identifying alloys that will have a relatively long useful life, such as ferritic steels [16], even the most optimistic estimates of the lifetime of the first wall of a fusion reactor is about six years. This raises the question of the disposal of the large amount of highly radioactive material that has to be replaced. Efforts have been made to estimate the amount of radioactivity that has to be disposed [17,18]. For this a knowledge of the reaction cross sections of all the constituents is needed for all transmutations that lead to radioactive products. This is a second reason for the interest in 14-MeV cross sections.

In some cases an activation cross section yields a gas production cross section, but not always. Because of the many reaction channels that are open, the same activity can often be produced in different reactions, and conversely starting from a given nuclide, reactions producing alpha particles, for example, can lead to different activities. Several approaches have been used to determine relevant reaction cross sections. Kneff et al. [19] have measured directly the amount of helium generated by the fast neutrons by using a mass spectrometer. This method works, however, only for helium, not for hydrogen, which is also of importance, and it does not provide information about the kerma, which is of importance both for estimating radiation damage from atomic displacements and for determining local heating by neutrons.

An extensive program of measurements of charged-particle production by 14-MeV neutrons was started at Livermore eight years ago. Similar measurements have been performed more recently at lower neutron energy at Ohio University. The problem in performing such measurements is that one needs to measure the energy distributions of all the charged particles produced, primarily protons, deuterons, and alpha particles, as a function of reaction angle, and that the measurement has to be performed in a high neutron flux where semiconductor detectors have a short life time

and a high background count rate. The procedure which was
developed [20] uses magnetic lenses to transport the charged
particles from a radiator near the neutron source to the detectors
which are several meters away and are shielded from the source.
ΔE-E silicon surface barrier detectors served to identify the
charged particles and to measure their energy. As an example of
the energy distribution of protons, Fig. 4 shows the proton
spectrum from nickel. [21] Just as in Rosen's early measurements of

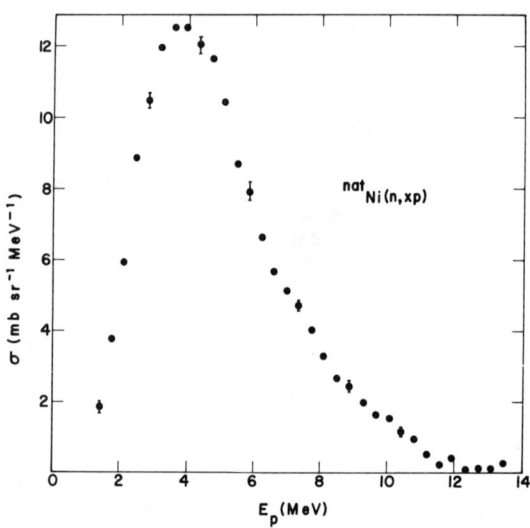

Fig. 4 Proton-emission cross section from natural nickel at 90°
with respect to the incident 15-MeV neutrons. (From
Reference 21).

secondary neutrons, there are many more high-energy protons than
would be expected from a statistical compound nucleus model. The
mechanism for their production was studied by observations at
different angles. This is illustrated in Fig. 5 for charged-
particle emission from ^{50}Cr. [21] The high-energy protons are peaked
forward and are attributed to preequilibrium processes, while the
angular distribution of the low-energy protons is isotropic. The
angular distribution of energetic alpha particles is also forward
peaked, and so is that of the deuterons. Fig. 6 shows proton
spectra integrated over emission angle. The curves are
calculations. [24] The statistical calculations shown by the dashed
curves reproduce all but the highest energy portion of the
spectra. The dot-dashed curves account for the preequilibrium
portion of the spectrum. The sum of the contributions is

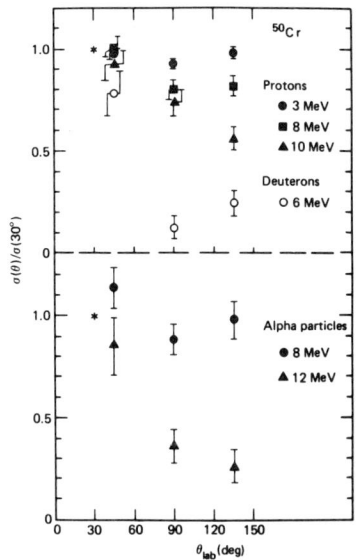

Fig. 5 Variation with angle of the cross sections for the emission of protons, deuterons, and alpha particles from the bombardment of ^{50}Cr with 15-MeV neutrons. The cross sections are normalized to unity at 30°. (From Reference 21).

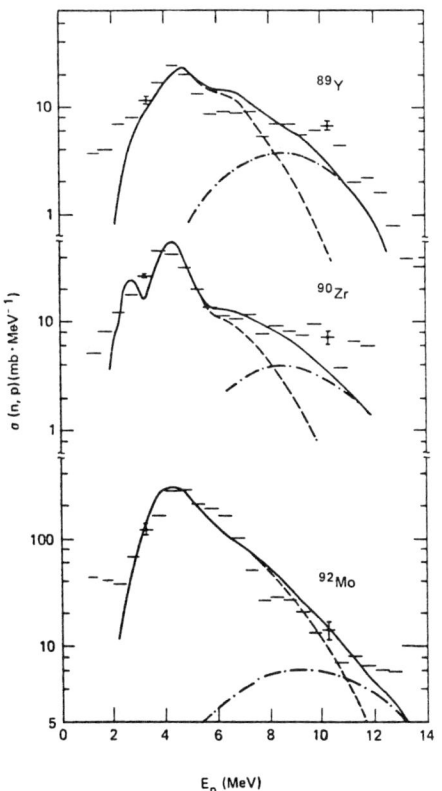

Fig. 6 Angle-integrated proton-emission cross sections for reactions induced by 15-MeV neutrons on targets of Y, ^{90}Zr, and ^{92}Mo. The dashed curves are a multistage Hauser-Feshbach calculation, the dot-dashed curves a hybrid-model calculation for preequilibrium emission, the solid curves the sum of the two. (From Reference 24).

represented by the solid curve. Addition of the preequilibrium reactions improves the fit to the data, but is not enough to account for all the observed high energy protons.

TABLE II
Cross Sections for 15-MeV Neutrons (mb)

Target	Proton Emission	Deuteron Emission	Alpha Emission	Helium Production
^{27}Al	400±60	19±8	121±25	144±7
^{46}Ti	670±90	9±4	98±18	
^{48}Ti	85±16	7±3	28±6	
^{51}V	91±14	7±3	17±3	18.7±1.4
^{50}Cr	830±100	13±4	94±15	
^{52}Cr	182±25	8±3	38±6	
^{54}Fe	900±110	10±4	80±13	91±7
^{56}Fe	190±20	8±3	41±7	46±3
^{58}Ni	1000±120	14±6	106±17	121±8
^{60}Ni	330±40	11±4	76±12	80±6
^{63}Cu	320±50	9±4	56±10	65±5
^{65}Cu	44±5	10±4	14±3	17±1
^{89}Y	98±12	10±3	8±2	
^{90}Zr	166±20	10±3	15±3	
^{93}Nb	51±8	8±3	14±3	17±5
^{92}Mo	967±116	22±7	36±7	31±2
^{94}Mo	124±15	9±3	28±6	22±2
^{95}Mo	84±10	8±3	24±5	17±1
^{96}Mo	64±8	6±2	18±4	12±1

The results of the cross section measurements for charged-particle emission induced by 15-MeV neutrons are shown in Table II for nuclides of particular interest to fusion technology. The first three columns of numbers are based on measurements with the magnetic lens spectrometer [21-24], the last column is based on helium collection.[19] For the cases where the same cross section was measured by these two very different methods the results agree quite well. Among the nuclides studied, the proton-emission cross section varies by more than an order of magnitude, while the deuteron-emission cross section is small and does not vary very much from nuclide to nuclide. The alpha-particle emission cross section, on the other hand, again varies by an order of magnitude between nuclides of interest. These measurements provide adequate information on the impact of the use of materials for fusion reactors from the point of view gas production by neutrons.

FUSION WASTE MANAGEMENT

The neutron cross sections of greatest interest to fusion technology are at present probably those for reactions that lead to long-lived radioactivities. Reference 18 lists 79 known radionuclides up to uranium with half-lives greater than five years. There may possibly be other not yet discovered long-lived radionuclides, such as products of neutron reactions with radioactive reaction products. Many of the known long-lived radionuclides can be produced in the neutron bombardment of structural materials or of impurities in these materials or of reaction products. In a fusion reactor some 500 tons of steel would be exposed to a neutron fluence of 10^{23} cm^{-2} in three years of operation. Hence quite small cross sections and concentrations of impurities of a few parts per million can produce large activities. Many of the relevant cross sections are very poorly known, and present estimates of activities that have to be disposed of when the radiation-damaged structures are replaced, are based on unreliable extrapolations. I am not aware of any program of measurements of the relevant cross sections, and I think that it is an area of applied neutron physics that deserves study.

I wish to thank G.L. Kulcinski for helpful suggestions regarding the needs in fusion technology and P.M. DeLuca for discussions of neutron dosimetry.

REFERENCES

1. L. Rosen and L. Stewart, Phys. Rev. 107, 824 (1957).
2. E.O. Lawrence, Radiology 29, 313 (1937).
3. R.S. Stone, Am. J. Roentgenol. 59, 771 (1947).
4. G.E. Sheline, et al., Am. J. Roentgenol. 111, 31 (1971).
5. M. Catterall, I. Sutherland, and D.K. Bewley, Brit. Med. J. 2, 653 (1975).
6. J.J. Broerse and J.J. Battermann, Med. Phys. 8, 751 (1981).
7. J.J. Battermann, A.A. M. Hart, and K. Breur, Brit. J. Radiol. 54, 899 (1981).
8. J.C. McDonald, J.S. Laughlin, and R.E. Freeman, Med. Phys. 3, 80 (1976).

9. ICRU Report 26 "Neutron Dosimetry for Biology and Medicine," Washington (1977).
10. K. Shibata, Report NEANDC 95/U (1983).
11. A. S. Meigooni, J.S. Petler, and R.W. Finlay, Phys. Med. Biol. 29, 643 (1984).
12. T.S. Subramanian, et al., Phys. Rev. C 28, 521 (1983).
13. R.C. Haight, S.M. Grimes, R.G. Johnson, and H.H. Barschall, Nucl. Sci. Eng. 87, 41 (1984).
14. P.M. DeLuca, H.H. Barschall, R.C. Haight, and J.C. McDonald, Radiation Research (in press).
15. H.H. Rossi and W. Rosenzweig, Radiology 64, 404 (1955).
16. R.W. Powell, D.T. Peterson, M.K. Zimmerschied, and J.F. Bates, J. Nucl. Mat. 103, 104, 969 (1981).
17. F.W. Wiffen and R.T. Santaro, Proc. Conf. on Ferritic Alloys, Snowbird, Utah (1983).
18. R.C. Maninger and D.W. Dorn, Fusion Technology (to be published).
19. D.W. Kneff, B.M. Oliver, M.M. Nakata, and H. Farrar, J. Nucl. Mat. 103, 104, 1451 (1981).
20. K.R. Alvar, H.H. Barschall, R.R. Borchers, S.M. Grimes, and R.C. Haight, Nucl. Inst. Meth. 148, 303 (1978).
21. S.M. Grimes et al., Phys. Rev. C 19, 2127 (1979).
22. S.M. Grimes, R.C. Haight, and J.D. Anderson, Nucl. Sci. Eng. 62, 187 (1977).
23. S.M. Grimes, R.C. Haight, and J.D. Anderson, Phys. Rev. C 17, 508 (1978).
24. R.C. Haight, S.M. Grimes, R.G. Johnson, and H.H. Barschall, Phys. Rev. C 23, 700 (1981).

CAN AN ATOMIC BEAM POLARIZED SOURCE BE IMPROVED FOR NEUTRON EMISSION EXPERIMENTS BY USING AN ECR IONIZER?

T. B. Clegg
University of North Carolina, Chapel Hill, N. C. 27514*
and Triangle Universities Nuclear Laboratory, Durham, N. C. 27706*

ABSTRACT

An electron-cyclotron-resonance source is considered as a potential ionizer for a $\vec{H}_o$ or $\vec{D}_o$ atomic beam. The advantages and possible problems are discussed.

Neutron emission experiments utilizing polarized beams have traditionally been difficult because polarized beam intensities on target are low. Furthermore, the short beam pulse timing needed for high-resolution time-of-flight experiments is more difficult to obtain than with unpolarized beams. In the common atomic beam polarized source[1,2] equipped with an electron bombardment (EB) ionizer,[3] these two difficulties might be removed if the ionizer could be improved.

The EB ionizer has a rather low absolute ionization efficiency for a beam of thermal polarized atoms of $\vec{H}_o$ or $\vec{D}_o$ of ~4%, and its output ion energy spread is large, perhaps 200 eV. The physical origin of these limitations is the large, radial space-charge electric field which both limits the stored ionizing electron density in the ionizer to n_e = 2 to 3 x 10^9/cm^3 and causes ions produced at different radii to have different electrostatic potential energies.

A possible appealing solution for both limitations is an electron-cyclotron-resonance (ECR) ionizer.[4] When used for unpolarized H_2 or D_2 gas, the ionization efficiency is typically ≥50% and the output beam energy spread is ≤5 eV. In addition, the ECR source is simple and has been shown to be very reliable. There is real concern,[2,5] however, that because the ECR frequency, ω_{ECR}, is equal to the electron spin-flip frequency ω_o, the polarized atomic beam might be depolarized during the ionization process. Investigation of this problem shows that this depolarization is not as likely to occur as might first be feared.[6]

An ionizer for an $\vec{H}_o$ (or $\vec{D}_o$) atomic beam source must operate at magnetic fields large compared to the 507G(117G) critical field produced by the nucleus at the site of the electron. This couples the electron spin and, via the hyperfine interaction, the nuclear spin to the external magnetic field direction at the moment of ionization. At B ≃ 1200G where an ECR source operates conveniently, this hyperfine coupling reduces substantially the probability for atomic transitions, in which the nuclear spin is flipped with that of the atomic electron, to 8% (for $\vec{H}_o$) and 0.6% (for $\vec{D}_o$) of their values at B = 0. Thus, resonant spin flip of the atomic electron before ionization may not destroy the <u>nuclear</u> polarization.

Can one determine, however, how quickly the atomic electron's spin might be reversed? The most likely process for electron spin reversal will be that of adiabatic passage. In order to determine whether this will occur in an operating ECR ionizer, the performance of a small ECR source built recently at Jülich for test purposes[8] was considered. For adiabatic passage to occur, the atomic electron should experience simultaneously a large and slowly changing magnetic field B and an r.f. magnetic field B_1 at frequency ω_o. It is necessary that these fields satisfy $dB/dt \ll \gamma_J B_1^2$ where γ_J is the electron's gyromagnetic ratio. For the test source at Jülich and the polarized, thermal atomic beam, it is found that $dB/dt \simeq 1250$ Tesla/sec and $\gamma_J B_1^2 \simeq 18$ Tesla/sec. Thus this adiabatic passage criterion is strongly violated, and it seems very unlikely that the driving r.f. field will cause spin-flip of the atomic electron.

The ECR plasma in the source is also characterized by an unknown r.f. plasma frequency distribution which depends strongly on the B-field, gas pressure, and injected r.f. power at frequency ω_o in ways which are not understood. Thus one can imagine that plasma frequency components could be present locally inside the source such that the above criterion for adiabatic passage is satisfied. The probability that this occur in an actual ionizer seems unfortunately not calculable at present.

What about the ionization efficiency? In a traditional ECR source gas atoms bounce around, passing several times through the ECR plasma. In an ECR ionizer the polarized beam must pass once through the plasma without striking walls where the atoms may depolarize and recombine to form background gas. Thus, comparing experience in various laboratories with higher-power and/or higher-pressure ECR plasmas, it seems reasonable that an electron density of $n_e \simeq 10^{10}/cm^3$ would be possible for injected r.f. powers of ~100 to 200 Watts. Comparing this figure with the electron densities of 2 to $3 \times 10^9/cm^3$ available in a tradiational EB ionizer, one indeeds expects a factor of 3 to 5 improvement over present ionization efficiencies. There is no doubt that the output ion energy spread would be very low, ≤ 5 eV.

These considerations make a test of an ECR ionizer on an atomic beam source both interesting and important.

REFERENCES

*Work partially supported by the USDOE Contract No. DE-AS05-76ER1067.
1. Proc. 4th Conf. on Intense polarized proton ion sources, eds. G. Roy and P. Schmor (Am. Inst. of Phys. New York, 1984) vol. 117.
2. W. Grüebler, in ref. 1, p. 1.
3. P. A. Schmelzbach, W. Grüebler, V. König and B. Jenny, Nucl. Instr. Meth. 186, 655 (1981).
4. R. Geller, B. Jacquot, and C. Jacquot, in ref. 1, p. 162.
5. S. Jaccard, in ref. 1, p. 55.
6. T. B. Clegg, V. König, P. A. Schmelzbach, and W. Grüebler, to be published.
7. A. Abragam and J. M. Winter, Phys. Rev. Lett. 1, 374 (1958).
8. H.-G. Mathews, H. Beuscher, and C. Mayer-Böricke, Physica Scripta, T3, 52 (1983).

ON PERFORMING ($\vec{p}$,n), ($\vec{d}$,n) AND ($\vec{n}$,n) MEASUREMENTS WITH A ROTATING MAGNET BEAM SWINGER*

F. S. Dietrich
Lawrence Livermore National Laboratory

T. B. Clegg
University of North Carolina, Chapel Hill, N.C. 27514
and Triangle Universities Nuclear Laboratory, Durham, N.C. 27706

ABSTRACT

Polarized beams of protons and deuterons may be used with a rotating-magnet beam swinger if a spin-precession solenoid is added before the swinger to precess the incoming spin axis.

Beam swingers followed by a <u>fixed</u> neutron detector allow neutron emission measurements versus Θ by using magnets to change the angle of the incident beam on target. Properties of such systems have been reviewed by Zafiratos.[1] In one common type of swinger installation, the magnet system rotates around the incident beam direction to accomplish measurements at various angles Θ. These systems have not been felt to be convenient for experiments with incident polarized beams, because the spin axis $\hat{s}$ would precess in the swinger magnets. This paper shows that this limitation may not be severe, since large values of $\hat{s} \cdot \hat{n}$ (where $\hat{n}$ is normal to the scattering plane) may in fact be obtained for incident $\vec{p}$ and $\vec{d}$ beams by placing a spin-precession solenoid just before the beam swinger.

For most neutron emission experiments, it is desirable that an incident polarized beam arrive on target with $\hat{s} \cdot \hat{n}$ as large as possible. Precession of the beam polarization in the swinger magnet could be avoided if $\hat{s}$ were rotated so that it is parallel to the B field upon entering the swinger, but this arrangement would be unsatisfactory because $\hat{s}$ arrives at the target in the scattering plane. Alternatively, a solenoid installed just before the swinger magnets can be used to precess the spin axis such that $\hat{s}$ remains perpendicular to $\vec{B}$ for all swinger orientations. As shown in the figure, then $\hat{s}$ precesses around $\vec{B}$ in the swinger and arrives at the target nearly perpendicular to the scattering plane (i.e. $\hat{s} \cdot \hat{n}$ large and constant vs. Θ and E). A simple calculation shows that for a net deflection of the beam in the swinger of 90°, the spin axis for a proton (deuteron) will precess by 252° (77°) so that $\hat{s} \cdot \hat{n}$ = 0.95 (0.97).

The additional solenoid must be capable of precessing the incident spin into the median plane of the rotating swinger

---

*Work performed under the auspices of the U.S. Department of Energy by the LLNL under contract number W-7405-ENG-48, and partially supported by the DOE by Contract # DE-AS05-76ER02408 TUNL.

magnet(s), i.e. through angles up to ± 90°. The ∫B·dℓ to accomplish this depends on the incident beam energy. Table 1 shows the required range of values.

Table 1. Value of ∫B·dℓ in kiloGauss-meters for 90° spin precession

Energy (MeV)	10	20	50	100
proton	2.56	3.62	5.72	8.10
deuteron	16.7	23.6	37.3	52.8

The B-fields are large enough that a superconducting solenoid will be necessary at the highest energies, particularly for deuterons. The field in a superconducting solenoid can be changed in times short compared to those for data-taking runs.

The spin will not arrive on target exactly perpendicular to the scattering plane. Thus, for experiments where a source reaction is used to produce outgoing polarized neutrons at 0°, the neutron polarization will be related to the incident beam polarization via a more complicated expression involving at least one more polarization transfer coefficient[2] than when $\hat{s}\cdot\hat{n} = 1$.[3] These coefficients should be measured for the likely source reactions,[4] $T(\vec{d},n)$ and $T(\vec{p},n)$.

1. C. D. Zafiratos, in The (p,n) Reaction and The Nucleon-Nucleon Force, eds. C. D. Goodman et al., Plenum, N.Y., 1980, p. 313.
2. G. G. Ohlsen, Rep. Prog. Phys. 35, 717 (1972).
3. P. W. Lisowski et al., Nucl. Phys. A242, 298 (1975).
4. R. L. Walter, in Neutron Sources for Basic Physics and Applications (ed. S. Cierjacks, Pergamon, Oxford, 1983) p. 259.

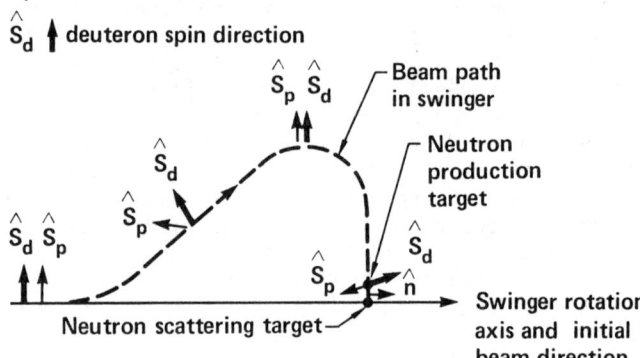

Illustration of the precession of the spin quantization axis for protons or deuterons in a movable-magnet beam swinger. The neutron flight path is perpendicular to the page.

SESSION DB

CONTRIBUTED PAPERS

NEUTRON, ALPHA AND TOTAL WIDTHS AND SPIN ASSIGNMENTS FOR RESONANCES IN ^{33}S+n FROM 10-400 keV

G. P. Coddens*
Universitaire Instelling Antwerpen,
Antwerp, Belgium

M. Salah
Minia University, El Minia, Egypt

J. A. Harvey and N. W. Hill
Oak Ridge National Laboratory, Oak Ridge, TN 37831

ABSTRACT

The S-33 neutron total cross section has been measured from 10 keV to 1 MeV. Resonance parameters (E_n, Γ_n, Γ_α, J^π) have been derived in the region from 10 keV to 400 keV.

INTRODUCTION

A good knowledge of the $\Gamma_\alpha/\Gamma_\gamma$ ratios for the compound nucleus ^{34}S is of interest in the study of the nucleosynthesis in the solar system. Recently Wagemans and Weigmann have measured the ^{33}S(n,α) cross section at a 30 m flight path of the Geel electron linear accelerator up to 1 MeV. A consistent set of resonance parameters can only be derived by combining these data with total cross section data.

EXPERIMENT

The T.O.F. measurements were performed on an 80-m flight path with a repetition rate of 800 Hz and 7 ns electron burst width. The energy resolution was about 0.1%. The ^{33}S sample was a 88.27% enriched powder pressed into 3 pellets with a total weight of 0.9732 g with $1/N_S$ = 28.212 at/b. The neutrons were detected with a NE110 scintillator.

DATA REDUCTION AND ANALYSIS

After proper corrections for deadtime and background, the data were analyzed using the computer code SAMMY[1] based on the R-matrix formalism in the Reich-Moore approximation. The program uses Bayes' theorem to derive the best values for the fitting parameters and keeps track of the covariance information in a rigorous way. Initial guesses for the resonance parameters and spin assignments were obtained based on the following considerations. The ground state of ^{33}S has J^π = 3/2$^+$. The ground state of ^{30}Si has J^π = 0$^+$. Therefore the α channel is only open to the ground state of ^{30}Si for resonances with J^π = 0$^+$, 1$^-$, 2$^+$, 3$^-$, 4$^+$, etc. The (n,α) reaction

*Neutron Scattering Project, IIKW.

has a Q-value of 3.492 MeV and a thermal cross section of 140±30 mb. The (n,α') reaction (Q=1.25 MeV) to the 2^+ first excited level of ^{30}Si and the (n,p) reaction (Q=0.533 MeV, σ_{th}=2±1 mb) to the $1/2^+$ ground state of ^{33}P are assumed to be negligible. The s- and p-wave resonances were identified from their characteristic interference with the potential scattering. The (n,α) data from Geel were valuable in the analysis which is complicated due to two circumstances: 1) both 1^- and 2^- p-wave resonances can be reached through 2 neutron entrance channels. This and the consequent possibility of having varying amounts of neutron widths in these 2 channels should be included for both spins. 2) The angular distributions of the α particles in the Geel (n,α) experiments depend on the resonance spin, which causes in several cases an uncertainty in the normalization of the data which were taken with a small angular aperture at 0° and 90°. As a matter of internal checking all ^{32}S resonances present in our data were analyzed as well, showing good agreement with the known values.

RESULTS AND DISCUSSIONS

As an example our fit below 100 keV is shown in Fig. 1. The most prominent features of our results are: 1) an apparent serious disagreement between our data and the Geel data for the 13.7-keV (30%) and 23-keV (60%) resonances which we could not attribute to an error in our data or data reduction (sample thickness, isotopic composition, background, etc.). 2) A large value for Γ_γ for the 17-keV 2^- resonance which can be explained from the large neutron binding energy in the compound nucleus ^{34}S* (11.422 MeV) which has low-lying positive parity states. 3) The fact that two neutron entrance channels were needed in order to fit the interference between the two 2^- resonances at 168.5 and 178.3 keV.

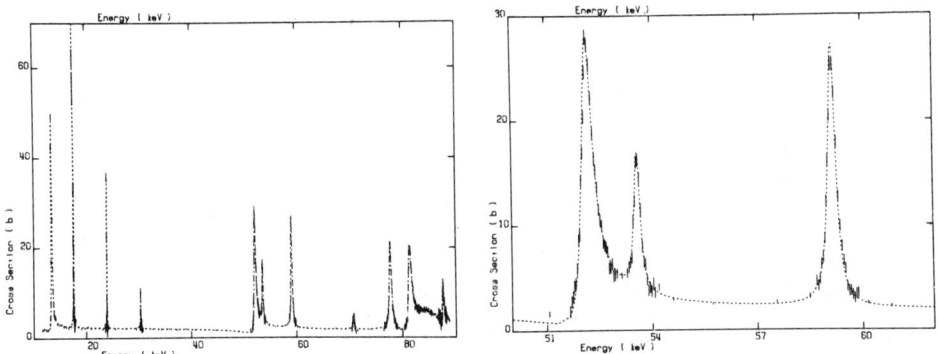

Fig. 1. The ^{33}S total cross section below 100 keV

This work was supported by the Belgian Nationaal Fonds voor Wetenschappelijk Onderzoek and by Martin Marietta Energy Systems, Inc. under Contract No. DE-AC05-84OR21400 with the U.S. Department of Energy.
1. N. M. Larson, ORNL/TM-9179 and N. M. Larson and F. G. Perey, ORNL/TM-7485.

$d_{5/2}$-SINGLE PARTICLE STRENGTH IN $^{48}Ca + n$

J. A. Harvey and C. H. Johnson
Oak Ridge National Laboratory, Oak Ridge, TN 37831

R. F. Carlton
Middle Tennessee State University, Murfreesboro, TN 37132

Boris Castel
Queen's University, Kingston, Canada K7L 3N6

ABSTRACT

The neutron total cross section of ^{48}Ca was measured up to 4 MeV and the data analyzed using an R-matrix code to obtain resonance parameters and potential scattering phase shifts. No s-wave resonances were observed and the small cross section (~ 0.5 b) at low energy requires a real well depth of 48 MeV. Three strong d-wave resonances (amounting to 45% of the single particle limit) were found in the 0.8 to 2.0 MeV energy region. Shell-model-in-the-continuum calculations agree with these observations.

INTRODUCTION

Neutron total cross section measurements of ^{48}CaO up to 1.4 MeV reported by Seibel et al.[1] established the existence of one strong $d_{5/2}$ resonance at 0.956 MeV and no s-wave resonances were found. Although several other small resonances were observed, their spins and parities could not be determined. Divadeenam et al.[2] calculated the energies and strengths of possible 2p-1h states and predicted that the lowest s-wave resonance would be above 1.5 MeV with the majority of the s-wave strength at 3 MeV.

EXPERIMENT AND DATA ANALYSIS

Neutron total cross section measurements of ^{48}Ca have been performed at ORELA over the energy interval from 10 keV to 4 MeV using a 200-m flight path. A $CaCO_3$ sample (1/N=126 b/a) enriched to 96% ^{48}Ca was used. Neutrons from the Ta target were used at higher energies while water moderated neutrons were used below 500 keV. The energy resolution $\Delta E/E$ was ~ 0.00011 $[1+16\ E]^{1/2}$ where E is in MeV. The experimental data after subtracting the oxygen and carbon cross sections are shown in Figs. 1 and 2. An R-matrix analysis using the code SAMMY[3] has been performed up to 4 MeV to obtain resonance parameters and potential scattering phase shifts for $\ell=0$, 1 and 2 wave resonances. The resulting fits to the total cross section are shown in Figs. 1 and 2.

RESULTS AND DISCUSSION

No resonances could definitely be assigned to s-wave neutrons. The non resonant cross section for low energy neutrons (<400 keV) is very small ~ 0.5 b (R' ~ 2.0 fm). This can be described by scattering from a real Woods-Saxon potential with r_0=1.21 fm, a=0.66 fm and V_0=48 MeV.

Three large $d_{5/2}$ resonances (at 0.960, 1.56 and 1.78 MeV) were identified up to 2 MeV whose combined strength (based on a radius of 5.27 fm) is ~ 45% of the single particle limit. The $d_{5/2}$ neutron strength function from 0.8 to 2.0 MeV is 0.60 (γ^2/D) or 14 x 10^{-4} ($\Gamma_n^{(2)}/D$)! The $d_{3/2}$ and p-wave strength functions are nearly 100 times smaller and the s-wave is ~ zero.

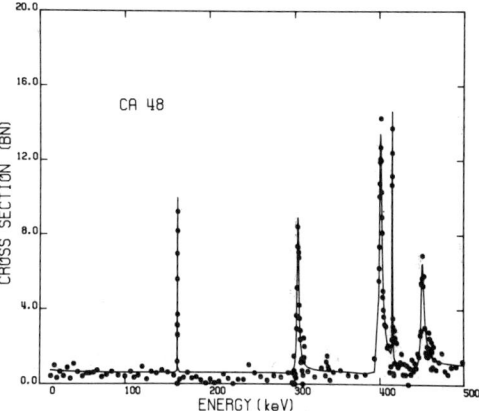

Fig. 1. σ_T (H_2O moderator)

We have done a shell-model-in-the-continuum calculation[4] of the ^{48}Ca+n system. The $p_{3/2}$ ground state and the $p_{1/2}$ first excited state at 2.03 MeV of ^{49}Ca exhausts most of the p-strength in agreement with (d,p) data and little remains at higher energies. The positive parity states are calculated as a combination of $1g_{9/2}$, $2d_{5/2}$ and $2d_{3/2}$ s.p. states and $p_{3/2}$ and $p_{1/2}$ neutrons coupled to the 3^- collective state in ^{48}Ca at 4.51 MeV. No $J=1/2^+$ states lie in the energy region from 0 to 3

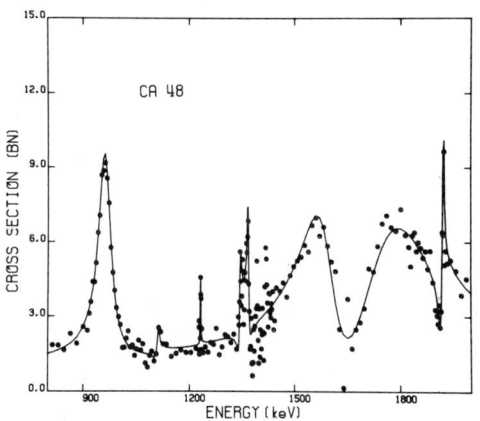

Fig. 2. σ_T (Ta target)

MeV above the neutron separation energy and four $J=5/2^+$ states are predicted in this energy region. They represent a total of 60% of the available s.p. strength, fragmented by the particle-core interaction. The remainder of the strength is predicted to lie within the 3 to 5.5 MeV energy region.

Research sponsored by the Division of Basic Energy Sciences, U.S. Department of Energy, under Contract Nos. DE-AC05-84OR21400 with the Martin Marietta Energy Systems, Inc. and DE-AS05-80ER10710 with MTSU.

REFERENCES

1. F. T. Seibel, E. G. Bilpuch, and H. W. Newson, Annals of Phys. 69, 451 (1972).
2. M. Divadeenam, W. P. Beres, and H. W. Newson, Annals of Phys. 69, 428 (1972).
3. N. M. Larson and F. G. Perey, ORNL/TM-7485, 1980 and N. M. Larson, ORNL/TM-9179, 1984.
4. J. A. Harvey et al., Phys. Rev. C 28, 24 (1983); R. F. Carlton et al., Phys. Rev. C 29, 1980 (1984).

HIGH RESOLUTION NEUTRON RESONANCE SPECTROSCOPY

R. Köhler[+], L. Mewissen[*], F. Poortmans[*], I. Van Parys[*], H. Weigmann[+]
[+]CEC, JRC, Geel Establishment, CBNM, B-2440 Geel, Belgium
[*]SCK-CEN, B-2400 Mol, Belgium

ABSTRACT

Neutron resonance spectroscopy in the MeV region at the Geel Linac is described. The first measurements performed with 1 ns burst width and a 400 m flight path done on ^{207}Pb, demonstrate the available resolution. The general purpose of these measurements are nuclear structure studies. Two examples are discussed. For ^{207}Pb, on which measurements have recently been started, the aim is the search for the expected fragmented giant magnetic dipole resonance[1]. Measurements on ^{28}Si have been concluded last year. Here the emphasis is on the location of a T=3/2 level and the determination of its isobaric spin impurity.

INTRODUCTION

The resolution of the neutron time-of-flight spectrometer GELINA has recently been improved by the installation of a post-acceleration pulse compression magnet[2]. Burst widths of 700 ps have been experimentally verified. The resulting resolution at 400 m is 100 and 700 eV at 1 and 5 MeV neutron energy respectively.

MEASUREMENTS ON ^{207}Pb

The first experiment with this improved resolution has been a transmission measurement on ^{207}Pb, using 200 and 400 m flight paths. The energy range covered was from 15 keV to 20 MeV. Analysis of the total cross section is under way. The good resolution allows to determine spin and parity of many resonances directly from the transmission measurement. As an example of the data, fig. 1 shows the experimental total cross section and a fit using the R-matrix code

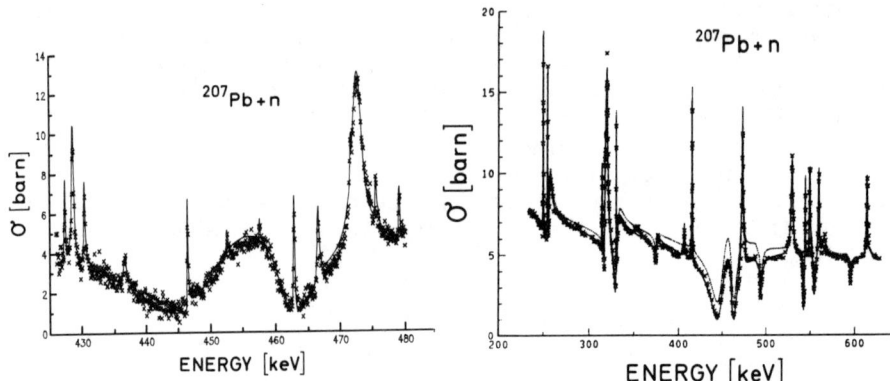

Fig. 1. Example of total cross section data with R-matrix fit.

Fig. 2. R-matrix fit to gross structures (crunched spectrum) with (full line) and without (dotted line) 0$^-$ resonances.

MULTI[3]. It turns out, that the two strong 0⁻ resonances between 400
and 500 keV, proposed by Seibel et al.[4] and questioned later on[5], are
necessary to have a consistent fit of the gross structure over a
large energy region (fig.2).

The strength function for s-waves including these two 0⁻ resonances shows a steep rise at ~450 keV indicating a doorway state. It sums up to about the same strength as the well known doorways for ^{206}Pb and ^{208}Pb at approx. the same energy (fig.3), so that one can assume the same doorway configurations to be involved.

For the search for the M1-strength a capture experiment at a
130 m flight path is presently under way. The covered neutron energy
range is from 3 keV to 4 MeV. Four BGO detectors are used under different angles with respect to the beam. Determination of spin and
parity of the resonances is done by the angular distribution of the
capture γ-rays and from the relative intensity of transitions
feeding the groundstate and the 3⁻ (2.61 MeV) level of ^{208}Pb

MEASUREMENTS ON ^{28}Si

Fig. 4 shows part of the total cross section data obtained for
^{28}Si, again together with a fit by MULTI. Resonances are indicated
on the figure. The 1/2+ resonance at 1200.8 keV neutron energy is
identified as a T=3/2 level in ^{29}Si, the isobaric analogue of the
1.41 MeV first excited state of ^{29}Al. The measured energy of this
level is in perfect agreement with the one obtained from the reaction
^{30}Si(^{3}He,a)^{29}Si [6]. The neutron width of this resonance yields its
isobaric spin impurity. It results in an average matrix element for
the mixing of this level with its T=1/2 neighbours of M=(22±5) keV.

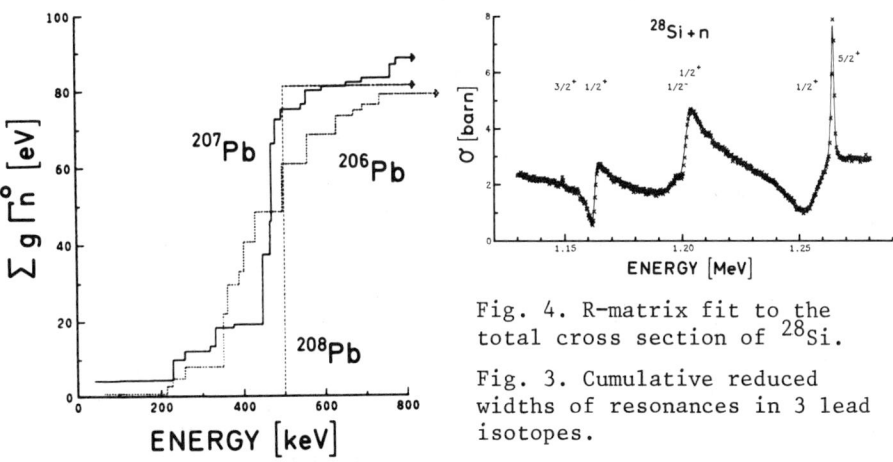

Fig. 4. R-matrix fit to the total cross section of ^{28}Si.

Fig. 3. Cumulative reduced widths of resonances in 3 lead isotopes.

REFERENCES

1. S. RAMAN, Neutron Capture Gamma Rays Spectr., Proc.Brookhaven 1979
2. D. TRONC et al., Nucl. Instr. & Meth. (submitted for publication)
3. G.F. AUCHAMPAUGH, LANL Scientific Report, LA-5473-MS,1974 (unpubl.)
4. F.T. SEIBEL et al., Ann. Phys. 69 (1972) 451
5. D.J. HOREN et al., Phys. Rev. C, 18 (1978) 722
6. C. DETRAZ and R. RICHTER, Nucl. Phys. A 158 (1970) 393

OPTICAL MODEL SCATTERING FUNCTIONS FOR LOW ENERGY NEUTRONS ON ^{86}Kr

R. F. Carlton
Middle Tennessee State University, Murfreesboro, TN 37312

J. A. Harvey and C. H. Johnson
Oak Ridge National Laboratory, Oak Ridge, TN 37831

ABSTRACT

Optical scattering functions are deduced from an R-matrix analysis and subsequent averaging of high resolution ^{86}Kr + n transmission data. These functions have been fitted by adjusting the well depths in an optical model with $r_0 = r_D = 1.21$ fm, $a_0 = 0.66$ fm and $a_D = 0.48$ fm. The well depths are $V_0 = 51.5$ MeV and $V_{SO} = 5.5$ MeV for $\ell=1$ and $V_0 = 48.5$ MeV for $\ell=0$. W_D is about 3.5 MeV for $s_{1/2}$, $p_{1/2}$ and $p_{3/2}$. Each well depth has an uncertainty of about ±1 MeV.

INTRODUCTION

Johnson, Larson, Mahaux and Winters[1] presented a method for deducing neutron optical scattering functions for individual partial waves from an R-matrix parametrization of high resolution total cross section data. Using that method we obtain scattering functions for ^{86}Kr and describe them with an optical model potential. (In Session V, Johnson compares our results with other nuclei.)

EXPERIMENTAL AND R-MATRIX ANALYSIS

Raman et al.[2] measured the transmission of a 99.5% enriched ^{86}Kr sample using the 80-m flight path at the ORELA facility. We repeated the measurement with the resolution of the 200-m flight path. The cross sections up to 400 keV are shown by points in Fig. 1. Between resonances the data are averaged over energy channels to reduce the fluctuations. These data permit J^π-assignments based upon the asymmetry produced by the resonance-potential interference. The curve is the multilevel R-matrix fit, which extends to 700 keV.

In Fig. 2 the staircase plots show the cumulative sums of p-wave reduced widths for a 6.4-fm boundary radius. The strength function, $\tilde{s}=<\gamma^2>/D$, is 0.058 for $p_{1/2}$ and 0.22 for $p_{3/2}$. For s-waves it is 0.022. Of equal importance for the model analysis is the external R-function which accounts for all levels outside the region from $E_\ell = 0$ to $E_u = 700$ keV. For each partial wave it is parametrized by $R^{ext}(E) = \tilde{R}(E) - \tilde{s} \ln[(E_u-E)/(E-E_\ell)]$ where $\tilde{R} = a+b(E-E_m)$ and E_m is the midpoint of the measurements. The fitted (a,b) with b in MeV^{-1} are (-0.13,0.15), (0.60,0.14) and (0.58,1.74) for $s_{1/2}$, $p_{1/2}$ and $p_{3/2}$.

Research sponsored by the Division of Basic Energy Sciences, USDOE under Contract Nos. DE-AC05-84OR21400 with Martin Marietta Energy Systems, Inc. and DE-AS05-80ER10710 with MTSU.

0094-243X/85/1240308-02 $3.00 Copyright 1985 American Institute of Physics

DISCUSSION

We averaged the scattering functions for each partial wave using the analytical approximation[1] with a width $2I=200$ keV. For p-waves the resulting compound cross section divided by g_J is about 2.5 times larger for $p_{3/2}$ than $p_{1/2}$ and the shape elastic is about 6 times larger for $p_{3/2}$ than $p_{1/2}$. We fit these by adjusting the well depths in a model with Woods-Saxon real well and surface-derivative spin-orbit and imaginary terms. (See abstract for geometric parameters.) The resulting model has $V_0 = 48.5$ MeV and $W_D = 3.5$ MeV for s-waves, and $V_0 = 51.5$, $V_{SO} = 5.5$, $W_D = 3.5$ MeV for p-waves. The uncertainties are each about ±1 MeV.

The spin-orbit term, which is required to fit the differences between $p_{1/2}$ and $p_{3/2}$, is consistent with other spin-orbit potentials of other nuclei. It places ^{86}Kr closer to the $3p_{3/2}$ size resonance than to $3p_{1/2}$ and it may explain why the $p_{3/2}$ strength function from Fig. 2 appears to be increasing in our region. The fact that the V_0 are about 3 MeV smaller than for some lighter nuclei is consistent with the N-Z difference for ^{86}Kr. The difference in V_0 between s-waves and p-waves suggests an ℓ-dependence consistent with results from other lighter nuclei.

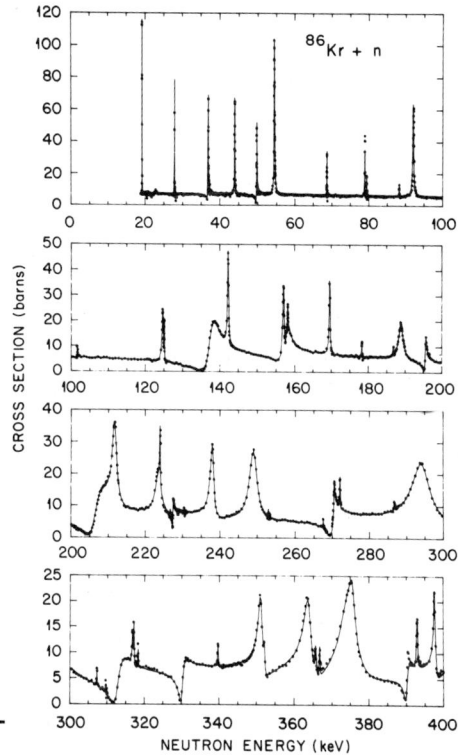

Fig. 1. Total cross section.

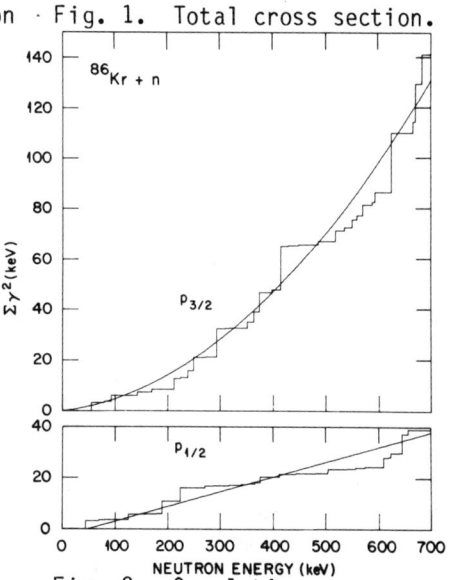

Fig. 2. Cumulative p-wave reduced widths.

REFERENCES

1. C. H. Johnson, N. M. Larson, C. Mahaux, and R. R. Winters, Phys. Rev. C 27, 1913 (1983).
2. S. Raman, B. Fogelberg, J. A. Harvey, R. L. Macklin, P. H. Stelson, A. Schröder, and K.-L. Kratz, Phys. Rev. C 28, 603 (1983).

ENERGY DEPENDENCE OF THE LOCAL OPTICAL POTENTIAL FOR NEUTRON-NUCLEUS SCATTERING

J.P. Delaroche,[*] P.P. Guss, G.M. Honore, C.R. Howell and R.L. Walter
Duke University and Triangle Universities Nuclear Laboratory[+],
Durham, NC 27706

D.C. Larson, D.M. Hetrick and J.A. Harvey
Oak Ridge National Laboratory, Oak Ridge, TN 37831

ABSTRACT

A systematic study of the energy dependence of the local neutron optical potential up to 80 MeV has been performed using coupled channels analyses of a large set of neutron scattering and reaction observables.

Several early attempts have been reported which established energy dependencies of the phenomenological neutron optical potential over a wide energy range. For instance, Bowen et al.[1] used total cross section σ_T for 10 MeV < E < 100 MeV and Becchetti and Greenlees[2] used σ_T and reaction cross sections for 1 MeV < E < 30 MeV. More recently, Rapaport et al.[3] combined σ_T and elastic scattering cross sections to obtain a global spherical optical potential for 7 MeV < E < 26 MeV.

In the present analysis we have extended the SPRT method[4] to include more observables, and used coupled channels formalism with deformed central and non-central potentials for some deformed and vibrational nuclei: ^{28}Si, ^{40}Ca, ^{58}Ni and ^{120}Sn. Three steps were followed. First, the new TUNL $\sigma(\theta)$ and $A_y(\theta)$ data from 8-17 MeV, the smoothed behavior of the total cross section σ_T from 10 keV to 20 MeV, and (when available) strength functions and the potential scattering radii at 10 keV were employed to develop the low-energy behavior of the potential depths and the geometry parameters. Second, published 20 to 40 MeV $\sigma(\theta)$ data were used to further constrain the surface and volume imaginary terms, $W_d(E)$ and $W_v(E)$, respectively. Lastly, the high accuracy $\sigma_T(E)$ data up to 80 MeV from ORELA were included (along with (p,p) data for consistency tests) primarily to determine the E-dependence of $W_v(E)$ and the energy at which $W_d(E)$ vanishes. Good overall agreement is obtained, especially for σ_T up to 80 MeV as seen in fig. 1. We find that the surface absorption W_d increases for neutron energies below approximately 15 MeV and decreases linearly above that energy, vanishing at about 70 MeV (except for ^{28}Si). The volume absorption W_v increases linearly with energy, taking on non-zero values at the energy at which W_d initially decreases. A more precise determination of the balance between W_d and W_v will require extensive measurements above 30 MeV for both differential elastic and inelastic neutron scattering.

[*]Permanent address: C.E. Bruyeres-le-Chatel, France
[+]Work partially supported by the USDOE, Contract No. DE-ACor-76ER01067

REFERENCES

1. P.H. Bowen et al., Nucl. Phys. 22, 640 (1961).
2. F.D. Becchetti and G.W. Greenlees, Phys. Rev. 182, 1190 (1969).
3. J. Rapaport et al., Nucl. Phys. A330, 15 (1979).
4. J.P. Delaroche, Ch. Lagrange and J. Salvy, Nuclear Theory IM Neutron Nuclear Data Evaluation (IAEA, Vienna, 1976), Vol. II, p. 251.

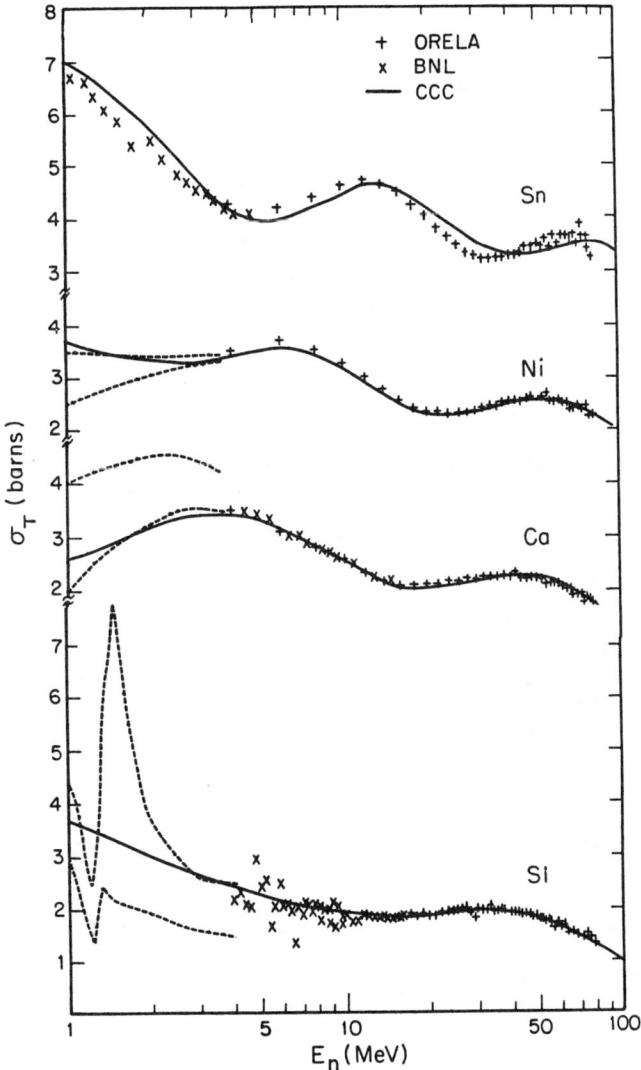

Fig. 1. Total cross section for ^{28}Si, ^{40}Ca, ^{58}Ni and ^{120}Sn. Comparison between measurements (dashed curves representing envelope of the data) and present calculations (solid curves).

THE IMAGINARY PART OF THE SPIN-ORBIT INTERACTION
FOR NEUTRON-NUCLEUS SCATTERING

R.L. Walter, W. Tornow[*] and P.P. Guss
Duke University and Triangle Universities Nuclear Laboratory
Durham, NC 27706

J.P. Delaroche
Centre d'Etudes de Bruyeres-le-Chatel, France

After the recent nuclear matter calculations of Brieva and Rook[1] for the nucleon-nucleus interaction, there has been a renewed interest in the phenomenology of the imaginary part $W_{so}(r)$ of the spin-orbit term. New $A_y(\theta)$ measurements[2] for (p,p) scattering from 80 to 180 MeV on four spherical nuclei support the negative sign for the strength W_{so} predicted in Ref. 1. On the other hand, for (p,p) scattering below 80 MeV, the magnitude and sign of W_{so} is poorly determined. Recently published[3,4] and unpublished studies at TUNL of $A_y(\theta)$ for neutron-nucleus scattering in the 10 to 17 MeV range has opened the question as to whether one needs a $W_{so}(r)$ term below 20 MeV. Spherical and deformed optical model analyses of the TUNL data for ^{40}Ca, ^{120}Sn and ^{208}Pb strongly suggest the existence of $W_{so}(r)$ with a depth $W_{so} \gtrsim +0.7$ MeV, where the positive sign indicates the same sign as the real spin-orbit potential depth V_{so}. (See fig. 1.) This sign is opposite to that predicted[1] for (p,p) scattering above 30 MeV. The results for $A_y(\theta)$ are shown in fig. 1 and correspondingly good fits were obtained for the other observables in these scattering analyses. Measurements of $A_y(\theta)$ above 20 MeV are needed to establish the energy dependence of W_{so} for neutrons and to investigate its isospin dependence, if any, through comparison with the proton results.

*On leave from University of Tübingen, Tübingen, FRG
+Work supported by the U.S. Department of Energy, Director of Energy Research, Office of High Energy and Nuclear Physics, under Contract No. DE-AC05-76ER01067.

REFERENCES

1. F.A. Brieva and J.R. Rook, Nucl. Phys. A297, 206 (1978).
2. P. Schwandt et al., Phys. Rev. C 26, 55 (1982).
3. J.P. Delaroche et al., Phys. Rev. C 28, 1410 (1983).
4. C.E. Floyd et al., Phys. Rev. C 28, 1498 (1983).

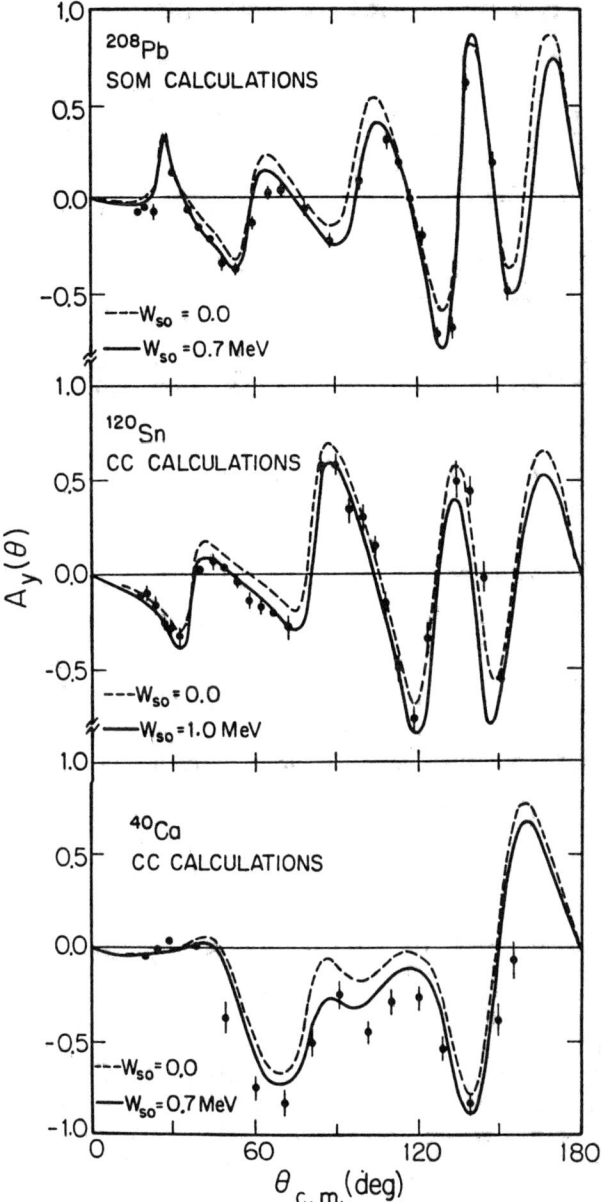

Fig. 1 Comparison between TUNL measurements and spherical (dashed curve) as well as deformed (full) curves.

NEUTRON SCATTERING ON DEFORMED NUCLEI*

L. F. Hansen, R. C. Haight, B. A. Pohl, C. Wong
Lawrence Livermore National Laboratory, Livermore, CA 94550

Ch. Lagrange
Centre d'Etudes de Bruyeres-le-Chatel, France.

ABSTRACT

Measurements of neutron elastic and inelastic differential cross sections around 14 Mev for ^{9}Be, C, ^{181}Ta, ^{232}Th, ^{238}U, and ^{239}Pu have been analyzed using a coupled channel (CC) formalism for deformed nuclei and phenomenological global optical model potentials (OMP). For the actinide targets these results are compared with the predictions of a semi-microscopic calculation using Jeukenne, Lejeune, and Mahaux (JLM) microscopic OMP and a deformed ground state nuclear density. The overall agreement between calculations and the measurements is reasonable good even for the very light nuclei, where the quality of the fits is better than those obtained with spherical OMP.

EXPERIMENTAL RESULTS AND CALCULATIONS

The elastic differential cross sections from Be, C, and Ta measured at 14.6 MeV (Fig. 1), and those from Th, U, and Pu, measured at 14.1 MeV (Fig. 4), were reported earlier.[1] New measurements were carried out for the differential elastic and inelastic (4.43 MeV level) cross sections from carbon for 13.6, 13.85, 14.04, 14.35, 14.66, and 14.80 MeV neutrons (Figs. 2,3). The relatively small energy spacings at which the data were taken were intended 1) to look for changes in the shape of the angular distributions around 14 MeV as suggested by resonance structure in the C total cross section; and 2) by subtraction to look for structure in the (n,α) cross section in this energy range.

The nuclei discussed in the present work are all characterized by large quadrupole deformations, $0.2 < \beta_2 \leq 1.2$. This suggests the need for deformed OMP when calculating the differential scattering cross sections to account for the strong couplings between the ground state (GS) and the low excited levels of the GS rotational band. The CC calculations have been performed with the code ECIS79 starting with phenomenological potentials with global OMP parameters optimized to fit neutron data.[2-4] For ^{9}Be and C, CC calculations were carried out with two OMP: a) the spherical potential of Ref. 2, by reducing the imaginary potential, W_D, by 25% to correct for coupling effects.[5] The fits to the data were obtained by a least squares analysis of the strengths of V_R and W_D. After search, the fits to the measured angular distributions were indistinguishable from those obtained with the spherical potential (Fig. 1, dashed lines) and consequently they have not been plotted. b) A second CC calculation was carried out to fit the elastic and inelastic (4.43 MeV) angular distributions for C between 13.60-14.80 MeV, using the deformed OMP given by Meigooni et. al.[3] This latter potential was obtained from fitting neutron scattering from C over a wide energy range. The CC calculations with this potential (discussed below), have given a much better fit to the data.

^{9}Be(n,n$_0$). The CC calculations include the GS (3/2-), the 2.429 (5/2-), and 6.660 (7/2-) excited levels. The value of β_2 is 1.2 and the Legendre multipole expansion of V_R and W_D include terms up to ℓ = 6 for all the calculations. The best fit to the Be cross sections, Fig. 1 (solid line was calculated with the deformed OMP that gave the best fit to the C data (Table 1-SetB).

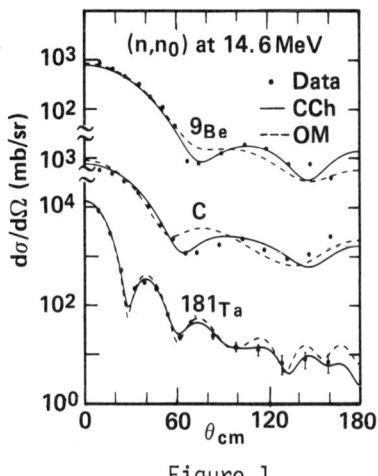

Figure 1

C(n,n$_0$) and C (n,n'). The measured angular distributions for the elastic and inelastic (4.43 MeV) at 13.60, 13.85, 14.04, 14.66, and 14.80 MeV are shown in Figs. 2 and 3 respectively. The data are in good agreement with measurements found in the literature[6,7] for some of these energies. Furthermore, the

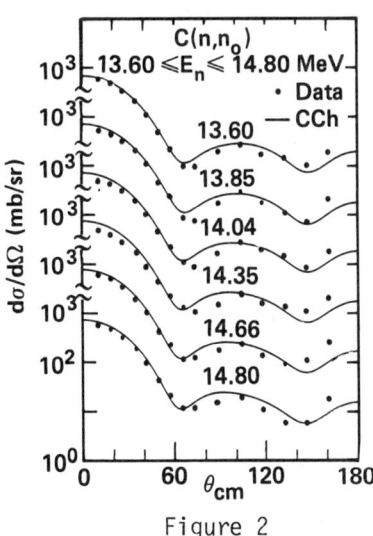

Figure 2

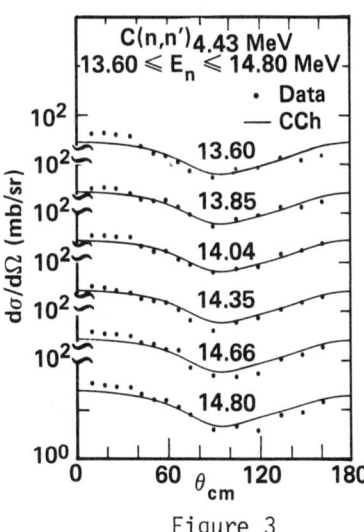

Figure 3

measurements[1] at 14.60 MeV are in very good agreement with the present one at 14.66 MeV. The CC calculations include the GS (0$^+$), the 4.43 (2$^+$) and 14.08 (4$^+$) excited levels. Starting from the Meigooni et al.[3] OMP a fine tuning of the parameters was necessary to obtain the fits shown in Figs. 1-3. The initial and final values of the parameters for set A and B are shown in Table 1. The quality of the fits to the 4.43 MeV angular distributions was most sensitive to the strength of the spin-orbit potential which is not deformed in the calculations. The fits are greatly improved by reducing the potential depth, V_{SO}, and increasing the diffuseness parameter, a_{SO}. The values of 5.50 MeV and 0.750, respectively, are in good agreement with those found from fits to neutron elastic and polarization data between 5-14 MeV.

Table 1. Values of the OMP parameters.

	V_R	r_R	a_R	W_V	W_D	r_D	a_D	V_{SO}	r_{SO}	a_{SO}
Set A	50.78-.34E	1.22	.478+.004E	0	.45E-1.7	1.25	.27	6.20	1.05	.55
Set B	52.45-.34E	1.22	.400	0	.44E-2.2	1.25	.35	5.50	1.15	.75

$\beta_2 = -0.53(A), -0.61(B); \beta_4 = 0.2(A \& B)$

^{181}Ta(n,n). The CC calculation shown in Fig. 1 for the neutron elastic scattering at 14.6 MeV includes in addition to the GS (7/2+), the 0.136 (9/2+) and 0.301 (11/2+) excited levels. The OMP^2 is the same used in the spherical OM calculations (dashed line), corrected for coupling. The value of β_2 is 0.260. The solid curve corresponds to the sum of the angular distributions calculated for the three levels, since the resolution of the measurements ($\approx$ 300 keV) did not resolve them. No parameter search was carried out for this calculation because the levels were unresolved. The quality of the fit was greatly improved by the inclusion of coupling among the GS and exited rotational levels.

^{232}Th, ^{238}U, ^{239}Pu(n,n). For ^{232}Th and ^{238}U the CC calculations, both with phenomenological and microscopic OMP, include the 0+, 2+, 4+, and 6+ states of the GS rotational band. For ^{239}Pu, levels of the GS (K=1/2) band are included up to the 9/2+ state. In Fig. 4 the measurements are compared with calculations carried out with the deformed potential of Ref. 4 (solid lines) and with two semi-microscopic calculations[8] to be discussed. The plotted curves correspond to the sum of the GS and excited levels differential cross-sections up to $E_{ex} \leq$ 300 keV. As for Ta, no search of parameters was carried out to optimized the fits to the data. In the semi-microscopic calculation of Lagrange et al.[8], the OMP is calculated by folding the JLM effective interaction with the deformed GS nuclear density. The differential cross sections are calculated using CC

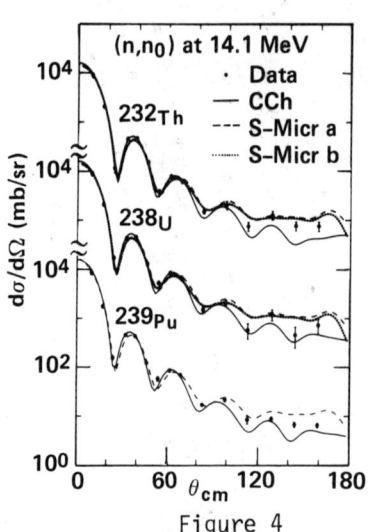

Figure 4

formalism and the adiabatic rotational model. The semi-microscopic potentials have been expanded in terms of Legendre polynomials up to $\ell_{max} = 8$, and complex coupling is used. The spin orbit potential is real and spherical. The calculations (a) and (b) (dashed and dotted lines respectively), differ in the calculation of the nuclear density. In (a), $\rho(r)$ is obtained from a Hartree-Fock-Bogolyubov and RPA calculation; while in (b) the nucleon density distribution has been obtained from fitting electron scattering data. (For details of these calculations and respective formalism see Ref. 8). The only free parameters in the two calculations are: the ranges of the real and imaginary parts of the effective interaction, $t_R = 1.2$ fm, and $t_I = 1.3$ fm; the normalization factors, $N_R = 0.94$ and $N_I = 0.82$, for V_R and W_D; and the strength of the spin-orbit potential, $V_{so} = 45$ MeV-fm^5. The difference between the two calculations is small, with (b) giving slighty better results at backward angles. Both, phenomenological and semi-microscopic calculations do poorly at larger angles, and possibly could be improved by a better representation of the spin-orbit potential.

CONCLUSIONS

This work shows conclusively the need for deformed OMP and CC formalism when calculating neutron scattering from strongly deformed nuclei. The calculations reproduce fairly well the measurements, even for nuclei as light as ^{9}Be. The fits to the C inelastic angular distributions (4.43 MeV level) need improvement. Possible effects of coupling of this level to other members of the GS rotational band or to some of the nearby vibrational levels need to be studied. The semi-microscopic calculations with effectively no free parameters (the values for t_R and t_I were obtained from ^{208}Pb data, while the magnitudes of N_R, N_I, and V_{so} were obtained from ^{238}U(n,n) at lower energies) give fits to the data of similar quality to those found with global OMP optimized for the actinide region. Neutron polarirization data would allow a better representation of the spin-orbit potential, to which the differential cross sections at large angles are sensitive.

REFERENCES

1. L. F. Hansen et al., UCRL 80316 Aug. 1984 (Submitted to Phys. Rev. C). L. F. Hansen et al. BAPS 29 (1984).
2. J. Rapaport et al., Nucl. Phys. A330, 15 (1979).
3. A. S. Meigooni et al., Phys. Med. Biol. 29, 643 (1984).
4. A. B. Klepalskij et al., INDC(CCP)-1511L March 1981.
5. L. F. Hansen et al., Phys. Rev. C25, 189 (1982).
6. G. Haouat et al., Nucl. Sci. Eng. 65, 331 (1978).
7. C. A. Pearson et al., Nucl. Phys. A191, 1 (1972).
8. Ch. Lagrange et al., J. Phys. G: Nucl. Phys. 9, 197 (1983).

*Work performed under the auspices of the U.S. DOE by the LLNL under contract number W-7405-ENG-48.

SEMI-MICROSCOPIC CALCULATIONS OF ELASTIC, INELASTIC, AND TOTAL NEUTRON SCATTERING BY ^{239}Pu

Ch. Lagrange* and D. G. Madland
Los Alamos National Laboratory, Los Alamos, N.M. 87545

M. Girod
Centre d'Études de Bruyères-le-Châtel, France

The purpose of the present work is first to investigate the capabilities of a semi-microscopic optical-model potential involving fewer adjustable parameters than phenomenological models and second to test the sensitivity of this potential to various rotational band assumptions. Our analysis of the model capabilities is performed using experimental elastic, inelastic, and total cross sections from ~10 keV to 15 MeV as well as neutron strength functions, and our sensitivity study is performed for the first three bands of ^{239}Pu: $1/2^+$ (0.0), $5/2^+$ (286 keV), and $7/2^-$ (392 keV). The two main ingredients of our calculations are the microscopic optical-model potential of Jeukenne, Lejeune, and Mahaux[1] (JLM) and nuclear densities from Hartree-Fock calculations performed using the density-dependent force D1.[2] Note that the Hartree-Fock density is calculated separately for each rotational band considered and that the coupled-channel calculations are performed for each band individually.

The present calculations differ from those previously reported[3] by the presence of an effective mass factor in front of the JLM imaginary potential. The parameters of the semi-microscopic model are mainly determined by adjustments with respect to the strength functions and total cross section over the full energy range. The consistency of the parameterization is tested by comparisons of calculated and measured elastic and inelastic angular distributions. Our final values for the ranges of the effective interaction are $t_R = 1.3$ fm and $t_I = 1.4$ fm. The renormalization factor of the real potential is $N_R = 0.935$ while that of the imaginary potential is $N_I = 0.46$ for 10 keV $\leq E_n \leq 4$ MeV, but is then rising linearly with energy to a value $N_I = 0.82$ at $E_n = 14$ MeV. The results of our calculations are summarized in Fig. 1 and Table I where they are compared to experiment and to the results obtained using the phenomenological coupled-channel potential of Ref. 4 with $\beta_2 = 0.20$ and $\beta_4 = 0.08$. The main effect of the rotational band assumption (not shown) is on the diffraction minima of the elastic angular distributions.

REFERENCES

1. J. P. Jeukenne, A. Lejeune, and C. Mahaux, Phys. Rev. C 16, 80 (1977)
2. D. Gogny, <u>Nuclear Self-Consistent Fields</u>, Ed., G. Ripka and M. Porneuf, North-Holland, Amsterdam, 1975.
3. Ch. Lagrange and M. Girod, J. Phys. G: Nucl. Phys. 9, L97 (1983).
4. D. G. Madland and P. G. Young, Proc. Int. Conf. on Neutron Phys. and Nucl. Data for Reactors and Other Applied Purposes, Harwell, UK, Sept. 1978, published by OECD, 1978, p. 349.

*Permanent address: Centre d'Études de Bruyères-le-Châtel, France.

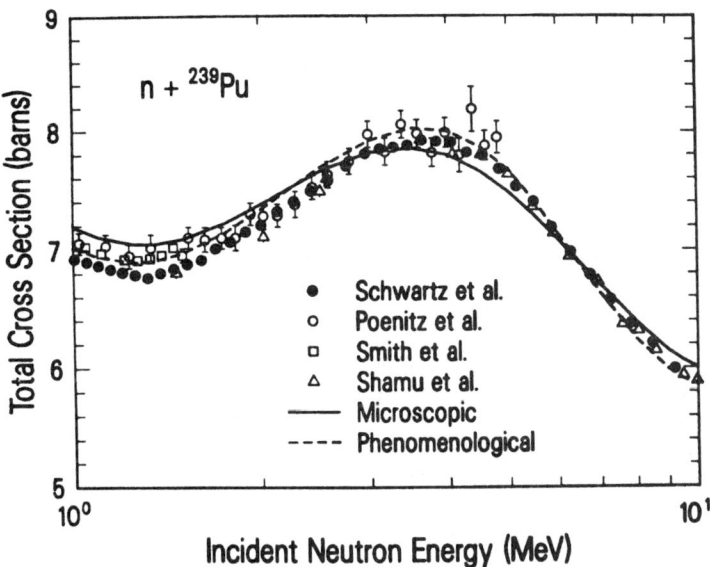

Fig. 1. Comparisons of calculated and experimental total cross sections as a function of incident neutron energy for ^{239}Pu.

TABLE I. CALCULATIONS FOR n + ^{239}Pu

E_n = 10 keV

Coupling Basis	$S_0 \times 10^4$	$S_1 \times 10^4$	R' (fm)	σ_{tot} (b)
$1/2^+,\ldots,9/2^+$ micro.	1.042	1.924	9.15	15.58
$1/2^+,\ldots,9/2^+$ phenom.	1.245	2.221	9.084	16.30
Experiment	1.3 ± .1	2.3 ± .4	9.6 ± .3	—
$5/2^+, 7/2^+, 9/2^+$ micro.	0.979	1.728	9.15	15.26
$7/2^-, 9/2^-, 11/2^-$ micro.	1.029	1.738	9.16	15.49

E_n = 3.4 MeV

Coupling Basis	σ_{el} (mb)	σ_1 (mb)	σ_2 (mb)	σ_3 (mb)	σ_4 (mb)	σ_{tot} (mb)
$1/2^+,\ldots,9/2^+$ micro.	4353	162.8	240.8	65.8	80.4	7853
$1/2^+,\ldots,9/2^+$ phenom.	4414	139.4	206.0	58.5	71.2	8010
Experiment	[4271 ± 581]	[380 ± 39]		—	—	8022 ± 100
$5/2^+, 7/2^+, 9/2^+$ micro.	4356	266.3	176.3	—	—	7755
$7/2^-, 9/2^-, 11/2^-$ micro.	4448	238.4	137.4	—	—	7776

SEMI-MICROSCOPIC INTERPRETATION OF FAST NEUTRON SCATTERING FROM ^{208}Pb

G. Haouat, Ch. Lagrange, J.C. Brient*, Y. Patin
R. de Swiniarski**, F. Dietrich***

SPNN, B.P. N° 12, 91680 Bruyères-le-Châtel, France

ABSTRACT

Angular distributions for elastic and inelastic neutron scattering on ^{208}Pb have been measured for the ground state and for excited states at 2.164 MeV (3^-), 3.198 MeV (5^-), 4.076 MeV (2^+) and 4.323 MeV (4^+) + 4.425 MeV (6^+), at incident energies between 7.5 and 15.5 MeV. Semi-microscopic folding-model calculations of the cross sections have been performed with the microscopic optical model of Jeukenne, Lejeune and Mahaux together with a Hartree-Fock-Bogoluybov ground state nucleon density and RPA transition densities The nucleon excitation strenghts deduced in this study will be compared to values obtained at higher energies and with other projectiles.

INTRODUCTION

In a recent paper Lagrange and Brient [1] have presented a semi-microscopic interpretation of nucleon elastic and inelastic scattering from ^{208}Pb in the energy range 7-61 MeV. Their coupled-channel (CC) calculations were achieved by using Jeukenne, Lejeune, Mahaux (JLM) effective potential [2] together with ground-state and transition nucleon density distributions obtained by Decharge [3] from Hartree-Fock-Bogoluybov (HFB) and RPA calculations. In the present work which in an extension of the previous one we investigate the sensitivity of the JLM predictions to various assumptions on the density dependence of the effective force.

EXPERIMENTAL CONDITIONS

Differential cross sections for neutron elastic and inelastic scattering from ^{208}Pb have been measured for the ground state and for excited states at 2.614 MeV (3^-), 3.198 MeV (5^-), 4.076 MeV (2^+) and 4.323 MeV (4^+) + 4.425 MeV (6^+), at incident energies between 7.5 and 15.5 MeV.

MICROSCOPIC OPTICAL MODEL ANALYSIS

The experimental data have been analysed by using the semi-microscopic model of Lagrange and Brient [1] in which the JLM imaginary effective force W_e (r) (see Ref.1 for notations) has been multiplied by an effective mass factor m*/m = 0.7 as proposed by Fantoni et al. [4].

* LPNHE, Université Pierre et Marie Curie, PARIS, France.
** ISN GRENOBLE, France.
*** Lawrence Livermore National Laboratory, USA.

This correction greatly reduces the difference between the theoretical and empirical imaginary potentials. The analysis has been performed with the values of Ref. 1 for the finite range parameters ($t_R = 1.2$ fm, $t_I = 1.3$ fm). Only the energy dependent normalization parameters N_R and N_I, and the spin-orbit potential parameter V_S were varied. The CC calculations have been made using a coupling basis (0^+, 3^-, 5^-, 2^+, 4^+, 6^+). Best fits to the data have been obtained for $V_S = 45$ MeV fm^5 and for the values of N_R and N_I presented in Table I :

Table I Normalisation parameters obtained from fits to the ^{208}Pb data

E_n (MeV)	7.5	9.5	11.5	13.0	13.5	15.5
N_R	0.940	0.925	0.958	0.983	0.983	0.983
N_I	0.520	0.590	0.683	0.752	0.752	0.752

The above calculations were achieved with a complex effective force $U_e(\rho,E) = V_e(\rho,E) + i W_e(\rho,E)$ where the local density ρ was taken at the projectile coordinate. We have performed further calculations in which we have used for the effective potential the arithmetic mean [5]: $1/2 \left[U_e(\rho(\vec{r_0}),E) + U_e(\rho(\vec{r_1}),E) \right]$, where $\vec{r_0}$ and $\vec{r_1}$ are the projectile and the target-nucleon coordinates, respectively. The two calculations which are presented in Fig.1 for 9.5 MeV neutron energy are very similar for the elastic scattering as well as for the inelastic scattering to the 3^- excited state.

SUMMARY REMARKS

The present work shows that the semi-microscopic folding model based on HFB and RPA nucleon density distributions seems to be a powerful tool for describing neutron elastic and inelastic scattering from ^{208}Pb.

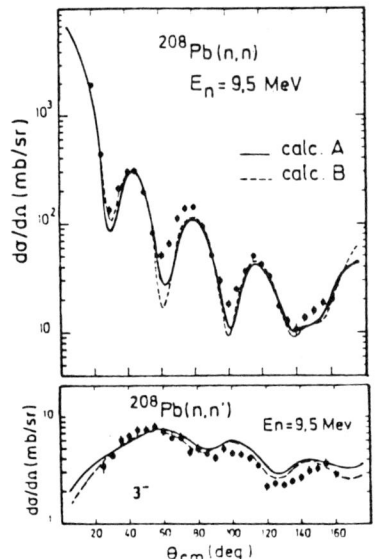

Fig. 1 : ^{208}Pb cross sections for 9.5 MeV incident neutrons
Calc.A : arithmetic mean.
Calc.B : local density at the projectile coordinate

REFERENCES

1. C.Lagrange and J.C.Brient, J. Phys. 44, 27 (1983).
2. J.P.Jeukenne, A.Lejeune and C.Mahaux Phys.Rev. C16, 80 (1977).
3. J. Dechargé, CEA Report CEA-N-2260 (1982).
4. S.Fantoni, B.L. Friman, V.R.Pandharipande, Phys.Lett. 104B, 89 (1981).
5. L.Rikus, N.Nakano and H.V.Von Geramb Nucl. Phys. A414, 413 (1984).

PHENOMENOLOGICAL MAPPING OF THE FERMI-SURFACE ANOMALY
WITH NEUTRON-NUCLEUS COLLISIONS

R.W. Finlay, J.R.M. Annand and J.S. Petler, Ohio University

F.S. Dietrich, Lawrence Livermore National Laboratory

Detailed new measurements of elastic and inelastic neutron scattering from ^{208}Pb (4-24 MeV) and ^{209}Bi (4-7 MeV) are analyzed together with total cross section [1,2,3], analyzing power [4,5] and earlier differential cross sections [6,7,8] in an effort to refine and extend the optical potential model for a heavy spherical nucleus over a wide energy range. The work proceeded in two steps. First, the ^{208}Pb data (7-50 MeV) were well described in terms of a conventional model with a real Woods-Saxon (WS) form factor, imaginary volume and surface-derivative WS form factors, and an energy-independent, real spin orbit potential; all with fixed values of the geometrical parameters even though there was strong evidence in the individual searches that the 7 MeV data required a rather different geometry for best fit. The energy dependence of the real well depth was given by V = 49.13 MeV - 0.31 E, which is shown as the lower solid line on Fig. 1.

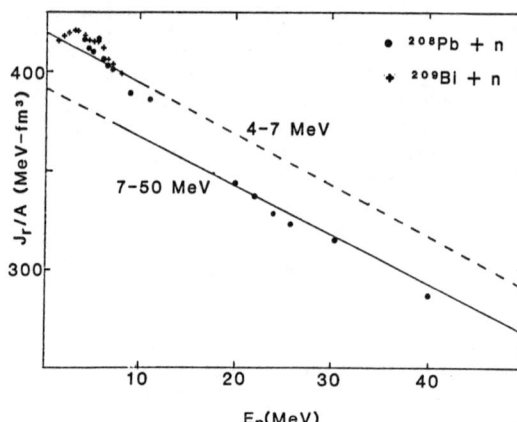

FIG. 1 Volume integrals for two different constant geometry potentials (lines) and for energy dependent geometry (points).

In the second step, the 4-7 MeV data for Pb-Bi were analyzed in a similar framework (with detailed Hauser-Feshbach calculations to account for compound elastic scattering), and good fits were obtained for both nuclei with V = 45.65 MeV - 0.28 E. This is shown as the upper straight line in Fig. 1. Unmistakably, the lower energy data group required larger values of real radius than did the higher energy group (1.265 vs 1.205 fm) which accounts for the vertical displacement of the nearly-parallel lines in the plot of J(real)/A vs E in Fig. 1.

All efforts to find a compromise geometry resulted in unacceptable deterioration of the quality of the fit in one or the other energy region.

Preliminary efforts to resolve the problem by invoking a real surface potential obtained from a dispersion integral were partly successful, so this effect was incorporated phenomenologically by allowing r_R, R_I and a_I to vary [9] linearly with energy between 0 and 24 MeV, and the well depths were searched again for best fits (which turned out to be very good). Resulting values for the real volume integrals are plotted as points (Pb) and pluses (Bi) in Fig. 1 and again in Fig. 2 where they are compared with a compilation of p-^{208}Pb parameters by Mahaux and Ngo [10] who emphasize the approximate symmetry of the potential anomaly about the Fermi energy ε_F (= -6 MeV for ^{208}Pb). Similarity of the trends in J(real)/A for both neutrons and protons is striking. A vertical displacement of the proton curve by 92 MeV-fm^3 (corresponding to the sum of twice the symmetry potential and the Coulomb correction) leads to dashed line through the neutron data.

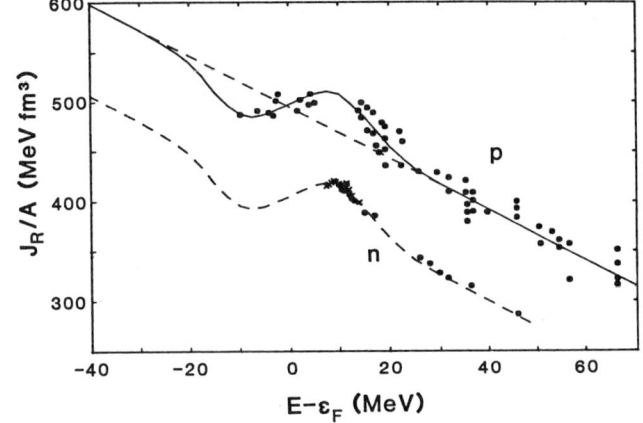

FIG. 2   Volume integrals for protons on Pb (●, ref. 10) and for neutrons on Pb (●) and Bi (x).

The use of neutron scattering data to fill in the gap in the potential systematics (where protons may not tred) is an interesting illustration of the importance of studying neutron-nucleus collisions.

REFERENCES

1. P.W. Lisowski et al., Proc. Symp. Neutron Cross Sections 10-50 MeV, BNL-NCS 51245, 301 (1980).
2. D.G. Foster and D.W. Glasgow, Phys. Rev. C3, 576 (1971).
3. A.B. Smith et al., Nucl. Sci. Eng. 73, 186 (1980).
4. G. Bulski et al., Proc. Int. Conf. on Nuclear Data, Antwerp (1982).
5. J.P. Delaroche et al., Phys. Rev. C28, 1410 (1983).
6. R.P. DeVito, Ph.D. dissertation, Michigan State University, 1979 (unpublished).
7. J. Rapaport et al., Nucl. Phys. A296, 95 (1978).
8. N. Olsson et al., Nucl. Phys. A385, 285 (1983).
9. C. Mahuax and H. Ngo, Nucl. Phys. A378, 1 (1983).
10. C. Mahaux and H. Ngo, Phys. Lett. 126B, 1 (1983).

NUCLEON INDUCED EXCITATION OF $K^\pi = 0^+$, 0^+_2, 1^- AND 3^- BANDS IN ^{12}C

Ali S. Meigooni, R.W. Finlay and J.S. Petler, Ohio University

J.P. Delaroche, Centre d'Etudes de Bruyères-le-Chatel

Elastic and inelastic neutron scattering to several excited states up to 15 MeV in excitation energy in ^{12}C has been studied at incident energies of 20-26 MeV [1]. These data, together with earlier measurements of the (p,p') reaction at 30-65 MeV, neutron total cross sections and reaction cross sections, have been analyzed in a coupled-channel framework using an optical potential with simple energy dependence.

^{12}C is considered to have a large, permanent, oblate deformation in the ground state, and the scattering cross sections for the ground state rotational band consisting of 0^+, 2^+ (4.44 MeV) and 4^+ (14.1 MeV) are calculated with the rotational model using the code ECIS79.

The rotation-vibration model is extended to include radius, diffuseness and potential depth *oscillations around a permanently deformed equilibrium shape* [2] in order to calculate scattering to other observed final states with ECIS79. The 0^+_2 excited state at 7.65 MeV is well described in terms of β-vibrations and diffuseness oscillations of a deformed nucleus (Fig. 1). The 3^- (9.64 MeV) and 4^- (13.4 MeV) states are treated as members of a $K^\pi = 3^-$ octupole vibrational band based on the deformed equilibrium shape. The 1^- (10.8 MeV) and 2^- (11.8 MeV) states are best described as members of the $K^\pi = 1^-$ band (Fig. 2).

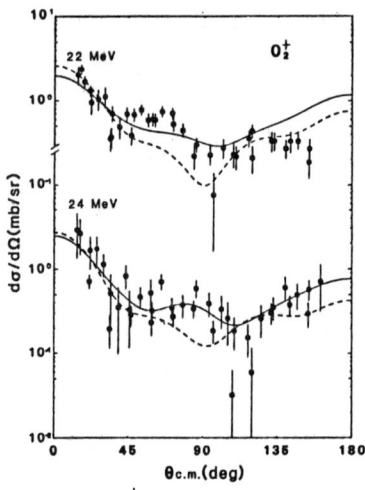

Fig. 1 0^+_2 state in ^{12}C -- β-vib. and a-osc. -- β-vib. and v-osc.

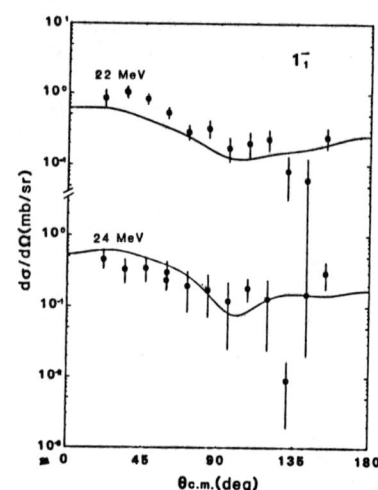

Fig. 2 1^- state in ^{12}C -- CC calculation with $K^\pi = 1^-$.

In all calculations, the strong 2^+ state at 4.44 MeV was coupled together with the excited band of interest, and more extensive coupling schemes were also calculated to test the sensitivity of the results. As the size of the coupling scheme increased, the description of the 2^+ (4.44 MeV) cross section at large scattering angle improved steadily with the 3^- state providing the next most important contribution.

A simple, real Coulomb correction was applied to the energy dependent optical potential and a deformed Coulomb field was added in order to calculate (p,p') scattering at 30-65 MeV [3,4,5]. Excellent agreement with the (p,p') data was obtained without further parameter variation. This is illustrated for the ground state rotational band in Fig. 3. Additional calculations in a large coupling basis (but with no new parameters) to the 0^+, 3^-, 1^- and 2^- states at 46 MeV that were remarkably good--better in every case than the fits presented in the original work [4]. No systematic improvement in the results could be obtained by the inclusion of an imaginary term in the Coulomb correction.

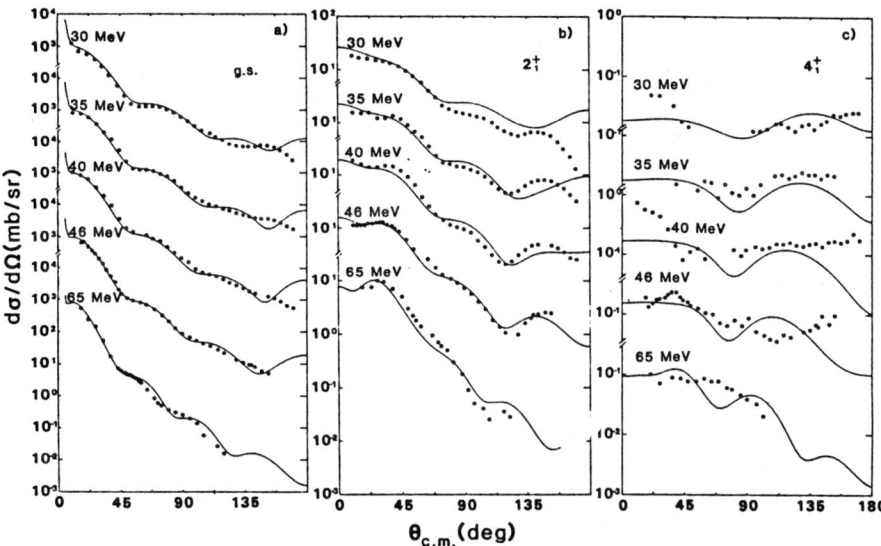

Fig. 3   Proton scattering from gs rotational band in ^{12}C.

Finally, calculations at 100 MeV were performed for the strong 2^+ and 3^- states, without reference to data, to show that coupled channel effects are not negligible at this energy.

REFERENCES

1. Ali S. Meigooni et al., submitted to Nucl. Phys.
2. Ali S. Meigooni and J.P. Delaroche, submitted to Nucl. Phys.
3. R. De Leo et al., Phys. Rev. C28, 1443 (1983).
4. G.R. Satchler, Nucl. Phys. A100, 481 (1967).
5. K. Hosono et al., Phys. Rev. Lett. 41, 621 (1978).

SESSION DC

WORKSHOP IN

EXPERIMENTAL TECHNIQUES

SUMMARY OF FACILITIES FOR EXPERIMENTAL STUDIES OF NEUTRON-INDUCED REACTIONS

Chris D. Zafiratos
University of Colorado, Boulder, CO. 80309

ABSTRACT

Experimental capabilities for measurements related to the topics of this conference were determined by mailing a questionnaire to an international mailing list of nuclear physics facilities. The responses show that a number of neutron-induced reactions can not be measured with present facilities.

INTRODUCTION

At the time of the Telluride conference on "The (p,n) Reaction and the Nucleon-Nucleon Force" (March, 1979) a brief survey was made to determine experimental capabilities for (p,n) studies and for (n,p) studies. The organizing committee of this conference felt that it would be useful to update and extend that survey to include all facilities related to the conference topics and to include facilities for the study of the associated (p,n) reaction.

EXPERIMENTAL NEEDS AND CAPABILITIES

A number of conference participants have pointed out the importance of (n,n') reactions near 60 MeV where the nucleus is unusually transparent to nucleon probes. Further, the presenttation by Roger Finlay showed that a facility for neutron elastic and inelastic scattering up to 75 MeV incident energy is a straight-forward extension of present experimental techniques.

Several talks stressed the importance of measurements near 200 MeV (where $V_{\sigma\tau}/V_\tau$ peaks). Other talks noted the importance of using neutron projectiles while measuring spin observables - in some cases on unnatural parity transitions. An experimentalist comes away with the feeling that what is needed is a facility that can determine spin transfers in interesting reaction channels such as $0^+ \rightarrow 0^-$ ($\vec{n},\vec{n}'$) at 200 MeV with $\Delta E = 10$ keV.

In order to determine present capabilities a questionnaire was mailed to over 200 nuclear physics facility addresses (international address list courtesy of the Cyclotron Laboratory at Michigan State University). The responses are part of this report. A summary of the results is given in Figure 1 where the circles indicate measurement capabilities extending to 50 MeV, 100 MeV and 200 MeV. The radial spokes indicate various reactions of interest. The heavy radial lines show present capabilities. Where an arrow points beyond 200 MeV, capability exists beyond 200 MeV. Where a bar is drawn (as at 200 MeV for ($\vec{p},\vec{n}$)) it indicates the

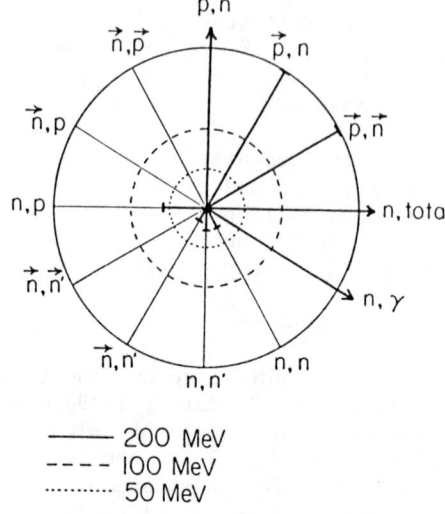

```
——— 200 MeV
- - - - 100 MeV
·········· 50 MeV
```

Fig. 1. The radial spokes indicate reactions of interest within the scope of this conference. The circles indicate energies of 50, 100 and 200 MeV. The heavy lines show present capabilities.

maximum energy for which that measurement can be made at present as determined by the survey.

The restricted energy range of neutron-induced reaction capabilities and the lack of information on spin observables is evident in Figure 1. There is a clear-cut need to extend our capability to observe (n, n') and (n,p), especially with polarized neutrons, to energies of 100 MeV or more.

Laboratory   Dept. Nuclear Physics, Australian National University
Name of respondent   Professor J.O. Newton
Accelerator Type (s)   14UD
Time structure   Minimum pulse width 50 ps

	Protons	Deuterons	Other
Energy range			
Maximum intensity			
Polarized intensity			

Neutron Facilities

Sources   T(p,n)   D(d,n)   T(d,n) ^{7}Li(p,n)   other

Intensity on
target (n/sec/sr)

Polarized intensity

Spectrometers: type, resolution, solid angle, completion date.

(n,n')

(n, charged particle)

(n,γ)

(p,n)

Other   (HI,xn) time of flight with up to 8 NE213 detectors.

Special features of your facility related to conference topics

Current, recent and proposed research related to conference topics

Study of shapes of neutron velocity distributions in HI,xn
reactions and measurement of pre-fission neutron multiplicities.

Laboratory __Brigham Young University Physics Laboratory__
Name of respondent __Gary L. Jensen__
Accelerator Type (s) __Van de Graaff__
Time structure _____

	Protons	Deuterons	Other
Energy range	1-3 MeV	1-3 MeV	
Maximum intensity	60 uamps	unknown	
Polarized intensity			

Neutron Facilities

Sources	T(p,n)	D(d,n)	T(d,n)	^{7}Li(p,n)	other
Intensity on target (n/sec/sr)		unknown		10^9	
Polarized intensity					

Spectrometers: type, resolution, solid angle, completion date.

_____

(n,n') _____

(n, charged particle) _____

(n,γ) _____

(p,n) _____

Other _____

Special features of your facility related to conference topics

Current, recent and proposed research related to conference topics

Laboratory	Centre d'Etudes de Bruyères-le-Châtel Service P2N		
Name of respondent	HAOUAT Gérard		
Accelerator Type (s)	Super EN Tandem and 4 MV HVEC		
Time structure	280-600 ps		

	Protons	Deuterons	Other
Energy range	1-14 MeV	1-14 MeV	
Maximum intensity	8 µA	6 µA	
Polarized intensity			

Neutron Facilities

Sources	T(p,n)	D(d,n)	T(d,n)	^{7}Li(p,n)	other
Intensity on target (n/sec/sr)	$\sim 10^8$	$\sim 10^8$	$\sim 10^7$	$\sim 10^6$	
Polarized intensity					

Spectrometers: type, resolution, solid angle, completion date.

(n,n') Time-of-flight spectrometer, $\Delta E \simeq 80-150$ keV, $\Delta \Omega = 210^{-4}$ sr, developed since 1972. Since 1984 use of a high energy R.F. Buncher

(n, charged particle)

(n,γ)

(p,n)

Other

Special features of your facility related to conference topics

The T.O.F. spectrometer is composed of six moving detectors placed 20° one for the other.

Current, recent and proposed research related to conference topics

-- Neutron scattering from ^{190}Os, ^{192}Os, ^{194}Pt at 8 MeV.
-- Neutron scattering on light nuclei between 7 and 15 MeV.
-- Neutron scattering from actinide nuclei between 5 and 8 MeV with an expected energy resolution of ~ 40 keV.

Laboratory LPC, Universite de Clermont, Aubiere, France
Name of respondent Professor Irigaray
Accelerator Type (s) Electrostatic
Time structure All Time

	Protons	Deuterons	Other
Energy range			
Maximum intensity			
Polarized intensity			

Neutron Facilities

Sources	T(p,n)	D(d,n)	T(d,n)	^{7}Li(p,n)	other
Intensity on target (n/sec/sr)		10^7	10^9		
Polarized intensity					

Spectrometers: type, resolution, solid angle, completion date.
NaI (Tl) Ge (Li) 20 keV and 2 keV

(n,n')

(n, charged particle) Activation Analysis

(n,γ) Activation Analysis

(p,n)

Other

Special features of your facility related to conference topics
Neutron as a probe of impurities in all kind of materials.

Current, recent and proposed research related to conference topics
Applications in biology and medicine.

University of California, Davis

Laboratory __Crocker Nuclear Laboratory__
Name of respondent __Paul Brady__
Accelerator Type (s) __Isochronous Cyclotron__
Time structure __1 nanosec pulses, Period ≃ 50 ns__

	Protons	Deuterons	Other
Energy range	5-70	5-44	α, ^{3}He, ^{14}N
Maximum intensity	20-40μA	40-80μA	
Polarized intensity			

Neutron Facilities

Sources   T(p,n)   D(d,n)   T(d,n) ^{7}Li(p,n)   other
Intensity on
target (n/sec/sr) ___ ___ ___ 3×10^{10}/MeV (16μA)
Polarized intensity ___ ___ ___ ___ ___

Spectrometers: type, resolution, solid angle, completion date.

(n,n') __MWC large solid angle, 20-65 MeV resolution ~ 1-2 MeV__

(n, charged particle) __Multiwire Chamber, large solid angle,__
__~ 1 MeV resolution__

(n,γ) _____

(p,n) __Future high resolution__

Other _____

Special features of your facility related to conference topics
(n,p) charge exchange reactions

Current, recent and proposed research related to conference topics
(n,p) charge exchange
(n,n') giant resonance and continuum studies
Planning high resolution neutron beam

Laboratory PHYSICS, EDINBURGH UNIVERSITY, U.K.
Name of respondent R.B. GALLOWAY
Accelerator Type (s) Van de Graff 500 keV
Time structure d.c.

	Protons	Deuterons	Other
Energy range	50-500 keV	50-500 keV	
Maximum intensity	150 µA	150 µA	
Polarized intensity			

Neutron Facilities

Sources	T(p,n)	D(d,n)	T(d,n)	^{7}Li(p,n)	other
Intensity on target (n/sec/sr)		10^6	2×10^7		
Polarized intensity					

Spectrometers: type, resolution, solid angle, completion date.

(n,n') assoc. part. time-of-flight, 250 keV, 10^{-4}, 1984
        recoil proton unfolding, 700 keV, 5×10^{-3}, 1983

(n, charged particle) _____

(n,γ) _____

(p,n) _____

Other Multi-detector scattering polarimeter, 1982
       reaction neutron polarimeter, 1979

Special features of your facility related to conference topics

Current, recent and proposed research related to conference topics
Analysing powers and cross sections for neutron scattering.

Laboratory Physikalisches Institut Universität Erlangen-Nürnberg

Name of respondent Eberhard Finckh

Accelerator Type (s) EN-Tandem, High Voltage Eng.

Time structure d.c.

	Protons	Deuterons	Other
Energy range	3-13 MeV	3-12 MeV	Heavy Ions, 6 MV Terminal
Maximum intensity	4 µA	4 µA	ca. 100 nA
Polarized intensity	300 nA	300 nA	--

Neutron Facilities

Sources	T(p,n)	D(d,n)	T(d,n)	^{7}Li(p,n)	other
Intensity on target (n/sec/sr)		3×10^8			
Polarized intensity		1×10^8			

Spectrometers: type, resolution, solid angle, completion date.

Proton recoil; E = 0.8 MeV; 6×10^{-3}; TOF using scintillating scatterer proton recoil

(n, charged particle) _____

(n,γ) _____

(p,n) _____

Other _____

Special features of your facility related to conference topics

Simultaneous measurement with 22 detectors, angular range 15°-170°. Collimated neutron beam. Measurement of σ_{tot}, $\frac{d\sigma}{d\Omega}$ and $A_y(\theta)$.

Current, recent and proposed research related to conference topics

Elastic and inelastic scattering from ^{28}Si, ^{40}Ar and ^{nat}Cr. n-D-breakup with two neutrons in coincidence, just started.

Laboratory	CBNM, JRC - CEC, Geel, Belgium	
Name of respondent	H. Liskien	
Accelerator Type (s)	7 MV CN-type Van de Graaff	
Time structure	width : 15/1.5/0.2 ns, distance 0.4/0.8/1.6 µs	

	Protons	Deuterons	Other (helium)
Energy range	1 - 7 MeV	1 - 7 MeV	1 - 7 MeV
Maximum intensity	5/50 µA	5/50 µA	5/50 µA ac/dc
Polarized intensity	-	-	-

Neutron Facilities

Sources	T(p,n)	D(d,n)	T(d,n)	^{7}Li(p,n)	other Be(d,n)
Intensity on target (n/sec/sr)	$9 \cdot 10^8$	$6 \cdot 10^8$	$2 \cdot 10^8$	$4 \cdot 10^8$	$2 \cdot 10^{11}$
Polarized intensity	-	-	-	-	-

Spectrometers: type, resolution, solid angle, completion date.

(n,n') TOF spectrometers

(n, charged particle) multi-angle telescope
multi-parameter parallel plate ionization chamber

(n,γ) -

(p,n) TOF spectrometers

Other

Special features of your facility related to conference topics

recent installation of a post-acceleration buncher yielding 210 ps pulses

Current, recent and proposed research related to conference topics

level structure, fission process

Laboratory   CBNM, JRC-CEC, Geel, Belgium
Name of respondent   K.H. Böckhoff
Accelerator Type (s)   150 MeV electron linear accelerator
Time structure   600 ps, 4 ns - 2 μs

	Protons	Deuterons	(electrons) Other
Energy range			40-150 MeV
Maximum intensity			100 A (600 ps)
Polarized intensity			/

Neutron Facilities

Sources	T(p,n)	D(d,n)	T(d,n)	^{7}Li(p,n)	other (electrons)
Intensity on target (n/sec/sr)					$2 \cdot 10^{13}$
Polarized intensity					/

Spectrometers: type, resolution, solid angle, completion date.

(n,n') _____

(n, charged particle) _____

(n,γ) _____

(p,n) _____

Other time-of-flight, 3 ps/m, 1965 - Modernizations '77,'82

Special features of your facility related to conference topics

high resolution neutron resonance spectrometry

Current, recent and proposed research related to conference topics

subthreshold fission, isobaric analogue states, level densities

Laboratory Indiana University Cyclotron Facility (IUCF)
Name of respondent D.L. Friesel/C.C. Foster
Accelerator Type(s) 2 DC Ion Source Terminals; 2 SSC Cyclotrons
Time structure 6 nSec Burst width, 33 nSec to 10 μSec Rep. Rate

	Protons	Deuterons	Other
Energy range	11 to 205 MeV	23 to 100 MeV	200 A^2/A
Maximum intensity	6.0 μA	2 μA	0.5 eμA
Polarized intensity	4.0 μA	0.16 μA	----

Neutron Facilities

Sources	T(p,n)	D(d,n)	T(d,n)	^{7}Li(p,n)	d($\vec{p},\vec{n}$)
Intensity on target (n/sec/sr)				10^5 (=1 MeV special)	
Polarized intensity					1×10^{10} (=18 MeV special)

Spectrometers: type, resolution, solid angle, completion date at 50 namp/p's

($\vec{n}$,n') Special purpose facility for Charge Symmetry Breaking Experiment
(n, charged particle) No Spectrometer, at present, ADHOL Facility with Neutron Collimator and proton sweeping magnet.
($\vec{n}$,γ) Possible for some purposes in Polarized Neutron Facility.
(p,n) Horizontal Beam Swinger Systems - 0° to 24° range, 0°, 24°, 45° up to 200 MeV flight paths.
Other

Special features of your facility related to conference topics
1. Low Energy Storage Ring is Injection Beam Line permits Beam pulse periods from 2 to 10 μSec at peak intensities of ~10μA
2. Electron Cooled High Energy Storage Ring presently under construction. Completion Date: Summer 1986.

Current, recent and proposed research related to conference topics
(p,n), ($\vec{p}$,n), ($\vec{p}$,$\vec{n}$) measurements
Preliminary (n,p) measurements
Charge Symmetry Breaking Experiment ($\vec{n}$,p), ($\vec{n}$,n)

Laboratory   Karlsruhe Nuclear Research Center (KFK) POLKA
Name of respondent   Hans O. Klages
Accelerator Type (s)   Cyclotron
Time structure $\lesssim$ 1 ns at 11 MHz

	Protons	Deuterons	Other (α)	^{6}Li
Energy range	26	52	104	156
Maximum intensity (μA)	10	10	5	50
Polarized intensity		30 μA (1985; 500 μA!)		

Neutron Facilities

Sources   T(p,n)   D(d,n)   T(d,n) ^{7}Li(p,n)   other
                                   (1985)
Intensity on
target (n/sec) $\sim 2 \times 10^6$
Polarized intensity 10^5 ($\sim 2 \times 10^6$; 1985)
(n/sec)

Spectrometers: type, resolution, solid angle, completion date.

(n,n') TOF with up to 20 scintillation detectors; ($\Delta T \sim 1$ ns);
$d\Omega \sim 8\text{-}20 \times 10^{-3}$ sr; in operation

(n, charged particle) ΔE(Si) + E(NaI); TOF (1 ns); $d\Omega \sim 5 \times 10^{-3}$
sr; in operation.**

(n,γ) _____

(p,n) _____

** (n, charged particle) reactions studies by telescope
techniques in vacuum chamber at five angles simultaneously.

Special features of your facility related to conference topics

Narrow collimated neutron beam (2-8 $\times 10^{-5}$ sr) continuous energy
spectrum of polarized neutrons ($E_{\vec{n}}$: 15-50 MeV), heavily shielded
target area.

Current, recent and proposed research related to conference topics

Neutron cross sections and spin observables in very light nuclear
systems, especially $\vec{n}$-d and $\vec{n}$-$\vec{p}$ interaction ($\vec{n}$,p) reaction will be
studied.

Laboratory University of Kentucky

Name of respondent Marcus T. McEllistrem

Accelerator Type (s) CN, Single ended Van de Graaff, 6.5 MV

Time structure DC, Terminal pulsed to 6 ns at 2 MHz, or 2.5 ns at 5 MHz; bunched to 0.6 ns at 2 MHz.

	Protons	Deuterons	Other
Energy range	≤ 6.5 MeV	≤ 6.5	He^+, ≤ 6.5
Maximum intensity	4 uA (pulsed)	4 uA	2 uA
Polarized intensity			

Neutron Facilities

Sources	T(p,n)	D(d,n)	T(d,n)	^{7}Li(p,n)	other
Intensity on target (n/sec/sr)	2.4×10^{10}	1.2×10^{10}			
Polarized intensity	none				

Spectrometers: type, resolution, solid angle, completion date.

(n,n') 1. modular shielded scintillators; operating resolution typically 50-80 keV at 3-7 MeV. Shielding update, 1981.

(n, charged particle)

(n,γ) 1. 4 in. diam. BGO scintillator; expected resolution of 5% at 20 MeV (develop). 2. 8 in.x10in. NaI scintillator, with anti-Compton shielding.

Other Ge(Li) and Hp Ge detectors, 100 to 120 cc, resolutions about 1.9 keV; shielding and detector implementation completed 1981.

Special features of your facility related to conference topics
Camac based, multi-parameter data acquisition system interfaced to LSI/11/23.
Computer for on-line monitoring of experiments. (under develop).

Current, recent and proposed research related to conference topics
Neutron scattering from deformed and shape-transitional nuclei.

Precision tests of spin-flip and polarization asymmetries in intermediate energy proton scattering, to assess isovector spin-flip interactions effective nucleon-nucleon interactions.

Neutron cross sections for Nuclear Astrophysics.

Laboratory	Lawrence Livermore National Laboratory
Name of respondent	Ivan Proctor
Accelerator Type (s)	Cyclograaff
Time structure	1ns @ 2.5 M Hz

	Protons	Deuterons	Other
Energy range (MeV)	1-27	1-20	
Maximum intensity	3µA	3µA	
Polarized intensity			

Neutron Facilities

Sources	T(p,n)	D(d,n)	T(d,n)	^{7}Li(p,n)	other
Intensity on target (n/sec/sr)*	$\sim 10^7$	$\sim 5 \cdot 10^7$	$\sim 10^7$	$\sim 5 \cdot 10^7$	
Polarized intensity	-	-	-	-	

Spectrometers: type, resolution, solid angle, completion date.

(n,n')   16 detectors at 11m, 1ns, 1.4msr, 1970
         1 detector at 26m, 1ns, 0.05msr, 1982

(n, charged particle)   Quadrupole, 30% acceptance,
         solid state detectors, 5msr, 1982

(n,γ) _____

(p,n)   same as (n,n')

Other _____

Special features of your facility related to conference topics
  * at 12 MeV incident, $\sim$0.5 MeV target thickness

Current, recent and proposed research related to conference topics

Laboratory: Lawrence Livermore National Laboratory
Name of respondent: Ivan Proctor
Accelerator Type (s): 100 MeV S band electron Linac
Time structure: $2\text{ns} \leq \tau \leq 3\mu\text{s}$, $f \leq 1440$ Hz

	Protons	Deuterons	Other
Energy range (MeV)			20-100
Maximum intensity			8A Peak
Polarized intensity			-

Neutron Facilities

Sources	T(p,n)	D(d,n)	T(d,n)	^{7}Li(p,n)	other
Intensity					10^{13} *
Polarized intensity					

Spectrometers: type, resolution, solid angle, completion date.

(n,n') _____

(n, charged particle) _____

(n,γ)   GeLi, 15 to 250m flight paths

(p,n) _____

Other _____

Special features of your facility related to conference topics
   * neutrons into 4π, $\tau \leq 100$ns

Current, recent and proposed research related to conference topics

Laboratory   Lawrence Livermore National Laboratory
Name of respondent   Ivan Proctor
Accelerator Type (s)   ICT
Time structure   DC → 2ns @ 2.5 M Hz

	Protons	Deuterons	Other
Energy range	-	400 keV	
Maximum intensity	-	20 MA	
Polarized intensity	-	-	

Neutron Facilities

Sources	T(p,n)	D(d,n)	T(d,n)	^{7}Li(p,n)	other
Intensity on target*	-	2.10^{10}	2.10^{12}	-	-
Polarized intensity	-	-	-	-	

Spectrometers: type, resolution, solid angle, completion date.

(n,n')   ~15m flight path at 28°, various detectors

(n, charged particle)

(n,γ)

(p,n)

Other

Special features of your facility related to conference topics
   * DC neutrons /cm^2/s at the source

Current, recent and proposed research related to conference topics

Laboratory __Los Alamos National Laboratory__
Name of respondent __C. Bowman__
Accelerator Type (s) __LAMPF/WNR/PSR__
Time structure __Microburst (0.4n-sec, 1µ-sec spacing, 50,000 Hz__
    PSR (270n-sec, 24 Hz)

Protons

Energy range   __800 MeV__
Maximum intensity __Microburst (6µa), PSR (100µa)__
Polarized intensity

Neutron Facilities

Sources   T(p,n)   D(d,n)   T(d,n)   Spallation
Intensity on   (1-300MeV)(10^{-2}-10^{3}ev)
target (n/sec/sr) 5×10^{13} 1.5×10^{14}
Polarized intensity

Spectrometers: type, resolution, solid angle, completion date.

(n,n') _____

(n, charged particle) __Total neutron induced reaction cross sections for particular charged particles including fission fragments__
(n,γ) __Angular distributions using several Ge detectors for (n,n'γ), (n,γo), (n,γ) etc.__
(p,n) _____

Other _____

Special features of your facility related to conference topics
World's most intense white sources for 1-300 MeV and .01 to 1000eV evergy ranges. The MeV and eV sources run simultaneously on separate targets.

Current, recent and proposed research related to conference topics
(n,n'γ), (n,γo), total cross sections, fission process

345

Laboratory   Los Alamos National Laboratory
Name of respondent   J. R. Tesmer
Accelerator Type (s) Vertical Van de Graaff
Time structure   Beams can be bunches to ∿1ns fwhm

	Protons	Deuterons	Other Tritons
Energy range	.5-7 MeV	.5-7 MeV	.5-7 MeV
Maximum intensity	10μA(dc)	10μA(dc)	10μA(dc)
Polarized intensity	---	---	---

Neutron Facilities

Sources	T(p,n)	D(d,n)	T(d,n)	^{7}Li(p,n)	other H(t,n)
Intensity on target (n/sec/sr)	*	*	*	*	9.4x10^7**
Polarized intensity					

Spectrometers: type, resolution, solid angle, completion date.

(n,n') _____

(n, charged particle) _____

(n,γ) _____

(p,n) _____

Other _____

Special features of your facility related to conference topics

*Typical values can be obtained from the literature.
**STP 1 cm thick target with 1μA tritons incident at 6 MeV.

Current, recent and proposed research related to conference topics

Laboratory __Los Alamos National Laboratory__

Name of respondent __J.R. Tesmer__

Accelerator Type (s)__Tandem Van de Graaff__

Time structure __Beams can be bunched to ~1ns fwhm__

	Protons	Deuterons	
Energy range	2-18MeV	2-18MeV	2-18MeV
Maximum intensity	10µA(dc)	10µA(dc)	10µA(dc)
Polarized intensity	150nA		150nA

Neutron Facilities

Sources	T(p,n)	D(d,n)	T(d,n)	^{7}Li(p,n)	H(t,n)
Intensity on target (n/sec/sr)	*	*	*	*	1.6×10^8**
Polarized intensity	—	—	—	—	—

Spectrometers: type, resolution, solid angle, completion date.

(n,n') _____

(n, charged particle) _____

(n,γ) _____

(p,n) _____

Other _____

Special features of your facility related to conference topics

* Typical values can be obtained from the literature

** STP 1cm thick target with 1µA of tritons incident at 15MeV

Current, recent and proposed research related to conference topics

Laboratory __CATHOLIC UNIVERSITY OF LOUVAIN__
Name of respondent __P. MACQ__
Accelerator Type (s) __ISOCHRONOUS CYCLOTRON__
Time structure __1 nsec bunch width__

	Protons	Deuterons	Other
Energy range	20-80 MeV	10-50 MeV	
Maximum intensity	25 µA	25 µA	
Polarized intensity	-	-	

Neutron Facilities

Sources	T(p,n)	D(d,n)	T(d,n)	^{7}Li(p,n)	other
Intensity on target (n/sec/sr)				2×10^{10}	
Polarized intensity				-	

Spectrometers: type, resolution, solid angle, completion date.

(n,n') _____

(n, charged particle) _____

(n,γ) _____

(p,n) _____

Other _____

Special features of your facility related to conference topics

Current, recent and proposed research related to conference topics
radiative neutron-proton capture at extreme c.m. angles
neutron-proton bremsstrahlung in the hard photon region.

Laboratory   Dept. of Physics, Univ. of Lund, Lund, Sweden
Name of respondent   Ingvar Bergqvist
Accelerator Type (s)   Pelletron tandem
Time structure   DC beam

	Protons	Deuterons	Other
Energy range	1 - 6 MeV	1 - 6 MeV	
Maximum intensity	5 µA	5 µA	
Polarized intensity	--	--	

Neutron Facilities

Sources	T(p,n)	D(d,n)	T(d,n)	^{7}Li(p,n)	other
Intensity on target (n/sec/sr)	$3 \cdot 10^6$	$2 \cdot 10^7$			
Polarized intensity	--	--			

Spectrometers: type, resolution, solid angle, completion date.

(n,n')

(n, charged particle)

(n,γ)   NaI anticompton spectrometer

(p,n)

Other

Special features of your facility related to conference topics

Current, recent and proposed research related to conference topics

Laboratory: Ohio University

Name of respondent: J. Rapaport, R. Lane, S. Grimes, R. Finlay and C. Brient

Accelerator Type(s): Tandem van de Graaff

Time structure: DC and Pulsed. $\Delta t \simeq 0.7$ ns (FWHM)
$F = 5$ MHz $\times (\frac{1}{2})^n$, $n = 0, 1, \ldots, 6$

	Protons	Deuterons	Other
Energy range	1-9 MeV	1-9 MeV	
Maximum intensity	100 µA DC (5 µA pulsed)	80 µA DC (4 µA pulsed)	
Polarized intensity			

Neutron Facilities

Sources	T(p,n)	D(d,n)	T(d,n)	^{7}Li(p,n)	other
Intensity on target (n/sec/sr)	6×10^8	3×10^8	2×10^8	2×10^8	
Polarized intensity					

Spectrometers: type, resolution, solid angle, completion date.

(n,n') Beam Swinger $\Delta E/E \simeq 0.013$ @ 13 m, 0.007 @ 30 m
$d\Omega = 5$ msr @ 13 m. Completed in 1981

(n, charged particle) Magnetic Quadrupole (7 msr, 1980) and Time-of-Flight (6.5 msr, 1983) Charged Particle Spectrometers for (n,z)

(p,n) Beam Swinger. See above (n,n')

Other (n,charged particle) ΔE-E telescope system, t.o.f. (8 msr); 4 arms; $\Delta\Theta = 20°$ (1984)

Special features of your facility related to conference topics

Current, recent and proposed research related to conference topics
Elastic and inelastic neutron scattering $4 < E_n < 11$ MeV, $18 < E_n < 26$ MeV ($7 \leq A \leq 209$).

Laboratory: Oak Ridge Electron Linear Accelerator (ORELA)
Name of respondent: J.A. Harvey
Accelerator Type(s): L-Bank Travelling-Wave Electron
Time structure: 5-40 nanosecond pulses, up to 1000 Hz

	Protons	Deuterons	Electrons
Energy range			100-170 MeV
Maximum intensity			20 amps of electrons in pulses
Polarized intensity			0

Neutron Facilities

Sources	T(p,n)	D(d,n)	T(d,n)	Ta(γ,n)	Be(γ,n)
Intensity on target (n/sec)				10^{14}	10^{12}
Polarized intensity				0	

Spectrometers: type, resolution, solid angle, completion date.

$\sigma_{el}(\theta)$ elastic scattering; scattering chamber 200 m

(n,n')

(n, charged particle)

(n,γ) Total energy for $\sigma(n,\gamma)$ $\frac{\Delta E_n}{E_n} \sim 0.2\%$; Ge(Li) for E_γ.

(p,n)

Other (n, total) ^{6}Li glass scin. $0.002 - 10^5$ eV, $\frac{\Delta E_n}{E} \sim 0.3 - 0.1\%$.
(n, total) NE110 scin. $10^4 - 8 \times 10^7$ eV, $\frac{\Delta E}{E} \sim 0.05 - 0.3\%$.

Special features of your facility related to conference topics
High resolution neutron resonance studies from 1 to 4×10^6 [eV].

Current, recent and proposed research related to conference topics
Determination of optical model potential depths, V and W, for
ℓ = 0, 1 and 2 for light and closed shell nuclides.

Laboratory Laboratoire National Saturne, SACLAY, FRANCE

Name of respondent Yves TERRIEN

Accelerator Type (s) Synchrotron

Time structure Pulsed beam by bursts of a few hundreds of milliseconds, every 1 to 3 seconds

	Protons	Deuterons	Other
Energy range	0.3 to 3 GeV	2.3 GeV	^{3,4}He, Heavy Ions
Maximum intensity	10^{12}	5×10^{11}	
Polarized intensity	10^{10}	10^{10}	

Neutron Facilities

Sources	T(p,n)	D(d,n)	T(d,n)	^{7}Li(p,n)	other Be(d,n)
Intensity on target (n/sec/sr)					5×10^6 at 0° in $\Delta\Omega \sim 10^{-6}$ sr
Polarized intensity					

Spectrometers: type, resolution, solid angle, completion date.

(n,n') _____

(n, charged particle) Ionization chamber, especially designed to detect low energy recoil protons (100 keV resolution)

(n,γ) _____

(p,n) _____

Other _____

Special features of your facility related to conference topics

Current, recent and proposed research related to conference topics

Cross sections and asymetries measurements for elastic scattering of 0.4 to 1.1 GeV neutrons on proton target.

351

Laboratory The Studsvik Science Research Laboratory, S-61182 Nyköping, Sweden

Name of respondent Nils Olsson

Accelerator Type (s) CN VdG

Time structure DC or pulsed 1 MHz, 0.2 ns pulse width

	Protons	Deuterons	Other
Energy range	1-6 MeV	1-6 MeV	
Maximum intensity	(DC 100 µA) 5 µA	(DC 100 µA) 5 µA	
Polarized intensity			

Neutron Facilities

Sources	T(p,n)	D(d,n)	T(d,n)	^{7}Li(p,n)	other
Intensity on target (n/sec/sr)	(pulsed) $<1.6 \times 10^9$	$<1.1 \times 10^9$	$<3.4 \times 10^8$		
Polarized intensity					

Spectrometers: type, resolution, solid angle, completion date.
  TOF spectrometer

(n,n') Flight path: 4 m; Angular range: -50° - +162°; Two detectors 5" x 2" NE213; Time Resolution 0.5 ns

(n, charged particle) _____

(n,γ) _____

(p,n) _____

Other _____

Special features of your facility related to conference topics
1 or 2 cm long gas targets for pressures ≤ 4 atm. Total resolution in scattering experiments at 22 MeV: 0.5 MeV

Current, recent and proposed research related to conference topics
Systematics studies of elastic and inelastic scattering at 22 MeV in the mass range A = 9 to A = 209.

Laboratory Swiss Institute for Nuclear Research (SIN)
Name of respondent Dr. M.A. Pickar, Uni Basel
Accelerator Type (s) Sector focussed cyclotron
Time structure Bursts 0.6-5.0 ns long, sep. by 50-150 ns

	Protons	Deuterons	Other, ^{3}He
Energy range (MeV)	10-72	10-65	15-165
Maximum intensity	20 µA	10 µA	2 µA
Polarized intensity	2 µA	1 µA	--

Neutron Facilities

Sources	T(p,n)	D(d,n)	T(d,n)	^{7}Li(p,n)	other
Intensity on					D(p,n)
target (n/sec/sr)					1.4×10^{10}/µA
Polarized intensity					1.4×10^{10}/µA

Spectrometers: type, resolution, solid angle, completion date.

(n,n') _____

(n, charged particle) MWPC+Plast. Scint.: $\Delta E/E \sim 7\%$;
$\Delta t \simeq 0.8$ ns; ~2mm resol. on tgt.; ~ 0.2° ang.resol.; $\Delta \Omega \sim$ 50msr

(n,γ) _____

(p,n) _____

Other _____

Special features of your facility related to conference topics
Circulating liquid deuterium production target; Production of both transverse and longitudinally polarized neutron beams; liquid ^{4}He active polarimeter; Polarized proton target.

Current, recent and proposed research related to conference topics
Measurement of the transverse polarization coefficient $K_y^{y'}(0°)$ for $D(\vec{p},\vec{n})2p$ at 30, 50 and 70 MeV.

Measurement of A_{zz} and A_{yy} for $\vec{n}\vec{p}$ elastic scattering.

Laboratory Cyclotron and Radioisotope Center, Tohoku University
Name of respondent H. Orihara
Accelerator Type (s) AVF Cyclotron
Time structure 350 p sec for 35 MeV protons

	Protons	Deuterons	Other
Energy range (MeV)	2.5 ~ 41	2.5 ~ 26	7 ~ 65
Maximum intensity (μA)	40	50	20
Polarized intensity			

Neutron Facilities

Sources	T(p,n)	D(d,n)	T(d,n)	^{7}Li(p,n)	other
Intensity on target (n/sec/sr)					
Polarized intensity					

Spectrometers: type, resolution, solid angle, completion date.

(n,n') _____

(n, charged particle) _____

(n,γ) _____

(p,n) TOF; $\Delta\tau$ ~ 0.9 nsec; $\Delta\Omega$ ~ 10^{-4} sr (44 m flight path); in operation; PDP-11/44 with CAMAC

Other _____

Special features of your facility related to conference topics

Neutron time of flight analysis system with a beam swinger covering $-10° \leq \theta \leq 165°$. Flight path up to 44 m is available. Data acquisition system is capable to handle signals from 16 neutron detectors with n-γ discrimination.

Current, recent and proposed research related to conference topics

Currently, (p,n) research at E_p = 35 MeV. In near future polarization measurements for emitted neutrons with liquid He detector.

Laboratory   Triangle Universities Nuclear Laboratory
Name of respondent   Richard Walter
Accelerator Type (s) 8 MV Tandem and Cyclotron Injector
Time structure   D.C. Beam; 1.5 ns Wide Pulsing (2 MHz)

	Protons	Deuterons	Other ^{3}He and ^{4}He
Energy range	2-31 MeV	2-24 MeV	5-24 MeV
Maximum intensity	3 µA pulsed	3 µA pulsed	0.5 µA DC
Polarized intensity*	150 nA D.C.	200 nA D.C.	

Neutron Facilities

Sources   T(p,n)   D(d,n)   T(d,n) ^{7}Li(p,n)   other

Intensity on target (n/sec/sr)   Use gas targets: Tritium--6 cm atm, Deuterium--up to 21 cm atm

Polarized intensity   Use D(d,n) Polariz. Transfer with Gas Target

Spectrometers: type, resolution, solid angle, completion date.
Two well-shielded time-of-flight spectrometers, maximum flight path 4 m and 6 m for 8 cm and 13 cm diameter NE213 scintillators, respectively. Two well-shielded 10" x 10" diam. NaI crystals, two-well shielded 3" x 3" diam. BGO detectors, all having anti-Compton shields.

(n,γ) _____

(p,n) _____

Other   Longer neutron flight paths/larger detectors under design.

Special features of your facility related to conference topics
* Pulsed Mode: 2 nS burst width, 100 nA protons, 150 nA deuterons.

Current, recent and proposed research related to conference topics

1) Emphasize 8-17 MeV neutron elastic scattering and inelastic scattering to low-lying excited states--measure $\sigma(\theta)$ and Ay (θ).
2) Study (p,n) and ($\vec{p}$,n), (p,n) and ($\vec{p}$,p) for isospin-consistent development.
3) Few-nucleon systems: ($\vec{n}$,n), ($\vec{n}$,np), (n,γ), ($\vec{n}$,γ) research.

Laboratory TRIUMF
Name of respondent W.P. Alford
Accelerator Type (s) Cyclotron
Time structure Normal - 2ns pulse at 23 MHz
Dedicated - 150 ps pulse at 4.6 MHz

	Protons	Deuterons	Other
Energy range	200-500 MeV		
Maximum intensity	100 na (limited by beam dump)		
Polarized intensity	100 na		

Neutron Facilities

Sources	T(p,n)	D(d,n)	T(d,n)	^{7}Li(p,n)	other
Intensity on target (n/sec/sr)				3×10^8	
Polarized intensity				3×10^8	

Spectrometers: type, resolution, solid angle, completion date.

(n,n') _____

(n, charged particle) QD medium resolution spectrometer
$\Delta p/p = 3 \times 10^{-4}$; 2.5 msr; 1984

(n,γ) _____

(p,n) QD above for recoil protons

Other _____

Special features of your facility related to conference topics

Overall resolution <1 MeV
System tests planned for early 1985

Current, recent and proposed research related to conference topics

(p,n) and (n,p) reactions 200-500 MeV

Laboratory   Gustaf Werner Institute (GWI), Uppsala University

Name of respondent   H. Condé

Accelerator Type (s) Cyclotron (frequency mod. or isochronous op.)

Time structure C.W.: ~20 MHz, 4-8 ns
                F.M.: ≤1 kHz, 10-50 µs (micro 20 MHz, 5-15 ns)

EXPECTED
PERFORMANCES:

	Protons	Deuterons	Other
Energy range	10-200 MeV	25-100 MeV	He3: 35-267 MeV
Maximum intensity	40 µA	40 µA	20 µA
Polarized intensity			

Neutron Facilities

Sources	T(p,n)	D(d,n)	T(d,n)	^{7}Li(p,n)	other
Intensity on target (n/sec/sr/µA/mm Li)				10^9	
Polarized intensity					

Spectrometers: type, resolution, solid angle, completion date.

(n,n')   T.o.F.-spectrometer (1985-86)

(n, charged particle)   Magnetic spectrometer with 4 MWPC, 0.5 cm/
%Δp/p at 100 cm, 30 msr (1985-86)

(n,γ)

(p,n)

Other   HESM: Magn. spectr., General facility for intermediate
energy reaction studies, 60 keV at 200 MeV, 1 msr (1986)

Special features of your facility related to conference topics
Storage ring with electron cooling (CELSIUS) under construction.
Expected performances, res. 10^{-4} for 10^{10} particles stored, 10^{-5}
for 10^8 particles, max. proton energy about 1 GeV

Current, recent and proposed research related to conference topics
1) (n,p) and (α,α'n)-reactions as probes for   giant
   resonance studies

2) σ_{tot} and σ_R-studies

Laboratory   Tandem Accelerator Laboratory, Uppsala, Sweden
Name of respondent   Leif Nilsson
Accelerator Type (s)   HVEC Tandem Accelerator
Time structure   Pulse Width: 1-2 ns, Frequency: 2 MHz

	Protons	Deuterons	Other
Energy range	2-14 MeV	2-14 MeV	
Maximum intensity	2 µA	2 µA	
Polarized intensity	−	−	

Neutron Facilities

Sources	T(p,n)	D(d,n)	T(d,n)	^{7}Li(p,n)	other
Intensity on target (n/sec/sr)	$5 \cdot 10^7$	10^8	$2 \cdot 10^7$		
Polarized intensity					

Spectrometers: type, resolution, solid angle, completion date.

(n,n') Liquid scint NE 213, diam 12", thickness 2".
        Resolution (FWHM) 1.5 ns. 1984

(n, charged particle) _____

(n,γ) NaI, diam 10", length 14". Resolution (FWHM) 2.6 % at
E_γ = 6.13 MeV. Plastic (NE 102) anticoinc shield. 1984

(p,n) _____

Other _____

Special features of your facility related to conference topics

Current, recent and proposed research related to conference topics
Studies of the (n,γ) reaction for E_n = 2-28 MeV.
Search for isovector E2 giant resonance in Ca, Y

Laboratory   University of Washington Nuclear Physics Laboratory

Name of respondent   William G. Weitkamp

Accelerator Type (s)   HVEC Model FN Van de Graaff (superconducting booster)*

Time structure   _____

	Protons	Deuterons	Other
Energy range	1-24 MeV (1-36 MeV)*	Same	
Maximum intensity	20 μA	Same	
Polarized intensity	0.1 (3) μA**	Same	

Neutron Facilities

Sources	T(p,n)	D(d,n)	T(d,n)	^{7}Li(p,n)	other
Intensity on target (n/sec/sr)					
Polarized intensity					

Spectrometers: type, resolution, solid angle, completion date.
QDDDDQ, 2×10^{-3} resolution, 12 msr solid angle, completed in 1983.

(n,n')   _____

(n, charged particle)   _____

(n,γ)   _____

(p,n)   _____

Other   1.3 m diameter graphite sphere neutron detector

Special features of your facility related to conference topics
* A superconducting quarterwave-resonator booster is under construction, completion expected in 1987.
** A cross-beam polarized hydrogen ion source is under construction, completion expected in 1985.

Current, recent and proposed research related to conference topics
(p,γ) and (γ,n) studies of giant resonances. Neutron induced reactions of interest in nuclear astrophysics. Analyzing power and depolarization measurements for proton inelastic scattering to the continuum.

Laboratory	Department of Physics, University of Wisconsin		
Name of respondent	Willy Haeberli		
Accelerator Type(s)	Tandem Van de Graaf (EN)		
Time structure	bunched and pulsed beam, nanosecond pulsing		

	Protons	Deuterons	Other
Energy range	1-12 MeV	1-12 MeV	Heavy ions
Maximum intensity	10 μA	10 μA	
Polarized intensity	1 μA	1 μA	

Neutron Facilities

Sources	T(p,n)	D(d,n)	T(d,n)	^{7}Li(p,n)	other
Intensity on target (n/sec/sr)					
Polarized intensity	1x10^8	3x10^8	8x10^7		

Spectrometers: type, resolution, solid angle, completion date.

(n,n') _____

(n, charged particle) _____

(n,γ) 10"x10" NaI crystal spectrometer with anticoincidence plastic scintilator shielding and 6Li$_2$CO$_3$ thermal neutrons shield.

(p,n) _____

Other _____

Special features of your facility related to conference topics
High polarization and easy reversal of neutron polarization by switching RF transition at the ion source (polarization sign of the incident charged particle beam is changed)

Current, recent and proposed research related to conference topics
Neutron-photon analyzing power at 25 MeV and 10 MeV
Neutron-proton radiative capture at 6.0 MeV and 13.4 MeV

AN ACCELERATOR SYSTEM FOR THE PRODUCTION OF AN INTENSE NEUTRON BEAM FOR RESEARCH*

D. L. FRIESEL

Indiana University Cyclotron Facility, Bloomington, Indiana

ABSTRACT

While the desire to perform neutron-nucleus reactions as an alternative means of studying the nuclear interior has always been present, a suitable facility dedicated to the production of an intense neutron source of sufficient quality to perform high resolution nuclear reaction studies over a broad energy range has not been developed. In addition to the technical problems that must be solved for the successful development of such a facility, part of the reason for this has been the availability of a large variety of charged-particle beams, which even today is being expanded by the development of several new heavy ion accelerators[1,2]. However, recent results from charge exchange reactions at intermediate energies[3] have intensified the desire to study neutron induced reactions. Some thoughts on the desirable properties of a neutron source suitable for nuclear research, and a discussion of the design of a facility to produce it are discussed below.

INTRODUCTION

The use of the neutron as a nuclear probe has long been recognized as a valuable complement to the large volume of high resolution charged-particle induced reaction data now available, and is indeed the subject of this workshop. In fact, if one had to choose between the two states of the nucleon as a projectile without regard for the technical or experimental difficulties involved, the neutron might be preferred. To obtain these data, intense, high-resolution neutron beams with the energy and time structure variability of currently available proton beams are needed. Recent advances in accelerator technology coupled with a renewed interest from the research community make it worth reconsidering the development of a new facility dedicated to the production of intense mono-energetic neutron beams for research.

The design of such a facility must begin with a list of the desirable beam properties which would permit a wide variety of experimental programs to be run over a broad energy range. As stated previously, a neutron beam with characteristics similar to presently available proton beams, like those typically produced at IUCF[4], would be ideal. A list of reasonable properties for a neutron beam is given in table I, which forms the basis for the design goal of the accelerator systems. In addition to the properties listed, one would

*Research supported by the NSF under grant NSF PHY 81-14339.

also like to have the option of selecting either a polarized or unpolarized neutron beam.

Since any neutron source is necessarily the product of some nuclear reaction, these requirements are transformed to a new set of requirements for a charged-particle beam which can perform that reaction. The one major difference required of the charged-particle beam from those listed in table I is the intensity. The (p,n) reaction on ^{6,7}Li, ^{9}Be, and ^{3}H is generally used as a source of mono-energetic unpolarized neutrons. The differential cross sections for

Table I Neutron Beam Design Characteristics

Neutron beam energy:	50-200 MeV (variable)
Flux density:	10^8 n/cm^{2}s
Energy resolution(ΔE/E):	2×10^{-3} fwhm
Time structure:	< 1 ns
Pulse Repetition rate:	20-1000ns (variable)

these reactions are nearly 25 mb/sr at zero degrees. Therefore, using one of these targets thin enough (200keV) to achieve the desired energy resolution, a proton beam of about 1 mA intensity is needed to achieve the desired flux density. While not all experiments require this intensity, an intensity range of from one to several hundred microamperes would be required to cover all neutron experiments of interest. The choice of a reaction mechanism for a polarized neutron source is not so clear. The D($\vec{d}$,$\vec{n}$) reaction produces a polarized neutron beam of poor resolution, while a ^{6}Li($\vec{p}$,$\vec{n}$) reaction produces a mono-energetic polarized neutron beam. The Lithium target is less desirable as an intense neutron source, however, because of its relatively low melting point ($\sim$186°c). Nevertheless, the production of any neutron beam with the above properties requires a variable energy light-ion particle accelerator capable of delivering at least several hundred microamperes of beam current.

CHOICE OF ACCELERATOR TYPE

In addition to intensity capability, the choice of which kind of accelerator is best suited for this purpose is determined by such considerations as the duty factor, energy resolution and time structure of the accelerated beams, and the ease with which they can be changed to meet the various research needs of the moment. Of the various types of intermediate energy particle accelerator designs, the cyclotron is uniquely well matched to these requirements. While linear accelerators have generally provided higher peak beam intensities than cyclotrons, the time structure variability and low duty factors make them less desirable as a neutron source. However, until recently, the use of any accelerator as a neutron source was restricted by their relatively low beam intensity capabilities. Beam intensities of a milliamp from an intermediate energy accelerator have yet to be achieved, but are now beginning to appear feasible. Several of the highest intensity intermediate energy cyclotrons now in operation or under construction[5,6,7] are listed in table II with

Table II High Intensity Cyclotrons Operating or Under Construction

Facility	Energy (MeV)	Ion	Maximum Extracted Current (goal)	(achieved)
Triumf	200-500	H	500 uA	170 uA
Phillips/SIN	72	H	100 uA	190 uA
SIN Injector II	72	H	2 mA	-
SIN Ring Cyclotron	590	H	2 mA	170 uA
South Africa NAC	50-200	H, A 40	100 uA	-

their achieved and projected maximum beam intensities. The most likely new cyclotron design to achieve an extracted beam current of a milliamp or more is the 72 MeV SIN injector II[8]. This is a fixed energy, separated sector cyclotron with an external ion source terminal and was to have begun operation in late 1983. By the time of this writing, it may have achieved the highest extracted beam intensity of any cyclotron to date. The separated sector (or ring) cyclotron[9] is the new idea that makes high intensity (1mA) cyclotrons possible. The only accelerator listed in table II that is not a ring cyclotron is the Phillips machine, which is now thought to be operating at its intensity limit. The problems encountered in the development of cyclotrons capable of delivering milliamp proton beams were explored extensively during the development of the SIN injector II[10]. These problems are briefly summarized below.

HIGH INTENSITY PROBLEMS IN CYCLOTRONS

The two major factors limiting the beam intensity in a ring cyclotron are longitudinal and transverse space charge forces and the design of the radio-frequency accelerating systems. There are many other limiting problems which, while solvable, are more engineering oriented and are dominated by economical constraints. The effects of space charge forces are velocity dependent and are most troublesome for low energy and tightly bunched beams[10], both of which are present in cyclotrons. Transverse space charge forces cause particles at the periphery of the beam to be repelled from the center. The resulting beam emittance growth then limits the intensity and dominates the design of the low energy transport beam line and injection system to the cyclotron. In the cyclotron, these forces reduce the original focusing frequencies V_r and V_z, which can have a significant effect on the design of the sector magnets, and cause the betatron oscillations during acceleration to be intensity dependent. Studies at the SIN facility indicate that intensity limits due to transverse space charge forces for the 860 keV terminal and acceleration column, and the 72 MeV Injector and 590 Mev ring cyclotrons are 10 mA, 5 mA, and 20 mA respectively[10,11]. The design study of the 860 keV beam line, using a computer code TRANSPORT which includes space charge effects[12], showed a need for an intensity dependent adjustment of the quadrupoles from about 2 to 40 mA. Longitudinal space charge forces can introduce energy spreading of the beam in the transport lines and the cyclotron during the acceleration process[13]. The consequence of this, an additional broadening of the beam emittance during acceleration which spoils the turn separation at extraction in the cyclotron, is unacceptable beam losses. The energy spread increases proportionally with the intensity and the square of the number of turns in the cyclotron, and inversely with the beam velocity. The dependence on the number of turns means that a high acceleration rate will reduce the effects of longitudinal space charge forces. This leads to some design constraints for the RF systems of high intensity cyclotrons.

The RF systems are the primary tools that can be used to overcome the effects of space charge forces in high intensity cyclotrons. High RF accelerating voltages are not only necessary to provide adequate turn separation and thus permit the efficient extraction of beam from the cyclotron, they are essential. With beam powers reaching 200 kW, the consistent loss of only a small fraction of beam could cause considerable damage to the cyclotron systems and raise radiation levels to intolerable levels. To obtain perfectly separated turns at extraction, the energy spread of the beam must be much less than the energy gain per turn. Furthermore, a high acceleration voltage is the key to minimizing longitudinal space charge effects, as described above. Additional RF flattopping cavities are also necessary to further improve the energy spread and increase the phase width of the beam during acceleration[14,15]. The larger allowable phase width eases the problems of longitudinal space charge forces by reducing the charge density of the accelerated beam bursts and making them less sensitive to RF phase fluctuations. The third harmonic flattop cavity on the SIN 590 MeV ring cyclotron permits a phase variation for single turn extraction of over 40 degrees. RF flattopping cavities typically operate on the third harmonic of the RF accelerating frequency at about 12% of the fundamental amplitude.

Another effect of high intensity beams on RF cavities that must be accounted for is beam loading. High intensity beams, when not exactly in phase with the RF accelerating voltage, exert a reactive load on the RF-resonator which could collapse the RF voltage. At SIN, the flattop cavity absorbs nearly 10 kW of power from the 170 uA circulating beam in the 590 MeV ring. The response of the amplitude and phase regulating systems determines the stability of RF systems with high beam loading factors. These feedback loops must be tighter and faster with increasing beam intensity. As above, the effects of beam loading are also minimized by increasing the power of the RF system.

The impact of high intensities on the operation of a cyclotron is also significant. Diagnostic and interlock devices must be non-intercepting and faster, and an increased reliance on closed-loop control systems will become necessary. In addition, increased protection systems and interlocks for personnel and equipment from the effects of unexpected beam excursions will be needed. Simple changes of intensity will require a careful retune of the beam lines and accelerator, and can be equivalent in effort to an energy change. A pulsed beam system permitting the initial tuneup of the beam at the desired intensity without damaging the accelerator is also necessary.

The limits of all the above problems will be explored in the near future by the SIN injector II development effort.

NEW ACCELERATOR FACILITY CHOICES

Once the choice of the accelerator type has been made, it remains to decide whether an existing facility can be upgraded to meet the requirements, or a new facility must be built. In the energy range of interest, 50-200 MeV, there is only one operating cyclotron facility with the potential to be upgraded to a high enough intensity to be useful as a neutron source in this country. The Indiana University Cyclotron Facility (IUCF) is a variable energy, separated sector cyclotron system with all the necessary beam requirements except intensity[4]. The maximum intensity proton beam achieved so far by this facility is 6 uA, which is more than two orders of magnitude less than needed. With a modest effort (by comparison with building a new facility) these accelerators could be modified to routinely deliver about 10 uA of unpolarized protons over the desired energy range. The effects of space charge at these intensities would not require an extensive rebuild of the RF systems or the cyclotrons themselves. Improvements to the ion source pre-injector, transfer beamline optics, inflection and extraction systems and in the radiation hardness of the cyclotrons would be needed, along with additional radiation shielding for the accelerator vaults and experimental areas. A neutron production station and experimental area would also need to be built for neutron-induced reaction studies. The development of this capability at IUCF would permit many of the desired experiments to be conducted. It is also conceivable that the K=200 separated sector cyclotron at Indiana could transmit over 100 uA of protons. However, this would require the construction of an entirely new injector cyclotron and ion source terminal, as well as an extensive rebuilding of the existing cyclotron, including the RF, inflection, extraction, and trim coil systems. A detailed study of the required upgrades to achieve either of these options has not been made, but are predicated on the achievements of the several existing high intensity intermediate energy cyclotrons of similar design. However, it is reasonable to assume that the cost of achieving the 100 uA operation would be a significant fraction of that needed to build a new facility which could deliver an order of magnitude more intense beams. The construction of this new facility, if beam intensities in excess of several hundred microamperes are actually needed, appears to be the preferable course.

NEW ACCELERATOR FACILITY DESIGN

From the above discussion, the design of a new accelerator facility based on a k=200, variable energy, separated sector cyclotron capable of delivering proton beams at intensities up to 1mA will be explored. Primary consideration will be given to the problems of space charge forces and how to minimize their effect in the proposed accelerator. An excellent example of how to build such an accelerator exists in the new SIN injector II machine. The only technical capabilities lacking in this machine for our purposes are its energy variability and maximum energy (72 MeV). Unfortunately, these are not features which can be trivially modified on the SIN design. The more demanding change in terms of accelerator design is caused by the energy variability constraint. Because a separated sector cyclotron is an energy multiplying device, the extracted beam energy is a multiple of the input energy, unless there is provision made for either a variable inflection and/or extraction radius. For the energy variation needed in this cyclotron, using a fixed energy injector with a variable radius inflection system would be difficult at the intensities envisioned. A fixed radius inflection system with a variable energy inflected beam offers the best hope of achieving efficient inflection and reproducible centered acceleration in, and therefore extraction from, the cyclotron. This is an important consideration because of the need to routinely obtain 99.9% extraction efficiency from the cyclotron over the entire energy range in a reasonable amount of setup time. A possible consequence of this is an energy dependent intensity maximum for the beams achieved. It is also desirable to reduce the number of stages of acceleration to a minimum. To inject and extract from multiple cyclotrons over an energy range of a factor of twenty at the IUCF facility has become routine, but would be cumbersome and time consuming at high intensity where 5% inflection and extraction losses are unacceptable. From this point of view, the ideal configuration would be to inject beam from an external ion source system into a single variable energy cyclotron. This would greatly simplify the daily operation of the facility and reduce the amount of time needed to make the various machine changes needed to meet the experimental demands. While this may not be the most economical solution to the problem, it will serve to illustrate the problems involved in obtaining a variable energy, high-resolution, intense neutron source, and is the configuration proposed here.

ION SOURCE PRE-INJECTOR

From the studies made at SIN, an external ion source with injection at several hundred keV is the preferred method of obtaining milliamp beams for injection. The newly developed RFQ accelerators, while promising, deliver beams with a 2 to 3% energy spread, which is too large to be injected into cyclotrons. The requirement of energy variability by a factor of four calls for a similar inflection energy variation. Recalling that the space charge forces are most pronounced at low velocity, an ion source terminal having a maximum potential of 3 MV would be used for the injection of protons to be accelerated to 200 MeV. 50 MeV protons would then require an inflection energy of 880 keV. It is impractical to consider the application of more than about 1 MV on an external ion source terminal in air, and a single ended Van de Graaff would not easily house an intense source nor transmit the beam because of accelerator tube loading. The source needs to be decoupled from the acceleration column so that unwanted ion species do not contribute to the tube loading or space charge effects.

The design of a possible 3 MeV ion source terminal housed in an

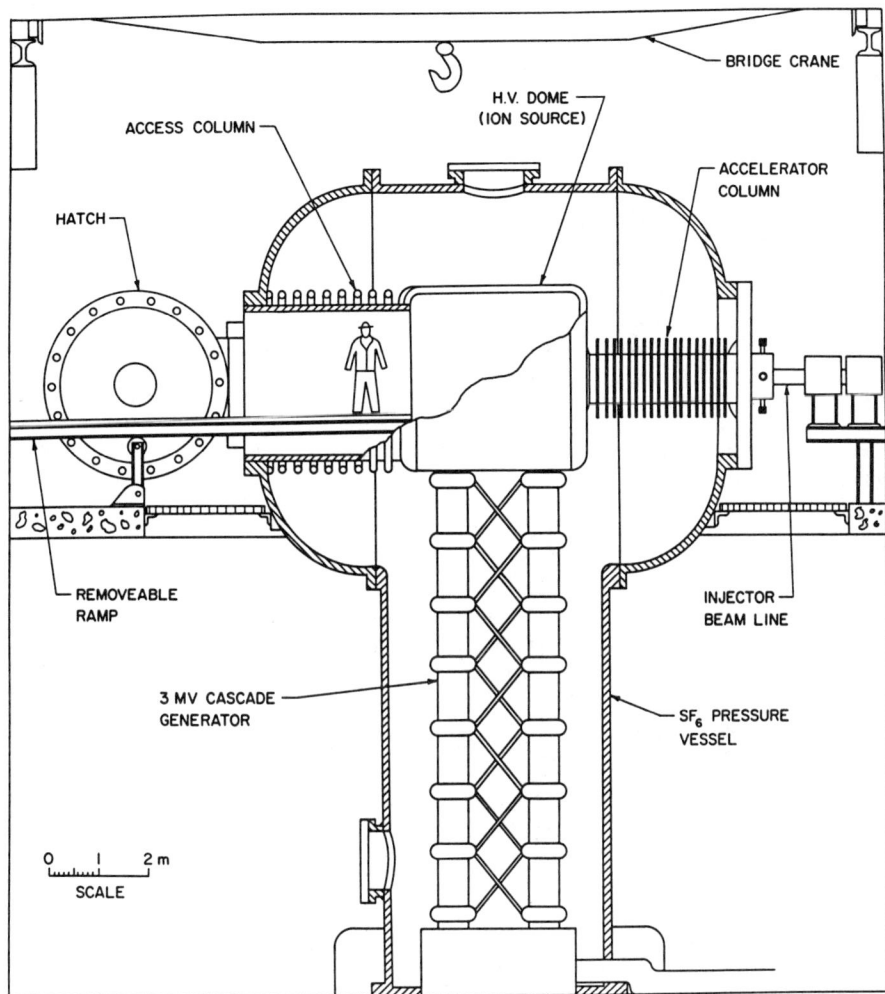

Figure 1. Sulfur-Hexafluoride insulated 3 MV ion source terminal

SF_6 atmosphere is shown in figure 1. The voltage is applied to the terminal via a seven-stage symmetrical cascade rectifier such as those manufactured by Emil Haefly & Co. of Basel, Switzerland. This manufacturer claims that even 10 MV pressurized gas-insulated dc power supplies with currents in the mA range are feasible[16] and has constructed a 2.5 MV, 200 mA air insulated rectifier with remote-controlled polarity-reversal. The polarity-reversal feature is an important consideration for obtaining a variable duty factor beam with a "stripper loop" in the transfer beamline[17]. A cylindrical 4 m diameter high voltage dome houses an ion source, 60 keV beam line, and associated support equipment. Access to the dome is provided by an additional 2.8 m diameter insulated column between the dome and the pressure vessel. During operation, the interior of the dome and the column are in equilibrium with the interior of the pressure vessel. When access to the dome is required, the SF_6 in the interior of the column and the dome is pumped into the pressure vessel and then let up to air. The volume of the dome and entrance column are small enough (46 cubic meters) to allow relatively rapid access to the ion source systems with no loss of SF_6. The total volume of the pressure vessel is about 400 m^3. A cusped field, single aperture ion source[18] and momentum analysis system are mounted in the dome to provide a 30 mA proton beam with a normalized emittance of about 0.5mm mrad. Ehlers type H- and polarized ion sources could be installed as well, but would have to be exchanged with the one another since room would not permit the installation of all sources simultaneously. A separate ion source terminal housing a new ECR[18] or other state of the art high intensity polarized H- source may be more efficient. Electrical power for the equipment in the dome would have to be provided by an alternator driven by a hydraulic or other system from ground potential.

LOW ENERGY TRANSFER BEAM LINE

The 3 MeV beamline to the cyclotron must provide a proper matching of the beam phase ellipses and dispersion trajectories to the acceptance of the cyclotron. At the intensities needed, the solutions to the optics problems are intensity as well as energy

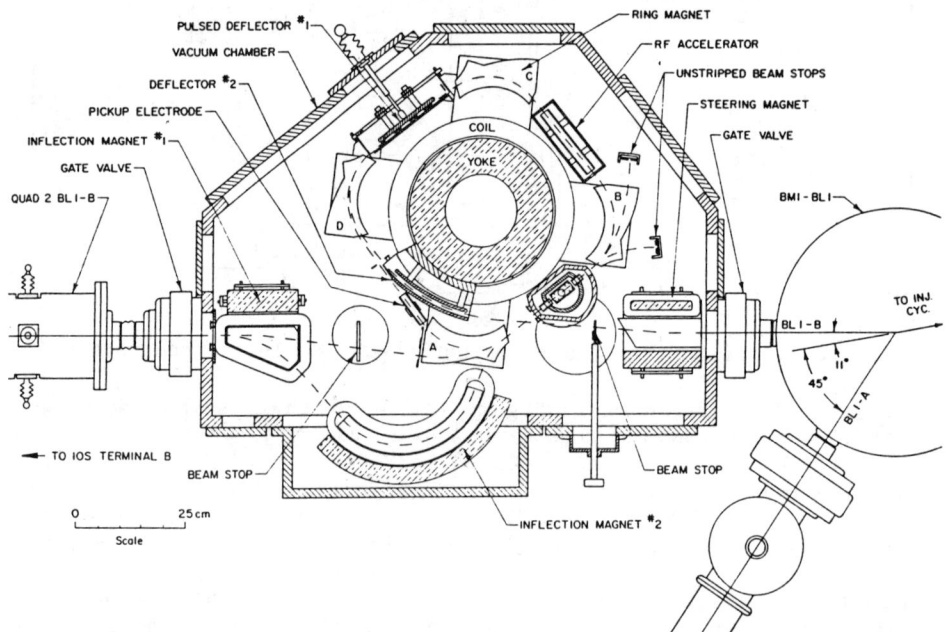

Figure 2. The IUCF 615 keV stripper loop design.

dependent and may require a physical reconfiguration of the beam line
between runs for the different energies. The detailed design of this
beam line may prove to be a challenging task. Space charge forces
argue against bunching the beam prior to entering the cyclotron, so
that the effectiveness of bunchers and choppers are questionable. The
dc beam will be injected into the cyclotron and the unaccelerated
portions stopped on high power collimators in the center of the
machine. The exception to this is caused by the need to have a
variable pulse repetition rate for the beams on target. A recently
suggested and tested device, called a "stripper loop", could be
installed in this beam line to provide the necessary pulse repetition
rate variability[17,20]. This device, a low energy storage ring into
which beam from the ion source is stacked by a charge exchange
reaction during the off time between beam bursts, uses an H⁻ beam and
therefore makes a reversible-polarity terminal necessary. The beam

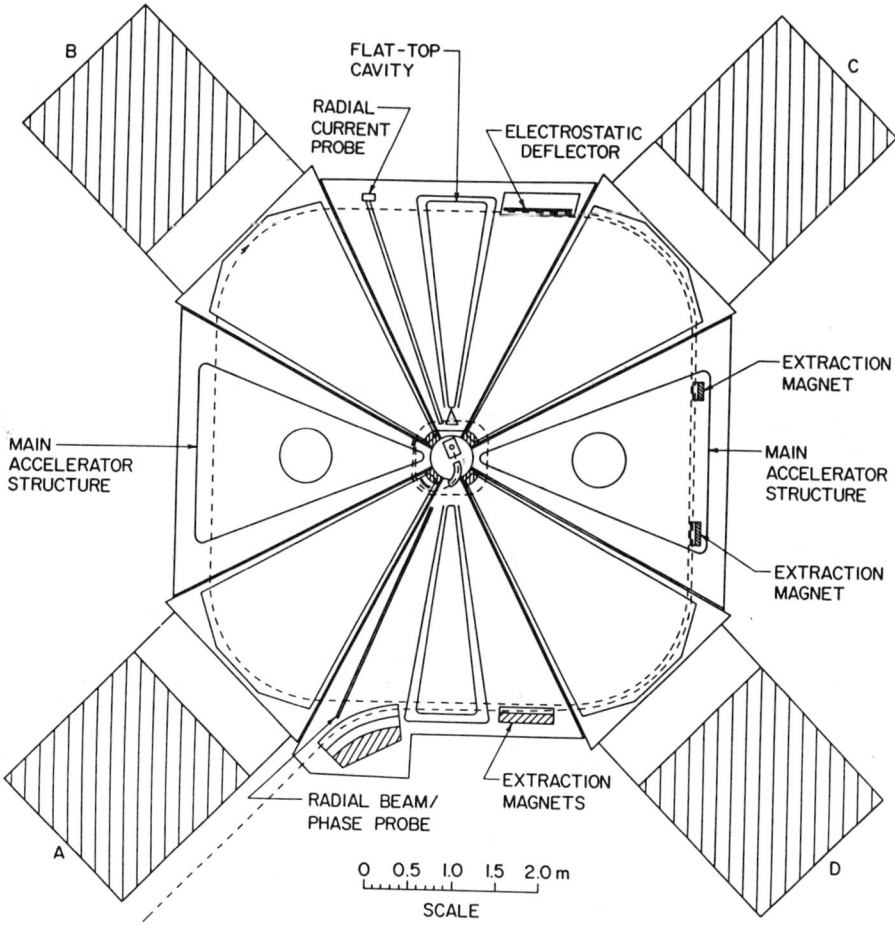

Figure 3. Plan view of the k = 200 cyclotron

intensity per burst is enhanced by the ratio of the beam off to beam on time. Calculations based on the performance of a similar loop at IUCF, shown in figure 2, indicate that a peak beam intensity gain of over 10 could be achieved. However, this gain may well be lost due to space charge effects, making the use of a simple chopper for this purpose as efficient. This aspect of the proposed design needs to be studied carefully. The use of the stripper loop is still important for polarized H⁻ beam operations, however, since these sources are more than two orders of magnitude less intense than their unpolarized counterpart.

SEPARATED SECTOR CYCLOTRON

The cyclotron takes the 3 MeV proton beam to a final energy of 200 MeV, which corresponds to an energy gain of about 67. The plan view and main characteristics for this machine are provided in figure 3 and table III. The proposed machine is quite similar to the IUCF main stage cyclotron in that it has the same magnet fraction and sector angle. The energy gain, however, is about 3 times larger, which makes the mechanical design of this machine much more difficult, particularly in the center region, where space for the inflection elements is tight. This can be helped by reducing the field of the magnets, but at extra cost. Comprehensive computer studies of the inflection and extraction regions, and the effects of high intensities on the focusing frequencies need to be made to determine the best sector angle and field strength combination. The configuration shown here only serves as a format for discussing the problems to be solved.

The proposed main RF accelerating cavities are also similar to the IUCF design but are expected to operate at a higher voltage to provide a larger turn separation at both the inflection and extraction radii. Also, two additional flattopping RF resonators

Table III. Cyclotron Parameters

Beam Properties: (goals)

Energy range	50-200 MeV
Beam intensity	1 mA
Beam quality	2π mm mrad
Energy spread (fwhm)	2×10^{-3}
Phase width; low intensity	6°
high intensity	15°

Sector Magnets:

Number of sector magnets	4
Sector angle	34°-35°
Maximum Field Strength	16.5 kG
Gap width	8 cm
Inflection radius (valley)	.39 m
Extraction radius (valley)	2.73 m
Extraction radius (hill)	3.30 m

RF Systems:

Accelerating freq. range	28-35 MHz
Peak Voltage; at inflection	150 kV
at extraction	300 kV
Flattop freq. range(h=3)	84-105 MHz

Internal Beam:

Orbit separation at inflection	2 cm
Orbit separation at extraction	.5 cm
Number of orbits	~300
Extraction efficiency	99.9%

are located in the remaining empty valleys. If possible, the flattop structures will be nested in the main accelerating deltas to allow for more free space in the remaining valleys for extraction and diagnostic systems.

However, efficient inflection of milliampere proton beams and quick access to the center region of the cyclotron are the primary problems to be solved to make this a practical accelerator. The design of the magnets and vacuum chambers will follow the example provided by the SIN II injector, where the two pole tips are welded into a single structure with stainless gap spacers. The trim coil

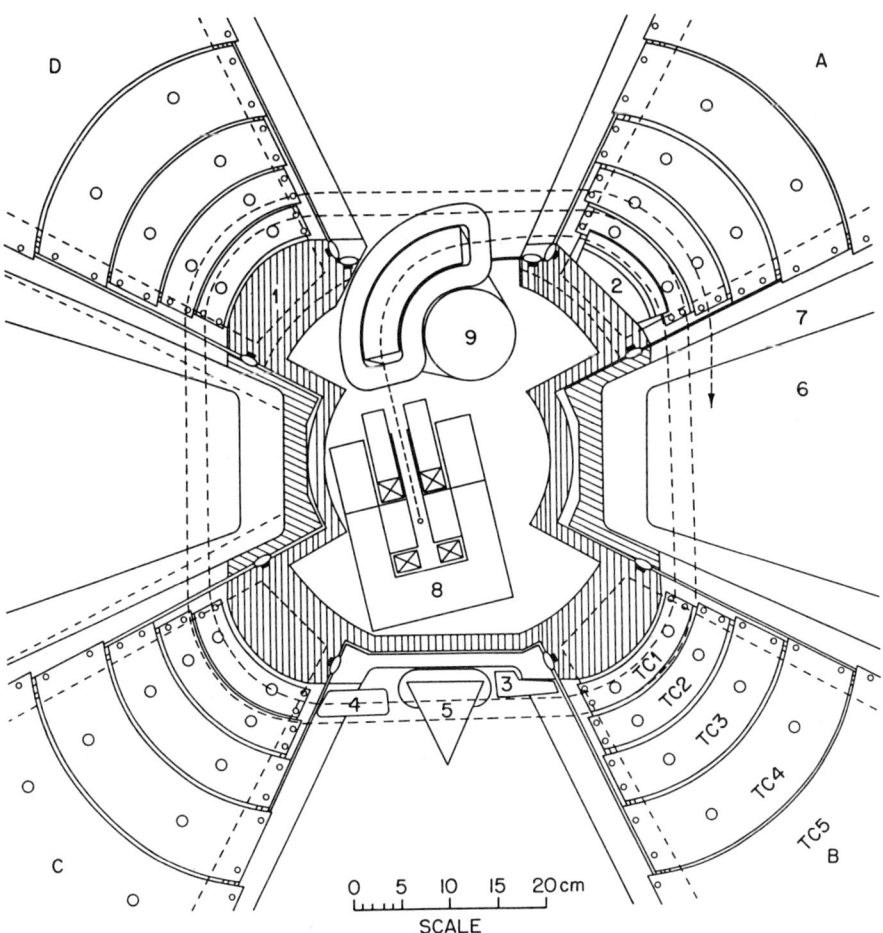

Figure 4. Center region of the 200 MeV cyclotron showing:
1. gap spacer, 2. electrostatic inflector, 3. high power beam stop, 4. vertical steering element, 5. compensating magnet, 6. RF delta, 7. Acceleration gap, 8. 90 degree bending magnet, 9. inflection magnet, and the first five trim coils

assemblies, which must provide the corrections to the field for saturation and the relativistic mass increase of the protons, would be individually bolted to the pole tips with special bolts designed to minimize the effects of the resulting magnet imperfections. The magnet vacuum chambers would then be welded directly onto the pole pieces with only enough of a gap to allow access to the trim coil lead connections. This has the added benefit that it would permit a smaller gap spacing than used in the Indiana machine. The main excitation coils of the magnets are mounted above and external to the vacuum chamber. The valley (RF) vacuum chambers would then slide between the magnet chambers using the inflatable seal design pioneered at SIN for quick access to both the RF systems and the center of the cyclotron. The center region for this machine is shown in figure 4, for the case of axial inflection. At the expected intensities, the use of electrostatic elements should be avoided. The system shown uses magnetic elements and a large gap, weak field (20 kV/cm) electrostatic inflector. The fringe field of the the inflection magnet will affect the first few orbits and corrections are made with a compensating magnet in the opposite valley. Further studies need to be made to improve upon this inflection system. The flattop resonators are not shown in figure 4, and cannot be made to interact with the accelerated beam until about the sixth or seventh turn if they are mounted in the free valleys. If mounted in the accelerating valleys, they can be made to interact with the first turn. Figure 5 is a schematic representation of the possible

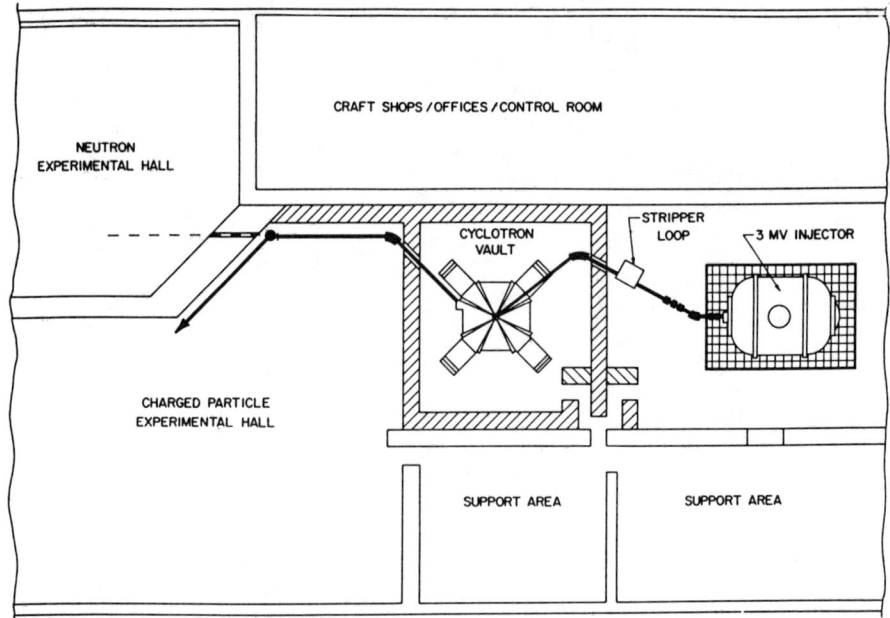

Figure 5. Layout of accelerator facility and experimental areas

orientation of the accelerators in the facility. The location of the stripper loop in the transfer beam line is also indicated. While no scale is given, the relative sizes of the ion source and cyclotron are correct.

SUMMARY

The tools needed to construct a high intensity cyclotron for the production of an intense, high-resolution neutron beam for reaction experiments are available. Because of the nature of the problems encountered with high intensity charged-particle beams and the variety of the type and length of the reaction experiments, extra money and effort spent in the construction of the facility which simplify its operation will be repaid by a more efficient and productive research program. If the scientific justification warrants the construction of such a facility, several design options need to be studied to minimize its cost. Whether it is necessary or more economical to build a multi-stage cyclotron system is still an open question, although from an operational standpoint, the simpler configuration is best. It may not be possible, however, to arrive at a single magnet design that will allow the acceleration of beams from 3 to 200 MeV over an intensity range from 1 uA to 1 mA. The question of whether a linear accelerator would be a suitable and more economical high intensity accelerator also needs to be studied carefully. Finally, in view of the difficulties to be overcome in the construction and operation of such an accelerator, the question of whether or not a high intensity cyclotron is required at all to perform the majority of interesting neutron-induced reaction physics experiments must be resolved. If beam intensities ranging from 1 to 10 uA are all that are required, then several facilities already exist which can, with modest improvements, be made to provide the necessary beams. The need is then reduced to constructing the necessary experimental apparatus at these facilities to perform the desired experiments. It would perhaps be useful to assemble several representative accelerator and neutron physicists for a summer study to determine the criteria for a neutron source suitable to answer the interesting intermediate energy physics questions, and the best means to achieve it. While the need for an intense neutron source is well established, its parameters (energy and intensity range) do not appear to be as well established.

REFERENCES

1. J. Ferme', and M. Gouttefangeas, Proc. Ninth Intl. Cyc. Conf, (les editions de physics, Les Ulis Cedex, France, 1981), p. 3.
2. P. Miller, Proc. Tenth Intl. Cyc. Conf, East Lansing, Mich, to be published, (1984).
3. J. Rapaport, "The Interaction Between Medium Energy Nucleons in Nuclei", AIP Conf. Proc. #97, ed. H. O. Meyer, (1983) p. 365.
4. R. E. Pollock, IEEE Trans. Nucl. Sci, NS-26, 1965 (1978).
5. U. Schryber, Proc.Tenth Intl. Cyc. Conf, East Lansing, Mich, to be published, (1984).

6. A. H. Botha, Proc. Tenth Intl. Cyc. Conf, East Lansing, Mich, to be published, (1984).
7. R. Baartman et al., Proc. Tenth Intl. Cyc. Conf, East Lansing, Mich to be published, (1984).
8. U. Schryber, Pron. Ninth Intl. Cyc. Conf, (les editions de physics, Les Ulis Cedex, France, 1981), p. 43.
9. M. M. Gordon, Annals of Physics, Vol. 50, No. 3, 571 (1968).
10. W. Joho, Proc. Ninth Intl. Cyc. Conf, (les editions de physics, Les Ulis Celex, France, 1981), p. 337.
11. Ch. Markovits, Proc. Ninth Intl. Cyc. Conf, (les editions de physics, Les Ulis Cedex, France, 1981), p. 525.
12. F. Sacherer and T. Sherwood, IEEE Trans. Nucl. Sci, NS-18, 1066 (1971).
13. M.M. Gordon, Proc. Fifth Intl. Cyc. Conf, (Oxford, 1969), p. 305.
14. M.M. Gordon, Particle Accelerators, Vol. 2, 203 (1971).
15. B. Bischof, IEEE Trans. Nucl. Sci, NS-26, No. 2, 2186 (1979).
16. G. Reinhold, Emil Haefly & Co. Ltd, post box 4028, Basil Switzerland.
17. D.L. Friesel, R.E. Pollock, T.E. Ellison and W.P. Jones, Proc. Tenth Intl. Cyc. Conf, East Lansing, Mich, to be published (1984)
18. M. Olivo, W. Joho and U. Schryber, IEEE Trans. Nucl. Sci, NS-26, No. 3, 3980 (1979).
19. D. J. Clark, Proc. Ninth Intl. Cyc. Conf, (les editions de physics, Les Ulis Celex, France, 1981), p. 231.
20. P.A. Smith, Nucl. Instr. & Meth. 166, 229 (1979).

A BEAM SWINGER FOR NEUTRON SCATTERING

C. D. Goodman
Indiana University, Bloomington, IN 47405

ABSTRACT

A conceptual design is presented for a system that provides for varying the angle of incidence of neutron flux on a scattering target. This makes possible the use of long flight paths for time-of-flight spectroscopy of scattered neutrons. The system is also applicable to measurements of neutron induced charged particle reactions. Special emphasis is placed on the energy range of 50-200 MeV.

INTRODUCTION

I have been asked to sketch out a conceptual design for a beam swinger system that could be used for neutron scattering experiments in the energy range of 50-200 MeV. I will begin by describing the individual elements required for a system for making neutron scattering measurements. Then I will discuss the possible layouts for combining the elements into an arrangement that will be convenient for measuring angular distributions of elastically or inelastically neutrons. The components of the system are a neutron source, a sweeping magnet to deflect the charged particle beam into a dump, the scattering target, a flight path, and detectors.

NEUTRON SOURCE

Let us assume that the neutrons are to be generated from a cyclotron such as the Indiana Cyclotron. The choice of source reaction depends on the desired neutron energy, the energy resolution, whether polarized neutrons are needed, and considerations of safety, complexity, and cost of the target. If one desires neutrons up to 200 MeV, one must use the (p,n) reaction at zero degrees as the source reaction. Using other source reactions in the Indiana Cyclotron would produce much lower energy neutrons. For the (p,n) reaction the possible targets are tritium, ^{6}Li, and ^{7}Li. Each of these targets has its own advantages and problems. Briefly, tritium has the largest production cross section but requires containment of the radioactive material in a high pressure gas cell, a cryogenic liquid cell, or as a metal hydride. Because of the biological hazard, a secondary containment is required in case of rupture of the primary containment.

The target technology is relatively simple for lithium, but the neutron spectrum from ^{7}Li(p,n)^{7}Be contains two lines of about equal intensity separated by about 0.47 MeV. The ^{6}Li spectrum, on the other hand, is dominated by a single line, but the production cross section is smaller.

Polarization is another consideration for source reactions. With the (p,n) reaction one can use polarized protons and rely on polarization transfer to create the polarized neutrons. Ideally one would like a 0^+ to 0^+ analog state transition for the source reaction. In that case the neutrons from the reaction would have the same polarization as the incident protons. Unfortunately, no target nucleus exists that provides a 0^+ to 0^+ transition energetically well separated from other states. The (p,n) reactions on tritium and ^{7}Li produce the neutrons through a sum of Gamow-Teller and Fermi interactions, and the polarization of the neutron beam is diluted.

We can estimate the expected polarization for a mixed F and GT transition from some simple considerations. The spin distribution of the neutrons produced in the reaction is the sum of the separate distributions for the Fermi and the Gamow-Teller components. Since the Fermi interaction changes only the isospin, the product neutron has the same spin as the incident proton, so the resultant spin distribution of the neutrons is the same as that of the protons. For the GT part, the spin flip probability is 2/3, so a beam of protons of 100% spin up would result in neutrons with a spin up intensity of 1/3 and a spin down intensity of 2/3. We need now to consider what the relative weightings of the two components will be.

The following relationship is useful for estimating the relative Fermi and Gamow-Teller components of the cross sections:[1]

$$\frac{d\sigma}{d\omega}(GT) / \frac{d\sigma}{d\omega}(F) = \left(\frac{55}{E_p}\right)^2 \frac{B(GT)}{B(F)} \qquad (1)$$

For the ground state transitions in tritium and ^{7}Li $B(F) = 1$. For tritium $B(GT) = 2.7$. For ^{7}Li(g.s.) $B(GT) = 1.25$ and for the 0.47 MeV state $B(GT) = 1.11$. For ^{6}Li(g.s.) $B(GT) = 1.62$. Using these values one can see that the polarization of the neutrons from a ^{7}Li target would be zero at about 49 MeV even if the protons were 100% polarized. The effect of the mixture of Fermi and Gamow-Teller components is to depolarize the neutrons strongly over much of the energy range under discussion. The neutrons produced in the ^{6}Li(p,n) reaction, on the other hand, come from a pure Gamow-Teller transition, and the polarization is diluted only by the spin flip probability distribution. We might also note for estimation purposes that the ^{7}Li(p,n) cross section summed over the two states is nearly constant over the 100-200 MeV energy range with a value of about 35 mb/sr(lab).

When the energy of the cyclotron is limited by the magnetic field and radius, the maximum deuteron energy is about half the maximum proton energy. Thus, with the Indiana cyclotron for neutrons below 100 MeV, the D(d,n)^{3}He reaction can be useful. At low energies the T(d,n)^{4}He is very clean because of the large positive Q value. The neutron facility at the Crocker Laboratory in Davis, California uses the ^{7}Li(p,n) reaction and the Edwards Accelerator Laboratory at Ohio University uses the T(d,n) reaction.

SWEEPER MAGNET

The purpose of the sweeper magnet is to direct the proton beam into a well shielded beam stop after it has passed through the production target. Although most of the beam will emerge from the target as protons, a small fraction of the beam will emerge as neutral hydrogen atoms that will not be deflected by the sweeping magnet. To remove these, a thin, charge-equilibrating foil and a second magnet should be placed between the production target and the scattering target.[2] It might be sufficient to use the fringing field of the sweeping magnet for this purpose if it extends far enough along the line to the scattering target. Since the neutrals are of low intensity, it is not necessary to bend them all the way into a shielded beam dump. It is sufficient to deflect them only far enough to miss the target.

For polarized experiments a magnet between the neutron production target and the scattering target can be useful for precessing the spin orientation of the neutrons. The use of a precessor magnet may be necessary, if the sweeper magnet does not bend in the scattering plane, to correct for the precession introduced by the sweeper magnet. For reasons of geometric convenience it may be desirable to make the sweeping bend perpendicular to the scattering plane, so this precession consideration should be kept in mind. If a precession magnet is part of the system, it can also be used to eliminate the neutrals.

SCATTERING TARGET

The next essential item downstream of the sweeper magnet is the scattering target. In charged particle experiments one usually supports the target in a relatively massive frame. Then the beam must be collimated well enough not to hit the target. Such collimation is difficult for neutrons, but is also undesirable. Rather one can take advantage of the relatively uniform flux that one gets from an uncollimated source. Then one must suspend the target on a mount with very little mass to minimize the background from the target frame. With this approach one not only eliminates the need for collimation in the the cramped space between the neutron source and the scattering target, but one also eliminates the need for a target of uniform thickness.

The quantity of interest in determining a cross section is the integral over the target area of the product of the number of nuclei per unit area and the neutron flux. If the neutron flux is uniform over the target area, one needs only to know the total mass of the target. This is in contrast to usual charged particle experiments where the beam flux is not at all uniform and one tries to make the target thickness as uniform as possible.

FLIGHT PATH

The next essential element in the system is the device by which we detect the neutron and measure its energy. Time of flight is a very effective means of measuring neutron energy in the 50-200 MeV range. That technique requires a flight path perhaps as long as 200 meters, and for that reason we consider beam swingers. Obviously a 200 meter flight path is practical only if it remains fixed. Once we accept the condition that the flight path remains fixed we can also take advantage of the situation to place massive, accurately aligned collimation in the flight path. Thus we can limit the spatial region that the detector sees to the target itself and a small area immediately around it. The effectiveness of the collimation determines how close to zero degrees we can go for the measurements. The limit is the angle at which the production target becomes visible to the detector.

DETECTOR

The detectors for (n,n) can be of the same kind currently being used for (p,n) at IUCF. These detectors, described elsewhere,[3,4] are meter long bars of plastic scintillator with photomultiplier tubes at both ends. These are most advantageously used in a configuration parallel to the flight path. The difference in time of the signals from the two ends provides a measure of the position in the detector where the scintillation originated. This information allows one to compensate for the time spread introduced by the large physical size of the detector. Phase compensation to correct for phase drifts of the proton bunches with respect to the cyclotron RF is also required. A phase reference signal can be derived from a non-intercepting beam pickoff or from a scattered proton detector.[5]

THE BEAM SWINGER

Beam swingers for (p,n) work are reviewed in reference 6. These are designed to change the angle of incidence of the charged particle beam on the scattering target. There are essentially two styles. In the Colorado design, that was adopted for the design of the Michigan State beam swinger that has recently been moved to Ohio University, the swinger magnets move mechanically around the target. In the ORNL-IUCF[5] design the magnets do not move physically and the beam is redirected by changing the magnetic fields. The Colorado-MSU-OU style beam swinger has been used for (n,n) work at fairly low energies. To use that style swinger for (n,n) one must put the production target and the proton beam stop inside the swinger, and these must move with the swinger. This design becomes very problematic above, say, 60 MeV because the neutron production in the beam stop becomes very intense and it becomes a stronger source than the source target. If the beam is stopped in ^{12}C, one can take advantage of the Q-value difference between ^{12}C(p,n) and ^{7}Li(p,n) to distinguish between beam stop neutrons and source neutrons, but in the higher energy range the difference isn't a large enough fraction of the total energy to be a much value.

The ORNL-IUCF design is also not well suited to (n,n) experiments. Again the neutron source and the beam stop have to be put inside the swinger. Since there is no vertical motion, mechanical support of shielding is simplified, but the beam stop would have to be placed very close to the scattering target, and the problems mentioned above apply again.

The arrangement I propose instead is to use a 90° vertical bend downwards for sweeping the proton beam into the dump as shown in fig. 1. The production target and sweeper magnet as a unit would then pivot around the dump, and the scattering target would be on the pivot axis. Pivoting the production target around the scattering target would change the angle of incidence of the neutrons on the scattering target.

The remaining problem is to direct the proton beam to the movable production target. This can be accomplished with a two magnet system related to the system used in the ORNL-IUCF swinger design. I would propose, however, to reduce the amount of iron required by making the second magnet move linearly along a track as shown in fig. 2. The first magnet bends the beam away from the original beam line. The second magnet directs the beam to the target. With two magnets the direction and the position of the beam at the production target can be set properly.

USES OTHER THAN (n,n)

The beam swinger arrangement provides a neutron flux of variable incidence angle on a scattering target. This can be a valuable resource for use with (n,p), or for that matter with any neutron induced reaction study. One of the problems with studying (n,p) with scintillation counters or with solid state detectors is that these would have to be placed close to the scattering target where the neutron flux is high. It is very desirable to use a magnetic spectrometer for measuring the proton energy. If the beam swinger is already available, it is not necessary to make the magnetic spectrometer movable. The swinger also opens up the possibility of transporting the scattered protons far from the target to a detector through a series of quadrupoles on a beam line. This would alleviate the background problem.

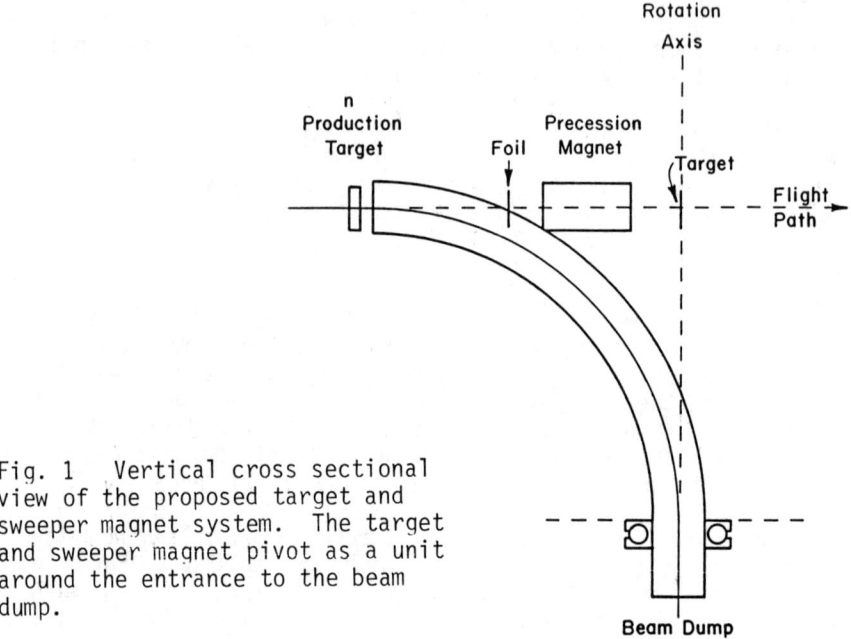

Fig. 1 Vertical cross sectional view of the proposed target and sweeper magnet system. The target and sweeper magnet pivot as a unit around the entrance to the beam dump.

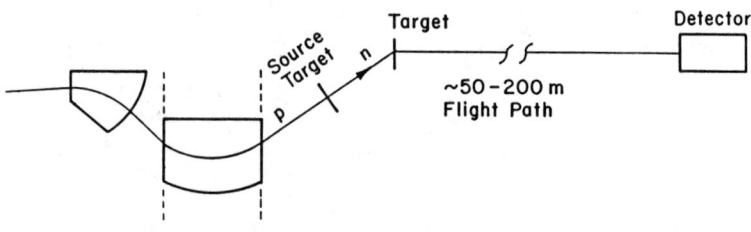

Fig. 2 Plan view of the swinger system. The second magnet in the swinger moves horizontally to accommodate the desired beam path to the target.

REFERENCES

1. T. N. Taddeucci, J. Rapaport, D. E. Bainum, C. D. Goodman, C. C. Foster, C. Gaarde, J. Larsen, C. A. Goulding, D. J. Horen, T. Masterson, and E. Sugarbaker, Phys. Rev. C 25, 1094 (1982); T. N. Taddeucci, in The Interaction Between Medium Energy Nucleons in Nuclei-1982, edited by H.O. Meyer, (AIP, New York, 1983), p. 228.

2. F. P. Brady, "Remarks on (n,p) Facilities," in these proceedings.

3. C. D. Goodman, J. Rapaport, D. E. Bainum, and D. E. Brient, Nucl. Instr. and Methods 151, 125 (1978).

4. C. D. Goodman, J. Rapaport, D. E. Bainum, M. B. Greenfield, and C. A. Goulding, IEEE Trans. Nucl. Sci. NS-25, 577 (1978).

5. C. D. Goodman, C. C. Foster, M. B. Greenfield, C. A. Goulding, D. A. Lind, and J. Rapaport, IEEE Trans. Nucl. Sci. NS-26, 2248, (1979).

6. C. D. Zafiratos, in The (p,n) Reaction and the Nucleon-Nucleon Force, edited by C. D. Goodman, S. M. Austin, S. D. Bloom, J. Rapaport, and G. R. Satchler, (Plenum, New York, 1980) p. 313.

7. D. A. Lind, R. F. Bentley, J. D. Carlson, S. D. Schery, and C. D. Zafiratos, Nucl. Instr. and Methods 130, 93 (1976).

FACILITIES FOR NEUTRON INDUCED REACTIONS

F. Paul Brady
Department of Physics and Crocker Nuclear Laboratory
University of California, Davis, California 95616

ABSTRACT

Facilities at Crocker Nuclear Laboratory for neutron induced reactions, e.g. (n,px) and (n,n'x), and our experiences in using them are described, along with future plans for their development and use. Comments on possibilities at higher energies are also included.

INTRODUCTION

Sector focusing cyclotrons have made possible intense beams (~ 100μA) of medium energy protons and deuterons in small time bunches which are well suited for the production of nearly "monoenergetic" neutron beams. In the past, neutron beams of medium energy ($25 \lesssim E \lesssim 100$ MeV) have been used largely for nucleon-nucleon experiments, and in general this has been the case at Crocker Nuclear Laboratory (CNL). However some years ago we began to explore the use of neutrons as a probe of the nucleus. In particular, we have measured the (n,p) reaction at E ~ 60 MeV for a range of nuclei, and considerable structure has been uncovered and identified.[1,2]

Here we describe the facilities and techniques we have used for these studies and our experiences in using them. We also describe new facilities we are developing for neutron induced reactions for energies up to 65 MeV. Some representative (n,px) data and preliminary (n,n'x) data are shown. Comments on possible future developments (at CNL and elsewhere) and on the possibilities at higher energies (near 200 MeV) are included.

FACILITY GENERAL DESCRIPTION

Fig. 1 shows the experimental layout of the cyclotron. With this arrangement the neutron source is located in the main vault and neutrons associated with the accelerator and the production are confined principally to the vault area whose walls are ~ 3m thick except for the lentil cut on the side in the neutron cave. The wall between the vault and the neutron experimental area is formed from 305 mm of steel and over the shortest distance 1.2 m of concrete which gives an attenuation perpendicular to the wall of ~ 10^4 for a neutron of 40 MeV. Additional shielding is placed around the Faraday cup (FC). The neutron beam is "dumped" in a shielded area (not shown) against the back wall of the neutron cave.

The shielding wall at top and bottom consists of very large blocks 3 m thick extending above and below the lentil opening containing the scattering chamber. The next set of large blocks (next to the top and bottom ones) have the lentil cut in them, and

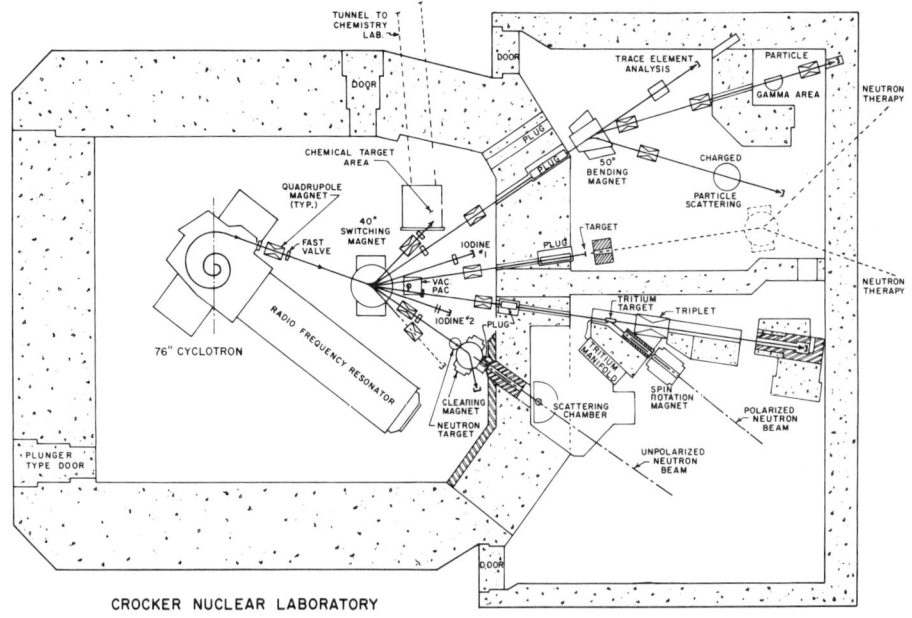

Fig. 1 CNL Cyclotron and Experimental Areas

these blocks sandwich the steel collimator assembly and smaller concrete blocks which surround the collimator assembly and completely fill the gap. This arrangement allows some flexibility in changing the shielding at and near beam level. The blocks can be manipulated by hand and we have replaced some of them with lead bricks. However it is our opinion that greater flexibility would be achieved by having the large blocks replaced by intermediate-sized blocks which can be more easily manipulated. This would allow greater changes in the experimental arrangement and greater flexibility in shielding than is possible now. One wants to optimize count rate and shielding according to the needs of each experiment.

The optimum arrangement (given more space) would be to have the neutron production take place, not in the cyclotron vault, but in a separate shielded area with flexible shielding arrangements between the production and experimental areas. This also has the added advantage of allowing one to enter the production area when the cyclotron is running other beams, and it also reduces the general background in that area for working there (e.g. changing or installing targets).

Fig. 2 shows details of the neutron production set-up. The proton beam is focused and centered on the ^{7}Li (99.8% ^{7}Li) target by quadrupole and steering magnets. An adjustable 4-jaw insulated carbon collimator is located just upstream from the ^{7}Li target. Current on the jaws is monitored and minimized and assists the operator in centering the beam on the target. This collimator is adjusted to be slightly larger than the beam spot on the ^{7}Li target so that the transverse location of the neutron source is fixed. The beam pick-off which provides a reference time signal is farther upstream from the 4-jaw. It is 200 mm in length and 25 mm inside diameter and is protected by a carbon collimator. Since carbon has a Q = -18.2 MeV it does not contribute neutrons to the high energy peak of the ^{7}Li(p,n)^{7}Be reaction.

^{7}Li (and other targets) of various thicknesses are installed in the target wheel which can be remotely positioned. The wheel is water or "antifreeze" cooled and oscillated so proton beams up to 25μA can be used without target melting or deforming. Ref. 3 gives more details of the target wheel and the facility.

After passing through the (^{7}Li) target the proton beam is cleared by a magnet into the FC. Neutrons produced at and near 0° are formed into a beam by brass or iron inserts in a steel collimator unit 1.55 m long. The neutron collimator points back to look at a shielding block behind the magnet of the cyclotron. This block should be a relatively cool area so that we expect the neutrons produced in the cyclotron to contribute very little to the spectrum coming out of the collimator.

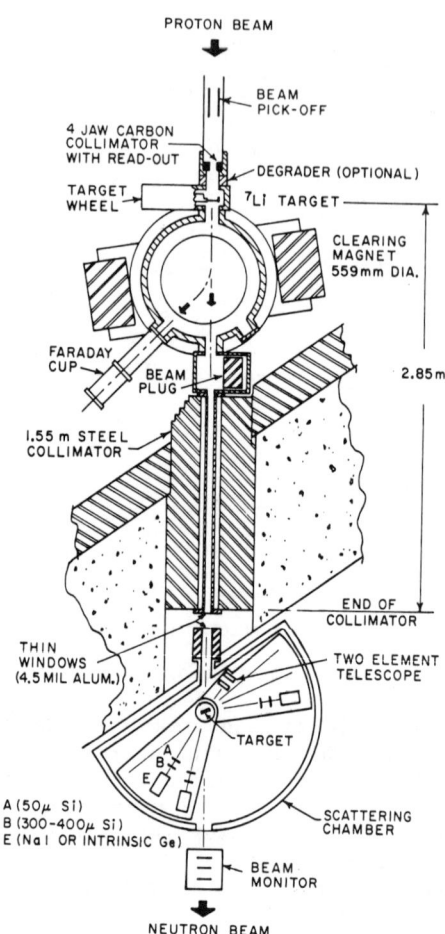

Fig. 2   Neutron production setup.

One of the drawbacks of our present arrangement is the proximity of the FC to the experimental area. It would be better to transport the proton beam to a more distant and well shielded area. In the present setup we surround as much as possible the FC with lead to reduce the gamma rays. Beam extraction efficiency from the cyclotron is large ~ 90% so the cyclotron per se should not be a large source of background neutrons and gamma rays. Gamma rays and neutrons can produce pile-up in charged particle detectors and where total energy detectors are used worsen the resolution as well as produce backgrounds at lower energies. Local shielding of the detectors as well as the sources is the most economical way of reducing backgrounds in the fairly large volume NaI and plastic scintillators that we use (250cc and 800cc respectively). Presently we are contemplating a more distant beam dump, but that might require a quadrupole as well as more bend of the protons in the magnetic field.

NEUTRON COLLIMATION

Fig. 2 shows the position of the collimator unit in the shielding wall. Small concrete and lead blocks are packed in around the steel collimator assembly. On the neutron cave side of the wall 5-10 cm of $(CH_2)_n$ followed by 20-30 cm of Pb is usually used.

Fig. 3 shows a cross section of a idealized collimator very similar to what we use. The x direction scale is greatly expanded relative to the y. In our case the proton beam spot is ~ 3 mm in diameter and its distance from the entrance of the collimator is ~ 1.3 m. This space is filled with the proton clearing magnet (not shown) whose vertical aperture is ~ 100 mm.

The neutron beam is defined spatially by a brass insert which slips into the steel collimator assembly. The insert opening is straight over the first half and is tapered to a wider opening over the last part. The insert is designed so that the collimating is done by the "throat" about half way down the insert and so the sides of the tapered section are not illuminated by neutrons from any part of the source target. With this geometry the tapered sides and exit of the collimator are not secondary sources of neutrons.

Of course there is a penumbra of neutrons from the straight section walls of the insert, but this is confined to a fairly narrow cone and need not hit the beam pipe and scattering chambers entrance walls if the latter are large enough. It is not always possible to have the (n,z) target frame (which is about 1 m beyond the end of the collimator) out of the neutron beam penumbra. However using multi-wire chambers for trajectory analysis one can usually separate events due to the target frame.

The borated paraffin is to thermalize and absorb neutrons whose energies are too low to undergo inelastic collisions. The lead is to attenuate gamma rays produced by neutron inelastic scattering and capture.

POLKA, the polarized neutron facility[4] at Kernforschungs-zentrum Karlsruhe (KfK) has a versatile collimator arrangement. A number of short annular rings are inserted into a tube. It is relatively easy to change the size and shape of the collimator aperture, and to use rings of various materials for example $(CH_2)_n$ to attenuate low energy neutrons. The whole shielding arrangement at POLKA is well designed and the result is a very low neutron and gamma background.

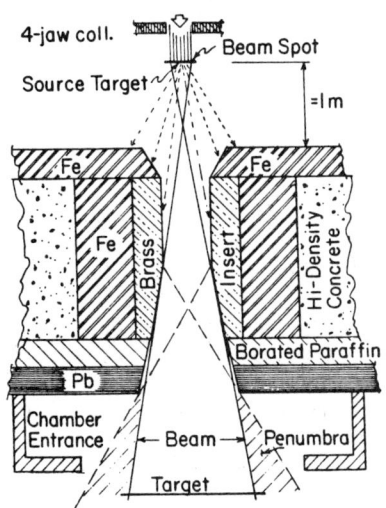

Fig. 3 An idealized n collimator

NEUTRON BEAM PRODUCTION

The 0° ^{7}Li(p,n)^{7}Be reaction is normally used to produce the CNL unpolarized neutron beam. (A polarized beam of 50 MeV, produced by the d+T → n + ^{4}He reaction at 30° lab, has also been used at CNL[5] (see fig. 1), mainly for few nucleon experiments.) The 0° cross section varies slowly with proton energy over the range of 40 to 70 MeV. The value (~ 33 mb/sr lab) includes that to the excited state at 0.43 MeV in ^{7}Be which constitutes around 30% of the total. Clearly this limits how narrow the neutron beam can be in energy. Typically a ^{7}Li target thickness of ~ 600 keV is used.

Other possible reactions have been studied.[6,8] ^{2}H(p,n)^{2}He cannot provide a high resolution beam, but ^{3}H(p,n)^{3}He does and has a substantial cross section (plus high nuclear density per MeV of energy loss).

Table I gives some cross sections measured at 50 MeV. These values have been extrapolated to higher energy (~ 200 MeV) using the distorted wave (DW) impulse approximation[9], β decay and EC matrix elements (measured or estimated) and distortion factors from DW to plane wave cross section ratios. Neutron intensities at 50 and 200 MeV are also estimated in Table I.

Table I. Peak lab cross sections and corresponding neutron intensities for (p,n) on the targets indicated at 50 and 200 MeV.

Target	σ(0°) (mb/sr)		Neutrons/s-μA-msr-MeV		Intrinsic Width (MeV)
	50	200 MeV	50	200 MeV	
^{2}H	30	37.5	4.3×10^6	1.5×10^7	~ 1.5
^{3}H	44	50	6.0	1.9	~ 0
^{7}Li	33	46	1.8	0.64	~ 0.5

It can be seen from Table I that ^{3}H gives almost three times the intensity that ^{7}Li (p,n) does for a given energy loss (beam energy width) and in addition has a very small intrinsic width so that neutron beams of a few hundred kilovolts are possible.

The d+T reaction is another possible source, particularly for polarized neutrons (due to polarization transfer from polarized deuterium beams) and for inelastic scattering measurements using TOF, where a large energy spacing between the d+T peak and the continuum is desirable. The neutron intensity per MeV is less for d+T than p+T due mainly to the larger energy loss of deuterons.

DETECTION SYSTEMS

(n,zx)

For studying (n,zx) reactions we used, to begin with, $\Delta E \cdot E$ telescopes in a scattering chamber as shown in fig. 2. (For cases where we wanted to detect all Hydrogen and Helium isotopes down to a threshhold of a few MeV, we used triple element telescopes with the first element being a thin (50 μm) Si(SB) detector.[10]) Even with 6 telescopes the data runs were long and statistics limited. Thus we turned to the use of multiwire chambers (MWC) combined with large area ΔE and E detectors, which allowed us to cover the forward angles down to ~ 5° degrees.

Fig. 4 shows the MWC system.[11] The MWC determine the trajectories of the charged particles coming from the target direction. This reduces backgrounds which comes from other directions. In addition the MWC allows one to map the response of both the large area ΔE and E detectors. The corrected response of the ΔE is very uniform and the corrected response of the E detectors allow one to achieve E resolutions of 1% for large area NaI and plastic scintillators. The plastic detector resolution, which includes some contribution from the proton beam energy width, is ~ 1.2% and the NaI ~ 0.9%.

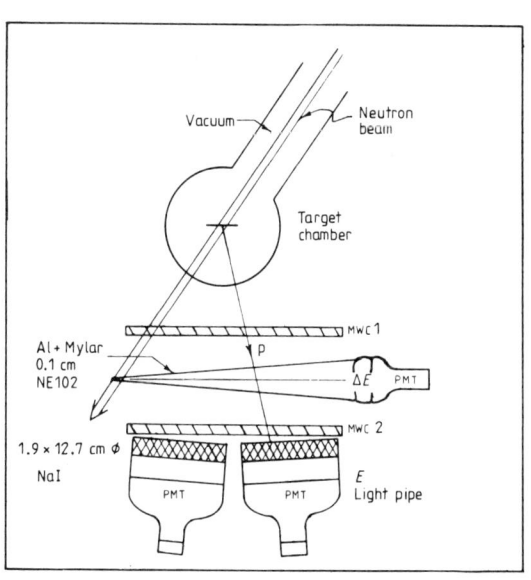

Fig. 4 MWC system for (n,z) measurements.

Energy resolution in (n,zx) reactions depends on a number of factors. To begin with one needs to use lots of proton beam to produce the neutrons via (p,n) and these beams at CNL have typically ΔE_p ~ 250 keV for 15μA at 65 MeV. (With phase slits near the center of the cyclotron one can improve this.) Then one needs a (p,n) neutron production target of finite thickness to produce adequate neutron intensity. A ^{7}Li target thickness producing ΔE_n ~ 600 keV essentially masks the effect of transitions to the 0.43 MeV state of ^{7}Be. Since the E detector resolutions are ~ 600 keV we usually choose a scattering target which gives rise to an energy loss dispersion of ~ 600 keV at 60 MeV. Then the overall resolution of the system is near 1 MeV.

Not shown in Fig. 2 is a remotely operated stripping foil in the middle of the clearing magnet. A small fraction of the protons which pass through the ^{7}Li (or other) target pick up electrons becoming neutral H^o's which can pass through the clearing field and down the neutron collimator (which is evacuated) to the target chamber. These H^o protons are useful for checking timing and energy resolutions of the system. This can be done with the full H$^+$ beam on a thin Ta target in the target wheel so that tests are made at full beam. We made some measurements of the ratio H^o to H$^+$ for several materials at 42.5 and 61.8 MeV.[12] These are shown in Fig. 5.

The first reaction we studied was ^{12}C(n,p)^{12}B whose cross section was needed for background subtraction in few nucleon studies with (CH$_2$)$_n$ and (CD$_2$)$_n$ targets. Fig. 6 shows a ^{12}C (n,p) spectrum taken with ΔE·E telescopes, and compared to ^{12}C (p,p'). This illustrates several of the features of the (n,p) reaction. The final ground state of the (n,p) reaction is the analog of the first T+1 state in the target. The analogs of T+1 target states have lower excitation in the residual nucleus by ΔE (Coul.) - Q$_{np}$.

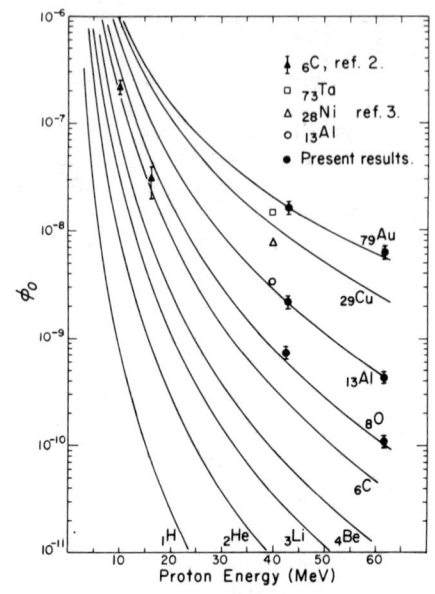

Fig. 5 Ratio of H^o to H$^+$ vs H$^+$ energy.

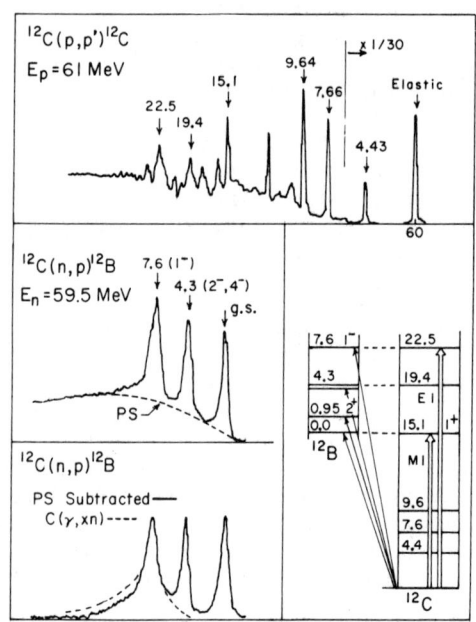

Fig. 6 ^{12}C(p,p') compared to ^{12}C(n,p).

Fig. 7 shows $^{12}C(n,p)^{12}B$ spectra taken with the MWC system. This is the data from E1, about half the data obtained in a run. The "ground state" peak is largely $\ell=0$ to the 1^+ state with some $\ell=2$ to the 2+ at 0.9 MeV state folded in. The next higher structure at 4.4 MeV is largely $\ell=1$ to the 2^- with some $\ell=3$ to the unresolved 4^-. The structure at 7.7 MeV is the dipole and part of the spin dipole resonance made up mostly of 1^- states. The feature illustrated in these spectra is that the angular distributions of these transitions varies slowly at this energy (60 MeV).

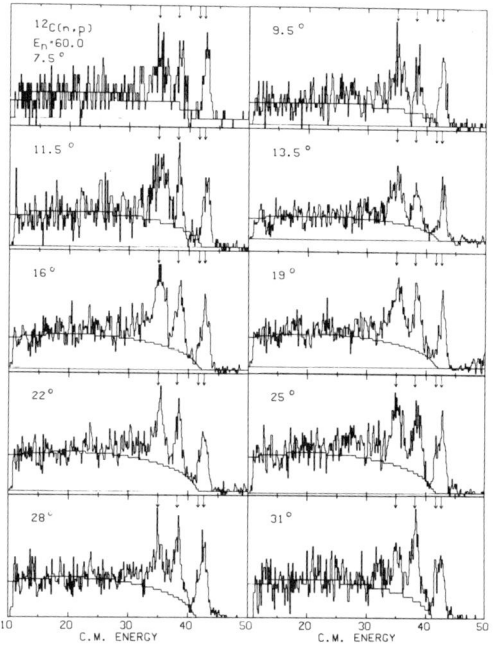

Fig. 7 $^{12}C(n,p)$ from MWC system (E1).

At higher energies (near 200 MeV) the angular distributions are much sharper because of the higher momenta and smaller distortion in the much smaller nucleon (optical model) potential. In addition the physical "background" continuum is largely due to direct one-step processes at 200 MeV[13], while at 60 MeV multistep processes must also be included.[14,15] The net result of these two effects is that the resonance signal to physical background continuum is considerably better at the higher energies.

$^{40}Ca(n,p)^{40}K$ has been measured with the MWC system at 65 MeV. Comparisons with (p,n) at 200 MeV, (π^-,π^0) at 165 MeV, and (h,t) at 197 MeV are being used to separate the dipole (GDR) and spin dipole components.[16]

It is difficult to measure (n,p) reactions at angles near 0° with these systems. The flood of neutrons and gamma rays in and near the neutron beam is too large. Thus we installed a magnet in the neutron beam near the exit of the collimator and detected protons from (n,p) reaction at and near 0° which are bent away from the neutron beam. In these feasibility studies the geometry was not optimum and the energy acceptance of the detector did not extend to the lowest proton (highest excitation) energies. This is always a problem. One wants to cover a large dynamic range of outgoing energies and this results in a large momentum (and hence bending) range. Despite these problems giving limited energy range (now resolved in the final design) we were able to identify what we believe to be the T+1 component of the giant isovector monopole resonance at $-Q_{np} \sim 23$ MeV.[17]

The final design of a detection system to allow (n,px) measurements over a wide range of angle (0-60° lab) and outgoing proton energy (20-65 MeV) is shown in Fig. 8.

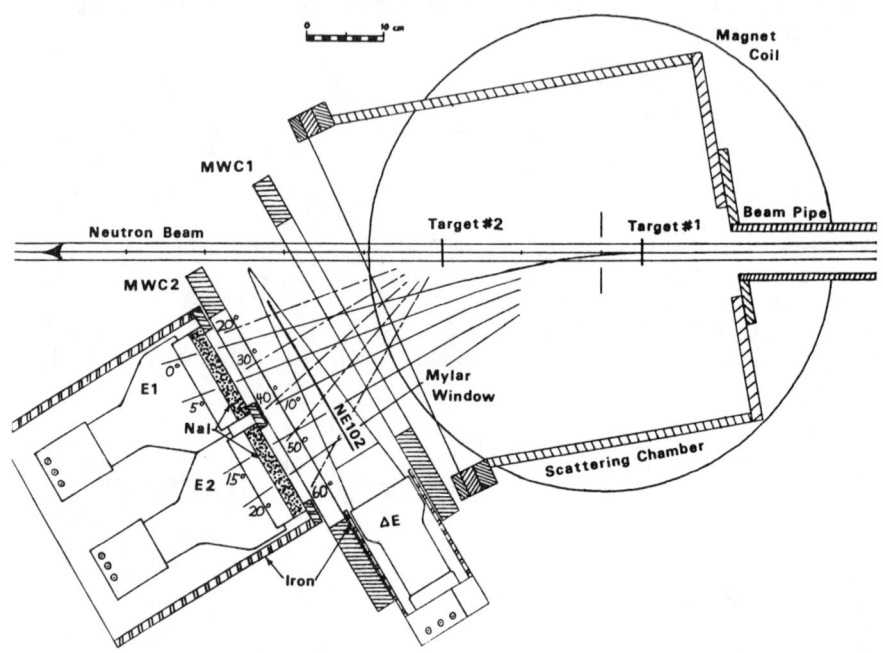

Fig. 8 Two target system for (n,px) measurements at $\theta \geq 0°$.

Charged particles, say protons, from target #1 pass through and are bent by the magnetic field. This allows measurements at and around 0°. Particles from target #2 do not pass through appreciable field and so are not bent. Trajectory analysis allows spectra from the two targets to be unambiguously separated. This system has been constructed and is being installed. An intrinsic Ge detector of relatively large area (55 mm) which can stop 70 MeV protons is being mounted for testing in the facility.

(n,n'x)

We have adapted our MWC system to make measurements of neutron inelastic scattering and continua over a wide dynamic range of outgoing neutron energies.[18] This is accomplished with the addition of a CH_2 (or other H) convertor as shown in Fig. 9.[17] A carbon convertor is used to subtract C events in CH_2.

The method of TOF is well adapted to (p,n) measurements where beams have small intensity and good time and energy spreads. For (n,n'x) spectrum measurements, the required high intensity primary beam (protons for (p,n) neutron production) and the energy width of the neutron production target increase the time dispersion so that longer flight paths and lower count rates result. Of more importance for (n,n'x) measurements is the fact that a neutron beam is not monoenergetic but has a lower energy tail. The TOF system cannot separate the elastic scattering of the beam tail neutrons from the lower energy (n,n'x) reaction neutrons due to the beam peak.

In fig. 9 the Veto is a third MWC. Calibration and normalization are carried out with C and CH_2 targets. The scattered neutron energy resolution is determined mainly by four factors: beam and H convertor energy loss widths, proton scattering angle uncertainty, and proton detector resolution. These combine to produce an overall energy resolution of ~ 2 MeV in the system as we use it now.

Fig. 10 shows preliminary data from a $^{nat}Fe(n,n'x)$ spectrum taken at 65 MeV. We also plan to measure Fe(n,p) so that the dipole and isoscalar quadrupole (E2) transitions can be separated more easily. Then a comparison with Fe(p,p'x) to E2 can be made to obtain the ratio of proton to neutron matrix elements.

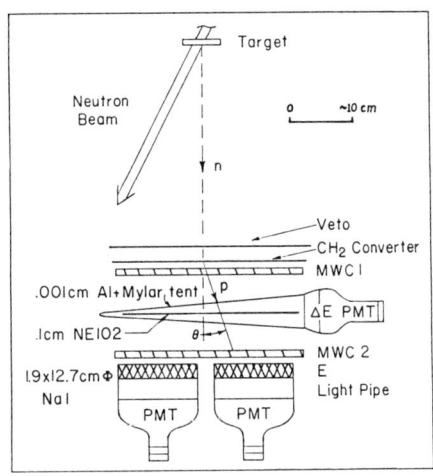

Fig. 9 MWC System for (n,n'x)

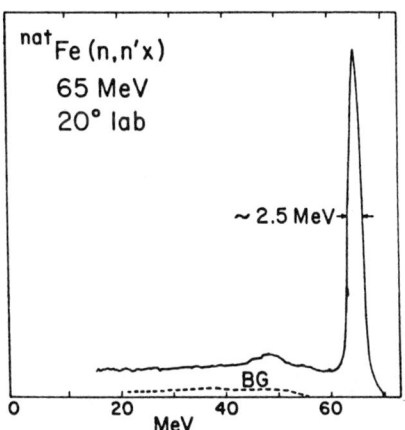

Fig. 10 $^{nat}Fe(n,n'x)$ at 65 MeV

FUTURE POSSIBILITIES

At CNL we are considering the installation of a tritium target system so that higher resolution neutron beams can be achieved with essentially the same intensity. We first have to verify that a high intensity (15μA) proton beam of good resolution (~100 keV) can be obtained. We can do this with the new Ge detector which is arriving soon.

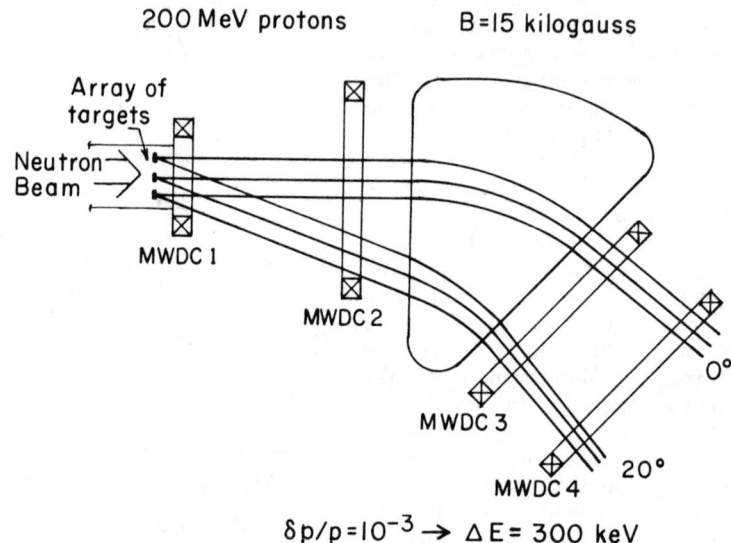

Fig. 11 Magnetic Spectrometer System.

Table I shows that the neutron intensity per MeV increases about a factor of three in going from 60 to 200 MeV. Detecting 200 MeV protons with good energy resolution does require a more elaborate system, such as that shown in Fig. 11. We envision that the neutron beam could have a large area so an array of targets (say 3x3) could be studied at one time. This is very feasible with MWC's, as are resolutions of a few hundred keV at angles down to θ=0°.

The physical advantages of higher energies are well known: spin mode selectivity, much smaller nuclear distortion, much better signal, σ(θ), to multistep physical background continuum. There are also the technical advantages of more neutron intensity per MeV (or smaller energy width in the neutron beam), and higher counting rate per MeV loss in the scattering target. Thus one gains nearly an order of magnitude in count rate (at a given E resolution) in going from 60 to 200 MeV.

This work is supported by the National Science Foundation under grant 84-03020 and the U.C. Davis Research Office.

The author would like to thank the following colleagues whose efforts made possible the facilities and data described here: Carlos Castaneda, Jim Drummond, Tim Ford, Jim Harrison, John Jungerman, Nick King, Bruce McEachern, Greg Needham, Juan Romero, John Ullmann and Mike Webb.

REFERENCES

1. F.P. Brady et al, Proceedings of Int. Conf. on Spin Excitations in Nuclei (Telluride). Edited by Petrovich, Brown, Garvey, Goodman, Lindgren, and Love.
2. F.P. Brady et al, Journal of Physics G: Nuclear Physics, Vol. 10 (1984) 363.
3. J.A. Jungerman and F.P. Brady, Nucl. Instr. and Meths. 89, 167 (1970).
4. H.O. Klages, H. Dobiasch, P. Doll, H. Krupp, M. Oexner, P. Plischke, B. Zeitnitz, F.P. Brady and J.C. Hiebert, Nucl. Instr. and Meths. 219:269 (1984).
5. A.L. Sagle et al, Nucl. Instr. and Meths. 129, 345 (1975).
6. C.J. Batty et al, Nucl. Instr. and Meths. 68, 273.
7. J.A. Jungerman et al, Nucl. Insts. and Meths. 94, 421 (1971).
8. J.L. Romero et al, Nucl. Insts. and Meths. 134, 567 (1976).
9. F. Petrovich, W.G. Love, and R.J. McCarthy, Phys. Rev. C 21, 1718 (1980).
10. T.S. Subramanian, J.L. Romero and F.P. Brady, Nucl. Instr. and Meths. 174, 475 (1980).
11. T.D. Ford et al, UCD-CNL 232 and Nucl. Insts. and Meths. (to be published).
12. J.L. Romero et al, Nucl. Insts. and Meths. 171, 609 (1980).
13. F. Osterfeld private communication and to be published.
14. S. Grimes, paper at this conference.
15. H. Feshbach, paper at this conference.
16. C.M. Castaneda et al, paper at this conference.
17. T.D. Ford et al, paper at this conference.
18. F.P. Brady et al, UCD-CNL 245 and Nucl. Insts. and Meths. (to be published).

NEUTRON SPIN TRANSFER MEASUREMENTS: WHY AND HOW

Terry N. Taddeucci
Ohio University, Athens, Ohio 45701
and
Indiana University Cyclotron Facility, Bloomington, Indiana 47405

ABSTRACT

The feasibility of measuring spin transfer observables is considered for nucleon-nucleus reactions involving a neutron in either the initial or final state. Measurements of transverse spin transfer for (p,n) reactions at 160 MeV are used as examples.

The information to be gained from the measurement of spin transfer observables has been stressed by several speakers at this conference.[1] Such measurements can provide unique information about the effective nucleon-nucleon interaction, the relativistic description of the optical potential, and various aspects of nuclear structure.

There are three nucleon-nucleus reactions that involve a neutron in either the initial or final state: (p,n), (n,p), and (n,n). Of these three, spin transfer data presently exist only for the (p,n) reaction. I will thus use the (p,n) reaction as an example, and will make an estimate of the feasibility for obtaining similar data for (n,p) and (n,n) reactions. Measurement of the full complement of spin transfer observables requires the use of beams polarized in three orthogonal directions, and measurement of the polarization of the outgoing nucleon in three directions for a range of scattering angles. For the purposes of this talk I will restrict the discussion to transverse polarization and a scattering angle of 0°. This restriction not only simplifies the discussion, but also applies to virtually all existing (p,n) spin transfer data.

The relationship between transverse spin observables is given by

$$(1 + p_i A)p_f = P + p_i D_{NN}, \qquad (1)$$

where p_i (p_f) is the incident (outgoing) projectile polarization, A is the analyzing power, P is the polarization function, and D_{NN} is the transverse spin transfer coefficient. At 0°, A=P=0, so Eq. (1) simplifies to

$$D_{NN}(0°) = p_f/p_i. \qquad (2)$$

The transverse spin flip probability S_{NN} is obtained from $S_{NN} = (1-D_{NN})/2$. Some simple estimates can be made for expected values of S_{NN} and D_{NN} by assuming a unique L-transfer and a central interaction.[2] These values are given in Table I. Improved

Table I. Values expected for transverse spin transfer observables for transitions of pure multipolarity (unique ΔL) and a central interaction.

ΔJ^π	D_{NN}	S_{NN}
1^+	$-1/3$	$2/3$
0^+	1	0
0^-	-1	1
1^- ($\Delta S=0$)	1	0
1^- ($\Delta S=1$)	0	$1/2$
2^-	$-2/5$	$7/10$

estimates that include the effects of noncentral interactions have been made,[3] and of course, full DWBA or DWIA calculations employing realistic effective interactions and optical potentials will give the best predictions. Despite the restrictive assumptions upon which it is based, Table I is still a useful guide. It should be noted that the values given for 0^+ and 0^- reactions are exact and are not approximations. Measurements[4] of $D_{NN}(0°)$ for Gamow-Teller type ($\Delta J^\pi = 1^+$) (p,n) reactions at 160 MeV are in very good agreement with the simple predicted value of $-1/3$.

The detection and polarization analysis of reaction-product protons is in general a two-step process. In the first step the energy (momentum) of the proton is determined by a magnetic spectrometer and wire-chamber/scintillator array. The proton is then scattered from a passive polarization-analyzing target (e.g., ^{12}C) into a second set of detectors that is operated in coincidence with the energy-analysis array. The energy and polarization measurements are therefore largely decoupled. Similar measurements on neutrons cannot be performed independently, however. A neutron must be detected by a nuclear interaction in a scintillator, and the polarization of the neutron is necessarily changed by the interaction. For the best possible energy resolution, the first interaction must also serve to determine the time-of-flight (TOF) of the neutron from the target. Thus, the energy and polarization analysis of neutrons must be performed simultaneously, with a single scintillator serving as both detector and analyzer.

Polarimeters employing liquid ^4He scintillators have been successfully used for measurements of neutron polarization at energies below 50 MeV.[5-7] Such designs are unsuitable for applications at higher energies, however, where neutron energies are determined by TOF measurements over long flight paths (>30m). The volume of scintillator required to achieve sufficient solid angle and efficiency at these distances makes liquid cryogenic scintillators (or even high pressure gas cells) impractical. Fortunately, ordinary hydrocarbon scintillators can serve as adequate polarization analyzers for energies above about 50 MeV. The primary analyzing reaction that takes place in such scintillators is $^1\text{H}(\vec{n},n)^1\text{H}$.

A schematic of the polarimeter used to obtain 160 MeV $(\vec{p},\vec{n})$ measurements is shown in Fig. 1. The polarimeter consists of two parallel planes of scintillator separated by a distance of 1 m. The planes are perpendicular to the incident neutron flux and each plane consists of three 15 cm x 15 cm x 102 cm blocks of NE102 plastic scintillator. Time signals derived from each end of the scintillators are used to determine the time and position of the neutron interaction in each plane. From this information the incident velocity v, the velocity of the scattered neutron v', and the polar and azimuthal scattering angles θ and ϕ are obtained. A valid event is defined at the hardware level as a coincidence between a single scintillator in the front plane and single scintillator in the back plane, with no signal from a thin intervening veto scintillator. Thus, coincidences caused by forward-going protons or cosmic ray showers are rejected. Additional restrictions upon the derived quantities v', θ, ϕ, and on the recoil pulse height in the analyzer are imposed in software in order to maximize the so-called figure-of-merit (FOM).

The figure-of-merit can be defined such that the statistical uncertainty in a polarization measurement is given by

$$\delta p_f = (I_n \text{ FOM})^{-1/2}, \quad (3)$$

where

$$\text{FOM} \equiv \varepsilon_c A_p^2, \quad (4)$$

I_n is the total number of neutrons passing through the face of the analyzer, ε_c is the coincidence efficiency, and A_p is the effective analyzing power (i.e., averaged over the geometrical acceptance) of the polarimeter. For the neutron polarimeter described above and in Fig. 1, the efficiency is typically $\varepsilon_c \simeq 2.5 \times 10^{-4}$ and the effective analyzing power is $A_p = 0.34 \pm 0.02$ for neutrons near 160 MeV. This analyzing power is comparable to that for proton polarimeters in the same energy region, while the efficiency is about a factor of 30 lower.[8]

The effective analyzing power of the above polarimeter was determined by observing neutrons from the $^{14}\text{C}(p,n)^{14}\text{N}(2.31 \text{ MeV})$ reaction. An energy spectrum for $^{14}\text{C}(p,n)$ is shown in Fig. 2. From Table I and Eq. (2) it can be seen that the neutrons produced in this $0^+ \to 0^+$ reaction have the same polarization at 0° as the incident protons. This calibration technique was suggested by R. Haight over a decade ago.[9]

The measured effective analyzing power of the polarimeter is shown in Fig. 3 for incident neutron energies of 117 MeV and 157 MeV. Also shown is the n-p analyzing power[10] at a center-of-mass scattering angle of 50° ($\theta_{c.m.} \simeq 2\theta_{lab}$). It seems likely that hydrocarbon scintillators should be suitable analyzers for neutron energies from several hundred MeV down to about 50 MeV. Measurements of neutron polarization near incident energies of 50 MeV have been made by Sakai et al. with a solid plastic scintillator.[11]

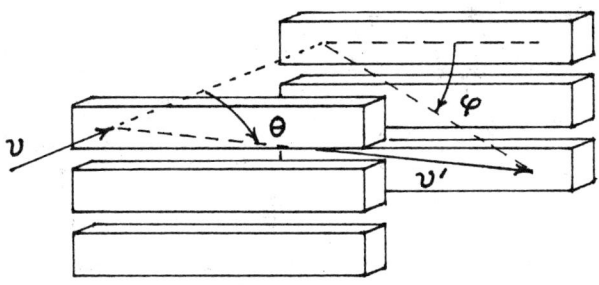

Fig. 1. Schematic of the neutron polarimeter used for (p,n) spin transfer measurements at 160 MeV (Ref. 4).

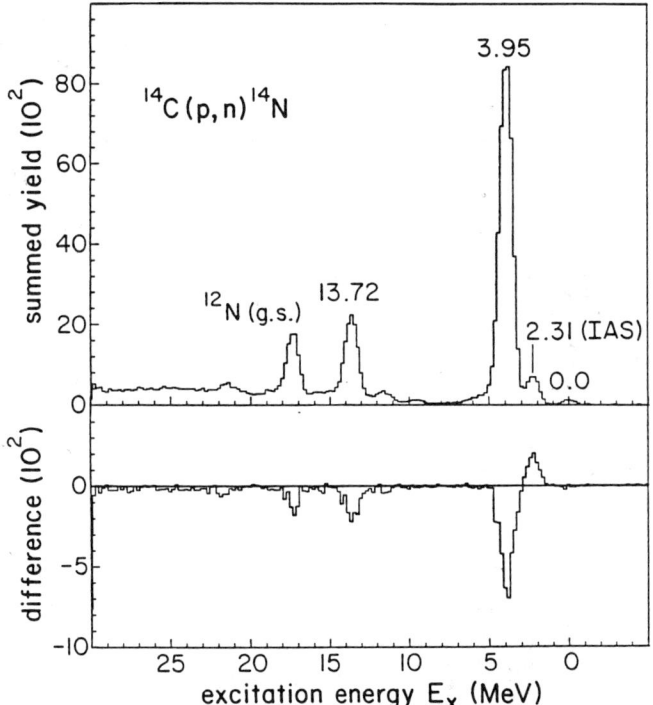

Fig. 2. Spectra for ^{14}C($\vec{p},\vec{n}$)^{14}N at 160 MeV. The summed yield (top) is proportional to $\sigma(0°,E_x)$, while the difference yield (bottom) is proportional to $\sigma(0°)D_{NN}(0°)$. Most of the peaks visible correspond to spin-flip transitions for which $D_{NN} < 0$. The $0^+ \to 0^+$ IAS transition has $D_{NN} = 1$ and therefore gives a positive peak in the difference spectrum.

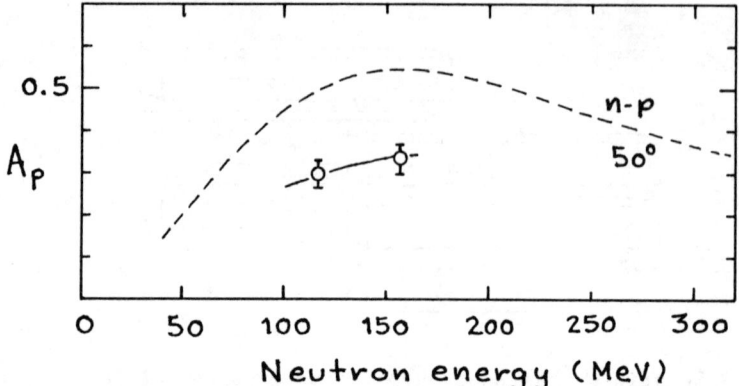

Fig. 3. The data points represent measured values of the effective analyzing power for the polarimeter of Fig. 1. The solid line connecting them represents the calculated energy dependence (including geometry effects) based on n-p scattering. Also shown is the n-p analyzing power for a c.m. angle of 50° (dashed line).

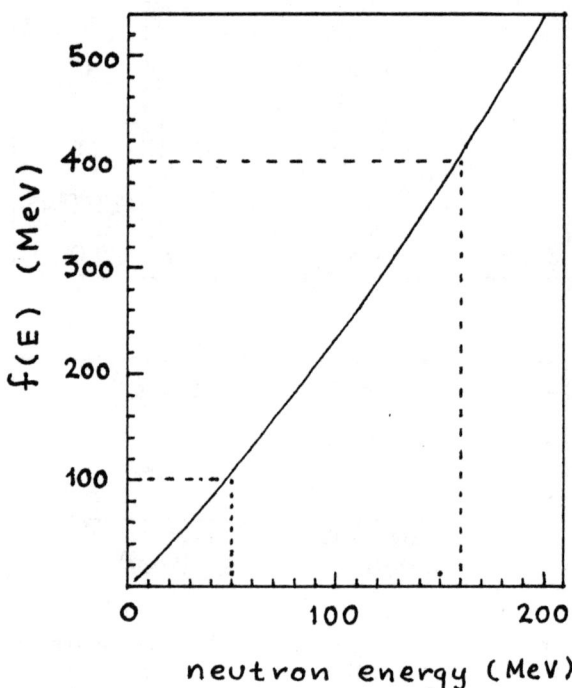

Fig. 4. Resolution function f(E) for neutron TOF measurements. See the text and Eq. (7) for details.

An important consideration in neutron polarimetry is the effective thickness Δz of the analyzing scintillator. It is desirable to make this thickness large for best efficiency, but energy resolution requirements put a limit upon this thickness and the resultant uncertainty in the neutron flight path ℓ. The energy resolution introduced by the flight time uncertainty can be expressed as

$$dE/dt = (p^3 c)/(m^2 \ell) \quad (5)$$

with the momentum p and neutron mass m in units of MeV. The flight path uncertainty contribution to the energy resolution is then obtained from

$$\Delta E = (dE/dt)\Delta t = (dE/dt)(\Delta z/v), \quad (6)$$

where v is the neutron velocity. This can be written as

$$\Delta E = E[(E/m)^2 - 1](\Delta z/\ell) \quad (7)$$

where $E = T+m$ is the total energy of the neutron. The quantity $f(E) = E[(E/m)^2 - 1]$ is plotted in Fig. 4 as function of neutron kinetic energy. An analyzer of 10 cm thickness at a flight path of 10 m will contribute about 1 MeV to the energy resolution for neutrons near 50 MeV. At 160 MeV, a thickness of 15 cm at 60 m gives about the same energy resolution.[4]

It is informative to make an "order-of-magnitude" estimate of the relative feasibility of spin transfer measurements for (p,n), (n,p), and (n,n). "Typical" parameters associated with currently existing accelerator facilities and polarimeters described in the literature will be used for this comparison. The results are shown in Table II. The amount of beam time required to make a measurement to a given precision is inversely proportional to the numbers in the third and fourth columns (L x FOM).

Table II. "Order-of-magnitude" estimate of the relative feasibility of spin transfer measurements for (p,n), (n,p), and (n,n). The figure-of-merit (FOM) is defined in Eq. (4).

Reaction	Luminosity, L	L x FOM	
	$I(enA) \times \mu(mg/cm^2)$	50 MeV	150 MeV
$(\vec{p},\vec{n})$	$10 \times 10^2 = 10^3$	1	10^{-1}
$(\vec{n},\vec{p})$	$10^{-3} \times 10^2 = 10^{-1}$	10^{-6}	10^{-4}
$(\vec{n},\vec{n})$	$10^{-3} \times 10^4 = 10$	10^{-2}	10^{-3}

Polarized beam currents delivered to the target are assumed to be on the order of tens of nanoamps (nA) for protons, and on the order of picoamp equivalents for neutrons. Target thicknesses for (p,n) and (n,p) measurements are generally on the order of hundreds of mg/cm^2, while (n,n) targets can be 100 times thicker (grams of material). Finally, the FOM for proton and neutron polarimeters is estimated for particle energies of 50 and 150 MeV based on information from four sources: for neutrons near 50 MeV the work of Sakai et al.[11] indicates that FOM $\simeq 10^{-3}$; for protons near 50 MeV the polarimeter described by Moss et al.[12] has FOM $\simeq 10^{-5}$; for neutrons near 150 MeV the polarimeter described in this talk has FOM $\simeq 10^{-4}$; and for protons near 150 MeV the work of Carey et al.[8] indicates that FOM $\simeq 10^{-3}$.

It should be kept in mind that the comparisons being made here are truly "order-of-magnitude" and that the assumed numbers can easily increase or decrease by factors of ten due to clever design or experimental constraints. In spite of this caution, the numbers under the last heading (L x FOM) of Table II tell a compelling story. It is clear that the measurement of spin transfer observables for (n,p) and (n,n) reactions is approximately 10^2 to 10^6 times harder than for (p,n), depending upon the energy and reaction. A typical (p,n) measurement currently takes from several to tens of hours of beam time. Thus, (n,p) and (n,n) spin transfer measurements are seemingly impractical with currently existing polarized neutron beam intensities. It should also be pointed out that no consideration has been made of how polarized neutron beams with good energy definition can be produced in the 50 to 150 MeV energy range. Such considerations may easily introduce additional (detrimental) factors of ten into the present comparison.

REFERENCES

1. See the contributions by B.C. Clark and J.R. Shepard.
2. W.D. Cornelius, J.M. Moss, T. Yamaya, Phys. Rev. C 23, 1364 (1981).
3. J.M. Moss, Phys. Rev. C 26, 727 (1982).
4. T.N. Taddeucci et al., Phys. Rev. Lett. 52, 1960 (1984).
5. L.P. Robertson et al., Nucl. Phys. A134, 545 (1969).
6. J.C. Hiebert et al., Phys. Rev. Lett. 37, 276 (1976).
7. R.C. Haight, J.E. Simmons, T.R. Donoghue, Phys. Rev. C 5, 1826 (1972).
8. T.A. Carey et al., Phys. Rev. Lett. 49, 266 (1982).
9. Gerald G. Ohlsen, Rep. Prog. Phys. 35, 717 (1972; p. 779.
10. R.A. Arndt and L.D. Roper, Scattering Analyses Interactive Dial-in (SAID) Program, phase shift solution SP84, Virginia Polytechnic Institute and State University (unpublished).
11. H. Sakai et al., J. Phys. G: Nucl. Phys. 10, L139 (1984).
12. J.M. Moss, D.R. Brown, W.D. Cornelius, Nucl. Instrum. Methods 135, 139 (1976).

A FACILITY FOR NEUTRON SCATTERING MEASUREMENTS AT 22 MeV

N. Olsson and B. Trostell
The Studsvik Science Research Laboratory, S-611 82 Nyköping, Sweden

ABSTRACT

The Studsvik fast neutron scattering facility has been improved with respect to energy resolution and signal-to-background ratio. The equipment can now be used for scattering studies in the 20 MeV region with a total energy resolution of 0.5 MeV.

INTRODUCTION

Measurements of neutron elastic and inelastic scattering from different nuclei have been performed at our laboratory for several years. Due to the limited energy resolution of the time-of-flight spectrometer, these studies were, however, restricted to the energy region below 8 MeV. Several elements should be available for a systematic study at 20 MeV, provided that the total energy resolution could be reduced to about 0.5 MeV. In the case of our spectrometer, with a maximum flight path of 4 m, this corresponds to a total time resolution of 0.75 ns. In order to reach this goal, considerable improvements have been made on the dominating sources of time spread, i.e. the ion pulse width, the neutron producing gas target, and the neutron detector time resolution. It was also important to maintain the signal counting rates at a reasonable level, while reducing the background from the direct target flux.

THE NEUTRON SCATTERING FACILITY

The pulsed CN VdG accelerator is used to produce 1-2 ns bursts of deuterons with an energy of 5 MeV. The repetition rate and average beam current is 1 MHz and 3 µA, respectively. The bursts are further compressed to about 0.2 ns by a post acceleration bunching system[1], working at a frequency of 90 MHz.

Neutrons of 22 MeV energy are produced with the T(d,n) reaction, using a new gas target and tritium gas handling system. The target cell is only 1 cm long, thus reducing this time spread contribution. The 4 mg/cm^2 Mo entrance foil, isolating the cell from the vacuum system, is sealed with 0.1 mm thick In O-rings. In order to maintain a high neutron yield, the target and filling system is capable of handling gas pressures up to 4 atm.

The two neutron detectors consist of Ø 5" x 2" NE 213 liquid scintillators, directly coupled to RCA 8854 5" PM tubes. The extremely tapered voltage dividers, similar to those described by Randers-Pehrson et. al.[2], are working at -2500 V and show very good linearity up to large pulse heights. The timing information is obtained from the DC coupled anode signals by using constant fraction discriminators, while the n-γ and threshold discrimination circuits work on dynode signals. With a threshold at the Compton edge of ^{60}Co, corresponding to 3 MeV neutrons, this results in a very good n-γ separation and an intrinsic time resolution of about 0.5 ns.

The detectors are housed in the old spectrometer shielding and are separated by 5 degrees. The spectrometer is remotely computer controlled and can be turned in the angular range from $-50°$ to $+162°$. The 15 tons heavy shielding and collimation system is fabricated from iron, borated paraffin, and lead. A hevimet (90% tungsten) shadow bar with a maximum length of 70 cm effectively shields the detectors from the direct target flux.

RESULTS

The performance of the new facility compared to the old one for the T(d,n) reaction with E_n=21.6 MeV is shown in table I.

Table I Contributions to the energy resolution (FWHM)

Sources of time spread (ns)		"Old facility"		"New facility"
Ion pulse width		1.5		0.2
Gas target length	(3 cm)	0.9	(1 cm)	0.3
Detector thickness	(5 cm)	0.4	(5 cm)	0.4
PM-tube & electronics		1.2		0.5
Neutron energy spread (0.1 MeV)	(3 m)	0.11	(4 m)	0.14
Total time resolution		2.2		0.75
Total energy resolution (MeV)		2.0		0.51

As can be seen the total energy resolution has been improved by a factor of 4.

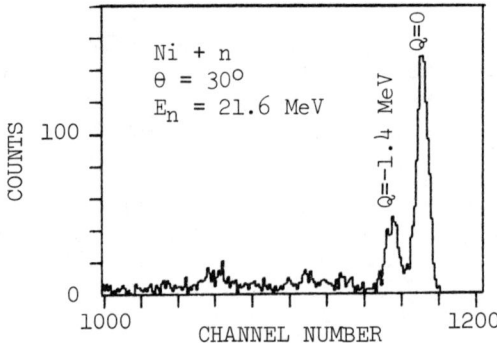

Fig. 1. Time-of-flight spectrum of neutrons from the Ni + n reaction.

The spectrometer is now in use taking data of elastic and inelastic neutron scattering at 22 MeV. A typical time-of-flight spectrum from the Ni + n reaction is shown in fig. 1. A sample-out background has been subtracted, and the separation of the first inelastic levels at E_x=1.4 MeV from the elastic one as well as the good signal-to-background ratio can be seen.

REFERENCES

1. N. Olsson, Nucl. Instr. and Meth., <u>187</u>, 341 (1981).
2. G. Randers-Pehrson, R.W. Finlay and D.E. Carter, Nucl. Instr. and Meth., <u>215</u>, 433 (1983).

PLANNING OF NEUTRON PHYSICS AT THE REBUILT CYCLOTRON OF THE GUSTAF WERNER INSTITUTE, UPPSALA UNIVERSITY

H. Condé
Gustaf Werner Institute, Uppsala University
Box 531, S-75121 Uppsala, Sweden

ABSTRACT

Experimental neutron physics in the energy region up to 200 MeV is being planned at the rebuilt cyclotron of the Gustaf Werner Institute, Uppsala University. The expected performance of the cyclotron including a ring CELSIUS for storing, cooling and accelerating ions injected from the cyclotron are given. The experimental neutron physics facility, using the direct beam from the cyclotron, is described. Thoughts about neutron physics at the storage ring are given. Finally, planned experimental arrangements for (n,p)-reaction studies and total cross section measurements are presented.

THE CYCLOTRON AND STORAGE RING PROJECT

A Swedish national accelerator center is being established in Uppsala, based on three different accelerators: the existing tandem van de Graaff, the synchrocyclotron under reconstruction and the CELSIUS ring for storing and cooling ions injected from the cyclotron.

The cyclotron will after reconstruction operate both as a synchroncyclotron and as a isochronous-three sector, variable energy-cyclotron with K = 200.

The layout of the cyclotron laboratory is shown in figure 1.

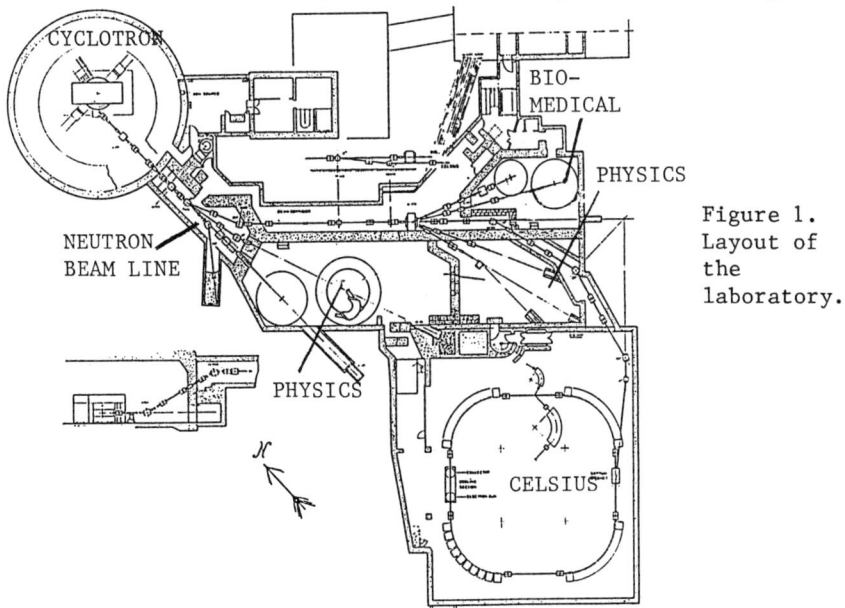

Figure 1. Layout of the laboratory.

The expected performances of the cyclotron and the CELSIUS ring are given in the facility summary presented to this meeting. Most of the buildings shown are situated below ground and closely surrounded by a number of other university buildings. The proximity to other houses has, in fact, been a major difficulty and explains much of the special features of the general layout.

NEUTRON PHYSICS FACILITY

Figure 1 shows the position of the beam line of neutron physics in the left corner of the layout.

Fast neutron beams will be obtained by (p,n) or (d,n) reactions in lithium- or beryllium targets in a well-shielded cave.

A water-cooled target-holder has been designed with place for four different targets remotely positioning controlled. The targets which are not in neutron production position are surrounded by lead for radiation shielding.

An H-magnet (10 cm pole-gap and an effective magnet length of 80 cm) will be used to clear the neutron beam from the proton beam passing through the target. The protons are bent $\sim 35°$ into a five meter long beam dump.

The neutron beam to the experimental hall (figure 2) is defined by three collimators with adjustable opening angles $< 10^{-4}$ sr. The collimator length is about 3 m and the distance from the neutron source to the experimental hall about 8 m. Furthermore, the beam line will have the filters for reducing the background of slow neutrons and gamma rays and a magnet for clearing the beam from protons produced by the secondary reactions. The choice of target

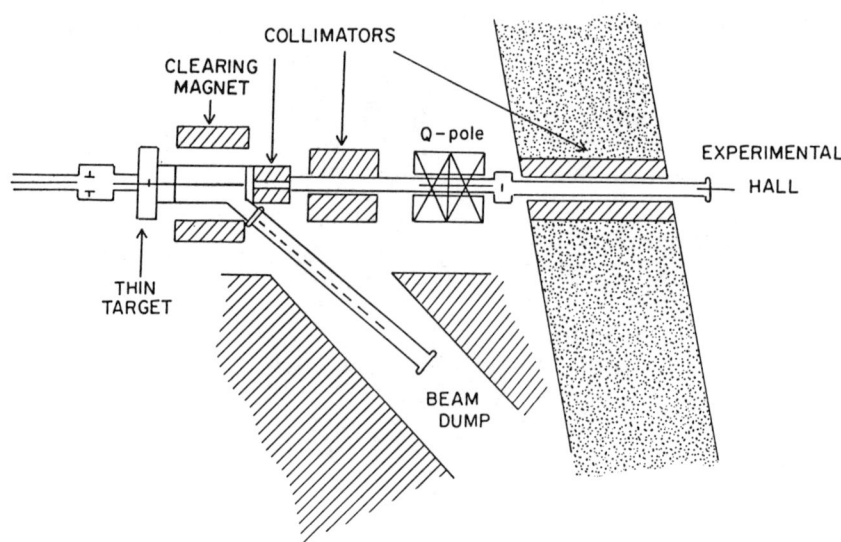

Figure 2. Neutron facility.

material and thickness determines the intensity and energy width of the "monoenergetic" neutron beam produced in the forward direction. With 1 μA protons (200 MeV) incident on a 1 mm thick lithium-target, the estimated neutron flux in a solid angle of 1 sr is 10^9 n/sec. The peak width is then of the order of 0.2 MeV/mm target. In the production of "white beams", thick beryllium or tantalum-targets are most suitable as production targets.

The maximum flight path for time-of-flight measurements is limited to 30 m by the experimental area available. Short beam pulses of the order of 1 ns separated by 1 μs can be obtained with arrangements inside the cyclotron at least for proton energies of 50 MeV and lower. An external sweeping system has to be added to obtain corresponding conditions at higher energies.

EXPERIMENTAL PROGRAM

The fast neutron beam facility will initially be used in the following two experiments:
1) Nuclear structure studies with the (n,p)-reaction.
2) Measurements of neutron-nucleus total cross sections.

Experimental equipment are being planned for the two experiments. The proton spectrometer in the first experiment will consist of a C-magnet with pole-gap 10 cm and with an effective length of 63 cm. The proton paths through the magnet will be detected by two 10 cm long multiwire-chambers before the magnet and two 30 cm long multiwire-chambers after the magnet. The wire-spacings are 1 mm and 0.5 mm, respectively. The magnet field will be measured and stabilized by an NMR equipment. The magnet dispersion is 0.3 which gives a space resolution of 0.5 cm/%Δp/p at 100 cm. Investigations are underway to minimize the total energy spread, which is mainly the sum of the energy spread of the cyclotron beam, the spread due to the neutron source and sample configurations and the energy spread of the magnetic spectrometer. The aim is to obtain a total energy spread of less than 1 MeV at 200 MeV. Measurements are initially planned on light nuclei and later on heavy nuclei (e.g. ^{208}Pb).

The neutron detector in the total cross section experiment consists of a cylinder of NE102A scintillator of 20 cm length and 10 cm diameter and viewed by a photomultiplier tube XP2040. An anticounter in front of the neutron counter consists of a thin scintillator disc with ⌀ 10 cm. The detector will be mounted on a rail so that it can be moved to cover various solid angles. The aim of the experiment will be to study the energy and mass dependence of the broad resonance structure of the total cross section between 20 and 200 MeV.

THOUGHTS ABOUT NEUTRON PHYSICS AT CELSIUS

Many neutron reaction studies, e.g. (n,p)-reaction studies, could largely be favoured by the improved energy resolution obtainable at the storage ring with electron cooling. However, several problems are connected with the geometrical arrangements of the experiments. The neutron output from suitable neutron sources ((p,n)-reactions in T and ^{6}Li) are peaked in the 0° direction. Thus, the neutron source has either to be put in the bend section of the ring or in the straight section together with the sample. In the first case the 0° neutron beam can be utilized though it might be more complicated to arrange with an experimental set-up in the bend section. In second case care must be taken so that the sample does not interfere with the beam, which exclude use of neutrons in the 0° direction. If e.g. the (n,p) experiment is made in the latter geometry, the magnets in the bend section might be used for proton spectroscopy as proposed by C. Gaarde at the CELSIUS workshop, Nov. 7-9, 1983, Uppsala.

Counting rate might also be a problem. Assuming a Li-target and a luminosity of 10^{32} cm^{-2}, s^{-1} the neutron intensity will be about 10^6 n, s^{-1}, sr^{-1} in the 0° direction ((Li(p,n)) $\sim$ 50 mb, sr^{-1}). This intensity might be low even if it is possible to tag the neutrons.

Improvements in neutron detector resolution are also needed for measurements on reactions with a neutron in the outgoing channel.

SESSION E

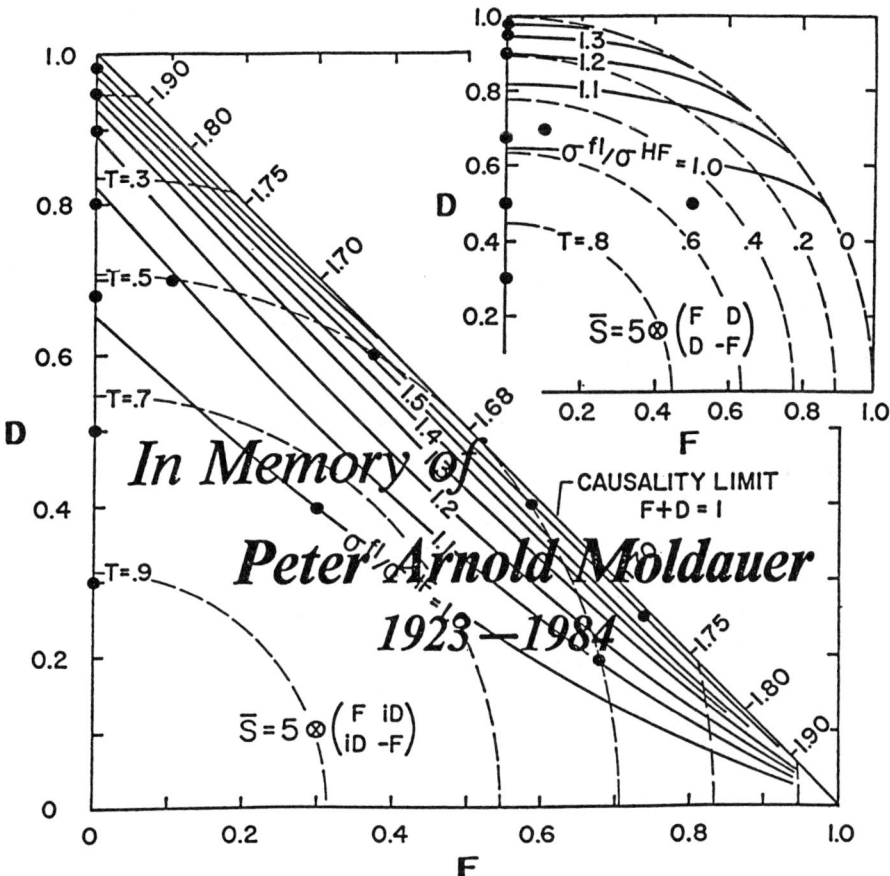

Peter Arnold Moldauer; Memorial Session
Introductory Remarks

Alan B. Smith
Argonne National Laboratory, Argonne, IL 60439

It is fitting and proper that the Organizing Committee has arranged this session in memory of Peter Arnold Moldauer. As his friend and co-worker for nearly three decades, it is my honor to assume the Chair.

While attending a professional conference at the International Center for Theoretical Physics, Trieste, Peter Moldauer was killed in an automobile accident.

Peter received the MA degree from Harvard and the PhD degree from the University of Michigan. His doctoral thesis dealt with the principles of relativistic quantum theory, a field in which he retained a life-long interest. After appointments at MIT and the University of Connecticut, he came to Argonne National Laboratory in 1957. A decade later he was appointed Senior Physicist. During his more than twenty-five years at Argonne, he was a visiting professor and guest lecturer at a number of prominent institutions here and abroad. Indeed, he was lecturing at the ICTP at the time of his death. He was a Fellow of the American Physical Society.

Peter was an internationally recognized authority on the theory and application of statistical nuclear processes, particularly those relevant to neutron-induced reactions. Thus this commemorative session is particularly appropriate to the present conference. His theoretical formulation of compound-nucleus fluctuations is widely employed in the interpretation of compound-nucleus reactions and, at the time of his death, he was extending these considerations to the neutron interaction with highly deformed nuclei. More generally, his scientific and personal interests were extremely broad and resulted in continuing contributions to a number of fields. He was very much interested in the foundations of quantum theory where his notable clarity of thought helped illuminate often murky areas. He had a gift for penetrating through the complicated mathematics to the underlying physics with perceptive judgements. He maintained a strong interest in the engineering application of nuclear physics, particularly the neutronic design of fast-reactor systems. He authored (with D. Okrent and S. Yiftah) the monograph "Fast Reactor Cross Sections", using fundamental theoretical concepts to predict nuclear data at a time when direct measurements were unreliable or unavailable. Looking back, one finds that those early predictions of reactor performance remain remarkably valid. For example, his calculated critical masses and spectral indexes are very similar to those obtained using the best of modern fast-neutron data. Peter provided valuable guidance and physical insight for experimental endeavors, often working closely with experimentalists. He made use of his physical understanding to construct extensive nuclear-model programs, guided their use, and made them the friends of the experimentalists. Some of these programs have been productively used by people at this meeting. He loved a good professional debate, and present here are both winners and losers of such engagements. Throughout his life he retained a deep interest in the social implications of nuclear science and he very much enjoyed the exercise of his appreciable artistic talents.

Peter Moldauer was a close personal and professional friend of numerous individuals the world over, many of whom are assembled here this evening to honor him with the type of professional commentary that he so much enjoyed.

SOME COMMENTS ON THE THEORY OF NUCLEAR REACTIONS

H. Feshbach
Massachusetts Institute of Technology, Cambridge, MA 02139

ABSTRACT

In part I, a simple derivation of the Hauser-Feshbach formula which avoids the calculation of the sum over resonances is presented. In part II, the description of a resonance in the scattering from a deformed nucleus in which the intermediate state is specified by the angle of the target is proposed. Part III contains a discussion of the bound states in the continuum which occur when the scattering potential is sufficiently absorbing as well as the derivation of a general theorem of great importance for the theory of direct multistep processes. Some experimental results are compared with the predictions of that theory.

My discussion this evening will consist of three items. The first is a remark which I have included because of its relationship to a long standing problem of interest to both Peter Moldauer and myself. It has to do with the derivation of the Hauser-Feshbach formula and it comes up to the course of my discussion of that formula in the second volume of Nuclear Theory to which I am now able to devote more time. To begin with let me remind you of some ancient formulas. The transition matrix $\mathcal{J}$ can be divided into two parts

$$\mathcal{J} = \mathcal{J}^{(P)} + \mathcal{J}^{(FL)} \qquad (1)$$

where $\mathcal{J}^{(P)}$ is the prompt amplitude including the direct and multistep direct processes given in the multi-channel optical model. $\mathcal{J}^{(FL)}$ is the fluctuation component generated by the compound nuclear reactions. The energy average of $\mathcal{J}$ is $\mathcal{J}_p$:

$$\langle \mathcal{J} \rangle = \mathcal{J}_p \quad ; \quad \langle \mathcal{J}^{(FL)} \rangle = 0 \qquad (2)$$

As a consequence the average cross section also separates

$$\langle \sigma \rangle = \sigma_p + \sigma^{(FL)} \qquad (3)$$

The equation for $\mathcal{J}^{(FL)}_{cc}$ where C denotes the channels is

0094-243X/85/1240412-15 $3.00 Copyright 1985 American Institute of Physics

$$J_{cc'}^{(FL)} = \frac{1}{2\pi} e^{i(\delta_f + \delta_c)} \sum_\lambda \frac{e^{i\phi_\lambda(c,c')} g_\lambda(c) g_\lambda(c')}{E - E_\lambda + \frac{i}{2}(\Gamma_\lambda + \Gamma_\lambda')} \quad (4)$$

where

$$\langle g_\lambda \rangle = 0, \quad \Gamma_{\lambda c} = g_\lambda^2(c), \quad \Gamma_\lambda = \sum_c \Gamma_{\lambda c}, \quad \sum_\lambda \Gamma_\lambda' = 0 \quad (5)$$

We immediately obtain

$$\sigma^{(FL)}(J\Pi) = \pi \lambdabar^2 \sum_{\lambda,\mu} e^{i(\phi_\lambda - \phi_\mu)} \frac{g_{\lambda f} g_{\mu f} g_{\lambda i} g_{\mu i}}{[E - E_\lambda + \frac{i}{2}(\Gamma_\lambda + \Gamma_\lambda')][E - E_\mu - \frac{i}{2}(\Gamma_\mu + \Gamma_\mu')]} \quad (6)$$

We now apply the random phase approximation to obtain

$$\langle \sigma^{(FL)}(J\Pi) \rangle_{R.P.} = \pi \lambdabar^2 \sum_\lambda \left\langle \frac{\Gamma_{\lambda f} \Gamma_{\lambda i}}{(E - E_\lambda)^2 + \frac{1}{4}(\Gamma_\lambda + \Gamma_\lambda')^2} \right\rangle$$

We assume that the number of exit channels is so large that fluctuations in the denominator are unimportant and thus the denominator can be taken to be a constant. Hence

$$\langle \sigma^{(FL)}(J\Pi) \rangle \longrightarrow \pi \lambdabar^2 \sum_\lambda \frac{\langle \Gamma_{\lambda f} \Gamma_{\lambda i} \rangle}{(E - E_\lambda)^2 + \frac{1}{4}(\Gamma_\lambda + \Gamma_\lambda')^2} \quad (7)$$

and finally

$$\langle \sigma^{(FL)}(J\Pi) \rangle \longrightarrow \pi \lambdabar^2 \langle \Gamma_f \rangle \langle \Gamma_i \rangle \sum \quad (8)$$

where

$$\sum = \sum_\lambda \frac{1}{(E - E_\lambda)^2 + \frac{1}{4}(\Gamma_\lambda + \Gamma_\lambda')^2} \quad (9)$$

One of the corrections to the steps Eq. (6) - Eq. (8) is given by the width fluctuation formula which Moldauer did much to develop. This is well known to this audience. I will not discuss it.

The point I want to make is that it is not necessary to evaluate Σ to obtain the H-F result. The value of that sum was at one time a considerable bone of contention. First note that Eq. (8) is equivalent to the Bohr independence hypothesis. To prove this sum Eq. (8) over all final states to obtain

$$\sigma_i^{(c)} \equiv \sum_f \sigma_{fi}^{(FL)} = \pi \lambdabar^2 \langle \Gamma \rangle \langle \Gamma_i \rangle \Sigma \tag{10}$$

so that

$$\sigma_f^{(FL)} = \sigma_i^{(c)} \frac{\langle \Gamma_f \rangle}{\langle \Gamma_i \rangle} \tag{11}$$

To obtain the H-F expression we need to establish a connection to the transmission coefficient given by

$$T_c = 1 - \sum_{c''} |S_{cc''}^{(OPT)}|^2 \tag{12}$$

This is again a point at which disagreement set in, a disagreement, as we shall see, of no importance. From Eq. (12) we have

$$T_c = \left\langle \sum_{\lambda c''} \frac{g_{\lambda c}^2 \, g_{\lambda c''}^2}{(E-\varepsilon_\lambda)^2 + \tfrac{1}{4}(\Gamma_\lambda + \Gamma_\lambda'')^2} \right\rangle$$

$$\rightarrow \langle \Gamma_c \rangle \langle \Gamma \rangle \Sigma$$

and

$$\sum_c T_c = \langle \Gamma \rangle^2 \Sigma$$

Hence from Eq. (8)

$$\sigma^{(FL)} = \pi \lambdabar^2 \frac{T_f}{\langle \Gamma \rangle \Sigma} \frac{T_i}{\langle \Gamma \rangle \Sigma} \cdot \Sigma = \pi \lambdabar^2 \frac{T_f T_i}{\langle \Gamma \rangle^2 \Sigma} = \frac{\pi \lambdabar^2 T_f T_i}{\Sigma T_c} \tag{13}$$

Note that it was not necessary to make an explicit statement about the relation between T_c and Γ/D; that relationship is simply not needed.

My second remark is made with respect to resonances involving deformed nuclei. This research was done in collaboration with Lutz Dohnert (1). In that situation one can expect that the potential energy between the target nucleus and projectile (which can also be deformed) will depend upon the orientation of the deformed nucleus as illustrated in Fig. 1. Greiner has indeed suggested that in the case of the collision of actinide nuclei

at certain orientations a pocket appears in the potential.
This pocket suggests the possibility of a resonance. Similarly
in Fig. 1, the potential at 90° might be able to support a
resonance. More generally it may be possible to generate a
resonance only for certain orientations. This is reflected in
the angular distribution at or near the resonance energy. The
proposal to be discussed below tries to make explicit use of
the orientation dependence to develop a more compact expression
with an explicit dependence on the most favorable orientation.
Moreover, it identifies the orientation angle most favorable for
resonance formation as the appropriate observable. The standard
observable (J,π) used for resonances are less appropriate as
many J's contribute to the final angular distribution.

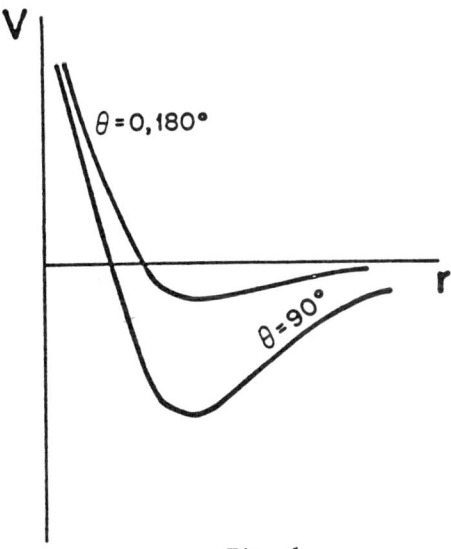

Fig. 1

The expression for the resonance transition amplitude $^{(R)}$
for the case of an isolated resonance is (1)

$$J^{(R)} = \frac{\langle \psi_f^{(-)} H_{PQ} \Phi_R \rangle \langle \Phi_R H_{QP} \psi_i^{(+)} \rangle}{E - E_R + \frac{i}{2} \Gamma_R} \quad (14)$$

In this formula $\psi_i^{(+)}$ and $\psi_f^{(-)}$ are the wavefunctions describing
the incident and final channels, Φ_R is the intermediate resonant
state which H_{PQ} is the interaction connecting the incident and
final channels with the resonant state. The resonance occurs
at an energy E_R and has the width of Γ_R.

For simplicity the wavefunction for the incident projectile is taken to be a plane wave, $\exp i\vec{k}\cdot\vec{r}$. The spheroidal target nucleus wavefunction consists of a factor specifying its internal state and a factor describing its orientation with respect to the incident direction $\vec{k}_i$. Initially, the target nucleus is taken to be in its ground state, $|0\rangle$. The second factor is taken to be a constant corresponding to the target having zero spin.

Thus, the initial wavefunction is

$$\phi_i^{(+)} = \frac{1}{\sqrt{4\pi}} \exp\{i\vec{k}_i\cdot\vec{n}\} |0\rangle \tag{15}$$

The factor $(4\pi)^{-1/2}$ describes the fact that the initial orientation of the target is arbitrary. Similarly, for elastic scattering we write

$$\phi_f^{(-)} = \frac{1}{\sqrt{4\pi}} \exp\{i\vec{k}_f\cdot\vec{n}\} |0\rangle \tag{16}$$

The resonant state consists, as stated above, of the projectile in a bound state, and the target in an excited state $|R\rangle$. The bound state projectile wavefunction is taken to be that of a single particle moving in a deformed harmonic oscillator well. Finally, the resonance is presumed to occur at or near a preferred orientation $\vec{\vartheta}_R$ of the target. We, therefore, have

$$\Phi_R = \frac{b\sqrt{a}}{\pi^{3/4}} N_s\, e^{-\frac{1}{2}s(\vec{\vartheta}-\vec{\vartheta}_R)^2} \exp\left\{-\frac{1}{2}a^2(\vec{\vartheta}\cdot\vec{x})^2 - \frac{1}{2}b^2(\vec{\vartheta}\times\vec{x})^2\right\} |R\rangle \tag{17}$$

The normalization factor N_S is given by

$$\frac{1}{\sqrt{\pi}}(1-e^{-4s^2})^{-1/2} \longrightarrow \frac{1}{\sqrt{\pi}}$$

The quantity s measures the angular width over which a resonance can occur.

Finally, we must specify H_{PQ}. This interaction must contain a factor producing the excitation to the internal state $|R\rangle$ as well as a space dependent factor which is orientation dependent. The assumed interaction has then the form (for simplicity we omit any spin dependence)

$$H_{PQ} = V_0 \exp\left\{-\frac{1}{2}\alpha^2(\vec{\vartheta}\cdot\vec{x})^2 - \frac{1}{2}\beta^2(\vec{\vartheta}\times\vec{x})^2\right\} |0\rangle\langle R| \tag{18}$$

The interaction H_{QP} has the same form except that $|0><R|$ is replaced by $|R><0|$. The coefficient V_0 is a constant measuring the strength of the coupling. The interaction is largest in the direction perpendicular to the symmetry axis or in the parallel direction according to whether α is greater or less than β.

With these assumptions, the evaluation of the matrix elements in the numerator of Eq. (1) is straightforward. One obtains

$$J_{fi}^{(R)} = \frac{1}{E-E_R+\frac{i}{2}\Gamma_R} \frac{V_0^2}{A^2} \frac{2\sqrt{\pi} a b^2}{(a^2+\alpha^2)(b^2+\beta^2)^2} exp\left\{-\frac{(\vec{k}_i \cdot \vec{\vartheta}_R)^2 + (\vec{k}_f \cdot \vec{\vartheta}_R)^2}{2(a^2+\alpha^2)}\right\}$$
$$\times exp\left\{-\frac{(\vec{k}_i \times \vec{\vartheta}_R)^2 + (\vec{k}_f \times \vec{\vartheta}_R)^2}{2(b^2+\beta^2)}\right\} \quad (19)$$

The width Γ_R has no angular dependence as can be seen from the following expression, valid when the resonance is isolated:

$$\Gamma_R = 2\pi \sum_a \left|\langle \psi_a^{(-)} | H_{PQ} | \Phi_R \rangle\right|^2 \quad (20)$$

Since we sum over all possible final states including, therefore, an integral over all directions of emission, Γ_R cannot depend upon $\vec{k}_f$.

The resonance amplitude thus has an angular dependence which is a function of the resonance angle ϑ_R which now becomes the important physical observable rather than the customary resonance angular momentum. At each angle the energy dependence is given by the same resonance formula dependent upon E_R. This energy dependence may be modified for broad resonances by the energy dependence of the partial widths. Of course, Eq. (19) will change when more realistic wavefunctions are used. However, it is to be expected that the qualitative features exhibited by Eq. (19) will remain.

The third comment brings to your attention a recent observation regarding the optical model (2) and the consequence for the statistical multistep direct theory. It was observed by Gal, Toker and Alexander that for sufficient absorption, solutions of the optical model Schroedinger equations whose amplitudes decay exponentially with distance from the interaction region exist. These are "bound states" in the continuum. The wavefunctions are normalizable and are orthogonal to other solutions. One can understand their presence by describing what happens to

the poles of the S matrix as absorption is increased. In Fig. 2
the dots indicate the position of these poles in the absence
of absorption in the complex k plane for a given orbital angular
momentum. These are in the lower half k plane, symmetrically
placed with respect to the Im k axis. Upon introducing absorption the positions of these poles rotate in a clockwise manner as
indicated. When the absorption is sufficiently strong, S matrix
poles in the third quadrant will move into the second quadrant.
the corresponding wavefunctions will be of the form $\exp i[-x+i\sigma]r/r$
or $e^{-ixr}/r \, e^{-\sigma r}$ exhibiting the aforementioned exponential decay.
The reader can readily verify these statements by considering the
simple ($\ell=0$) problem

$$\frac{d^2\psi}{dr^2} + [k^2 + i\lambda \, \delta(r-r_0)]\psi = 0 \tag{21}$$

involving a delta function absorption. Moreover one can also see
that these states will not give rise to a resonance since the
S matrix for this problem is:

$$S = e^{-2ikr_0} \frac{(\lambda/k - 1) + i \cot k r_0}{(\lambda/k + 1) + i \cot k r_0} \tag{22}$$

For λ and k positive the real part of the denominator cannot
approach zero as k varies. As a consequence these bound state
poles do not have any great effect on the scattering.

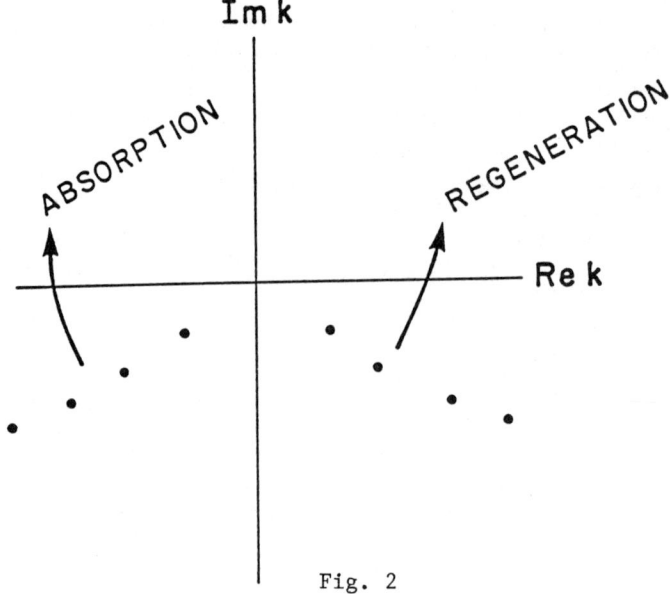

Fig. 2

The situation changes sharply for λ < 0, the regeneration case. Then resonances can occur near k = |λ|. The reason for this difference lies in the fact that for the absorptive case the bound state S matrix poles are in the second quadrant corresponding to incoming waves while for the regenerative case they are in the first quadrant corresponding to outgoing waves and these can give rise to resonances.

The regenerative case makes its appearance in multistep reactions. This can be readily seen in the second order distorted wave Born approximation considered for exampty by Robson. The amplitude has the form

$$J^{(2)} = \langle \phi_f^{(-)} v \frac{1}{E^{(+)} - H_{OPT}} v \phi_i^{(+)} \rangle \qquad (23)$$

In this expression $\phi_i^{(+)}$ and $\phi_f^{(-)}$ are the initial and final distorted waves with indicated boundary conditions. The quantity v is the perturbing potential inducing a transition of the system from the initial state $\phi_i^{(+)}$ or into final state $\phi_f^{(-)}$. The propagator $(1/E-H_{OPT})$ involves the complex potential of the optical model. One attempts to evaluate $J^{(2)}$ by expanding the propagator in a spectral series:

$$\frac{1}{E^{(+)} - H_{OPT}} = \int \frac{d\vec{k}}{(2\pi)^3} \sum_\lambda \frac{\chi_\lambda^{(+)}(\vec{k}) \rangle \langle \tilde{\chi}_\lambda^{(+)}(\vec{k})}{E^{(+)} - E_\lambda(\vec{k})} \qquad (24)$$

The function $\tilde{\chi}_\lambda(\vec{k})$ satisfies

$$(E - H_{OPT}^\dagger) \tilde{\chi}_\lambda = 0 \qquad (25)$$

It is, therefore, a solution of the optical model Schroedinger equation with a regenerative potential. The functions $\chi_\lambda^{(+)}(\vec{k})$ satisfy the optical model Schroedinger equation with an absorptive potential:

$$(E - H_{OPT}) \chi_\lambda = 0 \qquad (26)$$

One important property of $\tilde{\chi}_\lambda^{(+)}$ can be derived by expanding $\chi_\lambda^{(-)}$ terms of the biorthogonal set $\chi_a^{(+)} \rangle \langle \tilde{\chi}_a^{(+)}$

$$\chi_\lambda^{(-)} = \sum_a \langle \chi_\lambda^{(-)} \chi_a^{(+)} \rangle \tilde{\chi}_a^{(+)}$$

or

$$\chi_\lambda^{(-)} = \sum_\alpha S_{\lambda a} \tilde{\chi}_a^{(+)}$$

where S is the S matrix. Inverting one obtains

$$\tilde{\chi}_a^{(+)} = \sum_\lambda (S^{-1})_{a\lambda} \chi_\lambda^{(-)} \qquad (27)$$

This relationship leads to an apparent difficulty in the case of strong absorption as first pointed out by Robson (3). In that case the magnitude of S can be very small and consequently the magnitude of S^{-1} correspondingly large. However, the propagator $(1/E-H_{OPT})$ has no such singularities as shown by Austern and Vincent (4). The S^{-1} term must, therefore, disappear once the sum and integral are performed as a consequence of interference. Obviously great care is needed in order to evaluate $\mathcal{J}^{(2)}$ if expansion Eq. (24) is employed. Approximations to the individual matrix elements may not be sufficiently accurate and such stratagems as the naive random phase assumption of the statistical model are likely to lead to incorrect results.

This difficulty is resolved if one energy averages. Such as energy average is required in any event because of the resonances exhibited by $\chi_\lambda^{(+)}$. Energy averaging destroys the interference effects which remove the 1/S factor. As a consequence, as there is no singularity in $(1/E^{(+)} - H_{OPT})$, the S^{-1} factor must be replaced by a less singular function. This is demonstrated in a paper to be published in Annals of Physics (3) in which one finds that

$$\langle \chi^{(+)} \rangle = \psi^{(+)} \qquad (28)$$

$$\langle \tilde{\chi}^{(+)} \rangle = \psi^{(-)} \qquad (29)$$

so that

$$\left\langle \frac{1}{E^{(+)} - H_{OPT}} \right\rangle = \int \frac{d\vec{k}}{(2\pi)^3} \sum_\lambda \frac{\psi_\lambda^{(+)}(\vec{k}) \rangle \langle \psi_\lambda^{(-)}(\vec{k})}{E - E_\lambda(k)}$$

where $\psi_\lambda^{(\pm)}$ are solutions of a Schroedinger equation involving a new optical model Hamiltonian.

As a consequence of these results it now becomes possible to employ the random phase approximation and so obtain the following result for the statistical multistep direct reaction cross section (5)

$$\left\langle \frac{d^2\sigma}{d\Omega_f dU_f} \right\rangle = \sum_{m,\nu} \int \frac{d\vec{k}_1}{(2\pi)^3} \cdots \int \frac{d\vec{k}_\nu}{(2\pi)^3} \left[\frac{dw_{m,\nu}(\vec{k}_f, \vec{k}_\nu)}{d\Omega_f \, dU_f} \right] \left[\frac{dw_{\nu,\mu}(\vec{k}_\nu, \vec{k}_{\nu-1})}{d\Omega_\nu \, dU_\nu} \right]$$

$$\cdots \cdots \cdots \left[\frac{dw_{2,1}(\vec{k}_2, \vec{k}_1)}{d\Omega_2 \, dU_2} \right] \frac{d\sigma_i(\vec{k}_1, \vec{k}_i)}{d\Omega_1 \, dU_1} \quad (30)$$

The single step DWBA cross section is to be added to the above to obtain the total contribution. The cross section, Eq. (30), is the convolution in which the system with incident momentum $\vec{k}_i$ is scattered into a state with momentum $\vec{k}_1$, followed by a scattering from $\vec{k}_1$ to $\vec{k}_2$. To obtain the two step contribution one is directed to integrate over all possible intermediate moments $\vec{k}_1$. Energy is conserved at each transition so that

$$E = U_\nu + \frac{\hbar^2}{2\mu} k_\nu^2$$

Here U_ν is the excitation energy of the target nucleus. The w's are transition probabilities

$$\frac{dw_{\nu,\nu-1}(\vec{k}_\nu, \vec{k}_{\nu-1})}{d\Omega_\nu \, dU_\nu} = 2\pi^2 \rho(k_\nu) \rho(U_\nu) \left| v_{\nu,\nu-1}(\vec{k}_\nu, \vec{k}_{\nu-1}) \right|^2 \quad (31)$$

where ρ's are density of states and $v_{\nu,\nu-1}$ are matrix elements

$$v_{\nu,\nu-1}(\vec{k}_\nu, \vec{k}_{\nu-1}) = \langle \psi_{\nu,\gamma}^{(-)} \, v_{\nu,\nu-1} \, \psi_{\nu-1,\alpha}^{(+)} \rangle$$

In Eq. (31), an average over the possible state γ and α in the ν'th and $(\nu-1)$st stages is performed. Finally

$$\frac{d\sigma_i}{d\Omega_1 \, dU_1} = \frac{2\pi m}{\hbar^2 k_i} \rho(k_1) \rho(U_1) \left| v_{1i}(\vec{k}_1, \vec{k}_i) \right|^2$$

Calculations of these formulas have been performed by Bonetti, Colli-Milazzo, Hodgson and their collaborators (6). These have been highly successful. We provide only one example here, namely the ^{120}Sn(p,n) reaction. The spectrum at 30°, 90° and 120° is shown in Figs. (3) - (5) while the angular distributions for the indicated residual nuclear excitation are shown in Fig. (6). We see that as the scattering angle increases the multistep contribution increases in importance. For example, at all excitation energies except at the lowest, the spectrum is dominated by contributions from the two and three step processes. The calculated angular distributions shown in Fig. (6) agree with experiment even at very large angles of emission. These calculations have been carried for a variety of elements for proton energies ranging from 14 to 64 MeV. At the lower energies it is necessary to include contributions from the statistical multistep compound processes which we do not have time to describe. It is remarkable that agreement with experiment is obtained with the same interaction potential v for all cases.

REFERENCES

1. L. Dohnert and H. Feshbach, Phys. Rev. C $\underline{30}$, 1358 (1984).
2. A. Gal, G. Toker and G. Alexander, Ann. Phys. (N.Y.) $\underline{137}$, 341 (1981).
3. D. Robson, Phys. Rev. C $\underline{7}$, 1 (1973).
4. N. Austern and C.M. Vincent, Phys. Rev. C $\underline{10}$, 2523 (1974).
5. H. Feshbach, A.K. Kerman, S. Koonin, Ann. Phys. (N.Y.) $\underline{125}$, 429 (1980).
6. R. Bonetti, M. Camnasio, L. Colli-Milazzo and P.E. Hodgson, Phys. Rev. C $\underline{24}$, 71 (1981);
   R. Bonetti, L. Colli-Milazzo and P.E. Hodgson, Phys. Rev. C $\underline{26}$, 2417 (1982);
   R. Bonetti. L. Colli-Milazzo and M. Melanotte, Phys. Rev. C, in press;
   L. Avaldi, R. Bonetti and L. Colli-Milazzo, Phys. Lett. B $\underline{94}$, 463 (1980).

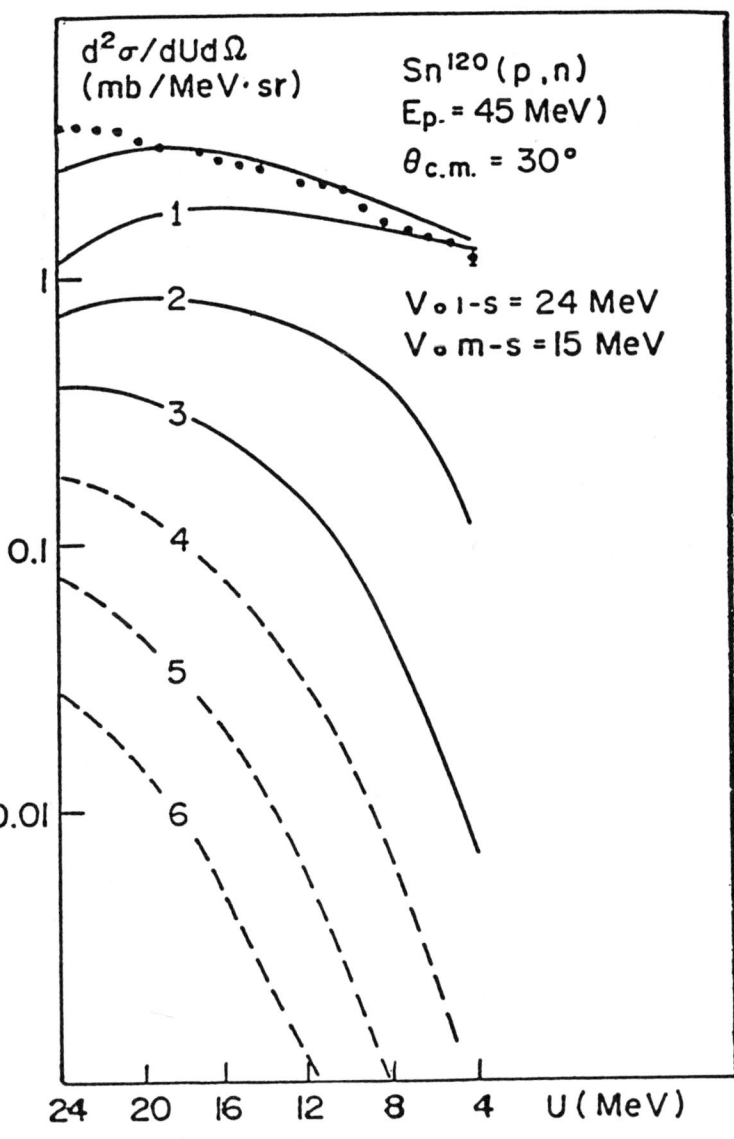

Fig. 3

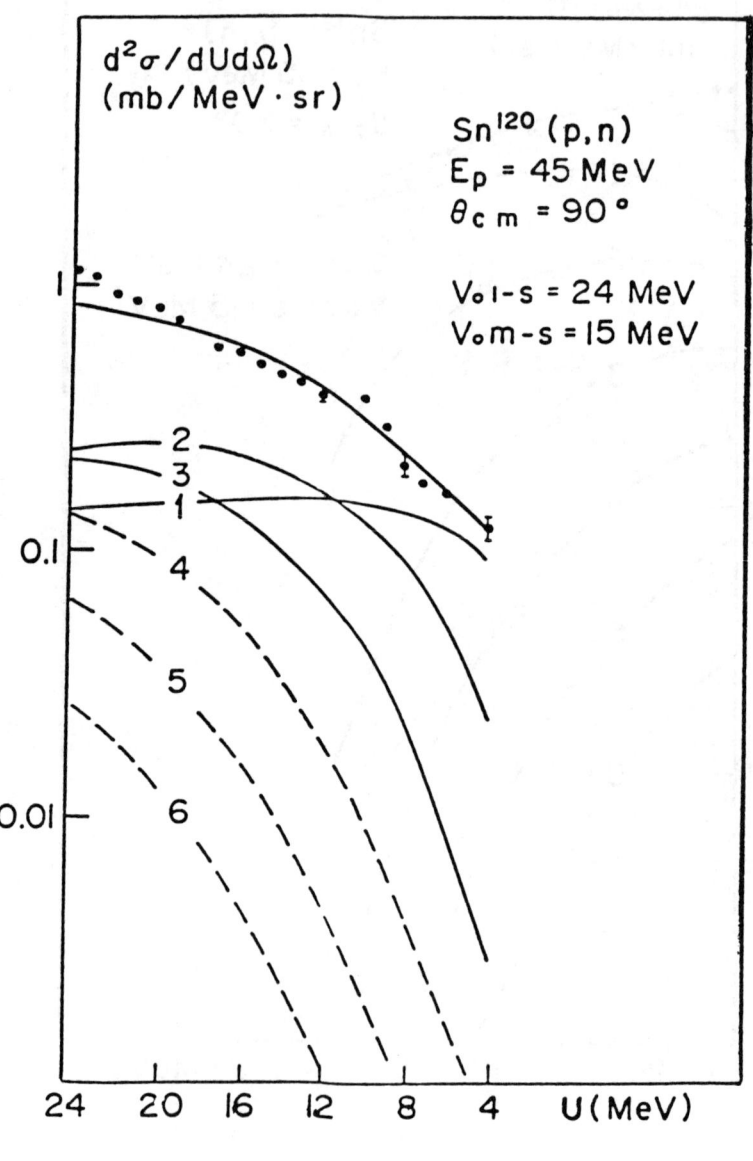

Fig. 4

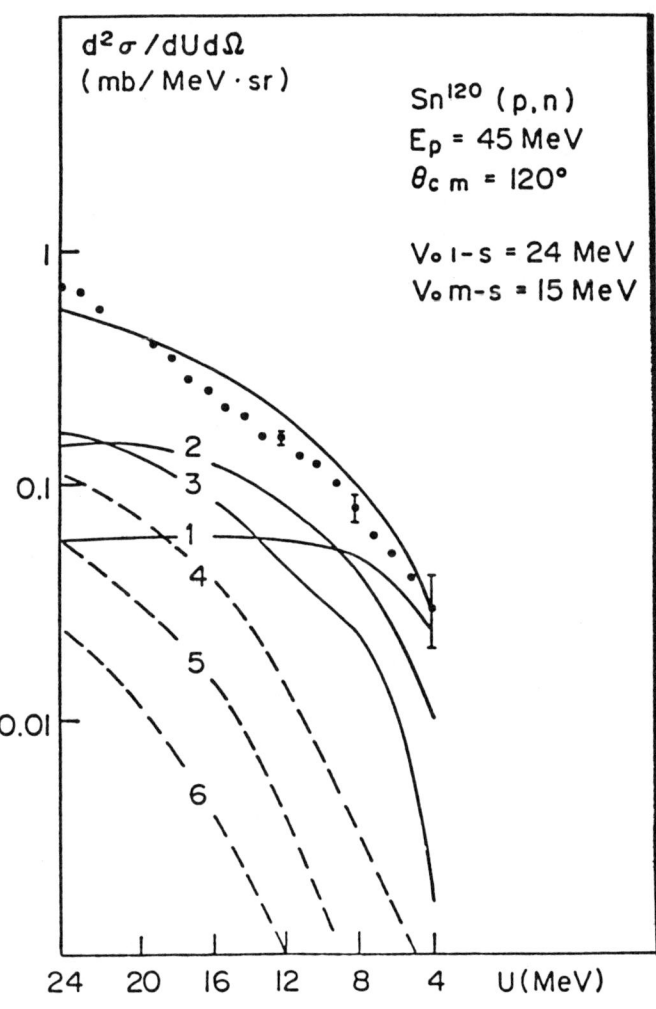

Fig. 5

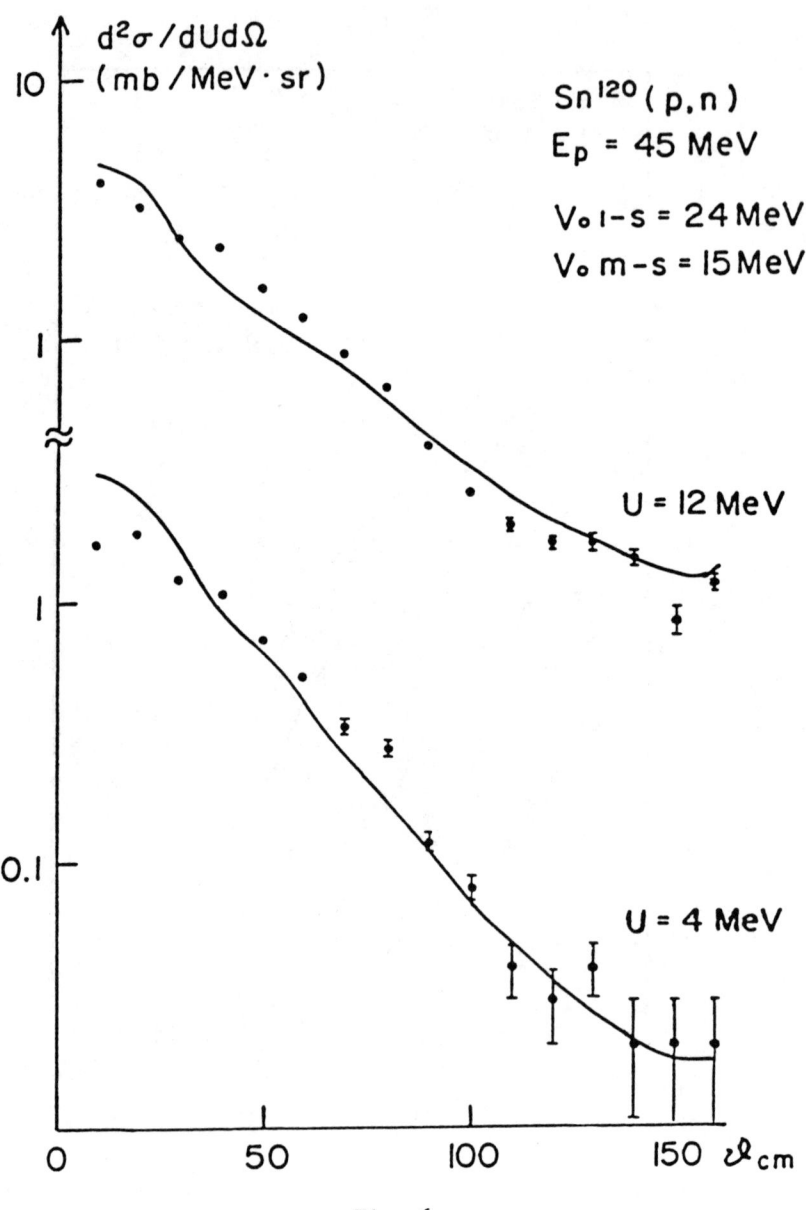

Fig. 6

STATISTICAL THEORIES OF NEUTRON
CROSS-SECTIONS OF THE ACTINIDES

J.E. Lynn
U.K.A.E.A., Harwell, Oxfordshire, OX11 ORA

ABSTRACT

The methods of treating resonance levels in statistical Hauser-Feshbach theory are reviewed. Special attention is paid to the fluctuations introduced by intermediate structure in the fission channels. The statistical theory is applied to a number of problems in the analysis of fission cross-sections of the actinides. It is shown how information on the structure of highly deformed nuclei may be obtained from such studies.

INTRODUCTION

The topic of the neutron cross-sections of the actinides is one that illustrates very well Peter Moldauer's special interests in nuclear reaction theory, in particular the fundamental statistical properties of the S-matrix and the problems of calculating average cross-sections. The actinides can well be considered as a special subject in neutron reaction theory because of the role that nuclear deformation has to play not only in the target and residual nuclei of inelastic scattering and radiative capture reactions, but also in the developing and extreme deformations of the fission process. Because of these two features there is a richness in reaction mechanisms and complexity in statistical phenomena that is not encountered elsewhere.
Section 2 of this paper reviews the basic problems of statistical reaction theory, specifically the difficulty of averaging over fluctuating cross-sections underlain by resonance poles with parameters that not only have wide statistical distributions but are also correlated in complicated ways. A desirable requirement of this averaging treatment is to bring out as nearly as possible the formal expression of independence of formation and decay embodied in the Bohr hypothesis and expressed simply in Hauser-Feshbach theory. From these more basic considerations we move on to Section 3 to consider the extra statistical difficulties imposed by intermediate structure, in this case specifically in the fission channels. This discussion is followed in Section 4 by some specific examples of the statistical treatment of fission cross-section problems, the emphasis being placed on what we can learn by neutron reactions about the fission process.

STATISTICAL REACTION THEORY

(a) Fluctuation of S-matrix parameters and transmission coefficients

The basic concept for describing neutron reactions of heavy nuclei up to medium energies is Hauser-Feshbach theory [1] embodying the Bohr hypothesis of independence of formation and decay of the compound nucleus. In its simplest form the cross-section for a reaction a→b through compound nucleus states of total angular momentum J can be expressed as

$$\sigma_{ab} = \pi \lambda_a^2 g_J T_a T_b / T \qquad (1)$$

where the T_c are transmission coefficients for formation (or decay) of the compound nucleus through channels c and T is the sum of transmission coefficients through all open channels. The transmission coefficient is normally expressed in terms of the S-matrix,

$$T_c = 1 - |<S_{cc}>|^2 \qquad (2)$$

This implies that the basic problem in statistical reaction theory is to deduce the transmission coefficients from a statistical representation of the microscopic properties of the S-matrix, in terms of compound nucleus states, and in these terms demonstrate the validity of eq. (1).

For relatively isolated resonance levels with widths Γ much smaller than level spacing D the Breit-Wigner formula shows that eq. (1) is exact at precise energies and also when the cross-section is averaged over the full extent of the resonance, but the statistical fluctuations of resonance partial widths imply that it cannot be precise when the cross-section is averaged over many resonances; it is thus common practice to introduce an extra multiplicative factor, the fluctuation factor S_{ab}, dependent on a and b, into the r.h.s. of eq. (1).

In the energy domains where resonances overlap ($\Gamma \gtrsim D$) much greater complications are introduced into the deduction of average cross-sections from the statistical properties of the levels of the S-matrix and Moldauer's work in particular was greatly concerned with these problems. Briefly, the situation is this. The expansion of the S-matrix in the complex energy (E) plane (2), is formally attractive:

$$S_{ab} = Q_{ab} - i \sum_m \frac{G_{m(a)} G_{m(b)} \exp[i(\xi_{m(a)} + \xi_{m(b)})]}{E - E_m^{(s)} + \tfrac{1}{2} i \Gamma_m^{(s)}} \qquad (3)$$

where Q_{ab} is a slowly varying background term, which is zero in the absence of direct reactions (a≠b), the $E_m^{(s)} - \tfrac{1}{2} i \Gamma_m^{(s)}$ are the poles of the S-matrix, corresponding to resonance terms, and the $G_{m(a)} \exp i\xi_{m(a)} G_{m(b)} \exp i\xi_{m(b)} (\equiv G_{m(a)} G_{m(b)})$ are their residues, which factorise into complex partial widths, amplitudes, ($G_{m(a)}$), etc. for each channel. But this expression does not automatically possess unitarity, and it can be shown that the attempt to impose unitarity upon it imposes correlations on its parameters; these greatly complicate the averaging process. On the other hand a K-matrix form of representation of the S-matrix at real energies E[3] (and the closely related R-matrix representations)

$$S_{ab} = \left[(\mathbf{1} - i\mathbf{K})^{-1}(\mathbf{1} + i\mathbf{K})\right]_{ab} \tag{4}$$

where $\hat{K}_{cd} = K_{cd} + \Sigma_\mu \Gamma_\mu(c) \Gamma_\mu(d)/(E_\mu^{(K)} - E)$ does possess unitarity but the energy averaging of this form is intractable. The R-matrix formalism[4] has the special attractiveness of relating the R-matrix poles directly to the solution of the Schrödinger equation (with imposed boundary conditions) for the internal compound nucleus of the reacting system;

$$S_{ab} = \Omega_a P_a^{\frac{1}{2}} \left[(\mathbf{1} - \mathbf{R}\hat{\mathbf{L}})^{-1}(\mathbf{1} - \mathbf{R}\hat{\mathbf{L}}^*)\right]_{ab} P_b^{-\frac{1}{2}} \Omega_b \tag{5}$$

where the Ω_c are channel phase factors, the $\hat{L}_c$ are the logarithmic derivatives of the outgoing channel wave-functions at the channel entrances, modified by the imposed boundary condition, B_c,

$$\hat{L}_c = S_c - B_c + iP_c = \hat{S}_c + iP_c \tag{6}$$

S_c being the level energy shift factor and P_c the penetration factor for the channel, and the R-matrix element is

$$R_{ab} = \Sigma_\lambda \gamma_{\lambda(a)} \gamma_{\lambda(b)} / (E_\lambda - E) \tag{7}$$

These pole parameters can not only be calculated directly for tractable physical situations (see e.g. Takeuchi and Moldauer[5]), but their statistical properties can be deduced directly from experimental data in the narrow resonance regime and extrapolated immediately to the overlapping level regime without concern for the change of penetrability with changing kinetic energy in the channels.

Despite the advantages of unitarity in the K-and R-matrix theories, and those of straightforward statistical properties, certainly in the latter, on physical grounds it would seem most appropriate to relate the transmission coefficients of Hauser-Feshbach theory to the residues of the S-matrix poles. Moldauer's attempt to do this[6] yields the following expressions

$$\sigma_{ab} = \pi \lambda_a^2 g_J \left[(2\pi/D) < |G_{m(a)}|^2 |G_{m(b)}|^2 / \Gamma_m^{(s)} > - M_{ab}\right] \tag{8}$$

$$T_c = (2\pi/D) < N_m |G_{m(c)}|^2 > - \Sigma_d M_{cd} \tag{9}$$

where

$$N_m = \Sigma_c |G_{m(c)}|^2 / \Gamma_m^{(s)} \tag{10}$$

and $M_{cd} = 2|\bar{S}_{cd}|^2 - \dfrac{2\pi i}{D} \left\langle \Sigma_{n \neq m} \dfrac{G_{n(c)} G_{n(d)} G_m^*(c) G_m^*(d)}{(E_m^{(s)} - E_n^{(s)}) + \frac{1}{2}i(\Gamma_m^{(s)} + \Gamma_n^{(s)})} \right\rangle$ (11)

The first term on the r.h.s. of eq. (8) is just that found in the simple treatment of the narrow-resonance case, and hence suggests that the second term may be small or vanishing owing to short-range level-level correlations causing an appreciable cancelling contribution from the second term on the r.h.s. of eq. (8). In fact, Moldauer[7] finds that strict cancellation of the M-terms does not occur, but, if the expression for the cross-sections (for many equally open channels) are recast in the form of the Hauser-Feshbach expression with width fluctuation factor explicitly included, his numerical experiments show that remainder terms related to the M-terms modified by statistical fluctuations and correlations of the partial widths $|G_m(c)|^2$ do effectively cancel. Thus, the Hauser-Feshbach formula for partial cross-sections of reactions proceeding through the compound nucleus appears valid, and the problem of using it reduces to finding statistical properties of the partial widths of the S-matrix poles.

The numerical results related to these statistical properties as found by Moldauer[7] are illuminating. For example, for 10 statistically equivalent uncoupled channels with transmission coefficients T = 0.75 the distribution of partial widths $N_\mu |G_\mu(c)|$ for any one channel can be described as a chi-squared distribution with $\nu \approx \frac{1}{2}$. The correlation coefficient for the partial width amplitudes for two different channels at a given pole is $\rho_{cd} \approx 0.34$. The M-terms are $M_{cc} \approx 0.64$ and $M_{cd} \approx 0.32$. From these figures we can deduce the average pole strength $<(2\pi/D)N_m|G_m(c)|^2>$ ≈ 4.3 (c.f. $T_c = 0.75$), and hence the comparative extent, for these broadly overlapping levels, that the M-terms do not vanish ($\Sigma_d M_{cd} \approx 3.5$). The effective ν-value that must be used in the chi-squared distribution in calculating the usual (narrow-resonance-based) width fluctuation factor that neglects channel-channel correlations is approximately 2 in this case. A series of calculations over a range of $T = \Sigma_d T_d$ indicates that this is a limiting value approached for full resonance overlap, and that for small values of the total transmission coefficient the expected Porter-Thomas distribution ($\nu = 1$) should be used. The asymptotic value of $\nu = 2$ can be expected intuitively from the fact that the $G_m(c)$ will be distributed uniformly around zero in the complex plane and real and imaginary components with gaussian distributions will contribute independently.

The highly-skewed nature of the distribution function of the pole channel widths and the positive correlation amongst them indicates that the total widths of the poles should also have a broad distribution. Moldauer[7] demonstrated this for a 20 channel case (with $T_c = 0.91$ for each channel). His results are qualitatively similar to a chi-squared distribution with $\nu \approx 4$, rather than the value of 20 that would be expected for 20 uncorrelated channels with Porter-Thomas distributions (as expected for the R-matrix parameters and the value of ≈ 10 for uncorrelated S-matrix pole channel widths. One way of viewing this kind of result is to see the broad distribution as a repulsion of poles in the complex energy plane, analogous to that of real energy eigen values of complicated systems on the real energy axis (leading to the Wigner distribution of spacings).

(b) Transmission coefficients from R-matrix parameters

Moldauer's work in this field, supported by others (e.g. Tepel et al[8]; see also the review by French et al [9])., has given a valuable basis of microscopic understanding and useful results allowing confident use of the Hauser-Feshbach formalism for calculating reaction cross-sections at excitation energies where resonant states overlap extensively. In this region also direct reactions can become significant, and the further modifications required to the statistical formalism have been discussed by Moldauer and others[10]. It is particularly valuable in this context for models of reactions that are of a phenomenological nature - optical models and coupled channel formalisms - used to calculate the appropriate transmission coefficients. Some complementing of this work is required if the approach to calculation of transmission coefficients is from the microscopic basis; the knowledge of strength functions and intermediate structure that can be gained from experimental study in the narrow resonance regime is most easily interpreted through R-matrix parameters.

In the narrow level regime the deduction of transmission coefficients from the average R-matrix partial widths is straightforward,

$$T_c = 2\pi <\Gamma_{(c)}>/D \qquad (12)$$

where $\Gamma_{(c)} = 2P_c\gamma_c^2$. At higher excitation energies no closed general relationship has been proved. Lane and Thomas[11] have proposed on the basis of one-channel and two-channel reduced R-matrix theory (in which the eliminated channels each satisfy the condition that $\Gamma_{\lambda(c)}/D \ll 1$) with a uniform picket-fence set of levels that

$$T_c = 2\pi(<\Gamma_{(c)}>/D)/\left|1 + \tfrac{1}{2}\pi<\Gamma_{(c)}>/D\right|^2 \qquad (13)$$

To achieve a general prescription that we can expect to be at least approximately valid we can use the observation that the sum of total widths $\Gamma_\lambda = \Sigma_c \Gamma_{\lambda(c)}$ calculated from the partial widths, for a set of R-matrix levels, is approximately equal to the sum of widths of the equivalent S-matrix poles, $\Gamma_m^{(S)}$. Then Simonius' relation[12]

$$\pi(1-T_c) = \exp(-2\pi<\Gamma^{(S)}>/D) \qquad (14)$$

leads us to the conjecture

$$(1-T_c) = \exp(-2\pi<\Gamma_{(c)}>/D) \qquad (15)$$

This relationship agrees numerically with eq. (13) within a few % up to values of $2\pi<\Gamma_{(c)}>/\bar{D} \approx 5$, and it seems possible that with physically realistic distributions of level parameters, eq. (15) should be valid up to much higher values of the argument.

We have attempted some numerical checks of the relationship (15) in a number of different physical situations. It turns out that for two-channel models in which the R-matrix reduced widths are selected randomly from Porter-Thomas distributions, and the mean widths in the channels bear a range of ratios, eq. (15) is realistic for values of $2\pi<\Gamma_c>/\bar{D}$ up to at least 5. Thus we can take eq. (15) as having general numerical validity for the actinides for channels with kinetic energy up to several MeV above any barriers they may contain.

STATISTICAL FEATURES OF INTERMEDIATE STRUCTURE IN FISSION CHANNELS

(a) Intermediate resonance modulation effects on average cross-section

Together with experimental data on the strength functions (for elastic and inelastic scattering) from neutron resonance data or optical model fitting at higher energies, the foregoing remarks comprise most of what we need to know for calculating and manipulating the neutron transmission coefficients for statistical model cross-section calculations. This is because there is no experimental evidence of intermediate structure resonance effects associated with neutron channels of nuclides in the actinide region. Indeed, Moldauer[13] has taken the view that much of what has been speculated as intermediate structure is simply the very considerable fluctuation implied by S-matrix statistics. With the important fission channels it is quite different; spectacular intermediate resonance effects are known and these are generally associated with the double-humped fission barrier associated with the actinides [14], or perhaps, for the lower-charge actinides, a triple-humped barrier. As a result, there are further important statistical modifications to be made to the Hauser-Feshbach formalism for fissionable nuclides.

At the excitation energies brought in by neutron bombardment the spacing of the class-II compound states[16] associated with the extended deformation of an actinide nucleus "caught" in the second well of the fission barrier is of the order of or less than a few keV for most actinides. The energy averaging interval adopted for statistical model estimates would normally be assumed to be considerably greater than this. Hence, there is a modulation of the fission probability $\Gamma_{(f)}/\Gamma$ within the averaging interval over and above the stochastic variation expected from the complexity of compound nucleus wave-functions. The modification of the fission probability from that due to a uniform picket-fence model of the fine structure resonances with the fission widths of the class-II intermediate states spread smoothly in Lorentzian manner is described by the expression[17]

$$\sigma_{af} = \pi\lambda^2 g_J T_a / \{1 + (T_I/T_f)^2 + (2T_I/T_f)\coth\left[\tfrac{1}{2}(T_A + T_B)\right]\}^{\frac{1}{2}} \quad (16)$$

where T_I is the transmission coefficient of the class-I compound states associated with normal deformation. At excitation energies above the barrier this differs little from the picket-fence value, but below the barrier (or, more strictly below both peaks of the double-humped barrier) it can deviate greatly from the standard statistical model, in which the transmission coefficient for fission through a double-humped barrier is described in terms of the transmission T_A, T_B through the two peaks, A,B, separately:

$$T_f = T_A T_B / (T_A + T_B) \quad (17)$$

A comparison of the standard model and modifications due to various conditions of intermediate resonance in the fission channel is shown in Fig. 1.

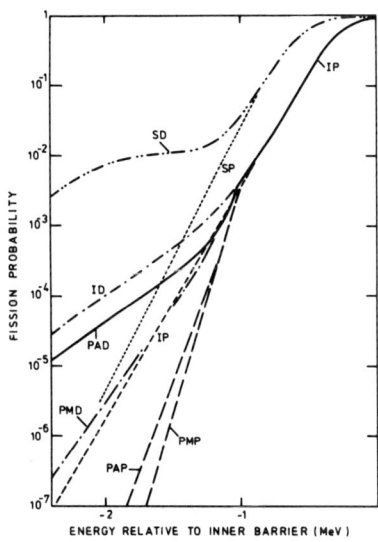

Fig. 1. Effect of intermediate structure on fission probability. SP-gross statistical model (prompt fission only; SD includes delayed fission). IP, ID - average over intermediate structure due to moderately weak coupling. PMP,D - very weak coupling, class-II state always mid-way between class-I states. PAP, D - same, but class-II state position is random average.

At excitation energies sufficiently far below barrier, intermediate resonances can become so narrow compared with the class-I ($\approx$ fine-structure) resonance spacing that perturbation treatments are valid. The two curves shown for this regime differ in the assumption of relative position of the class-II states and the nearest class-I levels, i.e. medium or averaged over the full range of class-I spacing. Also at low excitation energies fission decay of the class-II state becomes sufficiently weak that it is dominated by electromagnetic de-excitation. The low energy components of the curves in Fig. 1 therefore consist principally of delayed fission from decay of the lowest class-II state i.e. the spontaneously fissioning isomer populated by the radiative de-excitation cascade.

(b) Modifications of width fluctuation factors

The averaging factors due to fine structure level width fluctuations must also be modified to take account of the modulating effect of class-II intermediate states. Comprehensive numerical calculations of the modified fluctuation factors are not available, but indications of the magnitude can be obtained from certain limiting cases. Thus the case of a single entrance channel and one fission channel, representing the only reaction channel, each with Porter-Thomas width distributions[18] shows that the minimum value of the fluctuation factor is 0.55 rather than the value of 0.5 that would occur if the intermediate structure did not exist. But although this difference is not large the value of the fluctuation factor that should be used depends quite strongly on the sharpness of the intermediate resonance, as shown in Fig. 2. It is evident that fission cross-sections can be overestimated by a factor of up to 1.8 if intermediate resonance structure is ignored in calculating the fine-structure width fluctuation factor.

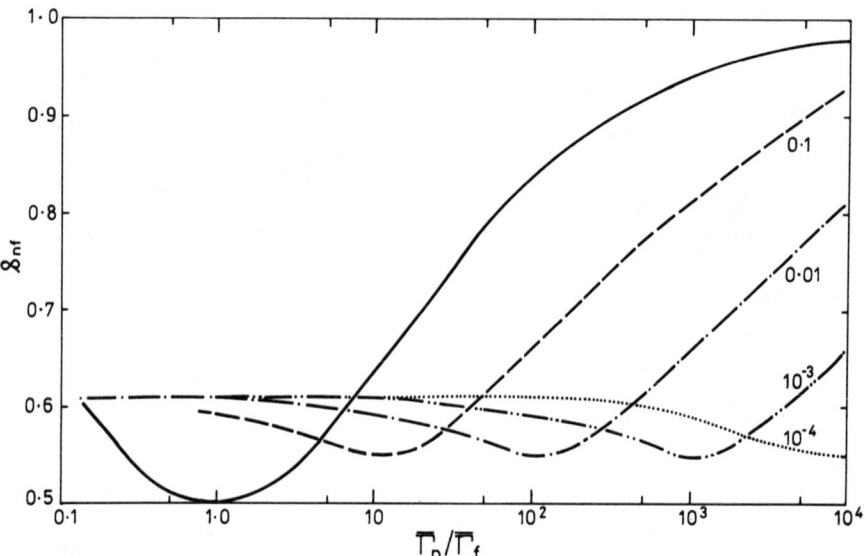

Fig. 2. Fluctuation factor for averaging fine structure states with Porter-Thomas distributions of neutron and fission widths through intermediate resonance with various values of width to spacing.

Normally the class-II states are themselves states of quite complex character. They are in fact states of a super-deformed compound nucleus with excitation energy only 2 to 3 MeV lower than the class-I states with which they couple. Hence we can expect their partial widths and related quantities, such as the coupling matrix elements to the class-I states to fluctuate in the same way as the class-I compound nucleus states. So, superimposed on the intermediate resonance modulating effect, and the fine structure level width fluctuation effect, we can expect yet another effect, the intermediate level width fluctuation factor. Integral expressions for evaluation of these fluctuation factors are given in ref.[19]. The minimum expected value for the fluctuation factor for a fission cross-section is expected to be of order $2/\pi$.

(c) Fluctuation properties of the intermediate state coupling width

An interesting facet of intermediate level fluctuation effects is the character of the distribution of the coupling widths. The matrix elements on which the coupling width is based can be formulated as a set of contributions from different channels across the first fission barrier hump connecting the primary and secondary wells of the deformation energy curve. But because the secondary well, with its characteristic class-II compound states. must be placed firmly within the internal region of any physically sensible separation of internal nucleus motion from channel motion these channels cannot retain their identity, and hence it is construed that the distribution of the class-II state coupling widths has the Porter-Thomas character, irrespective of the number of channels, in the sense of number of different internal degrees of freedom, that may effectively contribute to it. Hence, it can be surmised that fluctuations in the fission cross-section on the energy scale of the class-II level spacing will still persist at a significant level even when the class-II resonances overlap. The effective number of degrees of freedom entering the chi-squared distribution function for the transmission coefficient T_A will be of the order of $T_A/2\pi$ when $T_A \gtrsim 2\pi$ i.e. to the average number of class-II levels contributing significantly to the cross-section at a given energy. This measure of fluctuation should be used in calculating the integral expressions mentioned in Section (b) above. It should also be noted that this implies a considerable degree of fluctuation in fission cross-sections at excitation energies considerably above the fission barrier where very appreciable overlap of class-II levels will occur.

(d) Vibrational resonances

In the lower charge actinides (thorium and protoactinium nuclides) the intermediate structure resonances that have been found in the fission cross-sections are on a much larger scale than those with higher charge. These are generally known as vibrational resonances, it being assumed that their strong fission properties are due to the state principally having the character of a β-vibration about a nuclear shape with very extended equilibrium deformation. The relative lack of damping of such states into a denser background set with more complicated wavefunctions has led to the conclusion that they must have relatively little effective excitation energy for internal modes of motion and hence located in a relatively shallow potential well. It has been speculated that this shallow potential well is not the secondary well of the fission barrier, which theory and experimental data on the higher actinides suggest is some 3 to 4 MeV deep at neutron threshold, but rather the tertiary well splitting the outer barrier peak found in some sets of calculations on fission barrier mapping[20]. Whatever their source these vibrational resonances are sufficiently prominent objects both featurally and structurally that cross-section theories must aim to accommodate them rather than average them out.

The existence of the vibrational resonances has additional statistical implications. If the inner barrier of the (now) multi-humped barrier is very low compared to the excitation energy available it can be imagined that the class-I compound states will couple directly with the vibrational state in the tertiary well. Alternatively, if the class-II states have not lost all vestiges of their individual character in being mixed indiscriminately into the class-I states, then we must consider an hierarchy in which the class-I states, and hence the compound nucleus fine-structure levels, pick up their fission widths first by coupling with the class-II states which are in turn coupled to the vibrational state.

In the former case the fission widths of the individual fine-structure resonances can be expected to have a Porter-Thomas distribution about the smooth profile of the vibrational resonance, and the usual frame-work of Hauser-Feshbach theory with standard width fluctuation factors can be applied. In the second case the additional fluctuation factor due to intermediate structure must be applied. There is no evidence that distinct class-II resonances appear in the cross-sections of the low-charge actinides, and the theoretical calculations generally suggest that the inner barrier peak is sufficiently below neutron threshold that such resonances would overlap. Nevertheless, even for overlapping class-II levels the fluctuation of the coupling width across the inner barrier could be significant for quantitative calculations, as suggested in Section (c).

(e) General

The foregoing discussion applies mainly to comparatively low energy neutron reactions where resonances do not have appreciable overlap. As such the assessments of intermediate structure effects can be used fairly safely for quantitative calculations up to perhaps 1 MeV neutron energy. For the overlapping resonance regime we still require a fully comprehensive set of numerical calculations of these fluctuation effects.

EXAMPLES OF APPLICATION OF STATISTICAL REACTION THEORY TO THE ACTINIDES

Experimental evidence for intermediate levels governing fission channel properties of the actinides is spectacular, but extensive quantitative knowledge of its detail is surprisingly incomplete. The detailed parameters and strength of coupling are known for only a small handful of class-II compound states. Spin and parity dependence of these effects are virtually unknown, and we have little information on the pattern of fission channels open to the intermediate states. Experimental work in this field is difficult and demanding, requiring high energy resolution, intense neutron sources and sophisticated fission particle detection techniques. But the information that could be achieved is very significant.

Above the resolved resonance range of neutron energy, the overall energy dependence and absolute cross-sections (to within 20% or so) of many fission cross-sections have been measured. From these, and related data initiated by charged particle and transfer reactions, an overall survey of the fission barrier properties of many of the higher actinides has been completed[21]. The stastical theory analysis of these cross-sections has demonstrated the collective enhancement of the density of intrinsic states at the deformations associated with the fission barrier peaks; these enhancements are numerically of the order of 4 to 5 at the inner barrier, attributed to the extra rotational bands associated with the non-axially-symmetric shape of the nucleus at this barrier, and a factor of about 2 at the outer barrier, attributed to the opposite parity rotational bands allowed by the mass-asymmetric shape at this later stage of the fission process. In the lower actinides a number of vibrational resonances have been observed and the extensive experimental data on at least one of these have been analyzed with considerable success. In this section we survey some of the more detailed analyses of certain actinide fission cross-section that have special features or difficulties associated with them. In most of these calculations the neutron transmission coefficients are based on the limited information on s-wave and p-wave neutron strength functions measured in the resonance region, these being used in eq. (15) for the calculation of even-l and odd-l transmission coefficients respectively. Transmission coefficients for electromagnetic radiation are based on the giant dipole resonance formalism normalised to the neutron resonance radiation width of ^{238}U.

(a) The fission cross-section of ^{238}U

The fast neutron-induced fission cross-section appears generally to have the classic energy dependence expected of the Hill-Wheeler form for the transmission coefficient. Indeed its fall-away from the barrier region to lower energies is almost too sharp. Statistical model calculations to fit this cross-section, indicate that there appears to be a dearth of low-spin barrier transition states; in other words the neutron channel centrifugal barriers appear to assist significantly in suppressing the fission cross-section at lower energies. Barrier parameters deduced from this kind of fit[21] indicate that the height of the inner barrier is $V_A \sim 1.6$ MeV, with a Hill-Wheeler penetrability parameters, $\hbar\omega_A \sim 0.8$ MeV, while the outer barrier has $V_B \sim 1.3$ MeV, $\hbar\omega_B \sim 0.52$ MeV.

In spite of these high barriers and the sharp fall of the fission cross-section below them a measurable fission cross-section has been observed in the s-wave resonance region. This is in the form of very sharp intermediate structure resonances. Analysis of the best information on the resonance parameters of the group of fission resonances lying about 720eV[22] indicates that the class-II state fission width $\Gamma_{\lambda II}(f) \approx 1.6$ meV and the coupling width

$\Gamma_{\lambda_{II}(c)} \sim 2.5\text{eV}$. These, with the observed class-II level spacing $\sim 1\text{keV}$, imply barrier heights for the ^{239}U system with total angular momentum and parity $J^{\pi} = \frac{1}{2}^{+}$ of $V_A(J^{\pi} = \frac{1}{2}^{+}) \sim 0.6\text{MeV}$, $V_B(J^{\pi} = \frac{1}{2}^{+}) \sim 1\text{MeV}$, both values being very much lower than the results of the fast neutron cross-section fitting.

Various hypotheses can be put forward to resolve this discrepancy. One is that the class-II fission width as measured is not prompt fission, but electromagnetic de-excitation followed by delayed fission from the ^{239}U spontaneously fissioning isomer. This would allow the outer barrier (but not the much lower inner barrier) to be close to the statistical model barrier. Apart from the discrepant inner barrier height, the main difficulty with this hypothesis is the fact that no spontaneously fissioning isomer with the right half-life (less than the time resolution 0.25µs of the neutron resonance measurements, which is one or two orders of magnitude lower than systematics would suggest) has been observed, yet an anomalous gamma-cascade from the quasi-class-II resonance at 721eV$^{(23)}$ is suggested by experiment (23).

It is also possible that the 721eV resonance observed in the fission cross-section is not identical with the resonance at similar energy observed in the total cross-section. If this is the case analysis of the group would infer that $0.02\text{eV} \lesssim \Gamma_{\lambda_{II}}(f) \lesssim 0.5\text{eV}$ and $\Gamma_{\lambda_{II}(c)} \lesssim 0.13\text{eV}$, with corresponding barrier heights $V_A(J^{II} = \frac{1}{2}^{+}) \sim 1.0\text{MeV}$, $V_B(J^{II} = \frac{1}{2}^{+}) \sim 0.75\text{MeV}$. The further hypothesis of delayed fission allows the outer barrier to rise to at least the statistical model value, while the inner barrier is at a rather more acceptable value than before. The calculated fission cross-sections up to 1.6 MeV neutron energy, based on this hypothesis are shown in Fig. 3 and appears very acceptable. Prompt fission alternatives are also shown and tend to run much too high at intermediate energies. The difficulty concerning the isomer observations remains as before, perhaps even more acutely, because the branching ratio to fission would now be unity.

(b) Vibrational resonance in the fission cross-section of ^{230}Th

Of the vibrational resonances in the fission cross-sections of the low charge actinides the most intensively studied is that at 720keV neutron energy in the cross-section of ^{230}Th. The available data comprise a rather high energy resolution measurement and several sets of, often discrepant, fission product angular distributions with moderate neutron energy resolution. A selection of these data are shown in Fig. 4.

Analyses of the data have all rested upon the assumption that the resonance is due to a β-vibrational state in a well at high deformation in the fission barrier, the associated intrinsic state of motion of the odd-mass nucleus being that of a single-particle neutron with spin-projection $K = \frac{1}{2}$ (this is dictated by the forward-peaked angular distribution of the fission products relative to the neutron beam direction) and an associated rotational band.

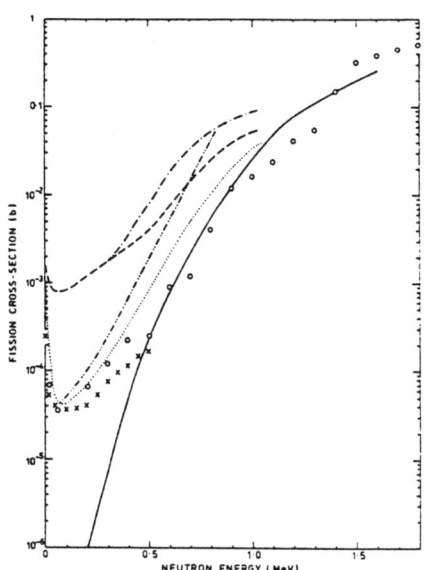

Fig. 3. The fission cross-section of ^{238}U. Circles are data points. Full curve is Hauser-Fechbach calculation using barrier transition states close to barriers at about 1.4 MeV. Dashed and dotted curves include $J^\pi = \frac{1}{2}^+$ barriers (with and without rotational bands) quoted in text. Crosses are modified $J^\pi = \frac{1}{2}^+$ barriers with delayed fission hypothesis.

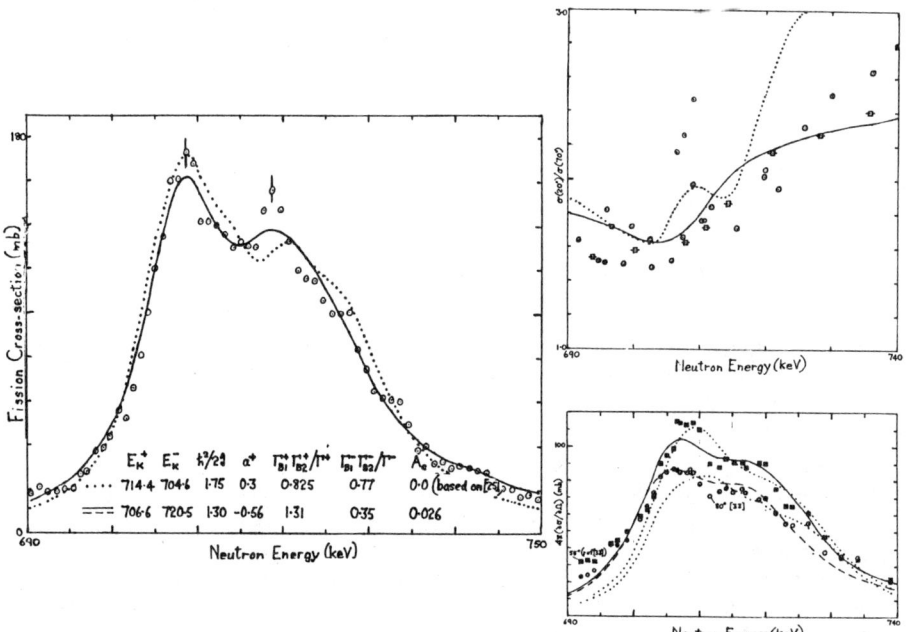

Fig. 4. Fission cross-sections and fission product angular distributions for ^{230}Th (n,f) Data from refs. (25,27,33). A_c is coriolis admixture coefficient $[(\hbar^2/2\mathcal{I})\langle K+1|j_\perp|K\rangle/(E_k - E_{k+1})]^2$

It has generally been found that a single vibration-rotation band is insufficient to adequately fit the data unless important modifications are made to the higher spin states within the band; these may arise from admixture with states from bands based on other single-particle configurations and the effects of Coriolis coupling in such admixing[24]. The key to a reasonable fit based on a "simple" model is the recognition that if the source of the vibration is a tertiary potential well in the fission barrier, the equilibrium deformation is strongly mass asymmetric along the axis of elongation, see Fig. 5. The high potential hill between the two positions of mass asymmetry implies that a second rotational band based on the same single-particle configuration but having opposite parity will be almost degenerate (at the band-head) with the first. The important Coriolis de-coupling parameter $a_{\frac{1}{2}}^{(-\pi)}$ associated with the second band will have opposite sign from that of the first $a_{\frac{1}{2}}^{(-\pi)} = -a_{\frac{1}{2}}^{(\pi)}$.

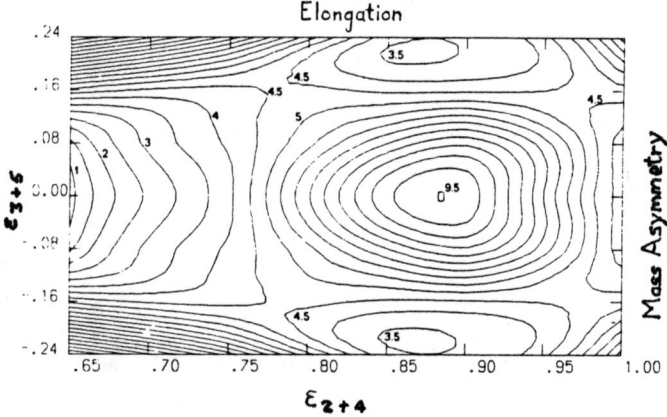

Fig. 5. Strutinsky type potential energy calculations of thorium nuclide as function of elongation and mass asymmetry, showing tertiary well(s) (after (20)).

Many authors (e.g. 25, 26) have carried out an extensive series of studies comparing this model with the available data. Blons et al's last preferred fit to the data is shown in Fig. 4; the angular distribution data of James, Grainger and Syme[27], which systematically differs from older data is included in this diagram. The barrier transmission coefficients for the two potential cols enclosing the tertiary minimum are similar, the only variation being due to the relative positions of states of the same spin and parity stemming from the different parameters of the rotational bands at each barrier. A more important source of variation could arise from the weakening of the potential hill at a given barrier between deformations of opposite mass symmetry; the opposite parity intruder band could then be appreciably higher than the normal parity band with consequent decrease in the barrier transmission coefficients of the two bands. The effect, which is intuitively expected to be stronger for the inner rather than the outer barrier, is included in the fits shown in Fig. 4. An effect for one barrier undoubtedly seems present, lowering these transmission coefficients for the odd-parity band by a factor of about 4.

The moment of inertia parameter that is deduced for these vibration-rotation bands is $\hbar^2/2\mathcal{J} \lesssim 2\text{keV}$ and the decoupling parameter for the even-parity band, which seems to be the basic Nilsson orbital is $a_{\frac{1}{2}}^{(+1)} \approx 0$. The former quantity suggests the very high deformation of the equilibrium shape of the vibration and is consistent with the tertiary well hypothesis. The second quantity gives additional information the specification of the Nilsson orbital of the single particle state in the band-head. Assuming the equilibrium deformation to be that of the tertiary well magnitude of $a_{\frac{1}{2}}$ suggests the orbital involved is $[631]\frac{1}{2}^+$.

One of the problems connected with analyzing the neutron-induced fission data of ^{230}Th, is the sparseness of definitive rotational state structure emerging from the natural line-width envelopes of the resonances. Higher states in the rotational sequence are more widely spaced, and these can be made visible in the (d,pf) reaction if measured with high energy resolution. Blons et al[28] have carried out some preliminary measurements of this kind, which seem to confirm their most favoured parametrisation, with observable spin states apparently to $J^\pi \approx 9/2^-$ and $11/2^-$.

(c) Intermediate resonances and other structure in the cross-section of ^{232}Th.

At higher neutron energies angular distributions of fission products from the neutron-induced fission of ^{230}Th fluctuate with neutron energy, but the cross-section structure is relatively featureless. By contrast the neutron fission cross-section of ^{232}Th is rich in structure at neutron energies between ~1.4 and ~2 MeV. The character of the cross-section appears to be resolvable into a rather smooth background cross-section of significant magnitude, on which is superimposed resonance groups about 100keV wide. Superimposed on this again is a much narrower structure of rather weak amplitude. Various qualitatively successful attempts have been made to fit the broad structures as vibrational resonances associated with single-particle neutron states with a range of total angular momentum and parity K^π (mostly with K>3/2 to explain sideways peaked fission product distributions), but the further attempt to recognise the narrower structure as rotational-band members of the vibrational resonances does not seem to scale correctly. The large background term is also not explicable within the simple model.

At lower energies[29] the fission cross-section has some interesting features (see Fig. 6), of which the most significant is perhaps the near-plateau region extending from 0.1 MeV to 0.6 MeV. When the energy behaviour of the neutron wave-number and centrifugal barrier terms are taken into account this appears to represent one or more resonance structure ~ 100-300 keV wide. Assuming that the resonance structures above 1.4 MeV are to be associated with the tertiary minimum of the fission barrier, these resonances would have to be about two orders of magnitude narrower to be one of their class. It is possible therefore that these are damped vibrational resonances associated with the secondary minimum. This implies the barrier height V_A is fairly close to the neutron separation energy. An attempted fit to the cross-section based on this hypothesis is shown in Fig. 6.

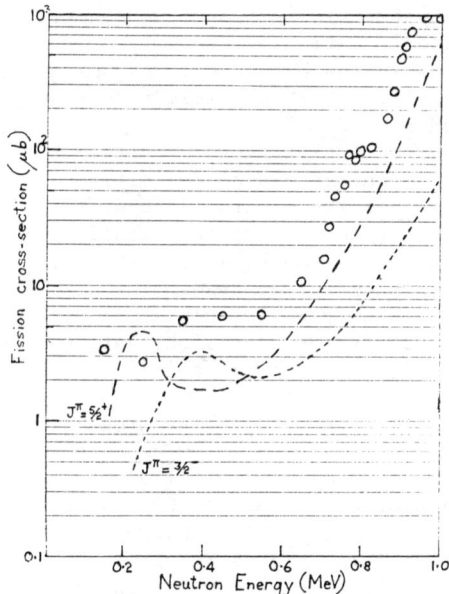

Fig.6. Neutron fission cross-section of ^{232}Th (28) up to 1 MeV. Indicative cross-sections for $J^\pi = 3/2^-, 5/2^+$ vibrational resonances with energies 0.37, 0.22 MeV, damping widths 0.15, 0.07 MeV and barrier B trans-energies 1.65, 1.34 MeV respectively are shown. Barrier A height is 0.4 MeV.

At higher energies the influence of class-II states can be revealed by fluctuations. We would estimate that at 1 MeV above the inner barrier the density of transition states of given (low) spin and parity is ~ 100 MeV^{-1}. This implies a fairly complete overlap of about 20 class-II compound states at any given energy, and an associated fluctuation level of perhaps 10 to 20%. We see this kind of fluctuation persisting in the ^{232}Th fission cross-section through the lowest vibrational resonances at 1.4 MeV. The modulation intervals of the fluctuation is ≈ 15 keV[30]. This is also qualitatively consistent with the above assumption of density of transition states and a secondary well 2.5 to 3.0 MeV above the ground state.

(d) Some observations on mass-distribution in neutron induced fission

In the actinides at low to moderate excitation energies the commonest mode of final division into fragments is asymmetric. Although this seems to bear some relationship to the quantitative theoretical calculations of potential energy of the nucleus as a function of deformation in the region of the fission barrier, these calculations showing the col at the outer barrier at mass asymmetric deformation, more careful study indicates that the final mass ratio distribution of the products is determined several MeV below the barrier. Particular shell effects in the potential energy surface approaching the scission point that are known to have major effects on the mass distribution are the $Z = 50$, $N = 82$ shells dominating the heavy mass peaks in the asymmetric division of uranium and plutonium nuclides and also the symmetric fission of the heavier fermium isotopes, and the $N = 50$ (spherical) and $N = 88$ (highly deformed) shells governing the even greater mass asymmetry of the lighter thorium nuclides. On this slope towards scission a complicated interplay of dynamic factors, including both inertia and viscosity, with the potential energy surface is at work, and simple conclusions are not easily forthcoming. Nevertheless, there are some striking features of mass distribution of the final products that suggest that potential energy features at the barrier are initiating them even though there may be considerable modification in the subsequent descent from saddle to scission.

In this context the well-known coexistence of symmetric and asymmetric modes of mass division in the fission of nuclides with mass $\sim 220 < A < \sim 230$ is remarkable. A recently discovered example of this kind of triple-humped mass yield peak is its occurence in the thermal neutron-induced fission of ^{229}Th[31]. It is a question of special interest therefore, whether the symmetric mode represents a special (high) valley in the potential energy landscape that is entered when the elongating nucleus has already descended a considerable way from saddle to scission, or whether it already has its own individual saddle point.

Some relevant information on this point has been found in studies of fission yields of actinium nuclides formed by charged particle transfer reactions[32]. The very rapid rise of symmetric fission persists 1 MeV or more beyond the equivalent rise in asymmetric fission. This rapid rise, which would normally be associated with a barrier tunnelling effect, is then followed in the case of symmetric fission by a slower rise. This latter rise is suggestive of the increasing share of phase space taken by symmetric fission as available internal energy of the system increases above the higher symmetric fission barrier. Study of the neutron-induced fission of a nucleus like ^{229}Th over a wide range of neutron energy with comparatively good energy resolution could add more definitive information to this topic. Firstly the hypothetical sub-barrier region for symmetric fission could be studied in much greater detail. Secondly, vibrational resonance phenomena will almost certainly be present in the fission cross-section of ^{229}Th. Absence of these in the symmetric fission yield would indicate a completely different route through the barrier region.

(e) Fission at higher neutron energies; second-chance fission

The systematic set of barrier parameters that have been deduced for the actinides from neutron reactions and related data in the region of the barrier parameters, together with reasonable extrapolations of level density data, are found to be capable of explaining quite well the fission cross-sections at higher neutron energies with the onset of multi-chance fission. One systematic discrepancy seems to remain; this is the sharpness of the rise in the statistical theory cross-section at each new threshold, particularly marked at second-chance fission.

One possibility to account for this is the distortion introduced by pre-equilibrium neutron emission. The statistical theory code can be modified to include in the residual nucleus after first neutron emission a slowly-rising level density component that represents the final state distribution for pre-equilibrium neutron emission. A calculated curve that contains this modification improves the fit . The extra level density component in this near-fit has a "temperature" of $\sim$ 2 MeV and accounts for some 30% of initial neutron emission. This is close to the expectation of pre-equilibrium theory.

CONCLUSIONS

The neutron cross-section of the actinides, especially those related to fission, are capable of yielding rich information on the structure of very heavy nuclides at very extended deformations. To extract this information measurements with high energy resolution seeking fine details of the fission process such as angular distribution and mass yields of fission products are required. At the same time sophisticated statistical analysis of the cross-sections is important, full account being taken of the various fluctuation effects that can arise from the hierarchy of fine-structure, intermediate-structure and gross resonance effects underlying the reaction process. The examples we have presented of applications of statistical theory to recent neutron-induced fission data show that there is still a great deal to be done and understood in this field.

REFERENCES

(1) Hauser, W. and Feshbach, H. (1952) Phys. Rev. 87. 366.
(2) Humblet, J. and Rosenfeld, L. (1961) Nucl. Phys. 26, 529.
(3) Humblet, J. (1970) Nucl. Phys. A151, 225.
(4) Wigner, E.P. and Eisenbud, L. (1977) Phys. Rev. 71, 29.
(5) Takeuchi, K and Moldauer, P.A. (1970) Phys. Rev. C2, 925.
(6) Moldauer, P.A. (1964) Phys. Rev. 135, B642.
(7) Moldauer, P.A. (1975) Phys. Rev. C11, 426.
(8) Hofmann, H.M., Richert, J, Tepel, J.W. and Weidenmüller, H.A. (1975) Ann. Phys. 90, 403 & 391.
(9) Brody, T.A., Flores J., French, J.B., Mello, P.A. Pandey, A. and Wong, S.S.M. (1981) Rev. Mod. Phys. 53, 385.

(10) Moldauer, P.A., (1975) Phys. Rev. 12, 744.
(11) Lane, A.M. and Thomas, R.G. (1958) Revs. Mod. Phys. 30, 257.
(12) Simonius, M. (1974) Phys. Lett. B52, 279.
(13) Moldauer, P.A. (1967) Phys. Lett. 18, 249.
(14) Strutinsky, V.M. (1967) Nucl. Phys. A95, 420.
(15) Möller, P. and Nix, J.R. (1974) in Physics and Chemistry of fission, Proc. of Rochester Conf. (IAEA, Vienna) Vol. 1, p.103.
(16) Lynn, J.E. (1968) Symp. on Nuclear Structure, Dubna, p.463.
(17) Lynn, J.E. and Back, B.B. (1974) J. Phys. A7, 395.
(18) Lynn, J.E. (1978) Neutron Physics and Nuclear Data, Proc. of Conf. at Harwell (OECD Nuclear Energy Agency, Paris) ed. G.D. James, p.941.
(19) Lynn, J.E. (1981) in Nuclear Fission and Neutron-induced Fission Cross-sections, ed. A. Michaudon (Pergamon, Oxford) p.157.
(20) Möller, P. (1980) in Physics and Chemistry of Fission, Proc. of Conf. Julich (IAEA, Vienna), vol.1, p.283.
(21) Bjornholm, S. and Lynn, J.E. (1980) Rev. Mod. Phys. 52, 725.
(22) DeSaussure, G., Olsen, D.K., Perez, R.B. and Difilippo, F.C. (1978) Oak Ridge Report ORNL/TM-6152.
(23) Browne, J.C.(1976), Proc. Int. Conf. on Neutron Int.,Lowell v2
(24) Lynn, J.E. (1983) J. Phys. G9, 665.
(25) Blons, J., Mazur, C., Paya, D., Ribrag, M. and Weigmann, H. (1984) Nucl. Phys. A414, 1.
(26) Boldeman, J.W., Gogny, D., Musgrove, A.R.D. and Walsh, R.L. (1980) Phys. Rev. C22, 627.
(27) James, G.D., Syme, D.B. and Grainger, J. (1983) Nucl. Inst. 205, 151.
(28) Blons, J., Fabbro, B, Hisleur, J.M., Julien, J., Mazur, C., Pattin, Y., Paya, D. and Ribrag, M. (1983) International winter meeting on nuclear physics, Bornio, Jan. 1983.
(29) Perez, R.B. de Saussure, G., Todd, J.H., Yang, T.J. and Auchampaugh, G.F. (1983) Phys. Rev. C28, 1635
(30) Auchampaugh, G.F., Plattard, S., Hill, N.W., de Saussure, G., Perez, R.B. and Harvey, J.A. (1981) Phys. Rev. C24, 503.
(31) Asghar, M. et al (1982) Nucl. Phys. A373, 225.
(32) Specht, H.J. (1973) Proc. Int. Conf. on Nucl. Physics, Munich (North-Holland, Amsterdam), vol. 2 p.311.
(33) Veeser,L.R., and Muir,D.W.(1981) Phys. Rev,C24,1541.

OPTICAL MODELS FROM LOW-ENERGY
s-, p- AND d-WAVE CROSS SECTIONS

C. H. Johnson
Oak Ridge National Laboratory, Oak Ridge, Tennessee 37831

ABSTRACT

From transmission measurements with good resolution at low energies one can obtain data on the optical model potential (OMP) for individual partial waves by first making a multilevel analysis to isolate the partial waves and then averaging for comparison to the OMP. For each J^π the averaging yields two quantities which are related to the amplitude and phase of the OMP scattering function or, alternatively, to the volume integrals of the real and imaginary potentials. Historically, the experimental averages have been represented by the s- and p-wave strength functions, S_0 and S_1, and the s-wave scattering radius R'. To make full use of data from modern time-of-flight facilities such as ORELA it is necessary to re-examine the averaging procedure in order to extend it upward both in energy and neutron ℓ-value. This averaging is discussed and applied to data on ^{30}Si, ^{32}S, ^{34}S, ^{40}Ca, ^{48}Ca, ^{60}Ni, ^{86}Kr and ^{208}Pb. The resulting OMP shows a systematic real potential with some indication of a parity dependence. The imaginary potential shows considerable fluctuations indicating the importance of nuclear structure at neutron energies below 1 MeV. A coupled channel OMP is also discussed for some of the nuclei.

INTRODUCTION

Most data for the neutron phenomenological optical model potential (OMP) have come from measurements at several MeV where many partial waves are effective. However, my subject is the deduction of model parameters for individual partial waves from multilevel analyses of high-resolution cross sections for neutron energies up to about 1 MeV. In 1963 Peter Moldauer wrote a paper[1] entitled "Optical Model for Low Energy Neutron Interactions with Spherical Nuclei." That paper provides valuable insight on the role of the OMP for low energy neutrons. Moldauer's primary experimental data were the s-wave and p-wave strength functions, S_0, S_1, and the s-wave potential scattering radius R'. In 1963 such data were obtained with lower energies and poorer resolution than are now possible. He also used total cross sections and angular distributions measured with poorer resolution at higher energies.

In the twenty years since Moldauer's paper the experimental techniques for high resolution transmission measurements have been improved such that there are now extensive tabulations[2] of S_0, S_1 and R'. The use of these data for parametrizing the OMP has continued. Delaroche, Lagrange and Salvy[3] refer to the use of S_0, S_1 and R', along with the total cross sections, as the SPRT method.

However, high resolution measurements provide far more information on the OMP than is embodied in S_0, S_1 and R'. For each partial wave the OMP scattering function embodies two quantities, i.e., an amplitude and a phase. In turn, these can be related to the real and imaginary volume integrals for the given partial wave. Thus, for $s_{1/2}$, $p_{1/2}$, $p_{3/2}$, $d_{3/2}$ and $d_{5/2}$ neutrons there are altogether ten OMP quantities. In principle these should be available from an experiment that is performed with good resolution such as to isolate and measure each partial wave. In fact, S_0 and R' do give the amplitude and phase of the s-wave scattering function, but only at low energies. Clearly, we need to extend the OMP analysis to higher energies and higher partial waves. That is the purpose of this study.

EXPERIMENTAL DATA

All measurements discussed here were obtained by several researchers at the neutron time-of-flight facility ORELA, i.e., the Oak Ridge Electron Linear Accelerator. Nearly all of the data come from transmission measurements using a burst width of about 6 nsec and a 200-m flight path. The resulting energy resolution, $\Delta E/E$, is better than 10^{-3} for neutron energies up to about 1 MeV. Some of the data are supplemented by differential scattering measurements. Only zero spin targets are included.

The total cross section is a sum over partial waves. For a given partial wave

$$\sigma_T = 2\pi k^{-2} [1 - \mathrm{Re}\, U(E)] \tag{1}$$

where the scattering function U is unitary for the one-channel cases studied here. We expand in the R-matrix formalism, which automatically embodies the unitary property;

$$U(E) = e^{-2i\phi(E)} \frac{1 + iP(E)\, R(E)}{1 - iP(E)\, R(E)} \tag{2}$$

where ϕ is the hard-sphere phase shift, P is the penetrability and R is the R-function. Each function depends on the chosen boundary radius a. In writing Eq. (2) we have set the boundary condition equal to the shift factor. The R-function includes a sum over the N resonances observed in the measured region from E_ℓ to E_u plus a parametric function which accounts for all levels outside the region;

$$R(E) = R^{\mathrm{ext}}(E) + \sum_{\lambda=1}^{N} \gamma_\lambda^2/(E_\lambda - E), \tag{3}$$

where γ_λ^2 is a reduced neutron width and E_λ is a resonance energy. Both terms in Eq. (3) are important not only for the fitting but also for the eventual OMP interpretation.

Using this formalism we fit the observed transmissions with the effects of finite resolution included. For each prominent resonance the correct ℓ and J are deduced from the resonance-potential interference pattern and from the peak height. Often correct assignments are not possible for very narrow resonances, but the OMP analysis is not sensitive to the narrower levels. For higher partial waves the effect of the centrifugal barrier makes all resonances narrow such that a meaningful OMP analysis is not possible. For some nuclei angular distributions were obtained to facilitate $J\ell$ assignments.

QUANTITIES FOR COMPARISON OF DATA TO THE OMP

One might assume that the higher partial waves could be described by an extended group of quantities like S_0, S_1 and R'. However, those quantities are appropriate only at very low energies. The scattering radius R' is normally defined as a low energy limit and is deduced from data by equating the off-resonance cross section to $4\pi R'^2$. The s-wave strength function is defined by

$$S_0 = \langle 2P_0 \gamma_\lambda^2 E_\lambda^{-1/2}\rangle/D \tag{4}$$

where D is the average level spacing and E_λ is a dimensionless quantity equal to the energy in eV. Often S_0 has been regarded as if it were independent of the boundary radius at all energies. Indeed, invariance does hold at low energies where $2P\gamma_\lambda^2$ is essentially the observed neutron resonance width Γ_n. However, at energies of a few hundred keV one finds from a multilevel analysis[4,5]

$$2P\gamma_\lambda^2 = \Gamma_n [1 + (P_0 R_0)^2] \tag{5}$$

where R_0 is the R-function for all but the considered level. Thus, S_0 is no longer invariant. This dependence on boundary radius is more severe for the p-wave strength S_1.

The variation of S_0 with the arbitrary boundary radius can also be demonstrated for the OMP. Following the usual procedure we expand the s-wave scattering function U in terms of a complex R-function $\overline{R}+i\pi s$ for given boundary radius and equate s to $\langle\gamma^2\rangle/D$ in Eq. (4). We find

$$S_0 = \frac{2}{\pi}\sqrt{\frac{1\text{ eV}}{E(\text{eV})}} \frac{1-|U|^2}{1+|U|^2 + 2\text{Re}(U \exp 2i\phi)} \tag{6}$$

Dependence on the boundary radius enters in the hard sphere phase shift ϕ or ka. Figure 1 shows S_0 calculated from Eq. (6) for a nucleus of mass 60 for an OMP model with given geometry (see later section) and with a real well depth $V_0 = 50$ MeV and imaginary depth $W_D = 5$ MeV. Curves are shown for several boundary radii. We see that S_0 is well defined only at zero energy.

To avoid this problem we compare directly the averaged

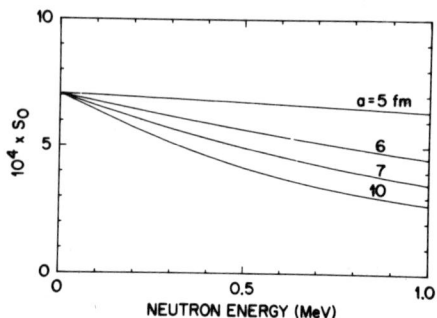

Fig. 1. OMP s-wave strength function for various boundary radii for A=60, V_0=50 MeV, W_D=5 MeV.

experimental scattering function for each partial wave to the OMP scattering function. Our vehicles for comparison are the compound and shape elastic cross sections. For the OMP for given partial wave we have

$$\sigma_c(E) = \pi k^{-2} g (1 - |U^{omp}(E)|^2) \tag{7}$$

and $\quad \sigma_{se}(E) = \pi k^{-2} g |1 - U^{omp}(E)|^2. \tag{8}$

For the corresponding experimental cross sections we replace U^{omp} by an appropriate average, which we designate U^{om}.

Parenthetically, we note that a poor resolution experiment with a thin sample yields an average of only the real part of U [see Eq. (1)]. Thus, that measurement does not give information on both the amplitude and phase of U^{om}.

NUMERICAL AVERAGING OF THE SCATTERING FUNCTION

For comparison to the OMP we wish to average the observed scattering function over the resonance structure. Ideally the resonances should be closely spaced such that

$$D << 2I << (E_u - E_\ell) \tag{9}$$

where 2I is the width of the averaging function and E_ℓ and E_u are the lower and upper energy limits of the data. However, the spacing D for a given partial wave is often so large that we satisfy only the inequality

$$D < 2I < (E_u - E_\ell) \tag{10}$$

Figure 2 illustrates the problem. The upper figure shows the neutron total cross section and multilevel fit[5] for neutrons on natural sulfur (95% ^{32}S). (Resolution effects were included in the analysis but not in the plotted curve.) The lower curve shows only the s-wave component of the multilevel curve. Even though the total cross section has many resonances, the given partial wave has relatively few. That is frequently the case. Since a multilevel analysis is practical only in a region where the total number of resonances is restricted, the number for each partial wave must be relatively small.

Recently, Johnson, Larson, Mahaux and Winters[6] (JLMW) discussed the problem of averaging. They define the

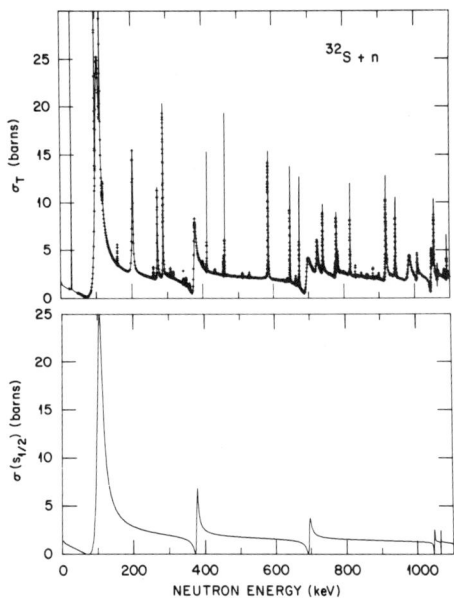

Fig. 2. Total cross section and s-wave component for sulfur (95% ^{32}S).

average of a function X to be

$$\langle X(\bar{E})\rangle_I = \int_{E_\ell}^{E_u} F_I(\bar{E},E')X(E')dE' \tag{11}$$

where the weighting function F is normalized to unity in the experimental region from E_ℓ to E_u and where the average energy $\bar{E}$ is calculated for X=E'. JLMW compared various F but here we use only their truncated Lorenzian;

$$F_I(\bar{E},E') = \frac{f(\bar{E})}{(\bar{E}-E')^2 + I^2} \tag{12}$$

where f is the normalizing factor. Figure 3 shows F for I=0.3 MeV, i.e., 30% of the experimental region. Such broad functions are needed to average over the sparse structure for s-waves in Fig. 2.

If the weighting function were not so broad relative to the neutron energy, we could average immediately by inserting the scattering function from Eq. (2) into the integral of Eq. (11). However, that procedure introduces a spurious absorption. To illustrate we consider a real potential (no absorption) such that the s-wave phase shift is real and given

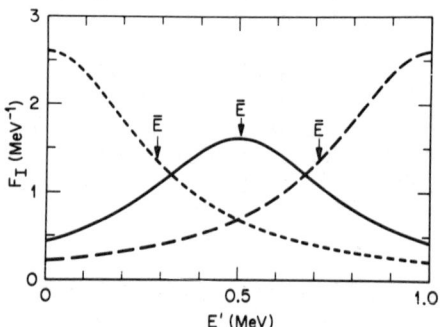

Fig. 3. Normalized weighting functions for I=0.3 MeV and $\bar{E}$=0, 0.5 and 1.0 MeV. Arrows indicate the average energies.

by the dashed line in Fig. 4. This curve is simply that for a hard-sphere of 5 fm radius. Averaging U, i.e., exp($-2i\phi$) yields the solid curves for I = 50 to 400 keV. The upper figure shows the real parts of the phase shifts for <U> and the lower figure shows the strength functions calculated by Eq. (6) from <U>. The S_0 are comparable to experimental values[2] in the valleys between size resonances, but they are spurious because there is no absorption in the model.

Figure 5 shows graphically the origin of the spurious absorption. The arrow on the circle represents the rotating unitary scattering function. Averaging over the large angle yields a vector of less than unit magnitude, as indicated by the enlarged view. Even though the difference from unity is small, it is serious when compared to the actual difference for real nuclei.

To solve the problem JLMW recommend the experimental scattering function be factored into resonant and background terms and averaged only over the resonant part

$$U^{om}(\bar{E}) = U_{bkg}(\bar{E})\langle U_{res}(E)\rangle_I. \tag{13}$$

The background is reasonably written as

$$U_{bkg}(E) = e^{-2i\delta(E)} \tag{14}$$

where
$$\delta(E) = \phi(E) - \tan^{-1} P(E) R^{ext}(E_m) \tag{15}$$

with R^{ext} being evaluated at the midpoint E_m of the energy interval. (A slightly better expression for δ is obtained by replacing R^{ext} in Eq. (15) by $\tilde{R}$, as defined below.)

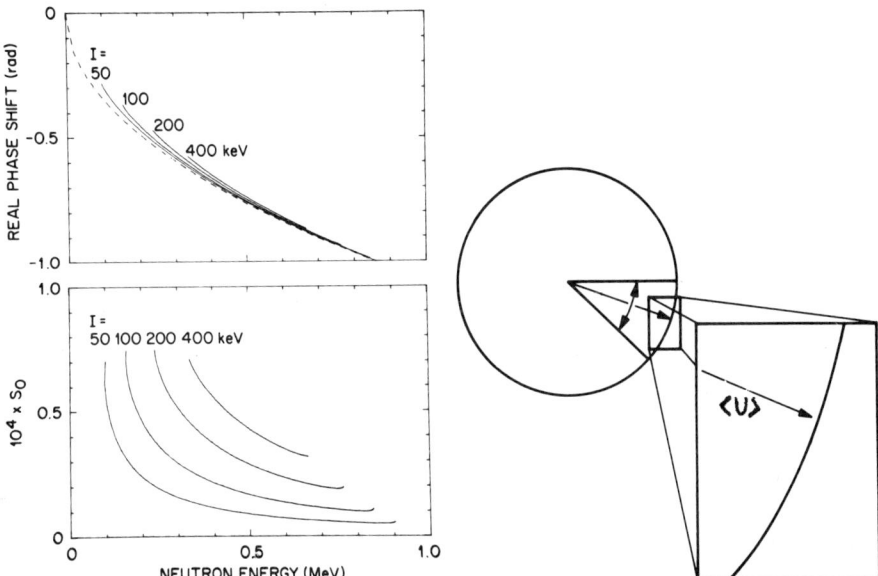

Fig. 4. Numerical averaging of only s-wave hard sphere scattering for various I. Upper: dashed curve for hard-sphere of 5 fm; solid for real part of phase shift of the average plotted versus $\bar{E}$. Lower: strength function from Eq. (6) for the averages.

Fig. 5. Graphical description of spurious absorption from averaging for a real potential.

To illustrate the use of Eqs. (11-15) for the definition of the average we have obtained U^{om} by numerical integration of an s-wave scattering function which Perey et al.[7] deduced from a multilevel analysis for 0-450 keV neutrons on ^{60}Ni. Figure 6 shows the resulting compound and shape elastic cross sections versus $\bar{E}$ for I=20, 50 and 90 keV. In the following discussion of this nucleus we use the curve for 2I=180 keV, which is about 12 times the level spacing. As expected the larger values of I produce a smoothing of the resonance structure.

ANALYTICAL APPROXIMATION TO THE AVERAGE

Since the integration of the scattering function cannot be done in closed form, the foregoing integration is difficult; it must be done numerically using carefully selected quadrature schemes. Therefore JLMW sought an analytical approximation to simplify the analysis. As a starting point they considered "Brown's[8] theorem." Brown pointed out that, if the scattering function were known for all energies from $-\infty$ to $+\infty$, then by contour integration with Lorentzian weighting one has

$$\langle U(E) \rangle_{I\infty} = U(E+iI), \quad (16)$$

where the subscript ∞ denotes the integration limits. This theorem is a powerful tool for theoretical works but JLMW found it can be quite inaccurate for the experimental case where U is known only in a limited region. One reason[6] for the failure is that, since the observed U senses only the <u>effects</u> of external poles, it can be parametrized by many forms of R^{ext}, none of which are likely to represent the actual poles and corresponding U in the external reagion.

To study this problem we consider a simple example, as illustrated in Fig. 7. For this hypothetical nucleus we find in the region ΔE a picket fence of 20 equally spaced levels, each with the same reduced width. The external R-function, R^{ext}, is adequately described by a continuation of the picket fence to $\pm\infty$, as illustrated in Fig. 7a. The corresponding R^{ext} is shown in Fig. 8. However, we find R^{ext} is also adequately described as in Fig. 7b with two small nearby poles and two

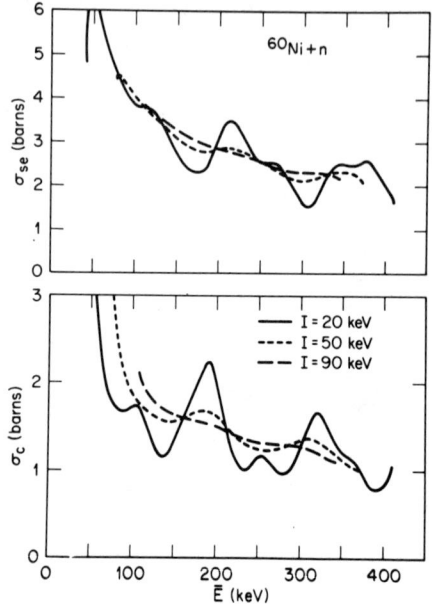

Fig. 6. Numerical average for s-waves on ^{60}Ni for three values of I.

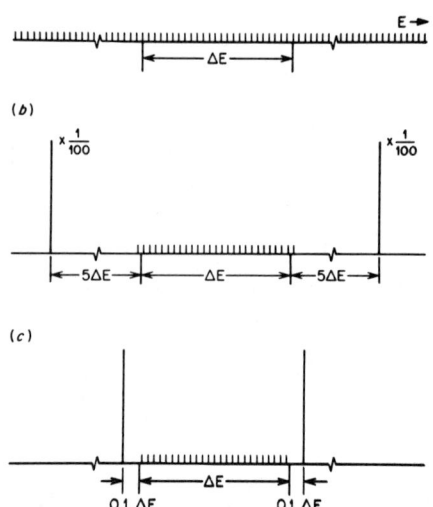

Fig. 7. A picket fence model of s-wave resonances with three equivalent parametrizations of the external R-functions.

distant very large ones or, as in Fig. 7c, by only two relatively large nearby poles. The corresponding R^{ext} are shown by dashed curves in Fig. 8.

By numerical averaging of our hypothetical scattering function using Eqs. (11-15) we find, as expected, nearly the same answer for all three parametrizations of R^{ext}. However, we find three different answers by evaluating $U(E+iI)$. We do find good agreement with the numerical average if we use the parametrization of Fig. 7a and substitute $E+iI$ only in the R-function.

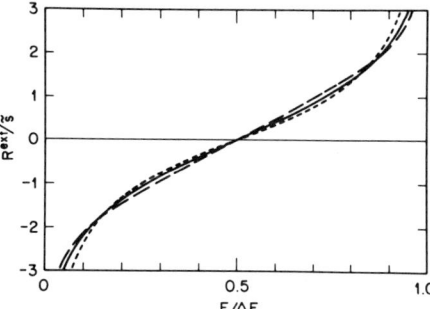

Fig. 8. External R-function for the three parametrizations of Fig. 7: solid, Fig. 7a; short dash, Fig. 7b; long dash, Fig. 7c.

The justification for this success in terms of a contour integration is presented by JLMW.

For the analytical approximation JLMW concluded R^{ext} should be parametrized by an expression that Johnson and Winters used earlier[5] for the analysis of the sulfur data in Fig. 1. It is

$$R^{ext}(E) = \tilde{R}(E) - \tilde{s} \ln [(E_u-E)/(E-E_\ell)] \tag{17}$$

where $\tilde{s} = <\gamma^2>/D$, (18)

and where $\tilde{R}(E)$ is a simple parametric function,

$$\tilde{R}(E) = \alpha + \beta E. \tag{19}$$

The effect of the log term in Eq. (17) is equivalent to the external picket fence in Fig. 7a. A good approximation to the averaged U is then achieved by substituting $E+iI$ in the R-function, except for the slowly varying part $\tilde{R}$. Thus we have

$$U^{om}(E) \simeq e^{-2i\phi(E)} \frac{1 + iP(E)[\tilde{R}(E) + i\pi\tilde{s} + R^f(E+iI)]}{1 - iP(E)[\tilde{R}(E) + i\pi\tilde{s} + R^f(E+iI)]} \tag{20}$$

where

$$R^f(E+iI) = \sum_{\lambda=1}^{N} \frac{\gamma_\lambda^2}{E_\lambda - E+iI} - \int_{E_\ell}^{E_u} \frac{\tilde{s}\, dE'}{E'-E+iI}. \tag{21}$$

We see that both the real and imaginary parts of the fluctuating term R^f tend to oscillate about zero and finally go to zero for large I. The net result is a smoothing similar to that shown in Fig. 6 for the numerical average.

To illustrate the success of this approximation we repeat the averaging for ^{60}Ni. In Fig. 9 the solid curve shows the original R^{ext} from Perey's[7] multipole parametrization, the short dash curve is our new parametrization from Eq. (17) and the long dash curve is

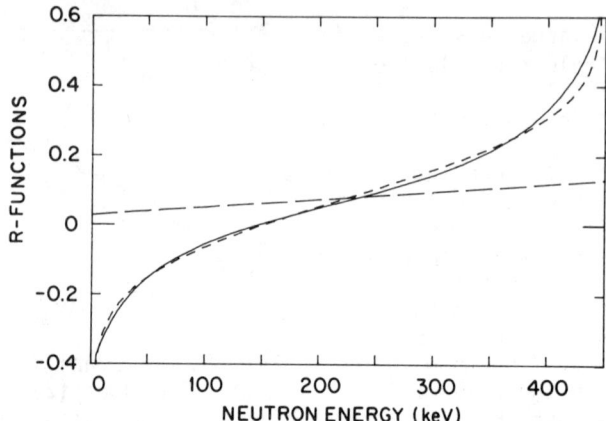

Fig. 9. External R-function and the $\tilde{R}$ (long dash) for 0-450 keV s-wave neutrons on ^{60}Ni. Boundary radius is 6 fm.

the corresponding linear $\tilde{R}$. Substituting in Eq. (20) we find an approximate U^{om} for calculating σ_c and σ_{se}. These are plotted by solid curves in Fig. 10 for I=90 keV. For comparison the corresponding numerical average from Fig. 6 is reproduced as a short-dash curve. The agreement is good.

The analytical approximation of Eqs. (17-21) is used for the rest of this report.

OPTICAL MODEL GEOMETRY

For the following OMP analysis we use a standard Woods-Saxon real potential with a surface derivative spin-orbit term and a Woods-Saxon derivative imaginary term. For each partial wave the experimental U^{om} or the resulting σ_c and σ_{se} can be fit by adjusting two parameters corresponding, qualitatively, to the volume integrals of the real and imaginary parts. For convenient comparison among various nuclei we fix the geometries and adjust the real and imaginary well depths, V_0 and W_D. For $\ell>1$ we also adjust the spin orbit depth V_{so}. Our geometry is

$$r_0 = r_{so} = r_D = 1.21\ A^{1/3}\ fm,$$

$$a_0 = 0.66\ fm\ and\ a_D = 0.48\ fm.$$

(22)

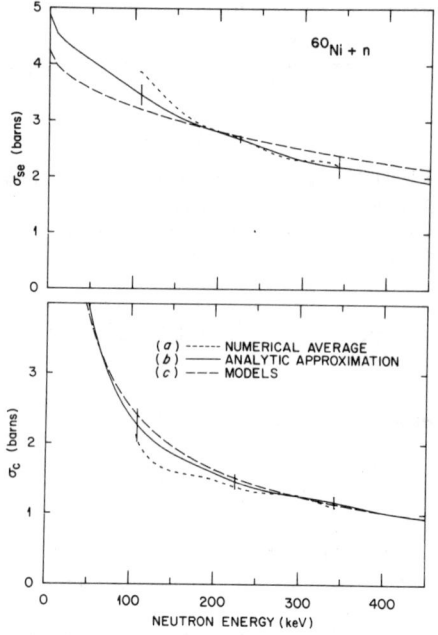

Fig. 10. Compound and shape elastic cross section for s-wave neutrons on ^{60}Ni

SPHERICAL OMP FOR ^{60}Ni

In Fig. 10 the long dash curve is the fit achieved by adjusting V_0 and W_D for s-waves. We find V_0 = 48 MeV and W_D = 29 MeV, each with a ±5 MeV uncertainty. The V_0 is consistent with other nuclei in our studies and it places the peak[2] of the 3s size resonances properly near A=55. However the W_D is much larger than any other in our studies.

Figure 11 demonstrates the roles of V_0 and W_D. The solid curves show σ_c and σ_{se} versus V_0 for various W_D at our midrange energy of 200 keV, and the dashed horizontal lines are the experimental values from Fig. 10, each with the small uncertainties indicated by the vertical bars in Fig. 10. The peak in σ_c near V_0 = 48 MeV is due to the $3s_{1/2}$ state. We see that σ_{se} with its small uncertainty restricts V_0 to near 48 MeV. With V_0 so restricted, the W_D has been increased to the large value of 29 MeV in order to spread the $3s_{1/2}$ state and bring σ_c down to the observed value. It might appear that a fit could

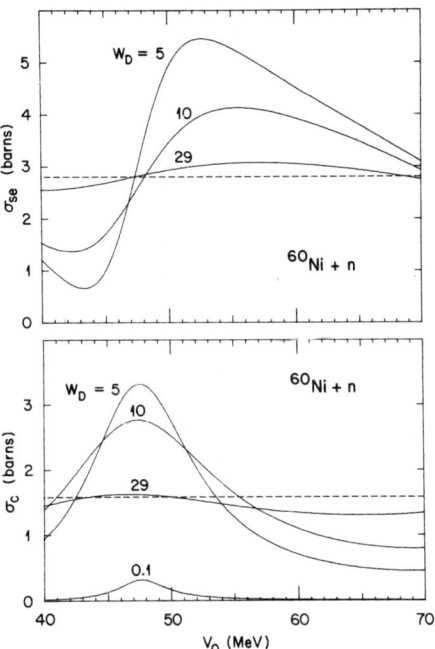

Fig. 11. Compound and shape elastic cross section for 200 keV neutrons on ^{60}Ni. Solid lines are for the OMP and dashed lines are experimental from Fig. 10.

be achieved with small $W_D \simeq 0.5$ MeV. However, the resulting σ_{se} has a strong energy dependence, in disagreement with the data of Fig. 10.

DEFORMED OMP FOR ^{60}Ni

The spherical model could be a poor approximation because ^{60}Ni is a vibrational nucleus. Therefore, we repeated the fit using the coupled-channel formalism.[9,10] The target channels are the 0$^+$ ground state and the 2$^+$ state at 1.33 MeV with β = 0.311. (The 3$^-$ state at 4.04 MeV has negligible effect.) Using this model we achieved the same fit as above with minor adjustments in the well depths, i.e., V_0 = 50 MeV and W_D = 24 MeV.

Figure 12 shows curves for this model plotted in like manner to Fig. 11. We see that the vibrational effects shift the $3s_{1/2}$ peak downward about 2 MeV and introduce two particle-core resonances. Arrows indicate the energies deduced for these resonances by simple addition of the 1.33-MeV 2$^+$ excitation to the bound 2d states in the potential, such as to equal the neutron energy of 200 keV.

The solid points indicate our chosen well depths. There is another solution, as indicated by the open circles, with V_0 = 54.5 MeV and W_D = 1.5 MeV. The smaller W_D is more consistent with other nuclei; however, the V_0 is too large for other considerations. Firstly, it places the 3s size resonance near A=44, in serious disagreement with the known peak[2] near A=55. Secondly, if the same well depth is assumed for p-waves, it predicts the bound 2p state to be about 3 MeV deeper than the known energies,[11] whereas the other solution (V_0 = 50 MeV) places the levels very close to the correct energies.

Recently, Guss et al.[12] measured neutron scattering on ^{60}Ni for 8, 10, 12 and 14 MeV neutrons and fit with a coupled-channel model including the requirements of fitting the low energy S_0, S_1, R' and σ_T, i.e., the "SPRT[3] method." Figure 13 is plotted in like manner to Fig. 12 for their potential geometry. The essential difference from Fig. 12 is that the resonance peaks are shifted upward due to the smaller real radius (r_0=1.165 fm versus 1.21 fm). The solid points indicate

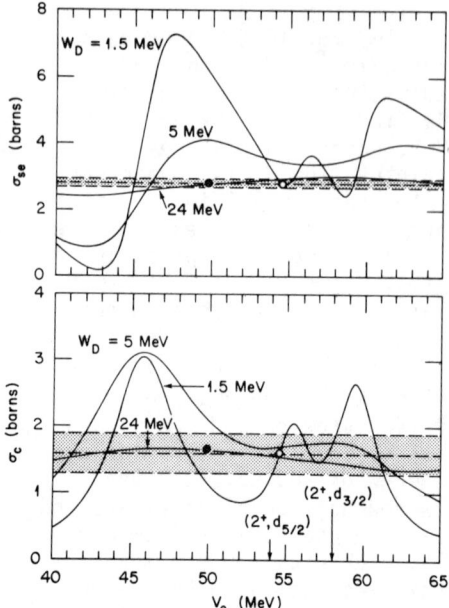

Fig. 12. Compound and shape elastic cross sections for 200 keV neutrons on ^{60}Ni. Solid Lines are for the coupled channel model. The dashed lines and shaded areas represent the observed cross sections with uncertainties from Fig. 10. Arrows indicate unperturbed energies of particle-core states. Solid points represent the accepted well depths and open points are rejected.

the solution of Guss, et al. That solution was influenced considerably by the low energy scattering radius[2] of R' = 6.7 ± 0.3 fm. The open circles represent the adjusted solution required to fit the present accurate evaluation of σ_{se}. With this solution (V_0 = 58 MeV) one must increase the slope for V_0 from the value of 0.46 MeV^{-1} of Guss et al. to about 0.62 MeV^{-1} to maintain consistency with their data at higher energies. Essentially this solution with their geometry[12] is the one we have discarded from Fig. 12 for our geometry. The choice is a matter of judgment. Further data on nearby nuclei are desirable.

CROSS SECTION AND OMP FOR ^{86}Kr

The total cross section of ^{86}Kr is being studied at ORELA and is the subject of a contributed paper by Carlton, Harvey and Johnson[13]. The analysis is nearly complete up to 700 keV. Since this nucleus lies near the 3p size resonance at A≈95, its resonance structure is dominated by p-waves. As an example Fig. 14 shows the structure for 340-440 keV neutrons. The upper figure shows the observed total cross sections with the fitted multilevel curve and the lower figure presents the p3/2 component without the effects of resolution broadening. Clearly p3/2 neutrons dominate the resonance structure.

Given the multilevel analysis we deduced average scattering functions for $\ell=0$ and 1 by use of the analytical approximation for $2I=200$ keV. In Fig. 15 the solid curves show the compound and shape elastic cross sections, divided by the g-factor, for p1/2 and p3/2. The dashed curves are preliminary fits achieved by adjusting the well depths of our standard model. (For the final fit we may increase the width of the averaging interval to give more smoothing over the p3/2 structure.)

The well depths are listed in the figure. We see that the imaginary potentials are about the same for the two partial waves and that the primary reason for J-dependence of the cross sections lies in the real spin-orbit term. This V_{so} is consistent with values deduced in the literature from polarization measurements.

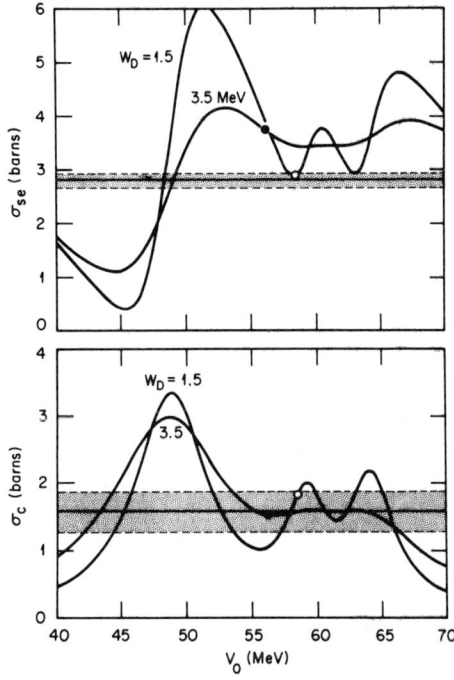

Fig. 13. Compound and shape elastic cross sections for 200-keV neutrons on ^{60}Ni. See Fig. 12 caption. Model geometry by Guss et al., ref. 12. Solid points are their solution and open circles are adjustments to present data.

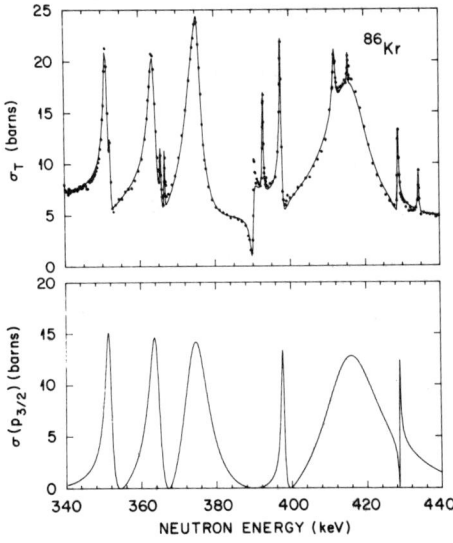

Fig. 14. Total cross section for ^{86}Kr. The upper figure shows the data and multilevel fit only from 340 to 440 keV and the lower curve is the p3/2 component.

The s-wave analysis gives V_0 = 48.5 MeV and W_D = 3.5 MeV. Thus, the imaginary potentials are about the same for all three partial waves but the real well depth is shallower for s-waves than for p-waves. This ℓ-dependence may not be very significant by iteslf but it is consistent with a similar trend that we observe in lighter nuclei.

CROSS SECTION AND OMP FOR ^{208}Pb

The ^{208}Pb nucleus is particularly interesting and has been studied extensively both experimentally and theoretically. At ORELA the total cross section has been measured and fitted to 1 MeV. Furthermore, angular distributions have been measured to aid in the J^π assignments. From the present viewpoint this nucleus is interesting because it is large enough to give substantial d-wave scattering.

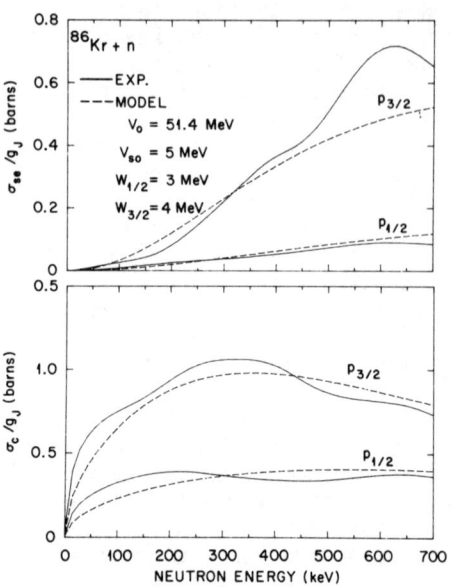

Fig. 15. Compound and shape elastic cross sections for ^{86}Kr. Solid curves are experimental for $2I = 200$ keV and dashed curves are for the model.

In Fig. 16 the upper curve shows the observed total cross section with the multilevel fit from 300 to 1000 keV and the lower curve shows the rather prominent $d_{5/2}$ component.

Given the multilevel scattering functions we averaged over the full 0-1 MeV region using averaging widths appropriate for each partial wave. We then fit σ_c and σ_{se} for each partial wave by adjusting the well depths for our standard geometry. These well depths are compared with the other lighter nuclei in the next section.

We also fit the cross section using the common geometry deduced by Finlay et al.[14] from measurements in the 7-40 MeV region. The fixed parameters are 1.205 fm, 1.105 fm and 1.283 fm respectively for r_0, r_{so} and r_D; 0.685 fm, 0.499 fm and 0.569 fm respectively for a_0, a_{so} and a_D and $V_{so} = 5.75$ MeV. In Fig. 17 the X-symbols show the resulting well depths for the first five partial waves. The real well depths are in very good agreement with the known bound single-particle states of ^{209}Pb and provide evidence for a parity dependence, with the potential being about 3 MeV deeper for odd parity than even parity. Of course the observed J-dependence for $\ell=1$ could be removed by increasing V_{so} for that partial wave. The fluctuations in W_D may result from the well-known nuclear structure effects in this doubly closed shell nucleus.

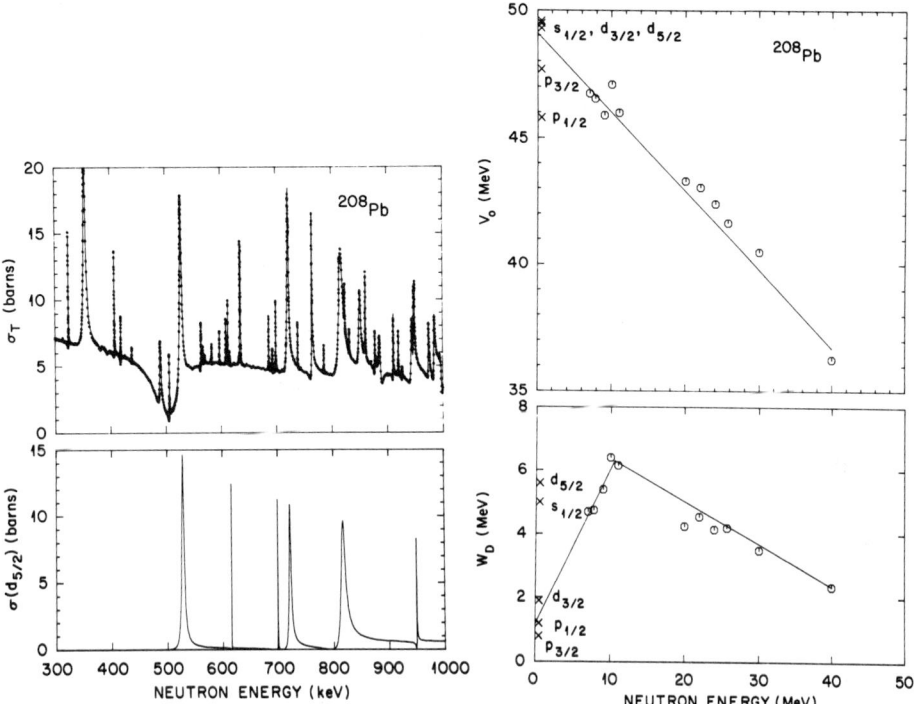

Fig. 16. Total cross section for ^{208}Pb. The upper figure shows the cross section and multilevel fit and the lower curve shows the $d_{5/2}$ component.

Fig. 17. Potential well depths for the model of Finlay et al., ref. 14. The X-symbols are from the present analysis for 0-1 MeV and the circles are from Finlay. The solid curves were suggested by Finlay for $E_n > 7$ MeV and have been extrapolated here to zero energy.

The open circles are from the report by Finlay et al.[14] and the solid lines are their suggested curves, with further extrapolation from 7 MeV to zero energy. Overall this figure shows a consistency in OMP parameters.

PRESENT SUMMARY OF V_0 and W_D for 0-1 MEV NEUTRONS

This paper is intended to be more a progress report than a review. Some potential parameters have been published for both the spherical[15,16] and coupled channel analysis.[17,18] The published[19] analysis for ^{206}Pb was performed with earlier techniques than reviewed here. But most of the work is still in progress and subject to modifications. Table I gives present results with appropriate footnotes.

Table I. Potential[a] Well Depths[b] from High Resolution Neutron Cross Sections.

Target	ℓ	V_o (MeV) sph	V_o (MeV) cc	W_D (MeV) sph	W_D (MeV) cc	V_{so} (MeV) sph	V_{so} (MeV) cc
^{30}Si[c,d]	0	48(1.7)	49.5	2.0(1.5)	1		3.3
	1	62(2.5)	51.5	4.5(2.5)	2		3.3
^{32}S[e,f]	0	51.5(0.4)	50.8	6.0(3.0)	4.0		5.3
	1	61.4(1.1)	50.8	2.7(1.5)	0.8	11(3)	5.3
^{34}S[c,d]	0	51.5(1.1)	50.5	3.0(2.0)	2.5		3.5
	1	58.5(1.2)	51	3.5(1.9)	2.5	6(3.5)	3.5
^{40}Ca[g]	0	53.6(1.0)	51	6.65(0.3)	7		
	1	56.1(1.0)	53	0.6 (0.3)	0.1	6(1)	4
^{48}Ca[h]	0	48(1)		0.0			
^{60}Ni[g]	0	48(5)	50	29(5)	24		5[i]
	1		50[i]		1.5		5[i]
^{86}Kr[j]	0	48.5(1)		1			
	1	51.4(1)		3.5(1)		5.5(1)	
^{206}Pb[k]	0	50.4		6			
	1	48.7		2		7.5	
	2	50.4		6		7.5	
^{208}Pb[g]	0	49.5(1)		6(2)			
	1	47(1)		1.5(1)		10(2)	
	2	49.5(1)		2-5(2)		7.5(1)	

a. Model geometric parameters are $r_o = r_D = 1.21$ fm, $a_o = 0.66$ fm, $a_D = 0.48$ fm.
b. Uncertainties for the spherical (sph) model are in parenthesis. The same uncertainties apply to the coupled-channel (cc) model.
c. See Ref. 16 for the spherical model.
d. See Ref. 18 for the coupled channel model.
e. See Ref. 15 for the spherical model.
f. See Ref. 17 for the coupled channel model.
g. Work in progress subject to modification.
h. Contributed paper, work in progress by Harvey, Johnson, Carlton, and Castel.
i. Assumed well depth.
j. Contributed paper, work in progress by Carlton, Harvey and Johnson.
k. See Ref. 19, Analysis by earlier method.

There are some trends in Table I. For the spherical model and A<100 the real well tends to be deeper for $\ell=1$ than for $\ell=0$. Coupled channels calculations remove some but not all of this difference. On the other hand the odd parity well is relatively deeper in the lead region. There is evidence for the isospin term for nuclei such as ^{48}Ca, ^{86}Kr and ^{206,208}Pb, each of which have relatively large N-Z differences. Altogether, the real well depths show consistent patterns. On the other hand the imaginary depths show large fluctuations with little apparent pattern at this time. Apparently W_D is sensitive to specific nuclear structure effects.

More data are needed with optical model analyses as reviewed in this paper.

ACKNOWLEDGMENTS

This work includes not only analyses cited in the references but also ongoing research with many colleagues. These include Profs. R. F. Carlton from Middle Tennessee State University, A. D. MacKellar at University of Kentucky, C. Mahaux at University of Liège, R. R. Winters at Denison University, and Drs. J. L. Fowler (deceased), J. A. Harvey, D. J. Horen, and N. W. Hill at ORNL. I am grateful to these colleagues for the privilege of presenting their preliminary results. Research was sponsored by the Division of Basic Energy Sciences, U. S. Department of Energy, under Contract No. DE-AC05-84OR21400 with the Martin Marietta Energy Systems, Inc.

REFERENCES

1. P. A. Moldauer, Nucl. Phys. 47, 65 (1963).
2. S. F. Mughabghab, M. Divadeenam, and N. E. Holden, Neutron Cross Sections, Vol. 1, Part A (Academic Press, 1981).
3. J. P. Delaroche, Ch. Lagrange and J. Salvy, Nuclear Theory and Nuclear Data Evaluation (IAEA, Vienna, 1976), Vol. II, p. 251.
4. A. M. Lane and R. G. Thomas, Rev. Mod. Phys. 30, 257 (1958).
5. C. H. Johnson and R. R. Winters, Phys. Rev. C 21, 2190 (1980).
6. C. H. Johnson, N. M. Larson, C. Mahaux, and R. R. Winters, Phys. Rev. C 27, 1913 (1983), and 29, 1563 (1984).
7. C. M. Perey, J. A. Harvey, R. L. Macklin, F. G. Perey, and R. R. Winters, Phys. Rev. C 27, 2556 (1983).
8. G. E. Brown, Rev. Mod. Phys. 31, 893 (1959).
9. T. Tamura, Rev. Mod. Phys. 37, 679 (1965).
10. J. Raynal, ECIS-79; (private communication), the present version was modified by R. L. Hershberger to print partial wave cross sections.
11. R. L. Auble, Nucl. Data Sheets 16, 1 (1975).
12. P. P. Guss, R. C. Byrd, C. E. Floyd, C. R. Howell, K. Murphy, G. Tungate, R. S. Pedroni, R. L. Walter, J. P. Delaroche, and T. B. Clegg, preprint (1984).
13. R. F. Carlton, J. H. Harvey and C. H. Johnson (contributed paper).
14. R. W. Finlay, J. R. M. Annand, T. S. Cheema, and J. Rapaport preprint (1984).
15. C. H. Johnson and R. R. Winters, Phys. Rev. C 27, 416 (1983).
16. R. F. Carlton, J. A. Harvey, and C. H. Johnson, Phys. Rev. C 29, 1988 (1984).
17. A. D. MacKellar and B. Castel, Phys. Rev. C 28, 441 (1983).
18. A. D. MacKellar and B. Castel, Phys. Rev. C 28, 1993 (1983)
19. D. J. Horen, J. A. Harvey, and N. W. Hill, Phys. Rev. C 24, 1961 (1981).

LEVEL DENSITY CALCULATIONS: PAST, PRESENT AND FUTURE

S.M. Grimes
Ohio University, Athens, OH 45701 U.S.A.

ABSTRACT

Nuclear level densities have been an area of both basic and applied interest for about fifty years. The largest fraction of the work in this field has been on the Fermi gas model. The current problems with this formalism will be discussed as well as the successes of this approach. A new technique which allows the inclusion of two-body forces is now available. Early results with this method are promising, but some unsolved questions remain. These will be examined with the goal of suggesting what further developments might be expected.

INTRODUCTION

In 1936 Bethe [1] formulated what has been called the Fermi gas model of the nucleus. He made the assumption that the protons and neutrons of the nucleus may be treated as non-interacting Fermions. With the use of statistical mechanics he was then able to derive an analytic form for the nuclear level density. It involved only one parameter a, which is proportional to the density of single particle states at the Fermi level of the nucleus. This form is still the most widely used level density expression some fifty years after its derivation.

An obvious limitation of the first formulation was that no difference was predicted between the level densities for even-A and odd-A nuclei. As is well known, even-A nuclei with even N and even Z have a much lower level density at low excitation than do odd-A nuclei, while the even-A nuclei with N and Z odd have a level density which is higher than for odd-A nuclei.

This difference is caused by pairing energy shifts and has been incorporated into level density formulas by introducing a second parameter, δ, which corresponds to an energy shift in the level density, since the energy u is replaced by u-δ in the formula. It is used as an adjustable parameter and, therefore, represents not only pairing but also closed-shell effects on the nuclear level density.

One difficulty with the use of the Bethe formula including two parameters is that the two parameters are coupled in a complicated way. Increasing the parameter a increases the level density as does reducing the magnitude of δ. Thus, by increasing both or decreasing both one can leave the level density unchanged at one energy but the slope with energy will be changed somewhat. Thus, introduction of a second parameter will require level density information to be available over a wider range of energies to constrain the energy dependence of the level density.

A second complication is the question of whether both shell effects and pairing shifts are adequately accounted for by simply shifting the excitation energy. Either effect could potentially modify the energy dependence in a different manner. Modifications to the original Bethe procedure now allow a test of the constant energy shift assumption. The effects of pairing can be studied with a quasi-particle formalism analogous to that used for superconductors [2]. As the thermodynamic temperature is changed, the number of paired particles changes, modifying the quasi-particle energies as a result. Because this change modifies the assumptions made in obtaining the Bethe formula, an analytic solution is not obtained. Numerical values of the level density are calculated as a function of excitation energy. It is found that at high energies (above about 7 MeV) the original analytic form provides a good fit to the calculations, but that for lower energies the form is not precisely that predicted by the Bethe expression.

Similarly, the consequences of shell structure in the single particle spectra can be investigated by utilizing realistic single particle energies. As in the case of pairing effects, this change results in a problem which can be solved numerically but not analytically. A summary of this formalism and some results using it has been presented by Huizenga and Moretto [3].

MICROSCOPIC FERMI GAS FORMALISM

To see what assumptions are made in evaluating level densities with this approach, we must examine the procedure involved in obtaining a level density. The Hamiltonian of the system is written

$$H = \sum_k e_k (a_k^+ a_k + a_{-k}^+ a_{-k}) - G \sum_{kk'} a_{k'}^+ a_{-k'}^+ a_k a_{-k} \quad (1)$$

where e_k is the energy of the kth doubly degenerate energy level and $a_{\pm k}^+$ and $a_{\pm k}$ are the creation and annihilation operators for particles with spin projections + or -. G is the pairing matrix element. The grand partition function of such a system can be written

$$\Omega = -\beta \sum_k (e_k - \lambda - E_k)$$
$$+ 2 \sum_k \ln[1 + \exp(-\beta E_k)]$$
$$- \beta \Delta^2 / G \quad (2)$$

where Δ is the energy gap, $E_k = \sqrt{(e_k - \lambda)^2 + \Delta^2}$, β is the reciprocal of the temperature and λ is the chemical potential (Fermi energy) of the system. For a given value of the temperature, one first solves the system of equations

$$\frac{2}{G} = \sum_{k_N} \frac{\tanh(\tfrac{1}{2}\beta E_{k_N})}{E_{k_N}}$$

$$\frac{2}{G} = \sum_{k_Z} \frac{\tanh(\tfrac{1}{2}\beta E_{k_Z})}{E_{k_N}} \qquad (3)$$

$$N = \sum_{k_N} \left[1 - \frac{(e_{k_N} - \lambda_N)}{E_{k_N}} \tanh(\tfrac{1}{2}\beta E_{k_N})\right]$$

$$Z = \sum_{k_Z} \left[1 - \frac{(e_{k_Z} - \lambda_Z)}{E_{k_Z}} \tanh(\tfrac{1}{2}\beta E_{k_Z})\right] \qquad (4)$$

where k_N and k_Z denote sums over neutron and proton orbitals and N and Z are the neutron and proton number of the nucleus respectively; this yields values of λ and Δ for each type of nucleon. The energy of the system is given by the expression

$$E = \sum_{k_N} e_{k_N} \left[1 - \frac{e_{k_N} - \lambda_n}{E_{k_N}} \tanh(\tfrac{1}{2}\beta E_{k_N})\right]$$

$$- \frac{\Delta_N^2}{G} + \sum_{k_Z} e_{k_Z} \left[1 - \frac{e_{k_Z} - \lambda_Z}{E_{k_Z}} \tanh(\tfrac{1}{2}\beta E_{k_Z})\right] - \frac{\Delta_Z^2}{G} \qquad (5)$$

Finally, the state density is given by

$$\rho(E,N,Z) = \frac{1}{(2\pi i)^3} \oint d\beta \oint d\alpha_N \oint d\alpha_Z e^s \qquad (6)$$

where $\qquad s = \Omega_N + \Omega_Z + \beta E - \alpha_N N - \alpha_Z Z$

and $\qquad \alpha_N$ and α_Z are $\beta\lambda_N$ and $\beta\lambda_Z$, respectively.

Use of the saddle point procedure then gives the following value for the integral

$$\rho(E) = \frac{e^S}{(2\pi)^{3/2} D^{1/2}}$$

where

$$D = \begin{vmatrix} \frac{\partial^2 \Omega}{\partial \alpha_N^2} & \frac{\partial^2 \Omega}{\partial \alpha_N \partial \alpha_Z} & \frac{\partial^2 \Omega}{\partial \alpha_N \partial \beta} \\ \frac{\partial^2 \Omega}{\partial \alpha_N \partial \alpha_Z} & \frac{\partial^2 \Omega}{\partial \alpha_Z^2} & \frac{\partial^2 \Omega}{\partial \alpha_Z \partial \beta} \\ \frac{\partial^2 \Omega}{\partial \alpha_N \partial \beta} & \frac{\partial^2 \Omega}{\partial \alpha_Z \partial \beta} & \frac{\partial^2 \Omega}{\partial \beta^2} \end{vmatrix} \quad (7)$$

The spin cutoff parameter σ is given by

$$\sigma^2 = \sigma_Z^2 + \sigma_N^2 \quad (8)$$

and

$$\sigma_N^2 = \frac{1}{2} \sum_{k_N} \frac{m_{k_N}^2}{\cosh^2(\tfrac{1}{2}\beta E_{k_N})} \quad \text{and} \quad \sigma_Z^2 = \frac{1}{2} \sum_{k_N} \frac{m_{k_Z}^2}{\cosh^2(\tfrac{1}{2}\beta E_{k_Z})}$$

and m_k is the angular momentum projection for the kth level.

We then use the usual relation between level and state densities [4] to obtain the following equation for the level density $\rho(E,J)$ of levels of spin J and energy E:

$$\rho(E,J) = \frac{(2J+1)}{2(2\pi)^{\tfrac{1}{2}} \sigma^3} \rho(E) \exp\left[-\frac{(J+\tfrac{1}{2})^2}{2\sigma^2}\right] \quad (9)$$

Obtaining the energy dependence of the level density simply requires incrementing the thermodynamic temperature a number of times and repeating the calculation for each value.

Thus, the calculation does not necessarily yield a level density of the form derived by Bethe, but will give numerical values of the level density at particular excitation energies. The additional assumptions needed to obtain the Bethe form are: 1) a uniform single particle state density (constant with energy), and 2) no pairing force. In fact, calculations with realistic single particle schemes and with a pairing force yield level densities which do not deviate substantially from the Bethe form as shown in Fig. 1. Calculated level densities with two single particle sets are fit with the traditional Fermi gas form; note the good agreement with the analytic form above about 6 MeV for both

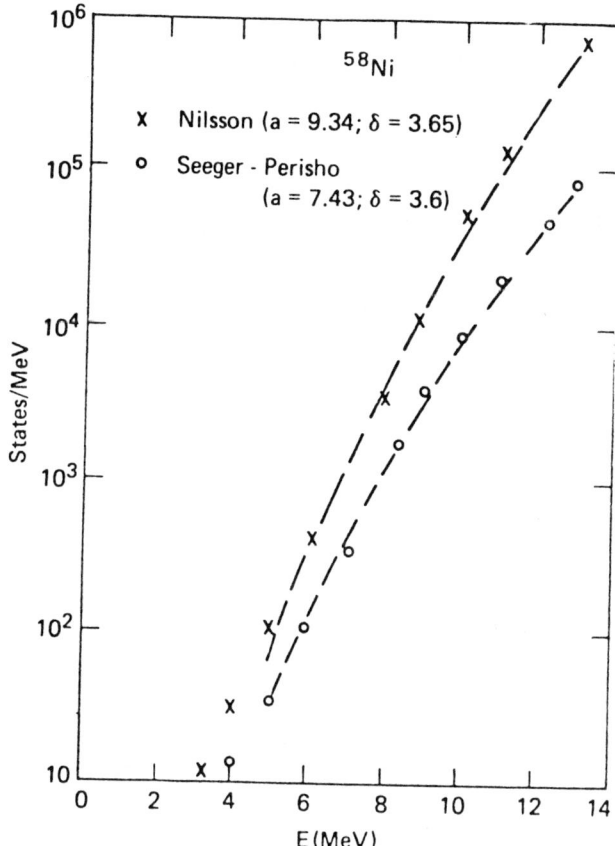

Fig. 1: Level density calculations for the nucleus ^{28}Si with the microscopic Fermi gas formalism. The x and o symbols denote the calculated values for the single particle energies of References 19 and 20, respectively. Note the good fit provided by a conventional Fermi gas expression at energies above 6 MeV.

single particle sets. Extensive comparisions of the results of
calculations with this model have been presented by Huizenga et
al. [5] and Grimes et al. [6]. For certain sets of single
particle energies, both Ref. 5 and Ref. 6 find good agreement
between calculations and data.

Some significant questions remain, however. The first
involves the mathematical convergence of the calculational
technique. Evaluation of the partition function is carried out
with the saddle-point procedure. To examine the validity of this
approach, we compare the Fermi gas calculation with exact values of
the level density for a system of four levels of spin 1/2 at 2, 4, 6
and 8 MeV containing four particles as shown in Fig. 2. This space
has only 70 states and is too small to be of physical interest,
but can be solved exactly (+ marks). The Fermi gas calculation
does well at the peak but has a tendency to give too large a
value on the average. Also shown for comparison is a moment
method calculation (discussed in Sect. III). It does better on
the average, but does not pick up the sharp rise between 18 and
20 MeV. Convergence tests in larger spaces are more difficult,
because exact results are not so easily obtained. Fig. 3 shows
results for a space consisting of 10 orbits of spin 1/2 equally
spaced between 2 and 20 MeV. Ten particles are placed in the
20 possible positions, giving a total of 184756 states. Comparison of the moment method with the Fermi gas results shows agreement to about 1% in the peak and within 10% down to 80 MeV. The
exact numbers are available for the first 10 MeV (60-70 MeV) and
they indicate that the lower values from the moment method are
more nearly correct than the Fermi gas results at low excitation.
It appears that the Fermi gas method is mathematically reliable
in situations where the level density is not too small.

A separate issue from the question of mathematical convergence is the question of physical correctness. The fundamental
assumption underlying the Fermi gas model is that the nucleons
are non-interacting. This assumption eliminates collective
effects completely from the calculation. For nuclei which are
deformed, this can cause a significant underestimation of the
level density. These levels are not missing from the calculation since our basis is complete, but they are found at too
high an energy. Huizenga et al. [5] have tried to remedy this
problem by adding additional levels at low energy to build up
rotational bands. While this technique does build the level
density at low energies, it produces a violation of unitarity.
The levels which are added in should be subtracted off at higher
energy, but there is no way to do this since we do not know from
what energy region they should be removed. This technique will
produce improved agreement at low energies but at high energies
will substantially overcorrect the level density, as bands are
built on levels which should have been removed from the spectrum.

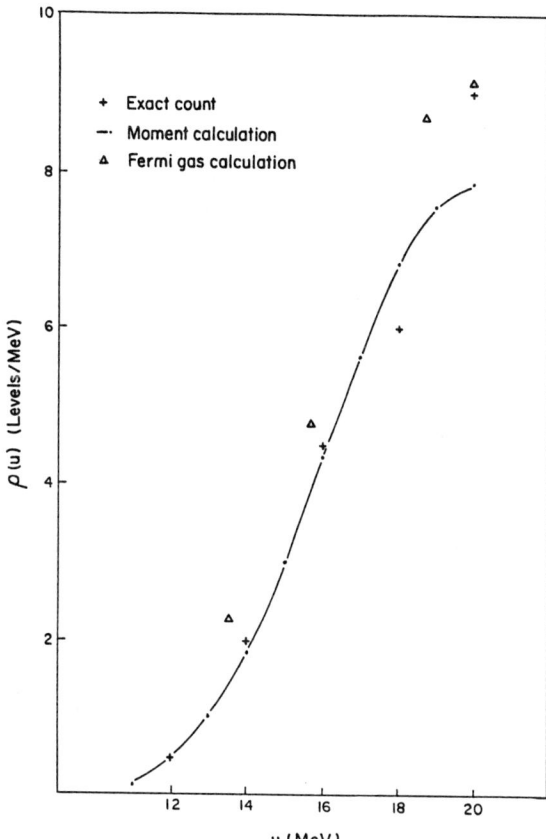

Fig. 2: Level density for the four orbit (doubly degenerate) system described in the text. The + marks denote exact values, while the Δ and dot-dashed symbols denote the result of microscopic Fermi gas and moment method calculations, respectively.

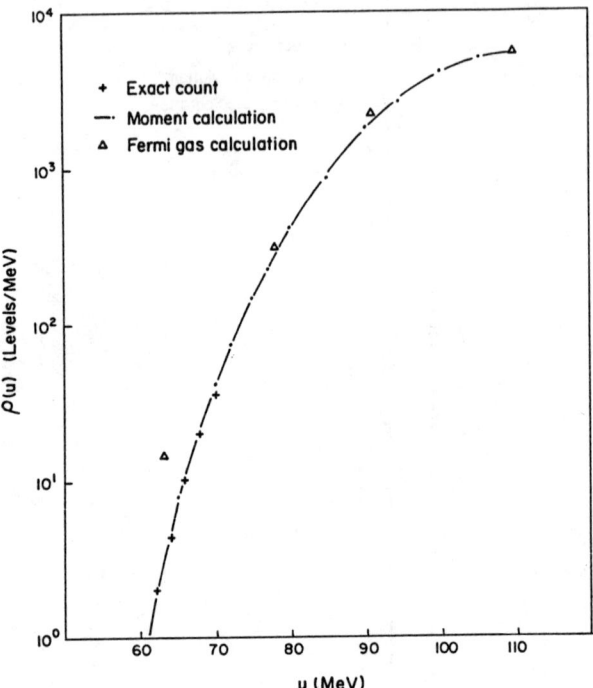

Fig. 3: Level density for the ten orbit system (doubly degenerate) filled with 10 particles as described in the text. The + marks denote exact values, while the Δ and dot-dashed symbols denote the result of Microscopic Fermi gas and moment method calculations, respectively.

A second problem with this approach is that it requires the deformation parameter to be known at all energies. If the parameter changes with energy or if bands with different deformation parameters are present at the same energy, the correction could become extremely difficult to make, both because of the large number of parameters involved and because of our lack of knowledge of their magnitude.

Finally, there is some evidence [7] that even in situations where the Fermi gas model correctly predicts the number of levels, some related parameters are not well described. The spin cutoff parameter is needed to give the distribution of levels as a function of J, while the positive-negative parity ratio is required to determine the relative fraction of the levels with a specified parity. Less experimental information is available for these two parameters than for the level density itself, but there appear to be difficulties associated with fitting these quantities. This may be correlated with subtle consequences of the two-body force: levels may be shifted by varying amounts depending on their spins and parities.

III. MOMENT METHODS

As the previous discussion indicates, it would be desirable to have a formalism which could directly incorporate the effects of the two-body force in a level density calculation. The theory of spectral distributions allows the inclusion of two-body forces in level density calculations in an enormous basis. It is routine to include two-body forces in shell model calculations, but these involve diagonalization of a large matrix and are, therefore, limited to a basis of 10^4 or less. Level densities are often needed in bases which have as many as 10^{10} levels or more. Even if we could diagonalize a very large matrix, this would not be an efficient way to proceed. We actually do not want all 10^{10} eigenvalues or eigenfunctions but only the distribution of levels with energy. French and collaborators [8] have developed a method which yields the level density (and spin cutoff and parity ratio) directly, without requiring a diagonalization. This formalism utilizes the assumptions that the Hamiltonian is two-body (actually a combination of one- and two-body parts) and that the distribution of levels with energy for such a Hamiltonian is nearly Gaussian. If we can calculate the total number of levels in our basis and also the average energy ($<H>$) and dispersion ($[<H^2> - <H>^2]^{1/2}$), we can make a Gaussian expansion of the level density.

We clearly depart from the typical Fermi gas approach in two ways: two-body forces are included and we deal with a finite (though large) basis. This latter difference may appear troublesome, since a finite basis obviously has no levels beyond a particular energy (that of the highest level). The Fermi gas form yields a level density which increases indefinitely with energy.

Actually, the limited basis assumption is physically more realistic [9]. For a nucleus such as ^{56}Fe with 56 nucleons and a binding energy of 8 MeV/nucleon, the nucleus will be totally dissociated into constituent nucleons at 8 x 56 = 448 MeV. For energies lower than this, many levels will have more than one nucleon in an unbound state and will have an extremely short lifetime. These are arrangements of 26 protons and 30 neutrons but probably do not live long enough to be considered excited states of ^{56}Fe. Thus, a level density which reaches a peak and then declines to 0 with increasing energy is actually more realistic physically than the traditional Fermi gas level density. The peak and region of negative slope, of course, will be at very high energies ($E \gtrsim 4A$ MeV).

Because these Gaussians span an enormous energy range, the shape of the predicted level density for moderate energies ($E \lesssim 30$ MeV) does not show dramatic effects of this cutoff. Indeed, it has been shown [7] that a Fermi gas form and a Gaussian are nearly indistinguishable over a wide energy range for proper choice of parameters. Fig. 4 illustrates this similarity. The Fermi gas form for $a = 4$ and $\delta = 0$ is fitted with Gaussians corresponding to two different dimensionalities with appropriate values for the average energy and width. Discrepancies between the two forms are extremely small over a 10-15 MeV range of energy. This similarity indicates that the general agreement of level density data with the Fermi gas form does not provide an argument against the spectral distribution approach.

Calculation of the parameters which enter the spectral distribution expansion is not difficult. The number of states in the basis will be the binomial coefficient

$$\binom{N_0}{m} = \frac{N_0!}{m! (N_0 - m)!}$$

where m is the number of particles and N_0 the total number of particle states available. We may obtain $\langle H \rangle$ and $\langle H^2 \rangle$ through use of a relation called the propagator theorem:

The expectation value of an n-body operator will be an n+1 order polynomial in the particle number.

As an example, consider the application of a one-body Hamiltonian to a system of 20 single particle states. By putting 10 particles in these states, we generate $\binom{20}{10}$ = 184756 states, and evaluating the average energy would seem to be difficult. According to the propagator theorem, the

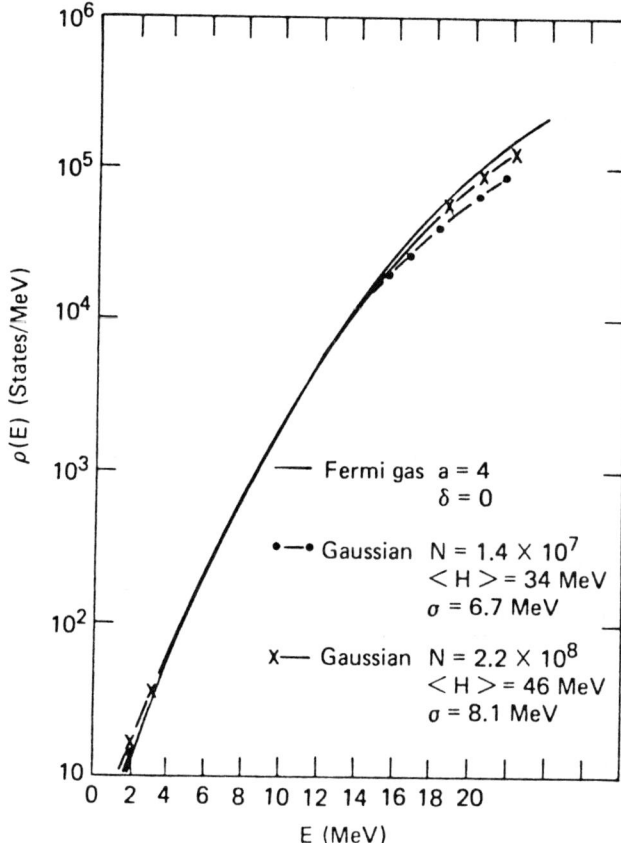

Fig. 4: Comparison of the conventional Fermi gas form with a Gaussian shape. The dot-dashed and x-dashed curves represent fits to the Fermi gas form for spaces of two different sizes. Note the close agreement in the energy range of 4 to 16 MeV.

average energy will be a two-term polynomial in the particle number. Thus,

$$\langle H \rangle = C_0 + C_1 m$$

If we define $\langle H \rangle$ to be 0 for 0 particles, then $C_0 = 0$. To evaluate $\langle H \rangle$ for a one particle system (= $\langle H \rangle_1$), we need a twenty term sum

$$\langle H \rangle_1 = \Sigma \varepsilon_i / 20$$

From the above form it is obvious that $C_1 = \langle H \rangle_1$. Thus,

$$\langle H \rangle = \left\{ \frac{\Sigma \varepsilon_i}{20} \right\} m$$

and we can evaluate the average energy for a system of 184756 states by multiplying two numbers together. For a one-body operator, the validity of the theorem can be appreciated immediately: we have the average energy of the system being given by the product of the average energy per particle times the number of particles.

For a realistic nuclear Hamiltonian, the situation is somewhat more complicated. The Hamiltonian is a two-body operator and $\langle H^2 \rangle$ is a four-body operator. Thus, for latter operator, a five term polynomial will be required. In addition, we want information about the energy dependence of the spin cutoff parameter, $\langle J_z^2 \rangle$, the average of the square of the z projection of the angular momentum. This requires, in addition to $\langle J_z^2 \rangle$, the expectation values of $\langle HJ_z^2 \rangle$ and $\langle H^2 J_z^2 \rangle$. These are two-body, four-body, and six-body operators, respectively.

The binomial coefficients $\binom{N_0}{m}$ are symmetric about a reflection through $m = N_0/2$, i.e. $\binom{N_0}{m} = \binom{N_0}{N_0 - m}$. Dimensionalities are, therefore, smallest for particle numbers which are small or near N_0. For example, the dimensionality for no particles or for $m = N_0$ will be one, for one particle or $N_0 - 1$ will be N_0, and for two or $N_0 - 2$ will be $\frac{N_0(N_0 - 1)}{2}$. It is, therefore, easiest to evaluate the moments for systems which are either nearly full or nearly empty. Evaluation of the moments of operators of higher rank is complicated not only by the complexity of the operator itself but also the fact that we need to evaluate it for systems with larger dimensionalities. It might appear that we face an additional problem in that H is not a diagonal operator in the shell model basis. Unitarity implies that the trace of an operator is the same in any representation, diagonal or not, so we are free to calculate the moments (normalized traces) in our non-diagonal basis.

For this approach to be useful, we need information on the degree to which the level density distribution is close to Gaussian. As has been previously discussed, the Gaussian form and the Fermi gas form are quite similar over an energy range of 10-20 MeV; thus, the good agreement of data with the energy dependence of such a form suggests that a Gaussian form would also be appropriate. From a theoretical standpoint, French [10] and Ginocchio [11] have shown that for a one-body Hamiltonian the level density will become asymptotically Gaussian as the number of orbits and particles increases. The comparisons in Figs. 2 and 3 utilize not just the first and second moments of the Hamiltonian, but also the fourth moment. Thus, the expansion is not strictly Gaussian but has small modifications ($\sim$ 5% except in the tail regions) due to the presence of a small component of the fourth Hermite polynomial needed to match the fourth moment. The small magnitude of this correction attests to the near-Gaussian shape of the level density in these systems.

It is harder to determine the appropriateness of the Gaussian form in general for a two-body Hamiltonian. Specific two-body Hamiltonians are known whose spectra are not approximately Gaussian. These are, however, believed to be exceptional situations and realistic Hamiltonians have been found [8,9] to give spectra which are very close to Gaussian. As was done for the systems shown in Figs. 2 and 3, one normally calculates higher moments to check on the extent to which the distribution is Gaussian.

Results [12] for the level density of ^{28}Si calculated with this approach have recently been reported. These were done with a basis which assumes an ^{16}O core and 12 active particles distributed in the $d_{5/2}$, $s_{1/2}$, $d_{3/2}$ and $f_{7/2}$ orbits. In Fig. 5 we show the comparison of such a calculation with the experimental points deduced from level counting [13] or from Ericson fluctuations [14-16]. Also shown is a level density proposed by Roeders et al. [17], based on the measurements of Put et al. [18]. Generally good agreement is seen between calculation and experiment, although the trend in the data indicates some inconsistencies between the various measurements.

To put these results into perspective, Fig. 6 shows a comparison of the same data with some Fermi gas calculations.

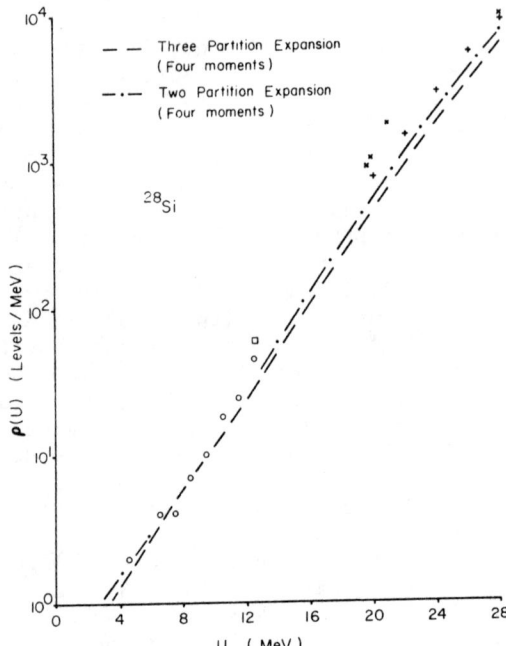

Fig. 5: Level density of ^{28}Si calculated with moment methods. The dots, circles, x and + symbols represent data, while the dot-dashed and dash lines indicate the calculated values. Deviations between the two lines are an indication of the numerical convergence of the calculation and represent the effect of re-partitioning the space.

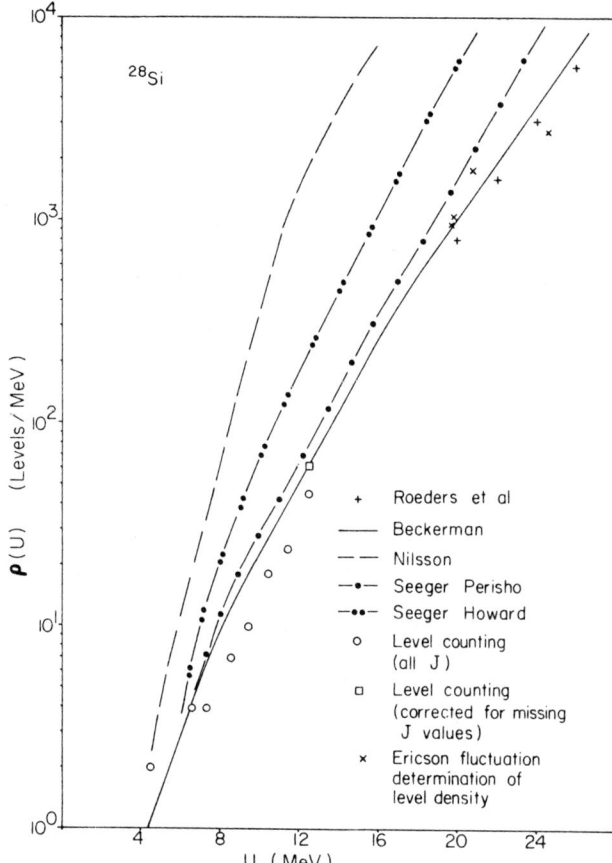

Fig. 6: Level density of ^{28}Si calculated with the microscopic Fermi gas formalism. The curves marked Nilsson, Seeger-Perisho and Seeger-Howard indicate the results of calculations with the parameters of Refs. 19-21, respectively. The curve labeled Beckerman is a fit to the data derived in Ref. 22.

The points labeled Nilsson, Seeger-Perisho, and Seeger-Howard, respectively, are points calculated using the microscopic Fermi gas formalism with the single particle levels proposed by Refs. 19-21, respectively. For this nucleus, the predictions from the various input sets vary widely with the Seeger-Perisho set providing the best agreement with the data. None of these is a fit to the data, so the range of values for the level density gives some estimate of how much variation can be expected from the use of global single particle sets. The solid line labeled Beckerman represents the fit of Beckerman [22] to the data; this then represents the result of using a and δ as fitting parameters. Moment method results can be seen to be roughly comparable in quality to the Beckerman fit. It should be pointed out, however, that some parameter adjustment was done in making the moment method calculation. Some reduction in the two-body force strength was required in order to adjust to the larger number of active particles. Hopefully, as more such calculations are done, the need for such adjustments will decrease as more knowledge about the input parameters accumulates. One significant remaining problem is the large demand for computer time ($\sim$ 1 hour) for these calculations.

IV. FUTURE DEVELOPMENTS

A recent conference at Brookhaven National Laboratory examined the present status of level density calculations and measurements. The proceedings [23] of this conference are testimony to the continuing interest in nuclear level densities for both basic and applied purposes. One point strongly emphasized in the summary report is that there is need for continuing effort in both experiment and theory in this area. The availability of more nuclear level density data would serve to stimulate efforts to refine present models and develop new ones.

Applied uses of nuclear level densities often require the level densities for many nuclei. Thus, models which require considerable computer time are not useful in these situations. As the proceedings indicate, continuing work [24] on understanding the systematics of the variation of a and δ in empirical Fermi gas fits and in improving our understanding of the rotational enhancement factor [25] (the factor by which a non-two-body approach underestimates the level density). Also underway is work on number-theoretical approaches [26] to calculation of nuclear level densities. These techniques are limited to one-body Hamiltonians at present, but those involved in their development feel that an extension to two-body forces may be possible.

The previous discussion of moment methods has indicated both strong and weak points to this approach. An important advantage is that a general two-body force may be included. This not only gives us a way of dealing with collective states, but provides a nice contact point with conventional shell model calculations. The much larger basis of the shell model calculations requires a renormalization of the two-body force strength because of the additional spreading possible (more strictly, the renormalization present in the small basis force parameters must be undone). Some discussion of this point is included in Ref. 12.

One further complication is the possibility that some of the nuclear properties change with excitation energy. To the extent that single particle centroids are modified by promotion of particles to higher orbits, the moments method should automatically take this into account (thereby eliminating one problem associated with the Fermi gas model). Other more subtle effects are not so easily accounted for. If, because of excitation, the nucleus were to expand substantially, both two-body and one-body parts of the potential could be appreciably affected. As yet there is no evidence that this is a significant problem, but the importance of this effect should be studied with temperature-dependent Hartree-Fock calculations.

V. SUMMARY

The field of nuclear level densities remains an active one. Both basic and applied areas have a need for level density information. The traditional Fermi gas model is still the most widely used, despite significant drawbacks. New approaches now allow the inclusion of two-body forces, with the consequent improvement of the calculations and interfacing of level density research with shell model investigations.

REFERENCES

1. H.A. Bethe, Phys. Rev. $\underline{50}$, 332 (1936).
2. J. Bardeen, L. Cooper and J.R. Schrieffer, Phys. Rev. $\underline{108}$, 1175 (1957).
3. J.R. Huizenga and L.G. Moretto, Ann. Rev. Nucl. Sci. $\underline{22}$, 427 (1972).
4. We follow the usual convention and denote as a state one of the 2J+1 degenerate components of a nuclear level of spin J.
5. J.R. Huizenga, A.M. Behkami, R.W. Atcher, J.S. Sventek, H.C. Britt and H. Freiesleben, Nucl. Phys. $\underline{A223}$, 589 (1974).
6. S.M. Grimes, J.D. Anderson, J.W. McClure, B.A. Pohl and C. Wong, Phys. Rev. $\underline{C10}$, 2373 (1974).
7. S.M. Grimes, Proceedings of the International Conference on the Theory and Application of Moment Methods in Many-Fermion Systems, edited by B.J. Dalton, S.M. Grimes, J.P. Vary and S.A. Williams, Plenum Press (1980), p. 17.

8. J.B. French and K.F. Ratcliff, Phys. Rev. C3, 94 (1971).
    F.S. Chang, J.B. French and T.H. Thio, Ann. Phys. (N.Y.) 66, 137 (1966).
9. S.M. Grimes, Proceedings of the Europhysics Conference on Neutron-Induced Reactions, edited by P. Oblozinsky, Slovak Academy of Science (1982), p. 79.
10. J.B. French, Proceedings of the International Conference on the Theory and Applications of Moment Methods in Many-Fermion Systems, edited by B.J. Dalton, S.M. Grimes, J.P. Vary and S.A. Williams, Plenum Press (1980), p. 17.
11. J.N. Ginocchio, ibid, p. 109.
12. S.M. Grimes, S.D. Bloom, H.K. Vonach and R.F. Hausman, Jr., Phys. C27, 2893 (1983).
13. P.M. Endt and C. van der Leun, Nucl. Phys. A310, 1 (1978).
14. P.P. Singh, R.E. Segel, L. Meyer-Schützmeister, S.S. Hanna and R.G. Allas, Nucl. Phys. 65, 577 (1965).
15. R.W. Shaw, Jr., A.A. Katsanos and R. Vandenbosch, Phys. Rev. 184, 1089 (1969).
16. K.A. Eberhard and C. Mayer-Boricke, Nucl. Phys. A142, 113 (1970).
17. J.D.A. Roeders, R.W. Put, A.G. Drentje and A. van der Woude, Lett. Nuovo Cimento 2, 209 (1969).
18. L.W. Put, J.D.A. Roeders and A. van der Woude, Nucl. Phys. A112, 561 (1968).
19. S.G. Nilsson, K. Dan Vidensk. Selsk, Mat.-Fys. Medd. 29, No. 16 (1955).
20. P.A. Seeger and R.C. Perisho, Los Alamos National Laboratory Report No. LA-3751, 1967 (unpublished).
21. P.A. Seeger and W.M. Howard, Nucl. Phys. A238, 491 (1975).
22. M. Beckermann, Nucl. Phys. A278, 333 (1977).
23. Proceedings of the IAEA Advisory Group Meeting on Basic and Applied Problems of Nuclear Level Densities, ed. by M.R. Bhat, Upton, N.Y. (1983).
24. See, for example, the papers by Maino and Menapace, Ramamuthy et al., Reffo, and Rohr in Reference 23.
25. G. Hansen and A.S. Jensen, Reference 23, p. 161.
26. A.M. Anzaldo Meneses, Reference 23, p. 91.

COMPLETE SOLUTION OF A MODEL IN STATISTICAL NUCLEAR REACTION THEORY

Martin R. Zirnbauer
Max-Planck-Institut für Kernphysik, Heidelberg, W. Germany

ABSTRACT

Weidenmüller's statistical model of nuclear reactions is solved exactly for all values of the transmission coefficients by using the method of integration over anticommuting variables.

The physics of nuclear reaction processes which proceed via the formation of an equilibrated compound nucleus (CN) is governed by two energy scales: the average spacing between neighboring CN resonances, D, and their width, Γ. All conventional approaches to the problem of evaluating CN cross sections analytically have been restricted to two extreme domains[1]: the limits of weak absorption (D>>Γ) and strong absorption (D<<Γ). By introducing mathematical techniques that are novel in the context of nuclear physics, the Heidelberg group has recently succeeded[2] in deriving a closed expression for CN cross sections which is valid for <u>all</u> values of D and Γ. This expression represents, in particular, the first analytical result for the intermediate regime where D $\approx$ Γ.

We consider a statistical model of nuclear reactions defined by the following parametrization of the S-matrix[3]:

$$S_{ab}(E) = \delta_{ab} - 2i\pi \cdot \sum_{\mu\nu} \tilde{W}_{a\mu}(E - H_{GOE} + i\pi W\tilde{W})^{-1}_{\mu\nu} W_{\nu b} . \qquad (1)$$

Here, $W_{\mu a}$ are fixed (i.e. non-statistical) matrix elements that couple reaction channels a (a=1,2,...,Λ) with CN levels μ (μ=1,2,...,N$\rightarrow\infty$), and H_{GOE} is taken from the Gaussian Orthogonal Ensemble of random matrices (GOE). The statistical assumption made by choosing a GOE Hamiltonian for the CN system accounts for the well-established statistical properties of CN resonance amplitudes and energies. This is to be contrasted with the K-matrix parametrization used earlier, which has difficulties in incorporating the observed energy level fluctuations.

The model defined by eq.(1) has been solved exactly[2] under the additional assumption that

$$\sum_{\mu} \tilde{W}_{a\mu} W_{\mu b} \propto \delta_{ab}, \qquad (2)$$

corresponding to the neglect of direct reactions. The method of solution involves such mathematical techniques as integration over anticommuting variables, the Hubbard-Stratonovich-transformation followed by a saddle-point approximation, and diagonalization of supermatrices. The final expression for the CN cross section can be expressed as

$$\sigma_{ab}^{CN}(E) = \overline{|S_{ab}^{fl}(E)|^2} = \int_0^\infty d\lambda_1 \int_0^\infty d\lambda_2 \int_0^1 d\lambda \cdot I(\lambda_1,\lambda_2,\lambda) \qquad (3)$$

$$\times \prod_{c=1}^{\Lambda} \frac{1 - T_c \lambda}{(1+T_c\lambda_1)^{\frac{1}{2}}(1+T_c\lambda_2)^{\frac{1}{2}}} \cdot \{\delta_{ab} \cdot |\overline{S}_{aa}|^2 \cdot f(T_a) + (1+\delta_{ab}) \cdot T_a \cdot T_b \cdot g(T_a, T_b)\}$$

where

$$S_{ab}^{fl} = S_{ab} - \overline{S}_{ab}, \quad T_a = 1 - |\overline{S}_{aa}|^2, \qquad (4)$$

and the horizontal bar denotes the average over the ensemble. The first factor is closely related to the invariant measure of the graded symmetry group underlying the model, while the second factor plays the role of a "symmetry-breaking" term. The last factor is a sum of two terms the first of which contributes to both elastic and inelastic scattering. The second term gives the dominating contribution, and it is this term that yields the Hauser-Feshbach formula with an elastic enhancement factor of 2 in the limit of strong absorption. An explicit expression for $\overline{S_{ab}^{fl}(E) \cdot S_{cd}^{fl*}(E + \varepsilon)}$ describing correlations between different S-matrix elements can be found in ref. 2.

The author has also carried out a similar calculation for the model which is obtained by replacing H_{GOE} in eq. (1) with a member of the Gaussian Unitary Ensemble (GUE). Ensembles with unitary invariance correspond to systems with completely broken time-reversal symmetry. The GUE result for σ_{ab}^{CN} differs from eq. (3) most notably by the absence of any enhancement factor for the elastic reaction channel. Evaluation of spectral correlation functions[4] has shown that the GOE-GUE transition occurs over a scale v/D where v^2 is the dispersion of the time-reversal non-invariant matrix element. Since this implies a very rapid transition, precise measurements of the elastic enhancement factor might provide a highly sensitive test of time-reversal symmetry in nuclei.

This work was performed in collaboration with J.J.M. Verbaarschot and H.A. Weidenmüller.

REFERENCES

1. C. Mahaux and H.A. Weidenmüller, Ann. Rev. Nucl. Part. Sci. vol.29, 1 (1979).
2. J.J.M. Verbaarschot, H.A. Weidenmüller and M.R. Zirnbauer, submitted to Phys. Rev. Lett.
3. H. A. Weidenmüller, Ann. Phys., in press.
4. A. Pandey and M. L. Mehta, Commun. Math. Phys. **87**, 449 (1983).

SESSION FA

ASTROPHYSICS

r and s-processes: Chronometers, Thermometers
and Neutron Dosimeters

R.R. Winters
Denison University, Granville, Ohio 43023

ABSTRACT

Recent measurements of neutron cross sections have provided new data which can be incorporated into models of nucleosynthesis to allow more reliable estimates of the duration and physical conditions under which the heavier chemical elements formed. Measurements of various neutron capture and scattering cross sections for the osmium isotopes have greatly reduced the uncertainty associated with the use of the Re/Os beta decay as a nucleosynthetic chronometer. In particular, measurements of the ^{189}Os(n,n') and ^{189}Os(n,γ) cross sections appear to support the use of the optical model and the Brink-Axel hypothesis to calculate neutron and gamma ray transmission factors needed to estimate the effect of the first excited state in ^{187}Os on neutron capture in a stellar environment. Our latest results imply that the laboratory cross section should be enhanced by about 20% as estimated by Woosley and Fowler and, hence, the age of the chemical elements is estimated to be (17 ± 3) byr. Branchings in the s-process path can provide information about the temperature and/or neutron density during nucleosynthesis. Recent measurements of the neutron capture cross sections of the samarium isotopes imply a branch at ^{147}Nd partly bypassing ^{148}Sm. Analysis of this branching yields an estimate for the s-process neutron density of $\sim 10^8$ cm^{-3}.

I. INTRODUCTION

As models of stellar nucleosynthesis have been developed since Burbridge, Burbridge, Fowler and Hoyle (1), quantitative tests of those models have demanded more and more accurately measured neutron cross sections. This paper will focus on the response to those data needs by those of us who measure the cross sections. Such focus must mean that I can only mention many of the very difficult and challenging problems abounding in this field of research. I apologize in advance if, in such a superficial treatment, I pass too quickly over some very difficult problems. Such exist in sufficient numbers to keep up busy for quite a while. Yet, after reflecting on Willie Fowler's (2) Nobel Lecture, I am optimistic about the progress made so far in understanding the origin of the chemical elements.

II. THE s-PROCESS $\bar{\sigma}_\gamma N_s$ CURVE

Early on, Seeger, Fowler and Clayton (3) presented a model of the s-process in which the ^{56}Fe nuclei which serve as the seed for the s-process were exposed to a superposition of exponentially decreasing neutron fluences. This phenomenological model was rather successful in describing the observed s-process current, $\bar{\sigma}_\gamma N_s$ (Maxwellian-averaged neutron capture cross section times s-process abundance) for those nuclides produced primarily by the s-process. During the past decade, this model has received a good deal of theoretical support. Iben (4) has produced models of intermediate mass (3-8$M_\odot$) stars in which nucleosynthesis takes place in a He-burning shell surrounding the core of a pulsating red giant star. In this model, the energy from He-burning increases the local temperature gradient in the shell until convection begins. Convection cells carry matter both into and out of the shell, effectively cooling the cells causing convection to cease. The cycle repeats each time the temperature gradient in the He-burning shell exceeds the critical value for convective transport. The duration of each thermal pulse is $\sim$ 20 years, with an interpulse interval of < 500 years. Ulrich (5) had already demonstrated that such thermally pulsing models lead to an exponential distribution of neutron fluences as in phenomenological models (3,6).

Using the notation of Ward and Newman (7), the abundance $N_s(A)$ of an isotope A under s-process conditions can be written

$$\frac{dN_s(A)}{dt} = \lambda_n(A-1)N_s(A-1) - [\lambda_n(A) + \lambda_{\beta^-}(A)]N_s(A), \quad (1)$$

Here $\lambda_n = \phi\bar{\sigma}_\gamma$ is the neutron capture rate which is proportional to the neutron flux ϕ and to the relevant Maxwellian-averaged neutron capture cross section $\bar{\sigma}_\gamma$. The λ_β are the relevant beta decay rates (at stellar conditions). Since we must write an equation such as (1) for each mass in the s-process path from the seed ^{56}Fe, equation (1) actually represents a coupled system of differential equations. In the classical s-process it is assumed that $\lambda_n \ll \lambda_\beta$, i.e. all beta decays are rapid relative to neutron capture. Recently developed models do treat branching, i.e. $\lambda_n \sim \lambda_\beta$, in the s-process. Assuming that most of the synthesis takes place near the base of the He-burning shell which remains at nearly constant temperature over most of the pulse, the neutron capture cross sections and beta decay rates are time independent, greatly simplifying the analysis. We assume that the ^{56}Fe seed are exposed to a distribution of neutron exposures where

$$\tau = \int dt \phi(t) \quad (2)$$

such that

$$dN(56) = N_{56} \rho(\tau) d\tau \qquad (3)$$

represents the number of ^{56}Fe seed having a neutron exposure between τ and $\tau + d\tau$. Then equation (1) becomes,

$$\frac{dN_s(A)}{d\tau} = \sigma(A-1)N_s(A-1) - \sigma(A)N_s(A). \qquad (4)$$

This set of equations can be solved analytically (Clayton and Ward (8)) and, for a two-component distribution of the form,

$$\rho(\tau) = \frac{f_1 N_{56}}{\tau_{01}} \exp(-\tau/\tau_{01}) + \frac{f_2 N_{56}}{\tau_{02}} \exp(-\tau/\tau_{02}), \qquad (5)$$

yields (neglecting alpha recycling in the lead-bismuth mass region),

$$\sigma(A)N_s(A) = \frac{f_1 N_{56}}{\tau_{01}} \prod_{i=56}^{A} \left[1 + \frac{1}{\sigma(i)\tau_{01}}\right]^{-1} + \frac{f_2 N_{56}}{\tau_{02}} \prod_{i=56}^{A} \left[1 + \frac{1}{\sigma(i)\tau_{02}}\right]^{-1}. \qquad (6)$$

The four parameters f_j and τ_{0j} are to be determined from measured $\bar{\sigma}_\gamma N_s$ products. The form of (6) makes clear the need for accurate determinations of both abundances and cross sections. In the past three years, three new evaluations (Anders and Ebihara (9), Palme, Suess and Zeh (10) and Cameron (11)) of abundances have become available. The evaluation of Anders and Ebihara is of particular importance since they include estimates of uncertainties for the abundances in their evaluation. The cross section data used in s-process modeling have overwhelmingly been taken from the work of R.L. Macklin and collaborators at the Oak Ridge National Laboratory (see eg. Allen et al. (12), Macklin and Winters (13), Kappeler et al. (14)), although others, especially Kappeler's group at Kernforschungzentrum, are making important contributions. Kappeler et al. (14) have used these data with their own modifications to determine the $\bar{\sigma}_\gamma N_s$ curve shown in figure 1.

The weak ($f_2 = 0.092\%$, $\tau_{02} = 0.24$ mb^{-1}) fluences (represented by the second term in equation (5)) are required to account for the steep decrease of the $\bar{\sigma}_\gamma N_s$ products from ^{58}Fe through ^{70}Ge and ^{76}Se to $A = 90$. The stronger exposure distribution, corresponding to the first term in equation (5), with $f_1 = 2.7\%$ and $\tau_{01} = 0.056$ mb^{-1}, provided a good description of the s-process current above $A = 90$. However, the systematic discrepancies

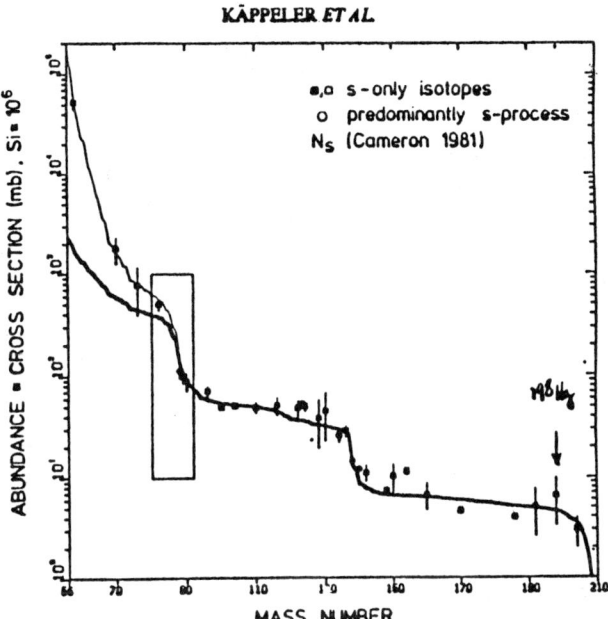

Fig. 1 The $\bar{\sigma}_\gamma N_s$ products vs. mass (from Kappeler et al. (14)). The two curves are the weak (A < 90) and the strong neutron exposures (equation (5)).

near A = 150 stimulated a series of cross section measurements of Nd (Mathews and Kappeler (15)), Dy (Beer et al. (16)) and Sm (Winters et al. (17)). These new results coupled with the abundance evaluation of Anders and Ebihara (9) have resulted in a revision of the $\bar{\sigma}_\gamma N_s$ curve in the higher mass region as shown in figure 2 taken from Winters et al. (17). Mostly due to the revised abundance, the mean fluence of the strong component is increased from 0.24 mb^{-1} to 0.295 mb^{-1}. The remaining problems appear to be for ^{134,136}Ba (below the $\bar{\sigma}_\gamma N_s$ curve by 20%) and for ^{128,130}Xe which have large uncertainties (see Beer et al. (16)).

This impressively successful model can be used to extract a good bit of information about the physical conditions in which nucleonsynthesis takes place.

Fig. 2 The $\bar{\sigma}_\gamma N_s$ products vs. A (Winter et al. (17)) for A = 150. The curve is calculated using the strong fluence term in equation (5).

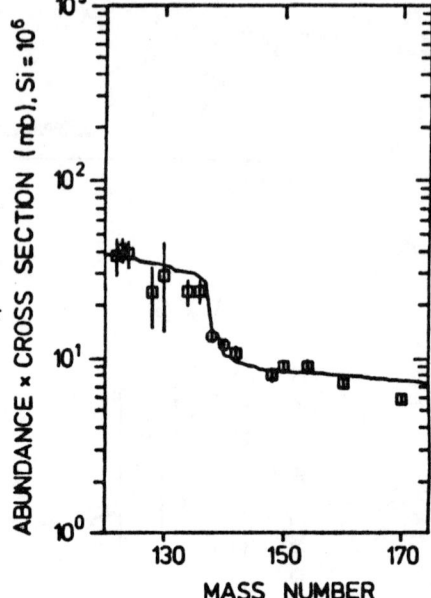

III. A NEUTRON DOSIMETER: ^{148,150}Sm

Clayton et al. (18) pointed out that in the plateau regions of the $\bar{\sigma}_\gamma N_s$ curve, the s-process current is essentially constant over small mass differences. In fact, one of the earliest tests of the s-process was provided by Macklin et al. (19) who measured the Sm capture cross sections and found for the ratio of currents

$$R = \bar{\sigma}_\gamma N_s(148)/\bar{\sigma}_\gamma N_s(150) = 1.02 \pm 0.06 \qquad (7)$$

The remarkably high precision of their result reflected only the uncertainty in the cross section ratios. However, a few years later Kononov et al. (20) remeasured these cross sections and found a result for ^{150}Sm about 50% smaller than that of Macklin et al. (19). The recent measurement of these cross sections by Winter et al. (17) confirm the result of Macklin et al. but yield R = (0.91 ± 0.03), significantly less than unity. There are a number of possible explanations for this result, e.g., failure to achieve equilibrium during the s-process flow through A = 150, or thermal neutron capture by ^{149}Sm (σ_γ(th) = 41,000 b) increasing the ^{150}Sm abundance in the meteoritic material from which solar abundances are derived. However, I believe the most likely explanation is that the s-process partially bypasses ^{148}Sm due to branchings at ^{147}Nd and ^{147,148}Pm (see figure 3).

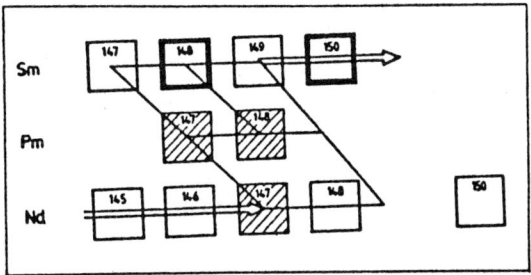

Fig. 3   The s-process paths through Nd-Sm.

If this conjecture is correct, the mean density of s-process neutrons can be obtained from the value of R. These branchings can be treated in the common s-process model (Kappeler et al. (14); Ward, Newman and Clayton (21)). Starting from ^{147}Nd, one follows the capture paths indicated in figure 3 and obtains

$$\bar{\sigma}_\gamma N_s(^{148}Sm) = \bar{\sigma}_\gamma N_s(^{146}Nd) \, \xi(^{147}Nd) \, \frac{f(^{147}Nd)}{1-f(^{147}Nd)} \, \xi(^{147}Pm) \, \zeta(^{148}Sm)$$

$$\cdot [\zeta(^{147}Sm) \, \frac{f(^{147}Pm)}{1-f(^{147}Pm)} + \xi(^{148}Pm) \, \frac{f(^{148}Pm)}{1-f(^{148}Pm)}] \quad (8)$$

and
$$\bar{\sigma}_\gamma N_s(^{150}Sm) = \bar{\sigma}_\gamma N_s(^{148}Sm) \, \zeta(^{149}Sm) \, \zeta(^{150}Sm)$$
$$+ \bar{\sigma}_\gamma N_s(^{146}Nd) \, \xi(^{147}Nd) \, \zeta(^{149}Sm) \, \zeta(^{150}Sm) \quad (9)$$
$$\cdot \{\zeta(^{148}Nd) + \frac{f(^{147}Nd)}{1-f(^{147}Nd)} \, \xi(^{147}Pm) \, \xi(^{148}Pm)\}$$

where

$$\zeta(^A Z) = [1 + 1/\bar{\sigma}_\gamma(^A Z)\tau_0]^{-1}$$

and

$$\xi(^A Z) = [1/(1-f) + 1/\bar{\sigma}_\gamma(^A Z)\tau_0]^{-1}$$

are the propagators for the s-process flow for the stable and unstable isotopes and $f = \lambda_\beta/(\lambda_\beta + \lambda_n)$ are the branching ratios at ^{147}Nd, ^{147}Pm and ^{148}Pm. The branching ratio is determined by the beta decay rates $\lambda_\beta = \ln(2)/t_{1/2}$ and the neutron capture rates

$\lambda_n = n_n v_T \bar{\sigma}_\gamma$. The latter rates are related to the s-process mean neutron density n_n, the thermal neutron velocity v_T, and the Maxwellian-averaged capture cross sections $\bar{\sigma}_\gamma$.

In addition to equations (8) and (9), one must consider that the ^{148}Pm abundance formed during the s-process decays afterwards and adds to the ^{148}Sm abundance. This fraction can be derived from

$$\bar{\sigma}_\gamma N_s(^{148}Pm) = \bar{\sigma}_\gamma N_s(^{146}Nd) \; \xi(^{147}Nd) \; \frac{f(^{147}Nd)}{1-f(^{147}Nd)} \; \xi(^{147}Pm) \; \xi(^{148}Pm) \quad (10)$$

The decay parameters of ^{147,148}Pm and of ^{147}Nd are needed in order to evaluate the s-process mean neutron density from these equations. These decay constants could be temperature dependent since low lying excited states are populated at s-process temperatures. Winter et al. (17) discuss these effects in detail, but I will assume that these effects are limited to whether or not the isomer and ground states of ^{148}Pm (see figure 4) are in thermal equilibrium.

Fig. 4   The isomeric state in ^{148}Pm.

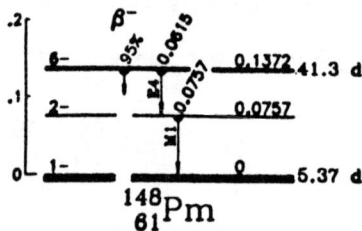

If thermal equilibrium is not attained, then we find

$$n_n = (1.0 \pm 0.4) \times 10^8 \; cm^{-3}.$$

Thermal equilibration between the isomer and ground state increases this result to $\sim 3 \times 10^8$ cm^{-3}. These results are in agreement with the analysis by Beer et al. (22) of the ^{151}Sm branching which yields

$$n_n = (1.3 \pm 2.8) \times 10^8 \; cm^{-3}.$$

Cosner, Iben and Truran (23) have studied the branching at ^{85}Kr in Iben's (4) thermal pulsing stellar model. They find a time-dependent neutron density $n_n(t)$ during a typical pulse given by

$$n_n \approx \frac{N_{22}(0)}{\Sigma \cdot \bar{\sigma}_\gamma \; N_i} \frac{1}{10 T_{min}} \exp(-4.6t/\Delta t_e)$$

$$\times \exp\left\{\frac{-\Delta t_e}{46 T_{min}}[1 - \exp(-4.6t/\Delta t_e)]\right\}. \quad (11)$$

where T_{min} is the lifetime of ^{22}Ne at the base of the He-burning shell at maximum temperature and Δt_s is the lifetime of the convective shell after maximum temperature. The abundance of the isotopes involved in the ^{147}Nd branching will freeze out when $\sigma_\gamma^{max} \tau(t) \sim 1$, where $\bar{\sigma}_\gamma^{max}$ is the largest cross section in the branch; in this case that of ^{149}Sm, $\bar{\sigma}_\gamma = 1511$ mb (see Winters et al. (17)). Thus freeze-out occurs for $\tau(t_0) \sim 6.6 \times 10^{-4}$ mb^{-1} where t_0 is given in terms of $n_n(t)$ by

$$\tau(t_0) = \int_{T_0}^{500} dt\, n_n(t) v_T$$

I find $t_0 = 7 \times 10^7$ sec corresponding to $n_n = 1.4 \times 10^8$ cm^{-3} surprising close to the value $(1-3) \times 10^8$ cm^{-3} determined from the ^{147}Nd, ^{148,149}Sm branching. Kappeler et al. (14) report similar calculations for ^{85}Kr and ^{170}Tm. A plot of $n_n(t)$ for a thermal pulse and these results are shown in figure 5 (a modification of figure 7 from Kappeler et al. (14)). The solid data points are the neutron densities derived from $n_n(t)$ and $\tau(t_0)$; the rectangles are the s-process mean neutron densities derived from the branchings. The figure suggests a correlation between the number of neutrons absorbed to produce a nucleus of mass A and the neutron density at which the abundance of A is fixed, i.e., the time after the thermal pulse begins.

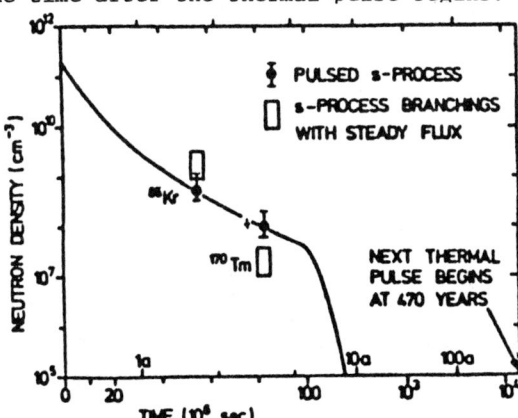

Fig. 5 Neutron density vs. time (solid line, equation (11)) for the thermal pulse in the He-burning shell (14). The full circles, bars and cross represent the model estimates and steady-flow analysis (14,16) of the branchings at ^{85}Kr, ^{170}Tm and in the Nd-Sm region.

IV. AN s-PROCESS THERMOMETER: $^{22}Ne(\alpha,n)^{25}Mg$

Assuming Iben's (4) thermal pulsing model as the site of the s-process, the neutron source is probably the $^{22}Ne(\alpha,n)^{25}Mg$ reaction. The number of neutrons per iron seed produced by this reaction should be at least sufficient to fuel the s-process in making the observed solar abundances from the ^{56}Fe seed nuclei. From the work of Kappeler et al. (14), a fraction $f_2 = 0.0097$ of the solar Fe abundance was exposed to the s-process neutron flux in synthesizing nuclei $A < 90$. If this same fraction of the lighter elements also served to absorb neutrons as they were mixed with the He-burning region, neutron balance requires

$$f_{\alpha n}\,^{22}Ne = n_c(^{56}Fe) + n_c(^{22}Ne) + n_c(^{25}Mg) + n_c(others) \quad (12)$$

where $f_{\alpha n}$ is the fraction of ^{22}Ne which undergoes the $^{22}Ne(\alpha,n)$ reaction in the He-burning shell. The quantities $n_c(A_{seed})$ are the number of neutron (per ^{56}Fe seed) capture beginning with seed of mass A building nuclides up through $A = 200$. Since ^{22}Ne is the most abundant nuclide (other than ^{4}He and ^{12}C) in the shell, its capture cross section, if large, would determine the balance. Almeida and Kappeler (24) measured the cross sections of $^{20,21,22}Ne$ finding all three to be very small, e.g., $\bar{\sigma}_\gamma = 0.9 \pm 0.7$ mb. These cross sections plus that of ^{25}Mg (Weigmann et al. (25), Macklin and Winters (13)) allowed Almeida and Kappeler (24) to calculate $n_c(A_{seed})$ and, from equation (12), $f_{\alpha n}$. Their results are shown in figure 6 which yields a value

$$f_{\alpha n} = (0.95 \pm 0.10),$$

i.e., 95% of the ^{22}Ne must participate in the source reaction in order to provide sufficient numbers of neutrons to fuel the s-process production of the solar abundances.

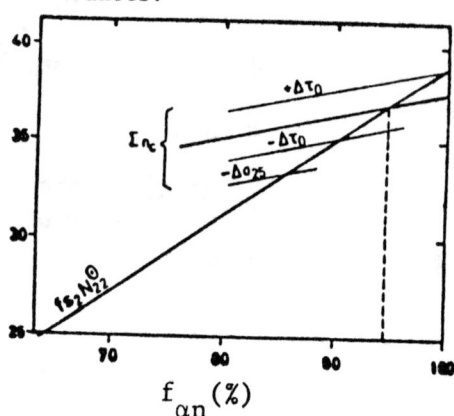

Fig. 6 (From Almeida and Kappeler, 1983). Right and left hand sides of equation (12). The value of Σn_c and uncertainties (from Almeida and Kappeler, 1983) determine the allowed range of $f_{\alpha n}$.

The rates at which this reaction and the competing (α,γ) reaction proceed are temperature dependent (Fowler et al. (26)). If the temperature is too low the (α,γ) dominates and too few neutrons are produced. In figure 7 (Fowler et al. (26)) is shown the ratio of the two rates as a function of temperature. Since

$$f_{\alpha n} > 0.85$$

and

$$f_{\alpha n} + f_{\alpha \gamma} = 1$$

we have

$$f_{\alpha n}/f_{\alpha \gamma} = \lambda_{\alpha n}/\lambda_{\alpha \gamma} > 0.85/0.15 = 5.7$$

This result implies $kT > 28$ keV, i.e., $T > 3.2 \times 10^8$ K.

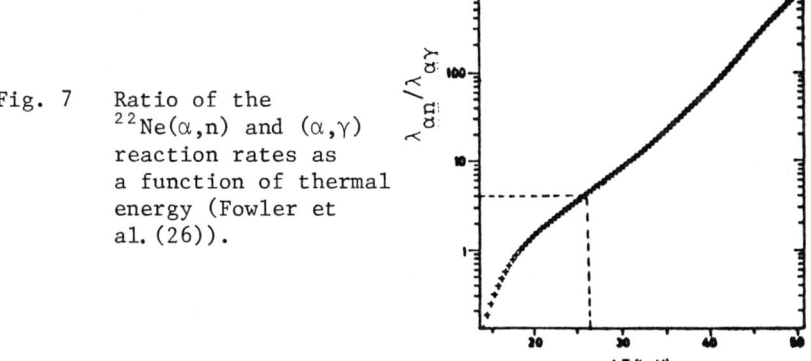

Fig. 7   Ratio of the ^{22}Ne(α,n) and (α,γ) reaction rates as a function of thermal energy (Fowler et al. (26)).

This lower bound on the s-process mean temperature agrees well with the analysis by Beer et al. (27) of the ^{176}Lu branching, 18 keV < kT < 42 MeV. Moreover, Cosmer et al. (23) find that more than 90% of the neutrons flux in the He-burning shell during a thermal pulse is produced while the temperature is about 27 keV.

V. A CHRONOMETER: Re/Os

The long laboratory half-life ($t_{\frac{1}{2}}$ = 44 byr.) of ^{187}Re and the fact that ^{187}Re is an r-process nucleus while ^{186}Os is made only by the s-process were first noted by Clayton (28) as characteristics making the Re/Os decay a good measure of the duration of the galatic nucleosynthesis. A great deal of work has been done over the past two decades in an attempt to improve the precision and accuracy of the clock. In figure 8 is shown the s-process path through the Os region.

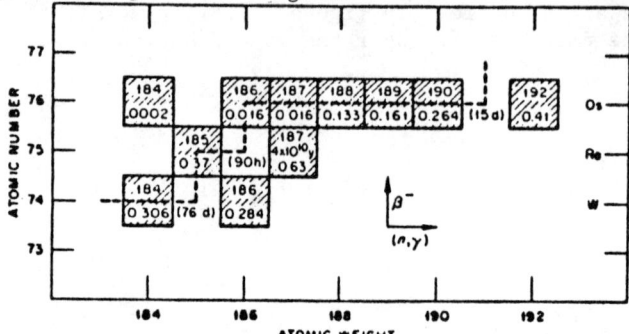

Fig. 8   s-process path in the Re-Os region.

While Arnould (29) and Beer et al. (27) have suggested branching at ^{185}W and ^{186}Re, the effect seems small (< 10%), so I shall use the conventional unbranched s-process throughout this discussion. The ratio of the abundance of ^{187}Os resulting from the beta decay of rhenium to the abundance of ^{187}Re can be written

$$\frac{^{187}Os(\beta)}{^{187}Re} = \frac{(^{187}Os/Os) - F_{67}(^{186}Os/Os)\langle\sigma_\gamma(186)\rangle/\langle\sigma_\gamma(187)\rangle}{(^{187}Re/Re)} \frac{Os}{Re} \quad (13)$$

The most recently available abundance and cross section ratio appearing in the right-hand side of equation (13) are given in Table I. The primordial (i.e. at time after r-process nucleosynthesis begins) value of the left-hand side of equation (13) can be written, assuming an exponential model $\exp(-\lambda_R t)$ for the rate of decrease of r-process nucleosynthesis, as

$$\frac{^{187}Os(\beta)}{^{187}Re} = \frac{\lambda_R - \lambda_\beta}{\lambda_R} \frac{1 - \exp(-\lambda_R \Delta)}{\exp(-\lambda_\beta \Delta) - \exp(-\lambda_R \Delta)} - 1 \quad (14)$$

In the model normally adopted in these discussions, it is assumed that the r-process has dropped to about 9% of its initial rate after a time Δ which determines λ_R. All quantities in equation (13) and (14) have been measured except F_{67} and Δ. The stellar enhancement factor F_{67} accounts for the stellar temperature effects on the neutron capture rates. Since ^{187}Os has an excited state only 9.7 keV above ground, that state is highly populated (48%) at kT = 30 keV, hence capture from that state affects the rate at which ^{187}Os is converted to ^{188}Os.

Table I Meteoritic abundance and ratio of 30 keV Maxwellian-averaged capture cross sections for the Re/Os chronometer.

	Present	Primordial [b]
^{186}Os/Os	0.0160 [a]	0.0160
^{187}Os/Os	0.0160 [a]	0.0122
^{187}Re/Re	0.626 [b]	0.646
Os/Re	(14.1 ± 2.1) [a]	(13.4 ± 2.0)
Ratio [d]	(0.478 ± 0.022) [c]	--

a) Anders and Ebihara (9)
b) Calculated using $t_{1/2} = 4.4 \times 10^{10}$ yr.
c) Macklin and Winters (13)
d) Ratio = $\bar{\sigma}_\gamma (186)/\bar{\sigma}_\gamma (187)$

The factor F_{67} has been calculated using the Hauser-Feshbach formalism by Holmes et al. (30) and more recently by Woosley and Fowler (31). The results depend critically on assumptions about the neutron T_{nn} and gamma ray T_γ transmission factors for the first excited state in ^{187}Os. Even though Woosley and Fowler attempted to normalize their calculations to extant capture cross section measurements (Winters et al. (32), Browne and Berman (33)) and to a preliminary result of a measurement (Winters et al. (34)) of the 30 keV (n,n') cross section, the allowed range of values for F_{67} was large,

$$0.80 < F_{67} < 1.2.$$

The 20% uncertainty in F_{67} dominates the small (5%) uncertainty in the ratio of Maxwellian capture cross sections. It was clear that additional measurements of the ^{187}Os(n,n') cross section and measurements of the ^{187}Os total cross section near 30 keV would serve to better define the neutron transmission factor of the excited state. Two additional measurements of the (n,n') cross section have now been reported (Macklin et al. (35) and by Hershberger et al. (36)). The results of these last two measurements are shown in figure 9. Winters and Macklin (37) have reported small adjustments to both capture cross section measurements bringing them into nearly perfect agreement. In addition, the total neutron cross sections for ^{187}Os has been measured by Winters et al. (38). With these results, Hershberger et al. (36) used a giant dipole form-factor (Johnson (39)) for the gamma ray transmission factor and an optical model (Table II) to calculate the particle transmission factors to obtain fits to measured cross sections as shown in Table III and figure 9. Clearly the agreement between calculation and measurement is acceptable but not perfect.

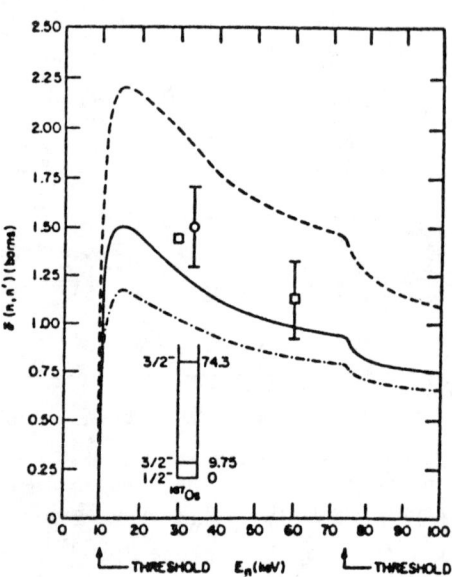

Fig. 9   Measured and calculated neutron inelastic cross sections for scattering to the $3^-/2$ state. The square point is the measurement at 60 keV by Hershberger et al. (36) and the circle is the result of a measurement at 34 keV by Macklin et al. (35).

Table II   The optical model and giant dipole parameters used by Hershberger et al. (36) in modeling the Os cross sections.

	(n,n)	(n,p)	(n,α)
V (MeV)	47.4 − 0.27 E	59.7 − 0.25 E	250
r_0 (fm)	1.22	1.25	1.24
a_0 (fm)	0.65	0.65	0.59
W (MeV)	9.97 + 0.45 E	6.19 + 0.45 E	27
r_D (fm)	1.26	1.26	1.24
a_D (fm)	0.47	0.50	0.59
GDR	E_g (MeV)	Γ (MeV)	σ_γ (mb)
	13.22	1.95	490

Table III Comparison of measured and calculated cross sections (in barns) at 60 keV. Calculations are using the optical and giant dipole models from Hershberger et al. (36) for ^{187}Os.

	^{187}Os Measured	^{187}Os Calculated	^{189}Os Measured	^{189}Os Calculated
$\bar{\sigma}_{tot}$ b)	12.8 a)	12.6	--	--
$\sigma_{n\gamma}$ b)	0.60 b)	0.78	0.71 d)	0.61
$\sigma_{nn'}$ b)	1.13 c)	1.44	0.45 e)	0.39

a) Winters et al. (38)
b) Winters and Macklin (37)
c) Hershberger et al. (36)
d) Macklin and Winters (41)
e) This work

Hershberger et al. (36) used their optical and giant dipole models to investigate the range of possible values of F_{67} and found

$$0.80 < F_{67} < 0.83 .$$

Fowler (40) had suggested one more check on the reliability of these calculations. As shown in figure 10, the spin, parity and Nilsson numbers of the ground and 1st excited states of ^{189}Os are reversed relative to those of ^{187}Os. Hence measurements of the neutron total, capture and inelastic scattering cross sections of ^{189}Os should provide additional constraints on the optical model/giant dipole parameters for ^{187}Os. Macklin and I have completed the measurement and Macklin has completed a preliminary analysis of the capture cross sections. Preliminary results of this work are shown in figure 11. Also shown in figure 11 are results calculations by Hershberger of the (n,γ) cross sections using their optical model for ^{187}Os (modified, of course, to correctly account for the low lying excited states in ^{189}Os). The agreement is remarkable.

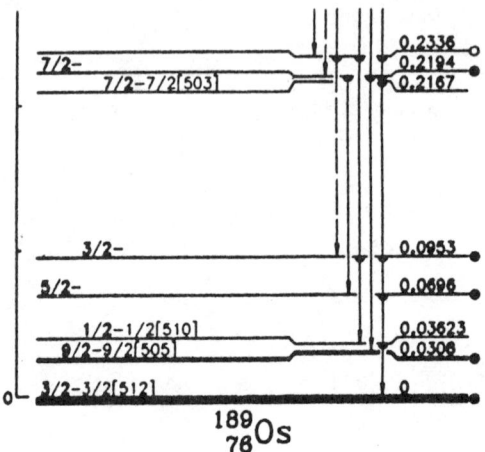

Fig. 10 Excited states of ^{189}Os from Lederer and Shirley (42).

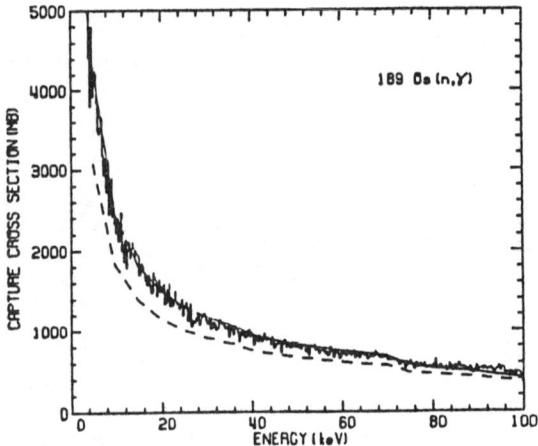

Fig. 11 Measurement (Macklin and Winters (41)) and calculated (Hershberger (43)) capture cross sections for ^{189}Os. The solid curve is a statistical model calculation without level fluctuation corrections; the dashed curve includes those corrections.

McEllistrem, Hershberger and I are now measuring the (n,n') cross section, but we have some preliminary results shown in figure 12. From the area in the peak corresponding to inelastically scattered neutrons, we estimate $\bar{\sigma}(n,n') = 0.45$ b at 75 keV.

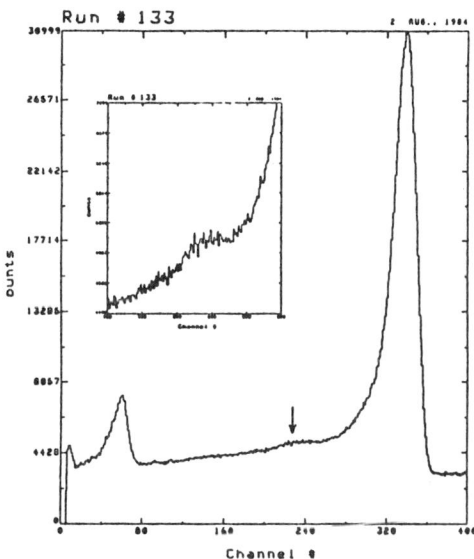

Fig. 12 Yields from a measurement (McEllistrem et al. (44)) of the ^{189}Os neutron inelastic scattering cross section at 75 keV. The insert shows the region of the peak corresponding to inelastically scattered neutrons.

In figure 13 are shown the results of a calculation by Hershberger (43) and our result for $\bar{\sigma}(n,n')$. The agreement is very good. Hence the optical/giant dipole parameters shown in Table III seem to provide an excellent description of the ground state and first excited state of ^{189}Os. F_{67} appears to be in the range

$$0.80 < F_{67} < 0.83 .$$

In figure 14 is shown the result of solving equations (13) and (14) for Δ using $F_{67} = 0.81$. The result is

$$\Delta = (11.0 \pm 2.5) \text{ byr} .$$

This leads to an age of the universe $A_u = 17$ byr. with an uncertainty of about ± 3 byr.

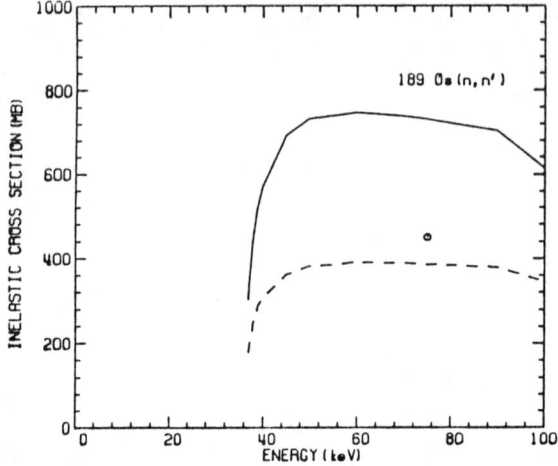

Fig. 13 Comparison of preliminary calculation (43) and measurement (44) (circle) of the ^{189}Os(n,n') cross sections. The solid and dashed curves are as in figure 11.

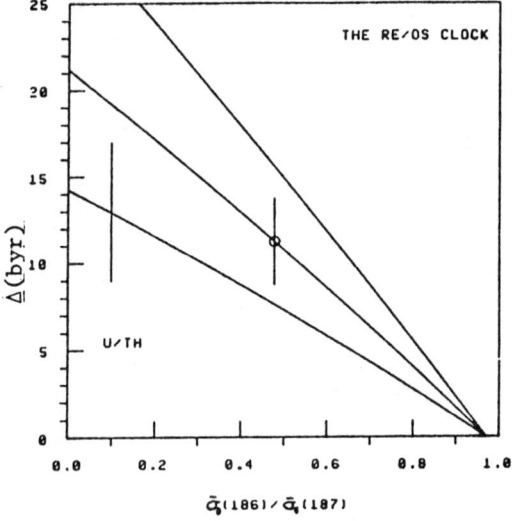

Fig. 14 Duration of galactic nucleosynthesis as a function of the Maxwellian-averaged cross section ratio $\bar{\sigma}_\gamma(^{186}\text{Os})/\bar{\sigma}_\gamma(^{187}\text{Os})$. The upper curve results from $\lambda_R = 0$, the middle, $\lambda_R \Delta = 1/0.43$ and the lower, $\lambda_R \to \infty$. Also shown is the result of a recent reanalysis (45) of the U/Th chronometer.

Recently Thielemann et al. (45) re-examined the U/Th chronometers taking into account the effects of beta-delayed fission. As shown in figure 14, that chronometer now yields $\Delta = (13 \pm 4)$ byr. or an age of the universe $A_u = 19$ byr. These results are comparable to the Hubble Time $H_0^{-1} = (19.5 \pm 3)$ byr. as determined by Sandage and Tammann (46) and are twice as large, $H_0^{-1} = (10.3 \pm 0.4)$ byr., as that determined by DeVaucoulers (47) and Aaronson et al. (48). A study by Fliche et al. (49) of the spatial distribution of the quasars leads those authors to suggest that a Lemaitre model with small positive curvature describes the universe rather than the Friedman model normally assumed in interpreting the Hubble constant as a measure of A_u. In that case, the shorter Hubble time implies $A_u = 18$ byr. What about the longer Hubble time?

There are problems with estimating the effects of Re/Os of temperature (Takahashi and Yokoi (50)) and electron capture (Arnould et al. (29)) on the beta decay rate. An excellent review of these problems is given in a paper by Yokoi et al. (51) who discuss not only Re/Os but also U/Th in the context of their model for the chemical evaluation of the galaxy. These authors tentatively conclude, primarily because of the uncertainty associated with estimating the initial galactic abundance ratio Os to Re, that perhaps neither chronometer can serve as a reliable measure of nucleosynthesis. It appears there remain problems in cosmochemistry of universal scale!

REFERENCES

1. E.M. Burbridge, G.R. Burbridge, W.A. Fowler and F. Hoyle, Rev. Mod. Phys. 29, 547 (1957).
2. W.A. Fowler, Rev. Mod. Phys. 56, 137 (1984).
3. P.A. Seeger, W.A. Fowler and D.D. Clayton, Astrophys. J. Suppl. 11, 121 (1965).
4. I. Iben, Astrophys. J. 196, 525 (1975).
5. R.K. Ulrich, in Explosive Nucleosynthesis, D.N. Schramm and W.D. Arnett, eds., Austin, Texas (1973).
6. D.D. Clayton and M.J. Newman, Ap. J. 192, 501 (1974).
7. R.A. Ward and M.J. Newman, Ap. J. 219, 195 (1978).
8. D.D. Clayton and R.A. Ward, Ap. J. 193, 397 (1974).
9. E. Anders and M. Ebihara, Geochim, Cosmochim. Acta 46, 2363 (1982).
10. H. Palme, H.E. Suess and H.D. Zeh, in Landolt-Bornstein, Neue Serie VI/Z, Chaps. 3,4 (1981).
11. A.G. Cameron, in Essays in Nuclear Astrophysics, C.A. Barnes, D.D. Clayton and D.N. Schramm, eds., Cambridge University Press (1981).

12. B.J. Allen, R.L. Macklin and J.H. Gibbons, Adv. Nucl. Phys. 4, 205 (1971).
13. R.L. Macklin and R.R. Winters, Nucl. Sci. Eng. 78, 110 (1981).
14. F. Kappeler, H. Beer, K. Wisshak, D.D. Clayton, R.L. Macklin and R.A. Ward, Astrophy. J. 257, 821 (1982).
15. G.J. Mathews and F. Kappeler, to be published in Ap. J. (1984).
16. H. Beer, F. Kappeler, K. Wisshak and R.A. Ward, Ap. J. Suppl. 46, 295 (1981).
17. R.R. Winters, F. Kappeler, K. Wisshak, A. Mengoni and G. Reffo, submitted to Ap. J. (1984).
18. D.D. Clayton, W.A. Fowler, P.E. Hull and B.A. Zimmerman, Ann. Phys. 12, 331 (1961).
19. R.L. Macklin, J.H. Gibbons and T. Inada, Nature 197, 369 (1963).
20. V.N. Kononov, B.D. Yurlov, E.D. Poletaev and V.M. Timokhov, Soviet J. Nucl. Phys. 26, 500 (1977).
21. R.A. Ward, M. Newman and D.D. Clayton, Ap. J. Suppl. 31, 33 (1976).
22. H. Beer, F. Kappeler, K. Wisshak and R.A. Ward, Ap. J. Suppl. 45, 295 (1981).
23. K. Cosner, I. Iben and J.W. Truran, Ap. J. Lett. 238, L91 (1980).
24. J. Almeida and F. Kappeler, Ap. J. 265, 417 (1983).
25. H. Weigmann, R.L. Macklin and J.A. Harvey, Phys. Rev. C 14, 1328 (1976).
26. W.A. Fowler, G.R. Caughlan and B.A. Zimmerman, Ann. Rev. Astr. Ap. 13, 69 (1975).
27. H. Beer, G. Walter, R.L. Macklin and P.J. Patchett, Phys. Rev. C 30, 464 (1984).
28. D.D. Clayton, Ap. J. 139, 637 (1964).
29. M. Arnould, Astron. Astrophys. 21, 401 (1972).
30. J.A. Holmes, S.E. Woosley, W.A. Fowler and B.A. Zimmerman, At. Data Nucl. Data Tables 18 (1976).
31. S.E. Woosley and W.A. Fowler, Ap. J. 233, 411 (1979).
32. R.R. Winters, R.L. Macklin and J. Halperin, Phys. C 21, 563 (1980).
33. J. Browne and B. Berman, Nature 262, 197 (1976).
34. R.R. Winters, F. Kappeler, K. Wisshak. B.L. Berman and J.C. Browne, Bull. Am. Phys. Soc. 24, 854 (1979).
35. R.L. Macklin, R.R. Winters and N.W. Hill, Ap. J. 274, 408 (1983).
36. R.L. Hershberger, R.L. Macklin, M. Balakrishnan, N.W. Hill and M.T. McEllistrem, Phys. Rev. C 28, 2249 (1983).
37. R.R. Winters and R.L. Macklin, Phys. Rev. C 25, 208 (1982).
38. R.R. Winters, R.F. Carleton, J.A. Harvey and N.W. Hill, Proceedings of the International Conference on Nuclear Data for Science and Technology, Antwerp, Belgium (1982).
39. C.H. Johnson, Phys. Rev. C 16, 2238 (1977).

40. W.A. Fowler, private comm. (1979).
41. R.L. Macklin and R.R. Winters, private comm., Sept. (1984).
42. C.M. Lederer and V.S. Shirley, eds., Table of Isotopes, John Wiley, New York (1978).
43. R.L. Hershberger, private comm., Sept. (1984).
44. M.T. McEllistrem, R.L. Hershberger and R.R. Winters, private comm., Sept. (1984).
45. F.K. Thielemann, J. Metzinger and Z. Klapdor, Phys. A $\underline{309}$, 301 (1983).
46. A. Sandage and G.A. Tammann, Ap. J. $\underline{256}$, 339 (1982).
47. G. DeVaucouleurs, Ap. J. $\underline{253}$, 520 (1982).
48. M. Aaronson, J. Mould and J. Huchra, Ap. J. $\underline{237}$, 655 (1980).
49. H.H. Fliche and J.M. Sourian, Astr. Astrophys. $\underline{78}$, 87 (1970).
50. K. Takahaski and K. Yokoi, Proceedings of the 4th International Conference on Nuclei Far from Stability, CERN-Report, $\underline{81-09}$, 351 (1981).
51. K. Yokoi, K. Takahaski and M. Arnould, Astron. Astrophys. $\underline{117}$, 65 (1983).

NEUTRON CAPTURE PROCESSES OF HEAVY ELEMENT SYNTHESIS

J. W. Truran
University of Illinois, Urbana, IL 61801

ABSTRACT

The elements in nature which are heavier than iron are now understood to have been formed primarily by neutron-capture reaction mechanisms. Sequences of neutron captures, interspersed by beta decays, allow the buildup of progressively heavier nuclei. Solar system abundances reveal signatures of two such processes, distinguished by the conditions that the appropriate neutron capture timescales are greater than (s-process) or less than (r-process) the typical beta decay timescales in the vicinity of the valley of beta stability. The r-process flow paths can proceed into the neutron-rich regions far off the valley of beta stability, under some conditions even approaching the neutron-drip line, depending upon the prevailing neutron density and temperature. The s-process flow path is constrained rather to follow along the valley of beta stability. Observations and theory confirm the identification of the site of s-process synthesis with red giant stars. The site of r-process synthesis has not been unambiguously established. Recent studies of isotopic abundance anomalies in meteorites and of abundances in old and correspondingly metal deficient stars in our galaxy provide potentially important clues to the nature of the r-process site. Current theories of neutron capture synthesis will be reviewed.

INTRODUCTION

Neutron capture processes have, from the outset, played an historically important role in the formulation of nucleosynthesis theory. Alpher, Bethe, and Gamow[1] early noted the existence of an approximate inverse correlation between neutron capture cross section and the relative abundance of heavy nuclei. The presence of significant abundance peaks associated with the positions of the neutron closed shells at N = 50, 82, and 126 is evident in solar system abundance compilations[2,3]. Alpher Bethe and Gamow sought an explanation for this correlation within the framework of cosmology. They recognized that free neutrons would constitute a significant fraction of matter at the extreme densities and temperatures which prevailed during the earliest stages of the cosmological big bang and anticipated that, during the subsequent expansion, the interaction of these neutrons with protons would lead via successive neutron captures to the production of progressively heavier elements. Modern calculations[4,5,6] reveal rather that the neutron to proton ratio emerging from these early stages, prior to the nucleosynthesis era, is approximately 1/7; for this ratio, subsequent nuclear transformations lead to the production of a ^{4}He mass fraction of order 25 percent but only trace amounts of heavy elements.

Neutron processes here again served to provide a critical clue to the nature of the alternative to cosmological nucleosynthesis: stellar nucleosynthesis. The realization that nucleosynthesis of heavy elements represents a continuing process in the interiors of stars followed upon Merrill's[7] detection of the presence of technetium in the atmospheres of red giant stars. Since the lifetime of the longest lived isotope of technetium is less than the ages of the very stars in which it has been observed, this serves to confirm that nuclear transformations <u>in situ</u> have produced the technetium.

The fact that stellar interior and supernova environments provide conditions compatible with nucleosynthesis was thus established. Subsequent investigations have revealed that the processes by which heavy elements are synthesized in these environments fall broadly into two categories. Most of the elements from carbon to nickel are believed to be formed by charged-particle dominated reaction sequences occurring either during relatively stable phases of nuclear energy generation in stellar interiors or in dynamically changing supernova environments. Alternatively, neutron capture processes are primarily responsible for the synthesis of the elements heavier than iron and nickel. These neutron capture processes are defined in the next section. The input nuclear parameters essential to studies of these nucleosynthesis processes are then briefly reviewed. Finally, specific models for s-process and r-process nucleosynthesis are presented and discussed. A more extensive review of nucleosynthesis has recently been provided by Truran[8] and a review of neutron capture processes in astrophysics by Mathews and Ward[9].

NEUTRON CAPTURE PROCESSES DEFINED

The formation of the bulk of the nuclear species of mass number $A \gtrsim 60$ occurs in nature by means of neutron capture processes. The abundance features in the heavy element region which are correlated with neutron shell closures strongly support this conclusion. These abundance peaks also reflect the operation of two distinct neutron processes - characterized by quite different neutron fluxes.

These two processes are distinguished on the basis of the relative lifetimes for neutron captures (τ_n) and beta decays (τ_β), where τ_β is a characteristic lifetime for beta unstable nuclei in the vicinity of the valley of beta stability. The s-process of neutron capture is defined by the condition $\tau_n > \tau_\beta$, which constrains the neutron capture path to remain close to the valley of beta stability. The peaks in the solar system abundance distribution at strontium (neutron number $N = 50$), barium ($N = 82$) and lead ($N = 126$) are attributable to this s-process: the peaks appear due to the fact that cross sections for neutron capture out of the closed shells fall quite precipitously.

The r-process of neutron capture is defined alternatively by the condition $\tau_n < \tau_\beta$, again noting that the τ_β here refers to typical decays in the vicinity of the valley of beta stability.

For this condition, it follows that successive neutron captures can proceed into the neutron-rich regions far off the valley of beta stability. This r-process neutron capture mechanism is expected to operate in an environment characterized by a more violent epoch of generation of neutrons. Following the exhaustion of the available neutrons, the capture products decay via beta decay toward the stable regions on a longer timescale. Since the neutron closed shell positions are encountered at lower mass numbers in the neutron rich regions, abundance peaks attributable to the r-process are realized, and are evident in solar system matter, several mass numbers below the s-process peaks at strontium, barium and lead. These double-peak features in the heavy element region are the signatures of the two distinct neutron-capture processes.

The neutron flux and timescale characteristics of the astrophysical neutron capture processes were identified in the early and defining papers on nucleosynthesis by Burbidge, Burbidge, Fowler, and Hoyle[10] and Cameron[11]. The neutron flux associated with the r-process is sufficiently high that the most neutron-rich isotopes along the r-process capture path typically have beta decay lifetimes < 1s. Lower limits on the lifetimes of critical nuclei involved in the s-process range rather from ~ 10-100 years. Analyses of the operation of these neutron capture mechanisms, together with experimental determination of critical neutron capture cross sections, have provided a quantitative means of disentangling the r-process and s-process contributions to solar system matter[12,13].

NUCLEAR PARAMETERS FOR STUDIES OF NEUTRON CAPTURE SYNTHESIS

The formation of the heavy nuclei (A > 60) in astrophysical environments occurs as a consequence of sequences of neutron captures interspersed by beta decays. Critical input to theoretical calculations of the nuclear transformations involved in these neutron capture processes includes both neutron-capture cross sections and beta decay rates.

Experimental cross section data is now becoming available for an increasing number of the neutron capture reactions which participate in the buildup of heavy nuclei[14-17]. These include particularly the neutron capture cross sections which are relevant to considerations of s-process nucleosynthesis. They are useful both as direct input to numerical calculations of the s-process in the context of realistic models of the appropriate astrophysical environments and as a means of disentangling quantitatively the s-process and r-process components of solar system abundances. The σN curve built upon the most recent estimates of the s-process abundances and experimental neutron capture cross sections is discussed by Mathews, Howard and Ward in these proceedings.

There are hundreds of neutron-capture reactions which participate in the processes by which heavy nuclei are produced in red giant and supernova environments. This is particularly true in the case of the r-process, for which the capture path can lie anywhere between the valley of beta stability and the neutron drip line. The absence of experimental information for these many cases makes it necessary to develop some theoretical model for the estimation of neutron capture cross sections. Calculations based upon the adoption of a Hauser-Feshbach[18] or equivalent expression for the energy averaged cross section have proven successful[19,20]. Holmes et al[20] provide the most complete compilation of neutron-capture reaction rates for stellar conditions. The need for additional experimental input, both to provide specific rates critical to our understanding of these processes and to serve as a guide for theoretical modeling, cannot be overemphasized.

The timescales for the buildup of heavy nuclei by neutron capture synthesis are substantially set by the beta decay lifetimes of the nuclei along the capture flow path. Two extremes are clearly encountered here. The s-process capture path is constrained to follow very closely the valley of beta stability; the beta decay lifetimes of many or most of the critical nuclei (in their ground states) are thus generally known from experiment. At the temperatures prevailing in the thermally pulsing helium shells of red giants, $2 \times 10^8 \lesssim T \lesssim 4 \times 10^8$ K, the low lying excited states of nuclei may be significantly thermally populated. Decays emanating from these states can in principle proceed far more rapidly than ground state decay. Electron and positron decay and electron capture rates for a large number of nuclei near the valley of beta stability which participate in the s-process reaction sequences have most recently been calculated from existing excited state data by Ward and Newman[21] and Cosner and Truran[22].

Beta decay rates utilized in the vast reaction networks involved in calculations of r-process synthesis have been determined in the context of several models. Takahashi, El Eid and Hillebrandt[23] have calculated rates based upon the gross theory of beta decay, while Klapdor et al[24] have provided rates based upon the Klapdor treatment of the Gamow-Teller resonance. Conclusions to be drawn regarding the nature and site of r-process synthesis, as we shall see, are extremely sensitive to these beta decay rates.

S-PROCESS NUCLEOSYNTHESIS

Observations and theory both substantially confirm the identification of red giant stars as the site of s-process nucleosynthesis. The seed nuclei upon which captures occur are largely provided by the primordial concentration of iron-peak nuclei in the stellar envelope. The neutron flux is believed to be provided by one of two neutron sources identified by Cameron[25,26]: the $^{13}C(\alpha,n)^{16}O$ and $^{22}Ne(\alpha,n)^{25}Mg$ reactions. The presence of both ^{13}C and particularly ^{22}Ne in the envelope matter may be readily understood in the context of stellar evolution: ^{13}C can be formed in abundance by $^{12}C(p,\gamma)^{13}N(e^+\nu)^{13}C$ when mixing processes bring protons into regions in which some helium burning to ^{12}C has previously occurred, while ^{22}Ne is formed during helium burning from ^{14}N, the residual of CN cycle hydrogen burning, by the reaction sequence ^{14}N$(\alpha,\gamma)^{18}$F$(e^+\nu)^{18}$O$(\alpha,\gamma)^{22}$Ne. Realistic calculations of s-process nucleosynthesis must necessarily be tied to stellar evolutionary models of the complex behaviors of intermediate mass red giant stars. Recent reviews of the s-process have been provided by Truran[27] and Ulrich[28].

Early studies by Schwarzschild and Härm[29], Sanders[30] and Ulrich[31] revealed that low and intermediate mass asymptotic giant branch stars, which contain active hydrogen- and helium-burning shells, provide a promising site for s-process nucleosynthesis. Iben[32,33] established that the operation of the ^{22}Ne neutron source is a natural consequence of the evolution of this stellar envelope environment. Truran and Iben[34] subsequently demonstrated that the s-process heavy element abundance pattern in solar system matter is produced as a forced consequence of the conditions which characterize these thermal pulses in more massive asymptotic giant branch red giants. Comparisons with observations[35] reveal, however, that the peculiar red giants which show s-process abundance enhancements have masses (and associated shell temperatures) which are too low for the ^{22}Ne$(\alpha,n)^{25}$Mg mechanism to operate. It has generally been assumed that the neutron source appropriate to these peculiar red giants is that provided by the ^{13}C$(\alpha,n)^{16}$O reaction. The successful operation of this source requires that there be a controlled mixing of protons into a carbon-rich region yielding ^{12}C$(p,\gamma)^{13}$N$(e^+\nu)^{13}$C. If too many protons are present, the ^{13}C$(p,\gamma)^{14}$N reaction acts both to deplete the ^{13}C abundance and to reduce its effectiveness as a neutron source: the ^{14}N thus formed competes favorably with heavy elements for neutrons, due to the large cross section for ^{14}N$(n,p)^{14}$C. Iben and Renzini[36] have demonstrated that an appropriate episode of mixing can in principle occur by semiconvection following thermal pulses in asymptotic giant branch stars of small core mass. This could possibly explain the s-process enrichments observed in the envelopes of low luminosity red giant stars. Further calculations of s-process neutron capture nucleosynthesis in red giant envelope environments are required before a definitive statement will be possible on this subject.

R-PROCESS NUCLEOSYNTHESIS

The astrophysical site for the operation of the r-process has not yet been clearly identified. The high neutron densities which are required would seem to suggest the association of the r-process with a violent event. Two possible environments have been examined in considerable detail. The shock heating of the helium layers of supernovae[37-41] provides a neutron flux, again attributable to such reactions as ^{13}C$(\alpha,n)^{16}$O and ^{22}Ne$(\alpha,n)^{25}$Mg. It would appear, however, that the typical neutron densities which can be achieved under these circumstances are not sufficient to produce the observed distribution of r-process heavy elements. The environments defined by the expansion and cooling of highly neutronized matter from the innermost ejected layers of supernovae or in neutronized jets released in the collapse of magnetized rotating neutron stars have also been shown to provide promising r-process sites[42-47]. These models are discussed in greater detail in the review by Truran[8] and in the article by D. N. Schramm in these proceedings.

REFERENCES

1. Alpher, R. A., Bethe, H. A., and Gamow, G., Phys. Rev. 73, 803 (1948).
2. Anders, E. and Ebihara, M., Geochim. Cosmochim. Acta 46, 2363 (1982).
3. Cameron, A. G. W., in Essays in Nuclear Astrophysics, eds. C. A. Barnes, D. D. Clayton, and D. N. Schramm (Cambridge University Press, Cambridge, 1982), p. 23.
4. Peebles, P. J. E., Astrophys. J. 146, 542 (1966).
5. Wagoner, R. V., Astrophys. J. 179, 343 (1973).
6. Yang, J., Turner, M. S., Steigman, G., Schramm, D. N., and Olive, K. A., Astrophys. J. 281, 493 (1984).
7. Merrill, P. W., Science 115, 484 (1952).
8. Truran, J. W., Ann. Rev. Nucl. Part. Sci. 34, 53 (1984).
9. Mathews, G. J. and Ward, R. A., preprint (1984).
10. Burbidge, E. M., Burbidge, G. R., Fowler, W. A., and Hoyle, F., Rev. Mod. Phys. 29, 547 (1957).
11. Cameron, A. G. W., Atomic Energy of Canada Ltd., CRL-41 (1957).
12. Seeger, P. A., Fowler, W. A., and Clayton, D. D., Astrophys. J. Suppl. 97, 121 (1965).
13. Cameron, A. G. W., Astrophys. Space Sci. 82, 123 (1982)
14. Macklin, R. L. and Gibbons, J. H., Rev. Mod. Phys. 37, 166 (1965).
15. Allen, B. J., Gibbons, J. H., and Macklin, R. L., Adv. Nucl. Phys. 4, 205 (1971).
16. Käppeler, F., Beer, H., Wisshak, K., Clayton, D. D., Macklin, R. L., and Ward, R. A., Astrophys. J. 257, 821 (1982).
17. Almeida, J. and Käppeler, F., Astrophys. J. 265, 247 (1983).
18. Hauser, W. and Feshbach, H., Phys. Rev. 87, 336 (1952).
19. Truran, J. W., Astrophys. J. 177, 453 (1972).
20. Holmes, J. A., Woosley, S. E., Fowler, W. A., and Zimmerman, B. A., Atom. Data Nucl. Data Tables 18, 306 (1976).
21. Ward, R. A. and Newman, M. J., Astrophys. J. 219, 195 (1978).
22. Cosner, K. M. and Truran, J. W., Astrophys. Space Sci. 78, 85 (1981).
23. Takahashi, K., El Eid, M. F., and Hillebrandt, W., Astron. Astrophys. 67, 185 (1978).
24. Klapdor, H. V., Oda, T., Metzinger, J., Hillebrandt, W., and Thielemann, F. K., Z. Phys. A. 299, 213 (1981).
25. Cameron, A. G. W., Astrophys. J. 121, 144 (1955).
26. Cameron, A. G. W., Astron. J. 65, 485 (1960).
27. Truran, J. W., Nukleonika 25, 1463 (1980).
28. Ulrich, R. K., in Essays in Nuclear Astrophysics, eds. C. A.

Barnes, D. D. Clayton and D. N. Schramm (Cambridge University Press, Cambridge, 1982), p. 301.
29. Schwarzschild, M. and Härm, R., Astrophys. J. 150, 961 (1967).
30. Sanders, R. H., Astrophys. J. 150, 971 (1967).
31. Ulrich, R. K., in Explosive Nucleosynthesis, eds. D. N. Schramm and W. D. Arnett (University of Texas Press, Austin, 1973), p. 139.
32. Iben, I. Jr., Astrophys. J. 196, 525 (1975).
33. Iben, I. Jr., Astrophys. J. 196, 549 (1975).
34. Truran, J. W. and Iben, I. Jr., Astrophys. J. 216, 797 (1977).
35. Scalo, J. M. and Miller, C. E., Astrophys. J. 246, 251 (1981).
36. Iben, I. Jr. and Renzini, A., Astrophys. J. (Letters) 263, L23 (1982).
37. Hillebrandt, W. and Thielemann, F.-K., Mitteilungen der Astron. Gesellschaft 43, 24 (1977).
38. Truran, J. W., Cowan, J. J., and Cameron, A. G. W., Astrophys. J. (Letters) 222, L63 (1978).
39. Thielemann, F. K., Arnould, M., and Hillebrandt, W., Astron. Astrophys. 74, 175 (1979).
40. Blake, J. B., Woosley, S. E., Weaver, T. A., and Schramm, D. N., Astrophys. J. 248, 315 (1981).
41. Cowan, J. J., Cameron, A. G. W., and Truran, J. W., Astrophys. J. 265, 429 (1983).
42. Cameron, A. G. W., Delano, M. D., and Truran, J. W., in Properties of Nuclei Far From the Region of Beta Stability, Vol. 2 (CERN, Geneva, 1970), p. 735.
43. Le Blanc, J. M. and Wilson, J. R., Astrophys. J. 161, 541 (1970).
44. Schramm, D. N., Astrophys. J. 185, 293 (1973).
45. Hillebrandt, W., Takahashi, K., and Kodama, T., Astron. Astrophys. 52, 63 (1976).
46. Hillebrandt, W., Space Sci. Rev. 21, 639 (1978).
47. Symbalisty, E. M. D., Schramm, D. N., and Wilson, J. R., preprint (1984).

A PARAMETRIC STUDY OF DYNAMIC S-PROCESS NEUTRON-CAPTURE NUCLEOSYNTHESES: NUCLEAR DATA NEEDS

G. J. Mathews, W. M. Howard, K. Takahashi, and R. A. Ward
Lawrence Livermore National Laboratory, Livermore, CA 94550

ABSTRACT

We summarize the stellar parameters which characterize the s-process σN curve in the framework of models which produce an exponential distribution of exposures by periodic time-dependent neutron irradiations. The optimum parameter space, defined by the best fit to the solar-system σN curve, is distinctively different than the classical smooth monotonically decreasing curve. Constraints are placed on the s-process environment. The needs for better neutron-capture data in various mass regions are highlighted.

INTRODUCTION

It has been clear for some time[1] that an exponential distribution of neutron exposures is required to fit the solar-system s-process σN curve (neutron radiative capture cross section times abundance) as a function of atomic mass. Ulrich[2,3] showed that an exponential distribution of exposures could be achieved in a single star which subjected initial seed material to periodic neutron exposures followed by dredge up of some fraction of the irradiated material to the stellar surface. This s-process scenario has been explored in a series of papers[4-7] based on a $^{22}Ne(\alpha,n)^{25}Mg$ neutron source for the s-process during thermal pulses of asymptotic giant branch stars. There is sufficient uncertainty in the stellar models, however, that a different approach is warranted, i.e. to utilize the observed solar-system s-process σN curve to define the constraints on any stellar model for the s-process. This is the subject of the present work. This study reveals that the s-process neutron-capture flow is probably considerably more complicated than the continuous chain assumed in the classical s-process. Deviations of the best fit from the observed solar-system values highlight needs for improved nuclear data.

S-PROCESS CALCULATION

In the classical s-process (without beta-decay branching) the abundance of an isotope is given simply by the solution to the set of coupled differential equations,

$$dN_A/d\tau = \sigma(A-1)N_{A-1} - \sigma(A)N_A , \qquad (1)$$

where τ is the time integrated neutron flux (i.e. neutron exposure) and σ the Maxwellian averaged neutron capture cross section.

At equilibrium, a single exposure would lead to a constant σN value for all isotopes in the s-process. The solar-system σN curve (see Fig. 1), however, requires an exponential distribution of neutron exposures, i.e. the probability, $\rho(\tau)$ for a given exposure is taken to be,

$$\rho(\tau) \propto \exp(-\tau/\tau_0) \ . \tag{2}$$

In a periodic s-process operating in a single star[2,3], the mean exposure, τ_0, is simply related to the average exposure per pulse, $\Delta\tau$, and the fraction of material, r, which remains after each dredge up, i.e. $\tau_0 = -\Delta\tau/\ln(r)$. Note, that the mixing fraction and exposure per pulse are not independent parameters.

In dynamic stellar environments, such as thermally-pulsing red giants, the simplicity of the classical s-process (Eq. 1) is lost due to a break down of the assumption that neutron captures are slow compared to beta decay. We therefore, compute the nucleosynthesis in a network:

$$\frac{dN(Z,A)}{dt} = N(Z,A-1)\Phi_n\sigma_{n,\gamma}(Z,A-1) + N(Z-1,A)\lambda_\beta(Z-1,A)$$
$$- N(Z,A)[\Phi_n\sigma_{n,\gamma}(Z,A) + \lambda_\beta(Z,A)] \ . \tag{3}$$

The time-dependence of the flux and temperature dependence of the cross sections and beta rates are included. In a few cases, positron, electron-capture, or alpha decay must also be added to Eq. (3). We utilize experimental neutron-capture cross sections when available[8-16]. Cross sections for unstable or unmeasured nuclei were taken from Hauser-Feshbach estimates[17] along with the temperature dependence of all cross sections. Decay rates from thermally populated excited states were calculated assuming thermal equilibrium and with appropriately choosen ft values[18]. The initial seed abundances were taken from solar-system values[19].

RESULTS

We have adjusted the parameters to minimize χ^2 for the fit to 23 s-only nuclei with $Z > 40$. Lighter nuclei were omitted because of possible contribution from other sources[9]. The most sensitive parameters in the fit are τ_0, and Φ_n, although the temperature, pulse shape, and interpulse period also enter.

Figure 1 is an example of a good fit ($\chi^2 \sim 1.2$) for a mean exposure of $\tau_0 = 0.31$ mb^{-1}, a constant flux and temperature ($\Phi_n = 3.2 \times 10^{16}$ cm^{-2}sec^{-1}, $kT = 30$ keV), with a long interpulse interval. The χ^2 quickly increases for higher fluxes, but increases more slowly as the flux is decreased. From this we conclude that the constant fluxes consistent with the observed σN curve for s-only nuclei must be in the range of $\Phi < 4 \times 10^{16}$ cm^{-2}sec^{-1}. The best fit value of 3.2×10^{16} cm^{-2}sec^{-1} implies a neutron density of $n_n = 1.3 \times 10^8$cm^{-3} for $kT = 30$ keV. This is consistent with previous estimates[3,14].

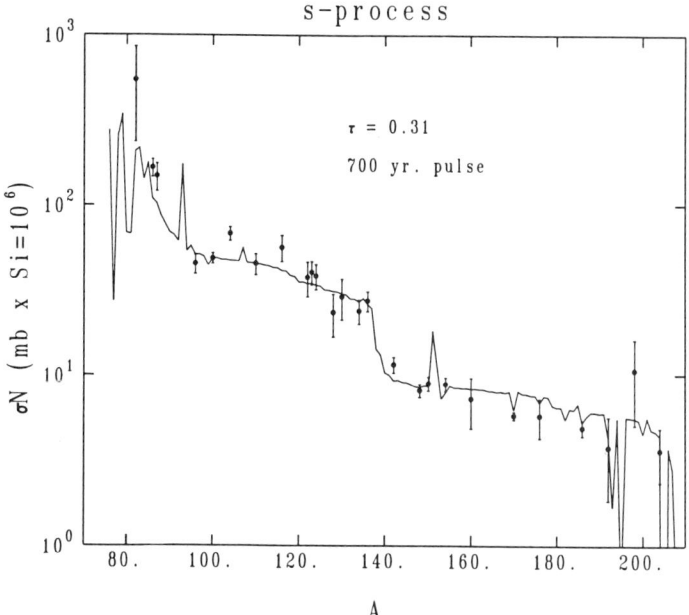

Fig. 1. Fit to the observed σN curve.

At the present time, the optimization of the fit with respect to the temperature and pulse shape is still in progress. The basic features of Fig. 1 are not significantly affected by these parameters however. The best fit for the solar-system σN curve always exhibits pronounced deviations from the smooth monotonically decreasing classical s-process curve. The reason for this is that many stable nuclei are actually produced as beta-unstable progenitors due to neutron captures on beta-unstable nuclei. If the progenitor has a larger (or smaller) cross section there will be a dip (or peak) in the σN curve when σN for the stable daughter is calculated. This is the reason for the pronounced peaks on Fig. 1 at ^{93}Nb (partially produced as ^{93}Zr) and ^{151}Eu (partially produced as ^{151}Sm).

Finally, we point out some nuclear data needs which are evident from this study. For one, there are still a number of stable nuclei along the s-process path for which neutron-capture cross sections are not measured. These include, ^{66}Zn, ^{72,73}Ge, ^{77}Se, ^{83,84}Kr, ^{99}Ru, $^{128-131}$Xe, $^{192-195}$Pt, and $^{198-201}$Hg. Amoung these isotopes ^{192}Pt and ^{198}Hg are particularly important s-only nuclei. The σN value for ^{192}Pt is substantially less than ^{198}Hg (see Fig. 1) due to branching at ^{192}Ir and ^{191}Os. The effect on the σN value of ^{204}Pb due to branching at ^{204}Tl and ^{203}Tl is also important, and would be much clearer if the uncertainties in the ^{204}Pb cross section could be reduced.

A second point is that there are several s-process only isotopes with measured cross sections which can not be brought into agreement (within one standard deviation) with the rest of the solar-system s-process σN curve for any variation of parameters. The most notable descrepancies are for ^{104}Pd, and ^{116}Sn. The uncertainty in the abundances for these nuclides is probably not very large[19]. Therefore, the nuclear data for these nuclei and neighboring isobars ought to be reexamined. There are also several s-only nuclei (like ^{160}Dy and ^{176}Hf) which may exhibit structure but the cross sections uncertainties are too large.

Work performed under the auspices of the U.S. Department of Energy by the Lawrence Livermore National Laboratory under contract number W-7405-ENG-48.

REFERENCES

1. P. A. Seeger, W. A. Fowler, D. D. Clayton, Astrophys. J. Suppl., 97, 121 (1965).
2. R. K. Ulrich, in "Explosive Nucleosynthesis", eds. D. N. Schramm and W. D. Arnett (Univ. Texas Press, Austin, 1973) p. 139.
3. R. K. Ulrich, in "Essays in Nuclear Astrophysics", eds. C. A. Barnes, D. D. Clayton, and D. N. Schramm (Cambridge Univ. Press, N. Y., 1982) p. 301.
4. I. Iben, Jr., Ap. J., 217, 788 (1977).
5. J. W. Truran and I. Iben, Jr., Ap. J., 216, 197 (1977).
6. I. Iben, Jr., and J. W. Truran, Ap. J., 220, 980 (1978).
7. K. R. Cosner, I. Iben, and J. R. Truran, Ap. J. Lett., 238, L91 (1980).
8. B. J. Allen, J. H. Gibbons, and R. L. Macklin, Adv. Nucl. Phy., 4, 205 (1971).
9. F. Käppeler, H. Beer, K. Wisshak, D. D. Clayton, R. L., Macklin, and R. A. Ward, Ap. J., 257, 821 (1982).
10. M. J. Newman, Ap. J., 219, 676 (1978).
11. H. Beer and R. L. Macklin, Phys. Rev., C26, 1404 (1982).
12. H. Beer, F. Käppeler, G. Reffo, and G. Ventorini, Ap. Space Sci., 97, 95 (1983).
13. R. R. Winters, F. Käppeler, K. Wisshak, A. Mengoni, and G. Reffo, (submitted to Ap. J., 1984).
14. H. Beer, F. Käppeler, K. Yokoi, and K. Takahashi Ap. J., 278, 388 (1984).
15. H. Beer and G. Walter, Astron. Ap. (1984 in press).
16. G. J. Mathews and F. Käppeler, Ap. J., 286, (1984 in press).
17. J. A. Holmes, S. E. Woosley, W. A., Fowler, and B. A. Zimmerman, Atom. Nucl. Data Tables, 18, 306 (1976).
18. K. Takahashi and K. Yokoi (to be published).
19. E. Anders and M. Ebihara, Geochim. Cosmochim. Acta, 46, 2263 (1982).

NEUTRON CAPTURE PROCESSES IN ASTROPHYSICS

Bradley S. Meyer and David N. Schramm
Department of Astronomy and Astrophysics
University of Chicago, Chicago, Illinois 60637

ABSTRACT

A brief discussion of the neutron capture processes of big bang nucleosynthesis, stellar and explosive nucleosynthesis, and neutron capture nucleosynthesis is given, followed by a more in depth review of the r(n)-process and its application to nucleocosmochronology.

I. INTRODUCTION

Study of neutron capture processes in astrophysics has led to important applications, including the development of nucleochronologies, the understanding of isotopic anomalies in meteorites, and cosmological constraints on particle properties. The neutron capture processes which lead to these applications include big bang nucleosynthesis, stellar and explosive nucleosynthesis, and the s- and r(n)-processes of neutron capture nucleosynthesis. This review will deal primarily with the r(n)-process and its application to nucleocosmochronology. First, however, we briefly discuss big bang nucleosynthesis and stellar and explosive nucleosynthesis. The s-process is discussed elsewhere in this volume.[1]

II. BIG BANG NUCLEOSYNTHESIS

Conditions were right in the early stages of the universe for the production of the bulk of present day abundances of ^{4}He[2] and deuterium,[3] as well as significant amounts of ^{3}He and ^{7}Li.[4] The calculated yields are all in remarkable agreement with observations, covering a range in abundances which differ by over nine orders of magnitude. This nucleosynthesis began once the temperature of the universe fell below about 10^9 K at a time about three minutes after the big bang. At that point, the neutron capture reac-

$$n + p \to D + \gamma$$

began to build up significant amounts of deuterium without destruction of that deuterium by the reverse reaction. With large amounts of deuterium then available, heavier elements were built up, including ^{4}He, ^{3}He, and ^{7}Li with abundances by number relative to hydrogen of about 0.1, 10^{-5}, and 10^{-10}, respectively. The lack of stable nuclei at mass numbers five and eight, the fact that the temperature was decreasing so that Coulomb barriers became increasingly impenetrable, and the fact that the denstiy was dropping made it difficult for big bang nucleosynthesis to proceed past mass number seven; hence, we say that the big bang produced elements with A$\lesssim$ 7. The heavier elements are made in stars where higher densities enable three-body reactions to bridge the mass five and eight gaps.

Because the abundance of ^{4}He produced in the big bang is so much greater than the abundances of the other elements produced, essentially all of the neutrons present at the beginning of nucleosynthesis subsequently went into ^{4}He nuclei. The net reaction, then, is

$$2n + 2p \rightarrow {}^4He + \gamma.$$

Clearly the amount of ^{4}He formed relative to hydrogen was dependent upon the neutron to proton ratio at the beginning of nucleosynthesis. Just before the beginning of nucleosynthesis, the universal expansion rate became greater than the weak interaction rates

$$p + e^- \longleftrightarrow n + \nu_e \,;\, n + e^+ \longleftrightarrow p + \overline{\nu}_e \,;\, \text{and } n \longleftrightarrow p + e^- + \overline{\nu}_e$$

Here the neutron to proton ratio could no longer be maintained at its equilibrium value; we say it froze out. The ratio could then only change by decay of the neutrons and by electron and positron captures until the start of deuterium production; thus, the final relative ^{4}He abundance is dependent upon the half-life $\tau_{1/2}$ for neutrons.

The final abundances of the big bang synthesized nuclei are also dependent upon the density in baryons in the early universe and on the number of neutrino species.[5] From the observed abundances, one concludes that the fraction of the critical density in baryons, Ω_b, is limited to $0.03 \lesssim \Omega_b \lesssim 0.15$ and that N_ν, the number of neutrino species is less than or equal to four.[3] It is possible to test this limit on the number of neutrino species by measuring the width Γ_Z of the Z^0 boson since the big bang model predicts $\Gamma_Z < 3.4$ Gev.[6] The current CERN measurements yield $\Gamma_Z \approx 2.6 \pm 2$ Gev. Colliding beam experiments in the near future should tighten up these error bars and begin to check the theory. This will be the first time cosmology has predicted something in fundamental elementary particle physics. Such a prediction might be compared to Hoyle's prediction of the ^{12}C excited state for stellar nucleosynthesis.

III. STELLAR AND EXPLOSIVE NUCLEOSYNTHESIS

We now consider the role neutron capture plays in stellar and explosive nucleosynthesis. We find that neutrons liberated during the synthesis of elements lighter than iron significantly affect the abundances of trans-iron nuclei. In the He burning zone, the reaction

$$^{22}Ne + \alpha \rightarrow {}^{25}Mg + n$$

provides neutrons which are rapidly captured by the trans-iron nuclei[7] or used for s-processing.[8] This s-processing yields significant production up to $A \approx 80$. Similarly, in explosive carbon burning, Wefel et al.[9] have shown that neutrons are liberated and can build up r-process like heavies such as Pd, Sm, Ba, and Gd in the same stars that can produce ^{26}Al and ^{16}O. Such processing also can produce the r-like nuclei between iron and $A \approx 80$. We conclude from the above processes that most of the nuclei from $A \approx 56$ to $A \approx 80$ are made as by-products of stellar and explosive nucleosynthesis, not by the main s- or r-processes, which work above $A \approx 80$. Thus the r/s classification of the isotopes below $A \approx 80$ by Burbidge et al.[10] is no longer appropriate. The main s-process which occurs in red giant shell flashes does not produce significant amounts of nuclei with $A \lesssim 80$ but does a great job for $80 \lesssim A \lesssim 210$.[11] Similarly, dynamic r-processes[12] do not do well for nuclei with $A \lesssim 80$ so it is nice that standard explosive processing fills in the iron to $A \approx 80$ region so well.

Apart from the role of actual synthesis by neutron capture, neutron induced reactions are important in explosive nucleosynthesis to probe the equation of state of the dense stellar core during the collapse which preceeds the explosive event. In particular, electron captures of the form

$$(Z,N) + e^- \rightarrow (Z-1, N+1) + \nu_e$$

lead to a lessening of the pressure support against gravity and, hence, to a sudden (explosive) contraction. The rates for these reactions can be probed in the laboratory by studying

$$(Z,N) + n \rightarrow (Z-1, N+1) + p$$

which involve the same nuclei and their energy levels. Such considerations are of great relevance to questions about the sites at which nucleosynthesis processes occur. We return to this problem later.

IV. NEUTRON CAPTURE NUCLEOSYNTHESIS

Nuclei of mass number 80 and greater are so heavily charged that the coulomb barriers other nuclei must overcome to fuse with the heavy nuclei are prohibitively large. It is evident, then, that neutron capture is the best means of building up heavier nuclei.

In neutron capture nucleosynthesis, a nucleus captures a neutron, thereby increasing its atomic mass. It will then either capture again or β-decay, depending on the relative rates for these two processes. For the s, or

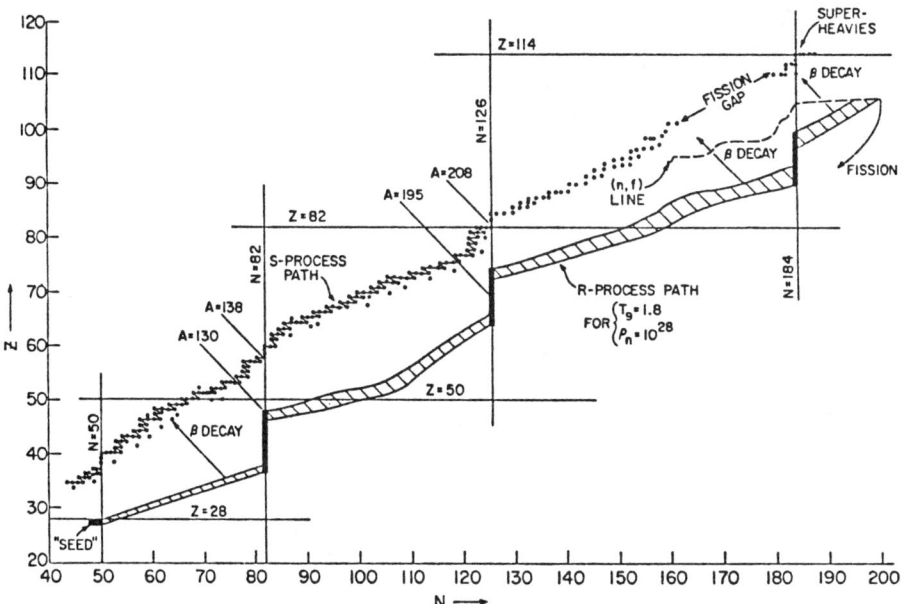

Figure 1. Neutron capture paths for the s-process and the r-process. The r-process was computed for initial conditions of $T = 10^9$ K and $n_n = 10^{28}$ cm^{-3} (Schramm and Norman[13]).

slow, process, the neutron capture rate is much slower than the β-decay rate so that a nucleus will decay back to the valley of β-stability before it captures another neutron. Thus, the s-process path follows the valley of β-stability, as shown in Figure 1. The s-process path ends where α instability sets in at Bismuth. In contrast, for the classical r, or rapid, process, the neutron capture rate is much faster than the β-decay rate. In this case, a nucleus continues to capture many neutrons until it gets out to neutron-rich material. Neutron capture continues until the (n,γ) rate is balanced by the (γ,n) rate. The nuclei are held in equilibrium there until β-decay allows them to begin capturing neutrons again. The resulting r-process path is shown in Figure 1. It is generally some tens of neutrons richer than the line of β-stability. The r-process terminates when the neutron density and temperature drop so that (n,γ) and (γ,n) reactions cease or when nuclei on the path fission thus preventing heavier nuclei from being built up.

The r-process abundance peaks occur near $A \approx 80$, 130, and 195, which correspond to where the path crosses the neutron magic numbers 50, 82, and 126, respectively. The s-process abundance peaks occur, for the same neutron magic numbers, at higher mass numbers. This is easily understood: the r-process nuclei encounter the neutron magic numbers at lower atomic Z, hence lower atomic mass, than do the s-process nuclei.

In the above picture of the classical r-process[10, 14] extremely high values of the temperature and neutron density are required. Blake and Schramm[15] showed, however, that less extreme values are required to generate the abundances attributed to the r-process if neutron captures are balanced by β-decay rather than by photodisintegration, i.e. (n,γ). This is the so-called n-process, and it is an important alternative to the r-process since it opens up alternative astrophysical sites. The n-process duplicates the r-process path so the same nuclei are produced.

V. ASTROPHYSICAL SITES FOR THE R(N)-PROCESS

Essentially all of the nucleosynthetic process now have reasonably well established astrophysical sites except for the r(n)-process. Let us now consider candidate astrophysical sites for the r(n)-process. Proposed sites require a large neutron density. These sites include:

1) The neutron-rich mass cut in a supernova explosion.[16] The site has the short-coming that the mass cut should have a neutronization gradient inward. The r-process peak, however, is observed to be quite distinct from the s-process peak, suggesting that there is highly neutronized and poorly neutronized material but no intermediate material, contrary to the expectation from a neutronization gradient.

2) Neutron processing during explosive C-(O,Si) burning.[17] This produces elements via the n-process. The problem with this site is that the calculations of Wefel et al.[9] indicate a destruction of actinides.

3) Black hole-neutron star collisions[18] or neutron star-neutron star collisions.[19] In these processes a jet of neutron material is expelled. This material then fissions and evaporates to form neutron-rich matter. This is not a classical neutron capture process since we are breaking large clumps of matter down rather than building up material from lighter elements, but it does give r(n)-process abundances and, notably, the actinides.[20] Such an

evaporation process may always produce the same path as opposed to capture processes where the path is more sensitive to temperatures and densities. This could yield the ubiquitous sharp, distinct abundance peaks discussed below. The problems with such sites are that they may be too exotic and that detailed calculations require a better understanding of large clumps of nuclear matter.

4) Explosive He-burning in supernova shocks[7] and novae.[21] Build up of nuclei here proceeds by the n-process. One problem with this site is that results are sensitive to parameters describing the shocks or stellar evolution, and these parameters are not well known. A second problem is that the events do not produce enough neutrons to make actinides.[22]

5) Magnetohydrodynamic jets resulting from rotating collapse described by Leblanc and Wilson[23] with magnetic fields and by Symbalisty[24] with and without magnetic fields. Such an event can produce material via the r-process.[25] The probability of the high rotation required for these events, however, remains to be determined.

All of the sites have problems; thus, we cannot claim one of the sites as *the* site for the r(n)-process. Indeed, we may have a combination of two or more of the sites producing the r-process abundances. For example, we might find that a common event such as neutron processing during explosive C-(O,Si) burning might combine with a rarer event such as neutron star-neutron star collisions to produce both the r-process abundances and the actinides. There might also be some other site or even synthesis processes we have not thought of that are better able to explain all of the abundances. We seem to be far from the answer so the search for the astrophysical site for the r(n)-process must continue. Wherever our search leads us, however, we must be guided by the fact that the observed r-process abundances peaks are sharp, revealing that the conditions under which most of these heavy nuclei were produced must fall in a narrow range. Such a convergence to a narrow path is a problem.

To find the site(s), we must be able to study the n or dynamical r-process accurately. This requires a knowledge of the β-decay rates, λ_β, and the neutron capture cross sections, $\sigma_{n,\gamma}$, for each nucleus affected by the r(n)-process. This is a heavy order since the r-process region covers about 5000 nuclei! Furthermore, most of these nuclei are far ($\Delta N \approx 10$) from the valley of β-stability so that no information exists for any but the low N nuclei. To estimate $\sigma_{n,\gamma}$ and λ_β far off the valley of β-stability is extremely difficult since one needs masses, deformations, levels, potentials, and so forth. Unfortunately, all one really knows is A, Z, and fits to theory at low N. Again we emphasize that much work is still required.

VI. NUCLEOCOSMOCHRONOLOGY

We now move on to consider the application of r(n)-process abundances to the study of nucleocosmochronology. The chronometers used to study the age of the galaxy are the r(n)-process nuclei ^{187}Re ($\tau_{1/2} \approx 43 \times 10^9$ years), ^{232}Th ($\tau_{1/2} \approx 14 \times 10^9$ years), and ^{238}U ($\tau_{1/2} \approx 4.5 \times 10^9$ years). Since these nuclei are radioactive, it is possible to use them to date the astrophysical production events in a manner similiar to ^{14}C dating of archaeological samples. Nucleocosmochronology is the use of observed or implied abundances to determine time scales over which nucleosynthesis of these nuclei occurred, which then determine or constrain the time scale for the duration of nucleosynthesis, the formation of the solar system, and the age of the galaxy and the universe. Cosmochronology began with Rutherford's use of uranium to determine the duration of nucleosynthesis.[26] The formalism has been further developed by Fowler and Hoyle[27] and Schramm and Wasserburg[28] and is reviewed by Symbalisty and Schramm.[29]

Let us review the principles of nucleocosmochronology. If two radioactive nuclei i and j are formed in a single event in the abundance ratio P_i/P_j, their abundance ratio a time t later is given by

$$\frac{N_i}{N_j} = \frac{P_i}{P_j} e^{-(\lambda_i - \lambda_j)t},$$

where λ_i and λ_j are the decay rates of nuclei i and j, respectively. The time t, then, is simply

$$t = \frac{\ln[(\frac{P_i}{P_j})/(\frac{N_i}{N_j})]}{\lambda_i - \lambda_j},$$

or

$$t = \frac{\ln R_{ij}}{\lambda_i - \lambda_j}$$

where

$$R_{ij} = (\frac{P_i}{P_j})/(\frac{N_i}{N_j}).$$

From the abundances of short-lived nuclei such as ^{26}Al, ^{129}I, and ^{244}Pu deduced to be present at solar system formation, it is possible to conclude that nucleosynthesis has occurred over all time.[30] For this reason, the single event model must be improved. To do this, we must look at the time history of nucleosynthesis. Nucleosynthesis has certainly varied over time due to evolution of the galaxy and to successive passages of spiral density waves over times like 10^8 years which induce the formation of massive stars. These massive stars explode in supernova events in times short ($10^6 - 10^7$ years) compared to the passage time of the density waves, thereby enriching the ambient medium with newly processed material. The long-lived chronometers ^{232}Th, ^{238}U, and ^{187}Re shed light on the total duration of nucleosynthesis; the intermediate-lived nucleus ^{235}U gives information on galactic evolution time scales; and the short-lived chronometers ^{244}Pu ($\tau_{1/2} \approx 82 \times 10^6$ years) and ^{129}I ($\tau_{1/2} \approx 16.9 \times 10^6$) reveal something of the nature of the density wave passage. Also, through the use of very short-lived chronometers like ^{26}Al and ^{107}Pd we can also study time scales on the order of 10^6 years during the formation of the solar system.

Figure 2 shows schematically a model of element production versus time. T is the beginning of collapse of the solar nebula since which time no significant new enriched material has been incorporated into the solar system material. Δ^{max} is the one event age prior to T, which for the longest-lived elements is the time interval from the mean age $\langle\tau\rangle$ of the elements to T. t_{ss} is the time since the beginning of the formation of the solar system. t_{ss} is known to be $\approx$ 4.6 x 10^9 years.

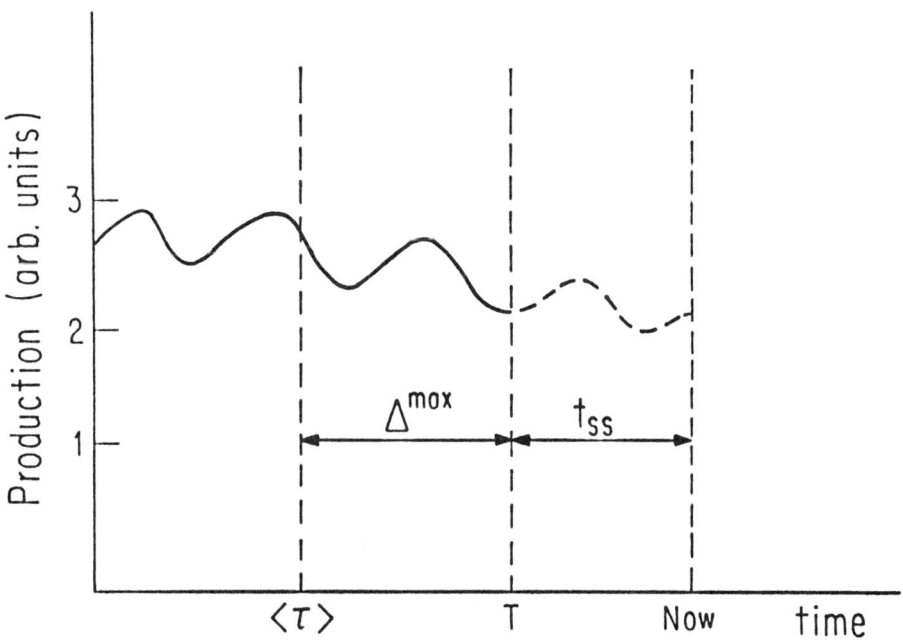

Figure 2. A schematic model of element production versus time.

The one event age can be calculated to be $(\ln R_{ij})/(\lambda_i - \lambda_j)$, where R_{ij} is $[(P_i/P_j)/(N_i/N_j)]$ and N_i and N_j are the abundances at T. It yields the mean age because sufficiently long-lived chronometers see nucleosynthesis as a single event.[28] To determine Δ^{max}, then, we require P_i/P_j, N_i/N_j, and λ_i and λ_j.

P_i/P_j, the production ratio of nuclei i and j, is obtained by adding up progenitors of the r-process abundances. Following Seeger and Schramm,[31] we assume equal abundances per isobar since there is no shell structure near typical progenitors for these nuclei[32] and correct for decay back effects forcing material out of the path of decay to the nucleus in question. We then merely add up progenitors. For example, for ^{232}Th, six progenitors decay to ^{232}Th once processing stops. These are (232), (236), (240), (244), (248), and (252). For ^{238}U, the progenitors are (238), (242), (246), and 1/10 of (250), or a total of 3.1. The other 9/10 of the (250) fissions. It is then clear that the production ratio Th/U is 6/3.1 or approximately 1.9. Wene[33] and Klapdor and Thieleman[34] have pointed, however, that delayed fission may eliminate one or more of the higher A progenitors of ^{232}Th. This effect would lower the production of ^{232}Th and, hence, require a longer Δ^{max} for a given observed present ratio of ^{232}Th/^{238}U. Note also that the Th/U pair is not a true long-lived pair since the ^{238}U half life is not $\gg T$; thus, corrections between Δ^{max} and the mean age should be expected.

The ^{187}Re/^{187}Os pair, first proposed for study by Clayton,[35] needs to be looked at somewhat differently from the ^{232}Th/^{238}U pair. ^{187}Os has no direct contribution from the r-process since it is shielded from β-decay from below by ^{187}Re. The present ^{187}Os abundance results from an s-process part $(^{187}\text{Os})_s$ and from a cosmoradiogenic part $(^{187}\text{Os})_c$ resulting from β-decay from ^{187}Re. The ratio R of ^{187}Re to ^{187}Os is

$$R = 1 + \frac{(^{187}\text{Os})_c}{^{187}\text{Re}}$$

The cosmoradiogenic abundance of ^{187}Os can be found from ^{186}Os, which is not shielded and, hence, whose abundance is not changing after nucleosynthesis:

$$\frac{(^{187}\text{Os})_c}{^{186}\text{Os}} = \frac{^{187}\text{Os}}{^{186}\text{Os}} - \frac{(^{187}\text{Os})_s}{^{186}\text{Os}}$$

or

$$\frac{(^{187}\text{Os})_c}{^{186}\text{Os}} = \frac{^{187}\text{Os}}{^{186}\text{Os}} - \frac{\sigma_{186}}{\sigma_{187}}\bigg|_{lab} \times f$$

since in the local approximation

$$(^{187}\text{Os})_s\, \sigma_{187\ 30\text{keV}} \approx (^{186}\text{Os})_s\, \sigma_{186\ 30\text{keV}}$$

where f is a factor allowing for the conversion from the lab cross sections to those appropriate for conditions inside stars, that is, at kinetic energies of ≈ 30 keV. This has been studied in detail by Winters[36] and others and is reported on elsewhere in this volume.

There are two wrinkles in the ^{187}Re/^{187}Os pair. The first is that, although the half-life for ^{187}Re β-decay is well known at lab temperatures, $\tau_{1/2} = 43 \times 10^9$ years, the value inside stars may be different. In the lab, β-decay goes to the continuum over 99% of the time.[37] In stars, however, the outer orbits are vacant so that β-decay can easily go to bound states.[38] This means a greater rate of decay, or, equivalently, a shorter half-life. A shorter half-life would lead to a shorter Δ^{max}. The question one must ask, however, is how much time does rhenium spend inside stars? From the existence of deuterium, which is easily destroyed in stars, we conclude that much matter never goes into stars. Also from the abundance by mass of the metals, $Z \approx 0.02$, not a lot of stellar processing has occurred. Thus, even though the ^{187}Re half-life inside stars may be much less than the half-life outside stars, the time ^{187}Re spends inside stars may be relatively small so that our basic conclusions may not require significant modification.

A second problem with the ^{187}Re/^{187}Os pair is that the local approximation may not be valid. Arnould, Takahashi, and Yokoi[38] have raised such doubts. They say that since some of the s-process ^{186}Re may actually neutron capture before β-decaying, we must question the use of the local approximation in solving for $(^{187}\text{Os})_c$. Clearly, more work needs to be done to resolve this question.

Our current best estimates for Δ^{max} from the Th/U and Re/Os pairs are

$$\Delta^{max}{}_{Th/U} = 5.1^{+1.0}_{-2.5} \times 10^9 \text{ years}$$

and

$$\Delta^{max}{}_{Re/Os} = 5.0 \pm 2.0 \times 10^9 \text{ years}.$$

A consistent range from the limits, then, is

$$3.6 \times 10^9 \text{ years} \lesssim \Delta^{max} \lesssim 6.1 \times 10^9 \text{ years}.$$

The mean age of the elements is $\langle \tau \rangle = \Delta^{max} + t_{ss}$, or for the above range

$$8.2 \times 10^9 \text{ years} \lesssim \langle \tau \rangle \lesssim 11 \times 10^9 \text{ years}.$$

Thus 8.2×10^9 years is an extreme lower limit to the age of the universe and rules out closed cosmologies ($\Omega > 1$) for $H_o > 80$ km/sec/Mpc.

The actual age of the universe we obtain from chronologies depends on the galactic evolution model assumed. In Figure 3, two simple, extreme models of the total duration of nucleosynthesis are shown. For the model in (a), $T = 2 \Delta^{max}$. For the model in (b), the value of T is very nearly Δ^{max}. We expect in general the true time variation to lie somewhere between these two models so that $T_{gal} \equiv$ age of galaxy $= T + t_{ss}$ corresponds to the range

$$\Delta^{max} + t_{ss} \lesssim T_{gal} \lesssim 2 \Delta^{max} + t_{ss}$$

The best detailed Galaxy evolution models[39] all tend to agree with the model in Figure 3(a) so that $T \approx 2 \Delta^{max}$. The best guess, then, for the age of the galaxy is $2 \Delta^{max} + 4.6 \times 10^9$ years or $T_{gal} \approx 15 \times 10^9$ years. This age value agrees with the 13.5×10^9 years $\lesssim T_u \lesssim 16.5 \times 10^9$ years range determined from cosmological ^{4}He and globular cluster ages.[40]

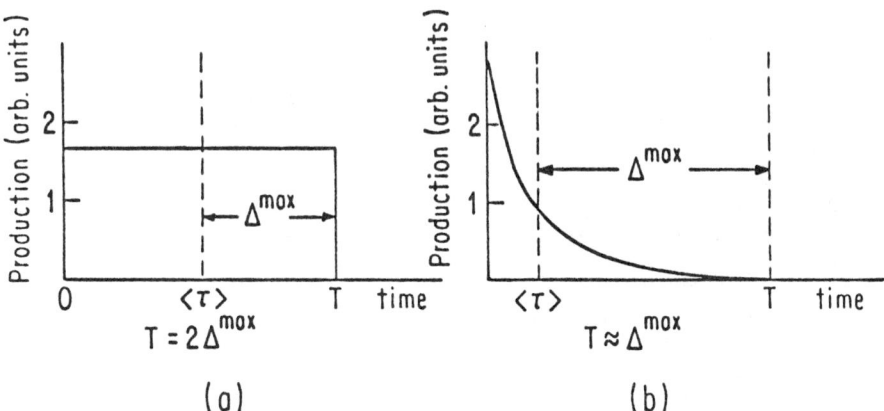

Figure 3. Two simple, extreme models of the total duration of nucleosynthsis.

VII. CONCLUSION

Neutron capture is extraordinarily important to astrophysics. It is relevant to nucleosynthesis and to the study of the age of the universe. Futhermore, although we have not discussed it in this review, neutron capture is relevant to the study of the origin of the solar system through exploration of the nature of meteoritic isotopic anomalies.[41] It may thus seem that nuclear astrophysicists already have more than enough work on their hands. We are greedy, however! We would like the detailed systematics of over 5000 nuclei so that we can better understand neutron capture processes and their applications.

ACKNOWLEDGEMENTS

We wish to acknowledge useful conversions one of us (D.N.S.) had with Ron Winters, Grant Matthews, and Jim Truran. This work was supported in part by NSF grant AST 8313128. One of us (B.S.M.) would also like to acknowledge the support of a National Science Foundation Graduate Fellowship.

REFERENCES

1. See articles by J. W. Truran and G. Matthews in this volume.
2. F. Hoyle and R. J. Taylor, Nature 203, 1108 (1964); P. J. E. Peebles, Astrophys. J 146, 542 (1966); R. V. Wagoner, W. A. Fowler, and F. Hoyle, Astrophys. J. 148, 3 (1967).
3. H. Reeves, J. Andouze, W. A. Fowler, and D. N. Schramm, Astrophys. J. 179, 909 (1973).
4. J. Yang, M. S. Turner, G. Steigman, D. N. Schramm, and K. A. Olive, Astrophys. J. 281, 493 (1984).
5. G. Steigman, D. N. Schramm, and J. E. Gunn, Phys. Letters B66, 202 (1977).
6. D. N. Schramm and G. Steigman, Phys. Letters 144B, 337 (1984).
7. J. W. Truran, J. J. Cowan, and A. G. W. Cameron, Astrophys. J. Lett. 222, L63 (1978); W. Hillebrandt and F.K. Thielemann, Astron. Astrophys. 58, 357 (1977).
8. R. G. Couch and W. D. Arnett, Astrophys. J. 196, 791 (1975); I. Iben, Jr., Astrophys. J. 196, 549 (1975).
9. J. P. Wefel, D. N. Schramm, J. P. Blake, and D. Pridmore-Brown, Astrophys. J. Suppl. Ser. 45, 565 (1981).
10. E. M. Burbidge, G. R. Burbidge, W. A. Fowler, and F. Hoyle, Rev. Mod. Phys. 29, 547 (1957).
11. See article by J. W. Truran in this volume.
12. D. N. Schramm, Astrophys. J. 185, 293 (1973).
13. D. N. Schramm and E. B. Norman, Proceedings of the Third International Conference on Nuclei Far from Stability, CERN Report 76-13, 570.
14. P. A. Seeger, W. A. Fowler, and D. D. Clayton, Astrophys. J. Suppl. Ser. 11, 121 (1965).

15. J. B. Blake and D. N. Schramm, Astrophys. J. 209 , 846 (1976).
16. M. D. Delano and A. G. W. Cameron, Astrophys. Space Sci. 10 , 203 (1971).
17. T. Lee, D. N. Schramm, J. P. Wefel, and J. B. Blake, Astrophys. J. 232 , 854 (1979).
18. J. Lattimer and D. N. Schramm, Astrophys. J. Lett. 192 , L145 (1974).
19. E. Symbalisty and D. N. Schramm, Astrophys. Letters 22 , 143 (1982).
20. J. M. Lattimer, F. Mackie, D. G. Ravenhall, and D. N. Schramm, Astrophys. J. 213 , 225 (1977).
21. F. Hoyle and D. D. Clayton, Astrophys. J. 191 , 705 (1974).
22. J. B. Blake, S. E. Woosley, T. A. Weaver, and D. N. Schramm, Astrophys. J. 248 , 315 (1981).
23. J. M. Leblanc and J. R. Wilson, Astrophys. J. 161 , 541 (1970).
24. E. M. D. Symbalisty, Astrophys. J., in press (1984).
25. E. M. D. Symbalisty, D. N. Schramm, and J. R. Wilson, Astrophys. J. Lett. (submitted) (1984).
26. E. Rutherford, Nature 123 , 313 (1929).
27. W. A. Fowler and F. Hoyle, Ann. Phys. 10 , 280 (1960).
28. D. N. Schramm and G. J. Wasserburg, Astrophys. J. 163 , 57 (1970).
29. E. M. D. Symbalisty and D. N. Schramm, Rep. Prog. Phys. 44 , 293 (1981).
30. T. Lee, Rev. Geophys. Space Sci. 17 , 1591 (1979).
31. P. A. Seeger and D. N. Schramm, Astrophys. J. Lett. 160 , L157 (1970).
32. D. D. Clayton, Principles of Stellar Evolution and Nucleosynthesis (New York, McGraw-Hill, 1968), p. 595.
33. C. O. Wene, Astron. Astrophys. 44 , 233 (1975).
34. F. K. Thielemann, J. Metzinger, and H. V. Klapdor, Astron. Astrophys. 123 , 162 (1983).
35. D. D. Clayton, Astrophys. J. 139 , 637 (1964).
36. See article by R. Winters in this volume.
37. R. D. Williams, W. A. Fowler, and S. E. Koonin, Astrophys. J. 281 , 363 (1984).
38. M. Arnould, K. Takahashi, and K. Yokoi, Astron. Astrophys. 137 , 51 (1984).
39. B. M. Tinsley, Astrophys. J. 198 , 145 (1975); R.J. Talbot and W.D. Arnett, Astrophys. J. 186 , 69 (1973); J.P. Ostriker and P.X. Thuan, Astrophys. J. 202 , 353 (1975).
40. E. M. D. Symbalisty, J. Yang, and D. N. Schramm, Nature 288 , 143 (1980); D. Kazanas, D. N. Schramm, and K. L. Hainebach, Nature 274 , 672 (1978).
41. G. J. Wasserburg, in Essays in Nuclear Astrophysics , edited by C.A. Barnes, D. D. Clayton, and D.N. Schramm (Cambridge University Press, New York, 1982), and references therein.

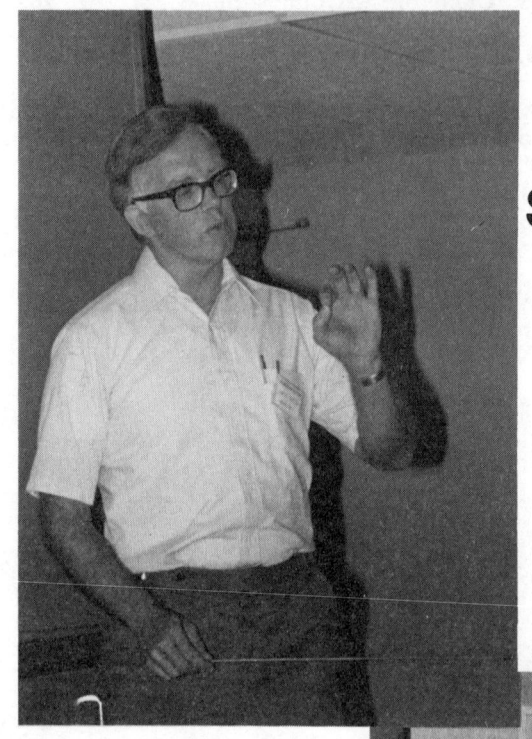

SESSION FB

SUMMARY

SUMMARY AND OUTLOOK--EXPERIMENTAL

Sam M. Austin[*]
National Superconducting Cyclotron Laboratory
and
Department of Physics and Astronomy
Michigan State University
East Lansing, Michigan 48824

INTRODUCTION

This Conference was devoted to recent progress and future directions of research in the study with neutrons of nuclear properties and structure; it also included a Workshop in Experimental Techniques. My talk will reflect this emphasis, summarizing both past work and opportunities for the future. I will also comment briefly on the technical capabilities and facilities needed to make this research possible. These comments are my personal views, but I hope that they might serve to focus discussion of future facility development.

An unusual feature of the Conference is that there are two summarizers: Claude Mahaux and me. We have not attempted to coordinate our comments, except that he will emphasize theoretical topics and I experimental topics; this reflects a point of view that real coordination is impossible anyway and that something that both of us find important probably warrants double emphasis.

ELASTIC SCATTERING

Much progress in the study of elastic neutron scattering has been made in the past few years, stimulated by the availability of cross section data at higher energies from Ohio University and MSU (temporarily) and of asymmetry data from TUNL. This work was described in detail at the Conference.

Howell presented the results of measurements of cross sections and analyzing powers on ^{32}S and ^{28}Si and their phenomenological analysis in terms of spherical and deformed optical models. The results shown in Fig. 1 give an indication of the high quality of the cross section data that can be obtained with modern facilities. Asymmetry data on many other nuclei were discussed by Walter. Cross sections above 26 MeV are available only for the few nuclei, mostly with N=Z, studied with the MSU facility and this facility no longer exists. Asymmetry data are not often available above 17 MeV.

These are significant limitations which should eventually be removed. Nevertheless, sufficient information is now available to extract the global features of the phenomenological optical model

---

[*] Research supported by the U.S. National Science Foundation

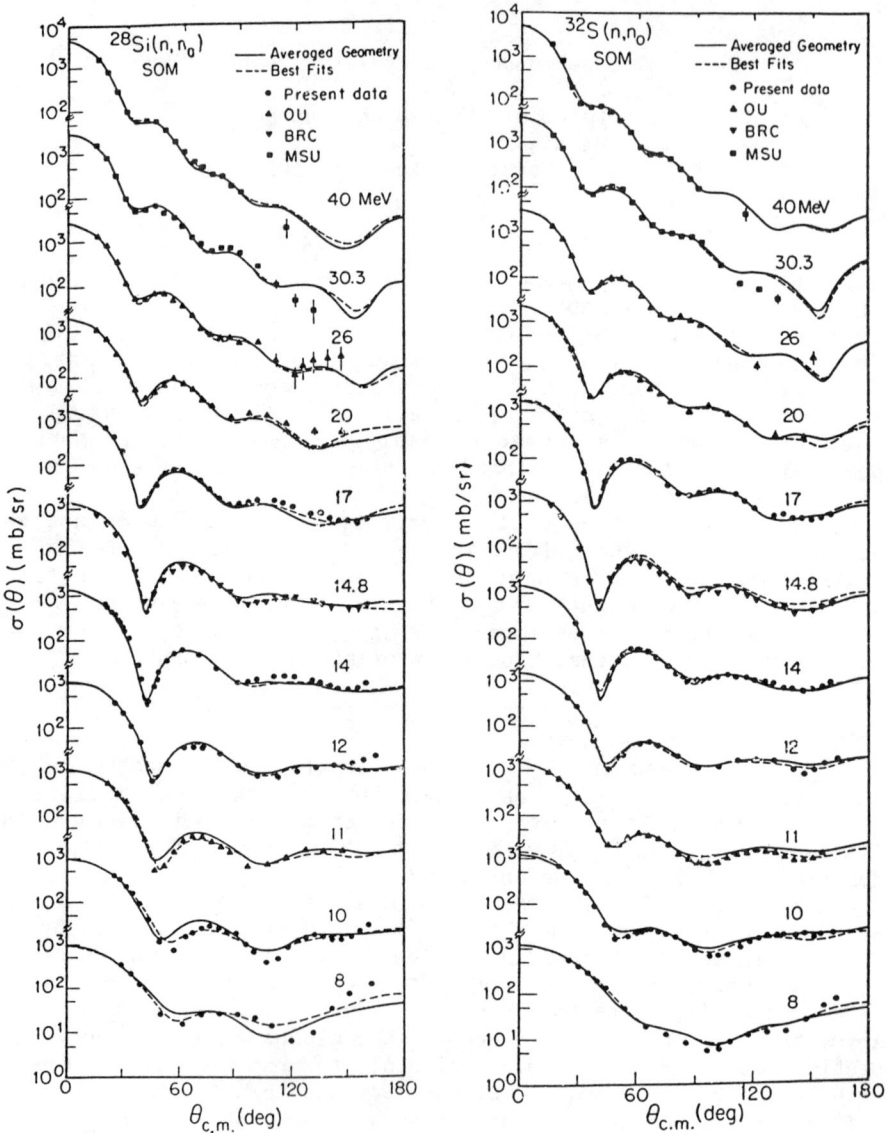

Fig. 1. Cross sections for the elastic scattering of neutrons from ^{28}Si and ^{32}S. Data are from TUNL, Ohio University, MSU and Bruyères-le-Châtel. (From the paper by C. Howell).

potential(OMP) for neutrons: the transition from surface to volume absorption is clear and there is evidence that the spin-orbit potential must contain an imaginary part of about 0.7 MeV(see paper DB6 by Walter, et al.). Global potentials have been obtained by Rapaport; more recent work was reported by Howell, Walter and Delaroche, et al. at this Conference. Dietrich, Von Geramb, Haouat, et al. and Lagrange, et al. showed that the data test microscopic or semi-microscopic descriptions of scattering and of the effective interaction.

A long term interest in nucleon scattering has been to determine the isovector or symmetry potential which describes the difference between nuclear(non-Coulomb) scattering of neutrons and protons from the same nucleus. The strength of this potential affects our understanding of all nuclear excitations in which neutrons and protons move differently(the best known of such excitations are the giant-dipole and Gamow-Teller resonances). In the Lane Model, the isovector potential U_1 is given by

$$U_1 = (U_p - U_n)/2\varepsilon \qquad (1)$$

where the U's are the optical potentials for protons and neutrons and $\varepsilon = (N-Z)/A$. The isovector potential $U_1 = V_1 + iW_1$, and the OMP $U = V + iW$, both have real and imaginary parts as noted. Thus it would appear to be straightforward to obtain U_1 given such neutron and proton potentials. However, an important problem remains: how to correct for the effect of the Coulomb interaction felt by the protons. While much progress has been made in understanding the Coulomb Correction(CC), much remains to be done. For this reason, our knowledge of the isovector potential is far from complete; since much future work will address this problem I will discuss it in some detail.

THE ISOVECTOR POTENTIAL AND THE COULOMB CORRECTION

If one assumes that nuclear forces are charge symmetric, then the Coulomb Correction for an N=Z nucleus is given by

$$\Delta U_{cc} = U_p(E) - U_n(E) \qquad (2)$$

The real part of the CC arises from the fact that the incoming proton is slowed by the Coulomb field and feels the (larger) attraction characteristic of lower energy particles in the local OMP. Reasonable estimates of dV/dE and of the average slowing down of the proton yield the usual real CC of $\Delta V_{cc} \approx 0.4Z/A^{1/3}$. Comparisons of real potentials for neutron and proton scattering verify this prediction with the experimental values of the coefficient generally lying between 0.4 and 0.5.

The origin of the imaginary part of the CC is more complex. Since the imaginary potential arises from second order processes, different excitation of levels in the intermediate system has important effects. These excitations are affected, for example, by different Q values for the (n,p) and (p,n) reactions, the inaccessibility of high-lying collective states because of the Coulomb repulsion, and the longer time the proton spends in the neighborhood of the nucleus. Osterfeld presented estimates of these effects for the case of ^{40}Ca and showed that the individual effects are large, but are energy dependent and almost cancel near 30 MeV. A comparison of available imaginary potentials for neutron and proton scattering by Rapaport verified this prediction for ^{40}Ca. However, it does not appear to apply generally. More recent work by the MSU group and work reported by Howell at this Conference indicates that the imaginary CC remains large at 30 MeV for ^{28}Si and ^{32}S. There is thus no general phenomenological picture that allows us to determine the imaginary CC from the available measurements on the N=Z nuclei. Theoretical calculations are seldom available and may not yet be reliable for such detailed predictions; Osterfeld's results underestimate the imaginary potential by roughly a factor of two. This situation greatly complicates the extraction of the imaginary isovector potential.

We have noted that U_1 can be obtained by comparing the OMP's describing neutron and proton scattering from the same nucleus. However, until recently neutron scattering data have been insufficient to follow this procedure. Instead values of V for proton scattering on a series of nuclei are plotted as a function of ε and, following Eq. 1, the slope of the curve yields V_1. Hodgson pointed out at this Conference that in fact the potentials are often not a smooth function of ε; they may break into families characterized by the isospin T or the nuclear charge Z. The reason for this behavior is not understood but might be related to the usual assumption that the geometrical parameters have a simple dependence on A. It may be possible to remove this fine structure by introducing shell effects and experimental information on nuclear radii and it will be interesting to find out if such procedures are possible. But it appears to me that comparison of neutron and proton potentials is a more reliable procedure and should be used whenever data are available.

The situation is still less clear for W_1 because the imaginary CC is not known independently. It may be possible to fix V_1 using Eq. 1 and then to determine W_1 and the imaginary CC simultaneously by requiring a fit to both scattering and the Lane-Potential-related (p,n) cross section. Sufficient data to carry out these procedures over a substantial energy range are available in only a few cases. It seems that direct inclusion of coupled channel effects may reduce the importance of the imaginary CC. (Hansen, et al.)

I note here that the much used potential of Patterson, Doering and Galonsky was obtained using CC's which in light of our present knowledge are inaccurate.

MODEL INDEPENDENT ANALYSES

At this conference microscopic theories of scattering were often compared directly to experimental data, a procedure that seems to have substantial disadvantages. A theory might predict a cross section which is statistically very improbable, but one is often unable to determine what is wrong: whether the imaginary potential is too weak or the real potential is too strong or the geometry is incorrect or what. It would seem more informative to make the comparison in terms of potentials: theoretical vs. phenomenological.

Another difficulty then arises: when one fits the scattering data to obtain an OMP, one assumes a potential shape, which might not bear a detailed resemblance to nature, and which might bias the result of the analysis in an unphysical and uncertain fashion. This problem can be avoided by adopting a model independent procedure, similar to that used for electron scattering. In this approach, the potential shape is expanded in an appropriate basis and the coefficients of the expansion terms are fixed by fitting the data. This yields an estimate of the potential and its uncertainty at each radius without assuming a specific radial form. I.e., one has an unfiltered report of the implications of an experiment.

An immediate question is whether the procedure will work as well for nucleons as it does for electrons. After all, three separate potentials, real, imaginary and spin orbit, must be fixed by the data, rather than a single charge density. However, the measurements span a similar range of momentum transfer, are of comparable accuracy and more types of data are available: $\sigma(\theta)$, $A(\theta)$ and eventually the spin rotation function. All in all, in seems likely that this approach will work. It has already been used for alpha scattering and by Tornow, et al. for neutron scattering near 11 MeV on ^{40}Ca. The resulting real potentials for this case differ significantly from a Woods-Saxon shape.

INELASTIC SCATTERING

During the past few years it has become clear that comparisons of scattering with different probes can yield information about the isospin structure of nuclear states if these probes interact differently with the protons and neutrons of the target nucleus. While these differences are well known for pions, the nucleons can serve a similar role and with comparable sensitivity. This sensitivity occurs because the central isovector part(v_T) of the two nucleon interaction is large for energies in the 40 MeV range yielding an interaction between unlike nucleons 3-4 times bigger than that between like nucleons. Thus neutron(proton) projectiles are sensitive mainly to protons(neutrons) in the target.

Madsen surveyed the results from comparisons of neutron and proton scattering carried out at bombarding energies below 26 MeV, mainly by the Ohio University group. In the case of single closed shell nuclei the ratio of neutron to proton matrix elements for low lying 2^+ states is found to be greater than N/Z(less than N/Z) for valence neutrons(valence protons). For open shell nuclei no definitive difference in the matrix elements is observed. These results are qualitatively understood in terms of schematic models. Alarcon, et al. presented the results of a comparison of neutron and proton scattering near 25 MeV which yields the relative sign of proton and neutron matrix elements for low lying 2^+ states in ^{34}S; the results disagree with the shell model for the second 2^+ state.

For the future, it appears likely that this technique will be extended to weaker, more complex states and to more highly excited states. A microscopic analysis based on effective charges in the shell model might yield information on the orbit dependence of effective proton and neutron polarization charges. Since reliable shell model wavefunctions are necessary, these studies might begin in the 1s0d shell as a test case and then move to heavier nuclei where they will provide unique information. When higher energy neutron sources are available, comparisons of neutron and proton scattering might be able to probe the isospin structure of the giant multipole excitations; preliminary work has been reported using pionic probes. Still more speculative would be the study of spin excitations, for example those to 1^+ states. Sensitivity to their isospin structure would depend on the relative size of the like-nucleon and unlike-nucleon interactions in the S=1 channel. While one expects the unlike nucleon interaction to be the weaker this has not yet been established experimentally.

OTHER STUDIES

Several studies somewhat outside the main line of the conference were very interesting. McEllistrem showed that scattering at low energies, a few MeV, is sensitive to the nature of the assumed coupling in deformed nuclei and may serve to determine this coupling.

Johnson described work of the ORNL group in which resonances observed in the total cross section for low energy(<1 MeV) neutrons from ORELA were analyzed to determine their spin and parity. Averages over these resonances then fix the optical model parameters in a given ℓ channel. One might not obtain a meaningful potential from a single nucleus, but systematic studies can yield interesting information. Results to date involving about eight nuclei with A = 30 to 208 yield real potentials in general agreement with those obtained from scattering data. The imaginary potential differs substantially from nucleus to nucleus; this is not understood in detail but is generally ascribed to nuclear structure effects. There is also some indication of ℓ-dependent effects; such effects have been invoked previously by Mackintosh and his collaborators to understand proton scattering data from ^{40}Ca and ^{16}O in the 25 MeV region.

Studies of elastic neutron scattering from ^{208}Pb and Bi^{209} in the few MeV region were described by Finlay. Data from Ohio University and Studsvik seem to show a significant anomaly in the real part of the OMP near the Fermi surface; the results are in good agreement with the proton data(after adjustment for the CC and the symmetry energy) where they overlap, and extend the systematics to very low energy where the proton data are dominated by Rutherford scattering.

Studies of the (n,p) reaction leading to giant resonances have been carried out over the past few years at Davis. While the available energy is low(65 MeV) and the signal to noise ratio is hence marginal, these data are our only source of information on these interesting reactions. Castaneda reported on the observation of the analog of the ^{40}Ca giant dipole resonance as observed in (n,p). Ford presented 0° data on several nuclei and apparently has observed the isovector monopole giant resonance for ^{90}Zr.

Finally, several interesting talks discussed the ways in which neutron reactions enter into astrophysical processes, particularly in the creation of the heavy elements and their use for astrophysical diagnostics. Truran, Mathews and Schramm discussed the nature of the slow and rapid neutron capture processes and their dependence on nuclear properties. Winters discussed measurements of radiative neutron capture and inelastic scattering which allow one to extract the age of the elements from the Re/Os chronometer pair. A value of 17×10^9 yrs for the age of the Universe results. This is compatible with values from the U/Th chronometer taking into account new information on the strength function for Gamow-Teller β decay obtained from (p,n) and other studies.

FUTURE DIRECTIONS

The rapid progress of neutron physics during the past few years has been based on the development of advanced detectors and ion sources, combined with pushing the present generation of accelerators close to their ultimate capabilities. It seems clear that new experimental facilities are necessary if the field is to advance. Much of the Conference was devoted to discussions of the future directions of neutron physics, so one could reasonably judge what facilities might be most useful, and to an evaluation of anticipated technical capabilities. A summary of these discussions follows.

SOME INTERESTING PHYSICS FOR THE FUTURE

This is meant to be a sampling of research opportunities I find interesting, to give a flavor of the field. When a topic has been discussed earlier, I simply list it here. Other topics are accompanied by a brief comment.

<u>The Nuclear Mean Field:</u> The nuclear mean field plays a fundamental role in our understanding of nuclear structure and reactions, yet our knowledge of the field and particularly its isospin nature is at best partial. Some research opportunities:
    1. Determine the global features of the OMP over a wide energy range by studies of scattering and model independent analyses.
    2. Determine the values and energy dependence of real and imaginary parts of the isovector potential from neutron-proton comparisons. An idea of what should be possible can be obtained from Fig. 2, where proton and neutron OMP's for ^{208}Pb are compiled. Subtraction of these potentials, using Eq. 1, should yield the magnitude, shape and energy dependence of U_1 (assuming the CC can be made accurately).

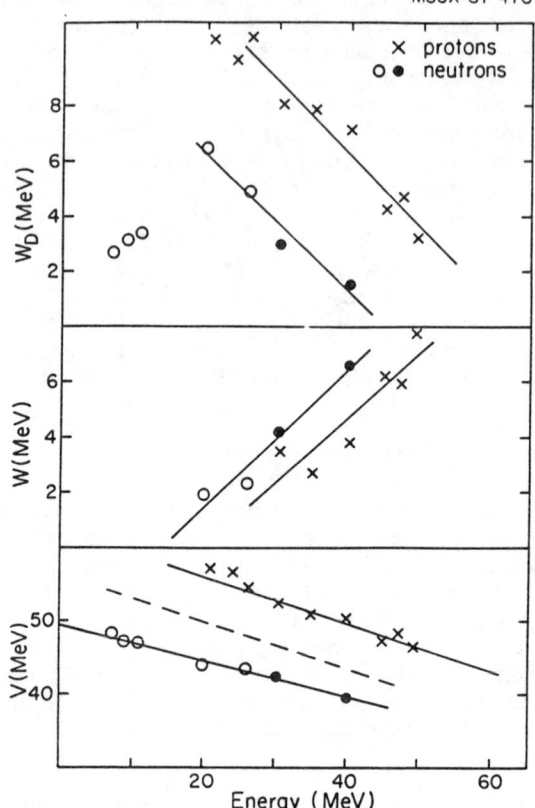

Fig. 2. Values of the OMP for proton and neutron scattering from ^{208}Pb. (R. Devito, et al.).

3. Develop a theoretical and experimental understanding of Coulomb effects and the Coulomb correction.

4. Determine or limit the amount of charge symmetry breaking(CSB) in the mean field. CSB is not ruled out by the two nucleon data at the percent level, and might be(at least partially) responsible for the Nolen-Schiffer(or Coulomb energy) anomaly. Present neutron and proton data on N=Z nuclei are marginally adequate to put limits on CSB at the requisite level; improved facilities covering a wider energy range could almost surely do so.

5. Elucidate the nature of relativistic effects in nucleon scattering, especially on the isovector and spin orbit parts of the field.

A general study of these phenomena will require a comparison of proton and neutron elastic scattering: cross sections, asymmetries and perhaps spin rotation functions in the 25 to 100 MeV range.

The Effective Interaction: Scattering provides the flexibility in choice of observables to study the effect of the nuclear medium on the interaction between two nucleons at different densities, momentum transfers and energies. Von Geramb presented evidence that the connection with theory will be particularly clean at energies near 60 MeV, since the difficult-to-treat exchange contribution is expected to be small there. A comparison of proton and neutron scattering will provide information on the isovector part of the interaction. Most of these studies will probably be done best in the 25 to 100 MeV range where the reaction mechanism is reasonably well understood and resolution will be adequate to resolve a variety of nuclear states. Studies of the spin dependent part of the effective interaction may be particularly interesting above 100 MeV where spin effects become dominant in certain reactions. Spin transfer (p,n) measurements such as those described by Taddeucci may be particularly important for this purpose.

Isospin Structure of Nuclear States: This section concerns spin-independent excitations.
1. Study the isospin structure of the giant isoscalar resonances.
2. Determine the isospin structure of non-collective states.
3. Measure transition densities of excited states, and determine their isospin structure. Carr showed that neutrons at energies near 60 MeV are particularly sensitive to the nuclear interior and are generally superior to pions for probing the interior(See Fig. 3).

For these experiments resolution will often be important. In addition one must limit the energy so that differential sensitivity to neutrons and protons remains large. An energy range up to 75 MeV, reduced when resolution is important seems desirable. Energies near 60 MeV may be especially useful.

Spin Excitations: These will mostly involve isovector excitations. Then, since $V_{\sigma\tau}/V_{\tau}$ increases with energy, higher energy, at least up to 200 MeV, generally yields greater sensitivity. On the other hand some of the studies listed require good resolution, which is hard to obtain at the high end of the range.

1. Study giant Gamow-Teller(β^+) strength using (n,p) reactions. If 2p-2h excitations are responsible for a significant part of the quenching observed in (p,n) reactions, then (n,p) transitions will not be entirely Pauli blocked and (n,p) cross sections will be able to provide information on the 2p-2h excitations. For N≈Z nuclei (n,p) strength will be concentrated in low lying states; the magnitude of this strength in nuclei near Fe has important implications for supernova evolution and explosive nucleosynthesis.

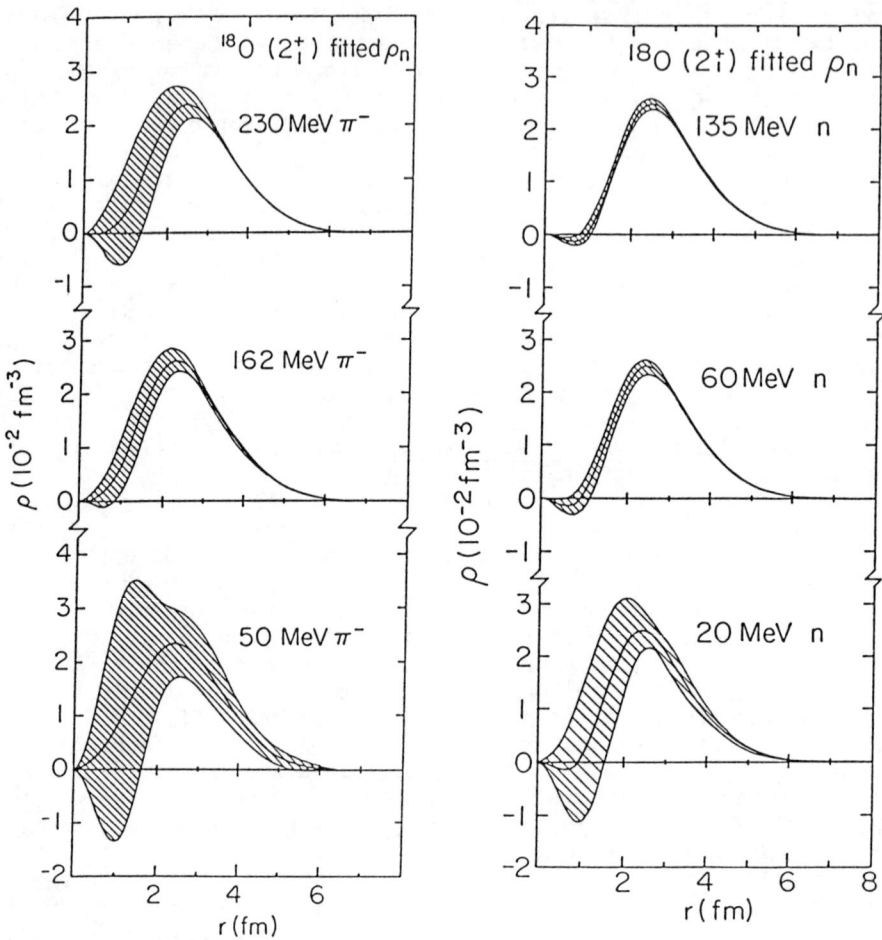

Fig. 3. Neutron densities and error envelopes for $^{18}O(2_1^+)$ determined from pion and neutron scattering pseudodata. (From the paper by J. Carr).

2. Study Gamow-Teller transitions to isolated states by (n,p) reactions--Strength to the $T_>$ states is especially sensitive to configuration mixing (see the paper by Brown) and should serve as a stringent test of shell model calculations.

3. Double beta decay--It has been pointed out that a combination of (p,n) measurements on a double beta decaying nucleus and (n,p) measurements on its daughter can yield upper limits on the rates for double beta decay. These are valuable since shell model calculations appear

to overestimate decay rates greatly. Brown applied shell model calculations to ^{48}Ca in a similar spirit.

4. $(\vec{p},\vec{n})$ measurements--Spin transfer is a specific probe of spin flip strength and Taddeucci showed it can yield otherwise unobtainable information about the reaction mechanism for charge exchange reactions and the location of spin-flip strength. Such measurements should help unravel the nature of spin transfer strength in the continuum above the giant Gamow Teller resonance.

5. High resolution (p,n) measurements--A dedicated facility would permit measurements with resolutions under 100 keV at energies up to 100 MeV. Such measurements would be useful for a variety of nuclear structure studies. For example, for the double beta decay studies mentioned earlier and for measurements of the longitudinal response function in $0^+ \to 0^-$ transitions as described by Orihara. These transitions should yield information about the importance of pion-condensate-precursor effects in nuclei.

It appears that an important program of spin-excitation studies leading to isolated states would be accessible with neutron energies in the 50-100 MeV range. Some important studies involving giant resonances require 100-200 MeV so as to increase the signal to noise ratio for these broad excitations.

EXPERIMENTAL POSSIBILITIES

During the Conference much discussion was devoted to new techniques and facilities, both at a Workshop on Experimental Techniques and in invited and contributed papers. The availability of facilities for carrying out various types of experiments was summarized by Zafiratos using the graph shown in Fig. 4; on this graph the circles are contours of energy and the solid lines indicate present capabilities. Compared to what is needed to carry out the program outlined above, present limitations are serious.

New facilities are coming on line at Studsvik and Uppsala. Olsson described improvements in the Studsvik facility which permit high resolution($\Delta E \approx 0.5$ MeV) scattering measurements up to 22 MeV. Conde discussed the capabilities of the rebuilt cyclotron at the Gustaf Werner Institute. This facility will have intensities in the 10 μamp range up to 200 MeV and will carry out a program of (n,p) reactions with a resolution of 1 MeV at 200 MeV energies. Further details are given in the Facility Summary attached to these proceedings.

Friesel discussed accelerators for the production of intense neutron beams with energies up to 200 MeV. He concludes that the IUCF facility could be upgraded to an

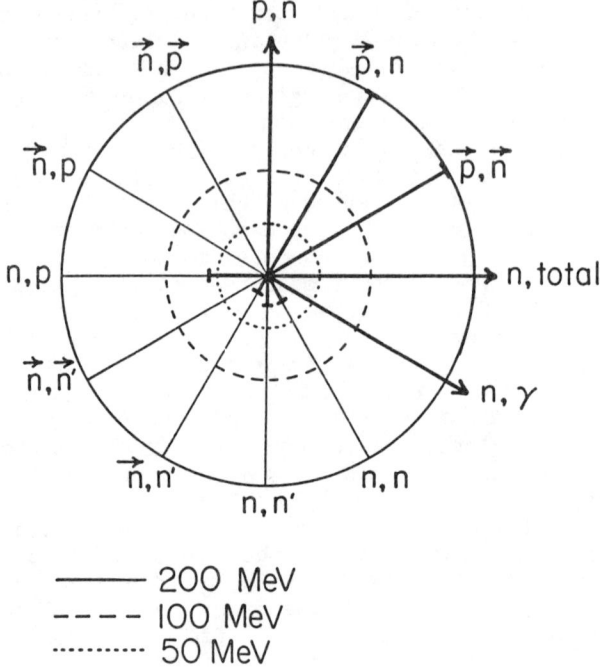

Fig. 4. Energies available for studying various reactions with present facilities. (From the paper by C. Zafiratos.)

intensity near 10 μamperes without extensive changes, but that an increase to 100 μamperes would be expensive. In general, construction of variable energy cyclotrons with milliampere beams is difficult, because of space charge effects and the requirement of high extraction efficiency to avoid radiation damage. It may be difficult to build a variable energy cyclotron for 1 milliampere beams; if this capability is required, Friesel concludes that other approaches should be examined.

Ancillary equipment was also discussed. Finlay showed that facilities employing rotating beam swingers such as that at Ohio University were well adapted to measurements at energies up to at least 75 MeV. Dietrich and Clegg pointed out that such beam swingers were consistent with the use of polarized beams. Goodman discussed beam swinger designs that would be useful at higher energies. Brady described the facility for 60 MeV (n,p) studies at Davis and techniques for working at higher energies. He concludes that studies at 200 MeV are feasible with beams in the 10 μampere range, noting that technical advantages yield "nearly an order of magnitude in count rate (at a given energy resolution) in going from 60 to 200 MeV." Taddeucci discussed the feasibility of spin transfer measurements and concluded that ($\vec{n},\vec{p}$) measurements are 10^4 to 10^6 times harder than ($\vec{p},\vec{n}$) measurements which with present techniques take typically a few to 10 hours per point.

SUMMARY

In listening to talks and participating in discussions at the Conference, I came away with a strong feeling that studies of neutron interactions with nuclei have progressed greatly during the past few years and have an exciting future in store. An important question for the field is the direction of facility development which will allow one to realize this future. It seems clear that an accelerator with energy in the range of 50 to 100 MeV and beam currents of ten or tens of μamperes would support a strong physics program. A more precise specification of energy and intensity is yet to be made and could be the subject of a technical workshop. However, such a facility will not be very useful for study of giant resonance phenomena—energies of well above 100 MeV are required. It seems that an upgrade of the IUCF to yield beam currents at the 10 μampere level might be cost effective. An investment in spectrometers for (n,p) and related studies is probably desirable and useful even at present IUCF intensities.

THEORETICAL CONTRIBUTIONS. HINDSIGHT AND OUTLOOK

C. Mahaux
Institut de Physique B5, Université de Liège,
Sart Tilman, B-4000 Liège 1 (Belgium)

One of the main aims of physics consists in the investigation of symmetry properties and of the accuracy of the corresponding conservation laws. In the case of nuclei, one basic symmetry derives from the charge independence of the nucleon-nucleon interaction which implies that isospin is a conserved quantum number. This symmetry is broken because of the existence of Coulomb forces. The amount of symmetry breaking can best be studied by comparing how nuclei react to a proton probe on the one hand and to a neutron probe on the other hand. For obvious technical reasons much more information is presently available for proton than for neutron probes. It is therefore essential that continuous and enhanced effort be devoted to the experimental and theoretical investigation of neutron-nucleus collisions.

The survey by C.R. Howell as well as several contributions illustrate the remarkable accuracy achieved in recent measurements of the differential neutron-nucleus cross section for bombarding energies between 8 and 40 MeV and of the analyzing power for energies up to about 20 MeV. These data should be extended to cover more targets and a wider energy range. Many speakers have rightfully insisted on the interest of the higher energy domain 50 MeV $<$ E $<$ 100 MeV. The results presented by Finlay et al. show, however, that the low energy domain (2 $<$ E $<$ 10 MeV) is also worth being studied in detail since it is not accessible to protons; neutron scattering therefore provides a privileged way of studying the Fermi-surface anomaly. This anomaly consists in the property that the volume integral per nucleon of the real part of the optical-model potential at low energies is not given by the linear extrapolation from the higher energy domain.

The optical model provides amazingly accurate fits to the experimental values of the elastic scattering and of the total neutron-nucleus cross sections. In the optical model, it is assumed that the incoming particle only feels a complex average potential which has a smooth dependence upon energy, mass number, neutron excess and spatial coordinates. This remarkable success of the optical model calls for a theoretical understanding of the properties of the empirical optical-model potentials. The theory should also provide a guidance for the relevant phenomenological parametrization of the optical-model potential. One of the striking features of the phenomenological analyses is indeed that equally good fits to the data can be obtained with the help of quite different optical-model potentials and even different wave equations.

Let us for simplicity restrict the discussion to the spherical case. Three main approaches have been used in the phenomenological analyses.

(a) <u>Local Schrödinger analyses</u>. Here, the neutron wave function $\psi(\vec{r};E)$ at energy E is computed from the equation

$$-\frac{\hbar^2}{2m} \nabla^2 \psi(\vec{r};E) + M(r;E) \psi(\vec{r};E) = E \psi(\vec{r};E) \quad , \quad (1)$$

where

$$M(r;E) = V(r;E) + i W(r;E) + [V_{SO}(r;E) + i W_{SO}(r;E)]\vec{\ell}\cdot\vec{s} \quad (2)$$

is a local-energy dependent optical-model potential. Some empirical properties of $V(r;E)$ have been surveyed by P.E. Hodgson. He particularly insisted on the property that for not too large variations of N and Z the real part of the proton optical-model potential appears to be a smooth function of N and Z separately rather than of the difference N-Z as usually assumed. He suggested that this could be an indication of a smooth change with A of the geometrical properties of $V(r;E)$ and that this change might be understood in terms of an approach where the matter density is folded with a density-dependent effective interaction.

(b) <u>Nonlocal Schrödinger analyses</u>. Here, the wave equation is taken of the form ($\vec{s} = \vec{r}-\vec{r}'$, $R = \frac{1}{2}|\vec{r}+\vec{r}'|$)

$$-\frac{\hbar^2}{2m} \nabla^2 \psi(\vec{r};E) + \int M(R;\vec{s}) \psi(\vec{r}';E) ds = E \psi(\vec{r};E) \quad . \quad (3)$$

Nonlocal potentials appear naturally in any microscopic approximation for the nucleon-optical model potential, e.g. in the Hartree-Fock approximation. In their pioneering work Perey and Buck have shown that in a limited energy domain good fits to the elastic scattering data can be obtained from the nonlocal Schrödinger wave equation (3), in which M has a simple dependence upon R and $s = |\vec{s}|$. Even if a larger energy domain is considered, it follows from the recent works by Kuo and collaborators that arbitrarily good fits could still be obtained with eq. (3). Then, however, M might be a complicated function of $\vec{s}$; this remark also applies to the case when some inelastic channels are strongly coupled to the elastic one.

(c) <u>Local Dirac analyses</u>. Here, the single-particle wave equation is assumed to be a Dirac equation of the following form

$$\{\vec{\alpha}\cdot\vec{p} + \beta[m+U_s(r;E)] - [\epsilon-U_o(r;E)] - i\beta\vec{\alpha}\cdot\vec{r} U_t(r;E)\} \Phi(r;E) = 0 \quad (4)$$

where $\vec{\alpha}$ and β are the Dirac matrices while $\Phi(r;E)$ is a four-component spinor and $\epsilon = E+m$ is the total energy. In her talk, B.C. Clark demonstrated that eq. (4) leads to very good fits even when it is simplified by setting U_t equal to zero and by assuming that $U_s(r;E)$ and $U_o(r;E)$ have Fermi form factors. These fits involve fewer parameters than the Schrödinger approaches. The strengths of the potentials U_s and U_o turn out to have opposite signs and to be very large. This implies that in this approach strong relativistic effects exist even for low energy nucleons. For instance, the spin-orbit coupling is then a purely relativistic effect. The possibility remains, however, that if one would not set U_t equal to zero good fits could be obtained with small strengths for the three potentials U_s, U_o and U_t. B.C. Clark insisted on the interest of gathering more detailed and more accurate experimental data on the analyzing power and of measuring the spin rotation function.

Which one of the three approaches should be favoured ? Should one use combinations of these, e.g. a nonlocal and energy-dependent Schrödinger potential ? There is no unique answer to these questions. It should be kept in mind that to some extent these various methods of analysis are interchangeable. For instance, one can construct a local Schrödinger potential $M(r;E)$ which approximately yields the same cross sections (but not the same wavefunction at small distances) as the nonlocal $M(R;\vec{s})$ provided that the nonlocality of $M(R;\vec{s})$ is not too complicated. A complicated nonlocality is attached to the coupling of the elastic channel to an inelastic channel. The success of the local Schrödinger phenomenology makes one hope that the cumulated effect of the coupling to many inelastic channels yields a simple type of nonlocality. It this is so nuclear matter approaches of the type discussed by H.V. Von Geramb could be accurate. This hope appears to be supported by the various tests reported by F.S. Dietrich. On the other hand, the nuclear structure calculations described by F. Osterfeld yield complicated nonlocalities for the imaginary part of the optical-model potential even at energies as high as 30 MeV in ^{40}Ca. This would exclude the possibility of constructing a simple local potential which would be equivalent to the calculated one. This would also imply that a few specific channels give large contributions to the absorption. In particular, the collective excitations of the target should be\explicitly taken into account, e.g. by a coupled-channel analysis of the data. This, however, does not appear to be required by the phenomenological analyses based on local Schrödinger potentials. This seems to leave one with the following two possibilities.

(a) The excellent fits obtained with local potentials are meaningless.

(b) The calculations based on the nuclear structure approach should be extended to include effects (coupling between the inelastic channels, absorption due to the feeding of the compound nucleus, composite particle emission, density-dependent interactions, ...) not yet included with sufficient accuracy. In particular, the feeding of the compound nucleus,which is responsible for the existence of an absorptive part at very low energy, is not contained in the nuclear structure approach.

Whether a Dirac optical-model potential must be adopted is an important theoretical question. It cannot be answered by analyses of the experimental data. Indeed, it is possible to construct various Schrödinger potentials which are equivalent to a Dirac optical-model potential. These Schrödinger-equivalent potentials can for instance be chosen to be local and energy-dependent, or to be energy-independent with a nonlocality similar to that of Skyrme-Hartree-Fock potentials, see e.g. the contribution by G. Rawitscher. Although phenomenological analyses cannot indicate whether a Dirac equation should be used, they may exclude some parametrizations of the Dirac potential. For instance, the Dirac analyses in which U_t is set equal to zero do not lead to the change of the sign of the imaginary part $W_{SO}(r;E)$ of the local equivalent Schrödinger potential at $E \approx 50$ MeV which appears to be required by the recent neutron analyzing power data reported by R.L. Walter et al. At present, one can only say that the Dirac phenomenology yields fits of remarkable quality between several tens and several hundreds MeV with a number of adjustable parameters which is smaller than that involved in many Schrödinger analyses.

All in all it appears necessary to continue to perform phenomenological analyses with the various approaches. Particular effort should be devoted to the refinement of the empirical local Schrödinger potentials. Indeed, the potentials obtained from the other two approaches can usually be transformed into local Schrödinger-equivalent potentials. These refined analyses should be performed in such a way that the main geometrical properties of the local Schrödinger potential are not fixed a priori. The feasibility of analyses of this type has been demonstrated by Friedman, Batty, Gils and others in the case of alpha-nucleus scattering and by MacIntosh, Brissaud and others in the case of nucleon-nucleus scattering. This is similar in spirit to the Fourier-Bessel analysis of the charge density distribution. Progress recently achieved by Fiedeldey, Lipperheide et al. on the inverse scattering problem at fixed energy also appear promising, especially for energies larger than several tens MeV.

The optical model amounts to reducing the many-body problem to a single-particle wave equation. In the nonrelativistic approach, this is usually performed by eliminating those basis configurations which correspond to closed as well as open inelastic channels. Difficulties are introduced by the need for antisymmetrization on the one hand and by the strong nature of the nucleon-nucleon interaction on the other hand.

These two problems can be rather successfully handled in the case of nuclear matter. Most available calculations use the Brueckner-Hartree-Fock approximation. Even at that level numerical uncertainties are involved in the various existing codes. Their outcomes should be compared with the same input nucleon-nucleon interaction. The size of some corrections to the Brueckner-Hartree-Fock approximation can and should be evaluated; only crude estimates exist, but A. Lejeune has described a recent progress in that direction.

The construction of the nuclear optical-model potential from the nuclear matter results requires the use of a local density approximation, of which several variants exist. The accuracy of these local density approximations is quite difficult to assess, especially for the imaginary part of the optical-model potential. Finally, the calculation of the effect of the Coulomb field and of neutron excess should be improved. In view of all these problems, it appears doubtful that the nuclear matter approach will ever yield an accuracy better than ten per cent for the real part of the optical-model potential and than twenty per cent for the imaginary part. One must therefore expect that some adjustments will be necessary. If no adjustment is needed this should be the result of the cancellation of errors.

The nuclear structure approach has the merit of including dynamical effects which are typical of the finiteness of nuclei, and the disadvantage of involving an effective nucleon-nucleon interaction as input. F. Osterfeld has evoked the possibility of deriving this effective interaction from a nuclear matter calculation in which the excitation of low-lying intermediate states would be excluded. The contribution of specific channels should be examined. If a few channels dominate, this may be an indication that the data should be analyzed in the framework of the coupled channel formalism. The optical-model potential obtained in this way could probably be more readily compared with the results obtained from the nuclear matter approach, possibly somewhat modified.

The microscopic theory of the Schrödinger optical-model potential shows that this operator is both nonlocal and energy dependent. We recalled that if the nonlocality is sufficiently simple, one can construct a local equivalent Schrödinger potential which yields approximately the same elastic scattering cross section as the nonlocal potential. Hence nonlocality and energy dependence cannot be disentangled by the analysis of elastic scattering data. However, phase equivalent potentials do not yield the same wave function at finite distance. This "Perey effect" influences the calculated values of inelastic scattering cross sections or of knockout reactions. Hence, the theory must provide an estimate of the nonlocality of the nucleon-nucleus potential. Another instance where it is necessary to distinguish between nonlocality and energy dependence is the theoretical evaluation of the corrections due to the Coulomb field and to neutron excess. This has been alluded to by F.S. Dietrich.

Fits to the nucleon-nucleus scattering data with the Dirac equation approach have been first attempted independently of theoretical support. A number of approaches (Walecka, Shakin, Horowitz et al.) now exist which aim at developing a relativistic many-body theory and at deriving a Dirac average nucleon-nucleus potential by introducing some mean field approximation in this many-body theory. This faces severe difficulties, especially if one wants to start from the properties of the elementary nucleon-nucleon interaction. The success of the Dirac phenomenology in accounting for several observations (nucleon elastic scattering, ground state density distribution, inelastic electron scattering, antiproton scattering, hypernuclei, ...) strongly encourages the pursuit of these theoretical attempts. Particular attention should be devoted to the analysis of data which involve antiparticles or the small component of the Dirac spinors. In particular, the antiproton-nucleus scattering data are of fundamental interest.

We now turn to the effective nucleon-nucleon interaction. In the nuclear matter approach surveyed by H.V. Von Geramb, this effective interaction v is defined by requiring that the optical-model potential M be given by (schematically)

$$M \psi(r;E) = [\int \rho(\vec{r},\vec{r}) v_d d^3r'] \psi(\vec{r};E)$$
$$+ \int \rho(\vec{r},\vec{r}') v_e \psi(\vec{r}';E) d^3r' \quad , \qquad (5)$$

where $\rho(\vec{r},\vec{r}')$ is the density matrix and v_d and v_e are the direct and exchange parts of the interaction. Here, we omit spin and isospin degrees of freedom. The effective interaction depends upon many variables. In the work by Brieva, Rook and Von Geramb, it is assumed that one can replace v_d and v_e by local functions which only depend upon E, $|\vec{r}-\vec{r}'|$ and the density $\rho(R)$, with $\vec{R} = (\vec{r}+\vec{r}')/2$ as above. The accuracy of this assumption had been tested by Negele for bound states but has apparently not yet been checked in the scattering case. Several checks can and should be performed. The exchange part of the interaction is usually handled in the framework of the Slater approximation. It is convenient to work in the momentum representation. For instance, this leads to the observation (H.V. Von Geramb) that the exchange contribution to the optical-model potential becomes small for energies close to 60 MeV. This is a useful property since the exchange contribution is difficult to handle accurately.

It is assumed that the effective interaction which appears in eq. (5) can be used to describe inelastic scattering. This allows the evaluation of the optical-model potential and of the effective interaction in a unified framework using the same input. The success of this consistent approach cannot be judged on the basis of only a few analyses, since the calculated cross sections also involve nuclear structure information with inherent uncertainties. Hence, it is not yet clear to what extent the measured inelastic scattering cross sections require the use of density- or energy-dependent effective interactions and, a fortiori, the particular energy-and density-dependence obtained from the nuclear matter approach.

J.R. Shepard described that at low energy the effective M3Y interaction constructed by the Michigan group is quite successful. At high energy, the effective interaction can be identified with the nucleon-nucleon transition operator. It would be of interest to investigate quantitatively if the nuclear matter approach smoothly bridges the gap between these two energy domains and interactions. J.R. Shepard pointed out that in a relativistic impulse approximation specific nuclear current contributions appear. These seem to lead to better agreement with polarization data at intermediate energy than the nonrelativistic impulse approximation does. Here again the question arises whether nuclear physics requires a relativistic framework.

Whether the experimental data yield detailed information on the effective interaction is one of the questions considered by J.A. Carr. He discussed the possibility of performing model-independent analyses of the inelastic scattering data. He investigated the following question. From a given expression for the transition amplitude and for the effective interaction (or for the transition density), can one determine an error envelope for the transition density (or for the effective interaction). This is an interesting attempt to extend model-independent analyses to the case of inelastic scattering.

One of the purposes of inelastic scattering is to obtain detailed information on the structure of nuclear excited states. This topic was discussed by J. Wambach, V.A. Madsen and B.A. Brown, with special emphasis on the isospin degrees of freedom, as fit to the scope of the present conference. J. Wambach focused on the comparison between calculated and experimental properties of the isovector vibrational modes. This yields information on the equation of state of asymmetric nuclear matter and on the density-dependence of the isovector component of the effective particle-hole interaction. The data appear to require that this component be strongly density-dependent. Wambach also discussed how the response functions calculated from the random phase approximation are modified when one takes into account the effect of excluded configurations, especially two particle-two hole configurations, and in the case of the giant Gamow-Teller state Δ particle-nucleon hole configurations. It now appears that the quenching of the Gamow-Teller strength due to Δ-h excitations is smaller than was hitherto generally believed.

The isospin structure of nuclear excitations can be obtained by combining (n,n') and (p,p') or (p,n) and (n,p) reaction data. For instance, B.A. Brown compared the information provided by (p,n) and (n,p) reactions concerning the Gamow-Teller excitations, and the relevance of these results for the problem of double beta decay. The isospin structure of the low-lying 2^+ states as emerging from (p,p') and (n,n') data has been discussed by B.A. Brown and by V.A. Madsen. B.A. Brown paid particular attention to theoretical interpretation based on microscopic shell-model calculations in the s-d shell. V.A. Madsen mainly surveyed the effect of core polarization on neutron and proton multipole matrix elements, and exhibited the existence of striking differences between closed shell and open shell nuclei.

The optical model had been triggered by the observation (H.H. Barschall et al.) of regularities in the dependence of the low energy average total neutron-nucleus cross sections upon the energy and the target mass number. We should therefore keep in mind that neutron probes can also yield information on the optical-model potential at low energy, in contrast to proton scattering which is dominated by the Coulomb field. This is illustrated by the analysis presented here by Finlay et al. of the scattering of low energy neutrons by ^{208}Pb, which shows some evidence for a Fermi anomaly.

For bombarding energies smaller than about one MeV, the neutron scattering cross section displays isolated resonances, as illustrated by several contributions presented by the ORELA group. The corresponding compound nuclear states have well-defined angular momentum and parity. These quantum numbers can often be determined experimentally. The contribution of individual partial waves to the total cross section can thereby be obtained. This may yield information on the dependence of the optical-model potential upon angular momentum and parity.

In the domain of isolated resonances, the optical model is related to the energy average cross section. As described by C.H. Johnson, caution must be exercised when calculating this average from analytic prescriptions because the experimental data only covers a finite energy range. For each partial wave the analysis yields two quantities, which are the analogs of the s-wave strength function and of the potential scattering length familiar in the case of scattering at very low energy. From these two quantities one can hope to determine information on the global properties of the real part of the spherical optical-model potential on the one hand and of the imaginary part on the other hand. Nuclear structure effects (deformation, doorway states, ...) can influence specific cases. Hopefully, average trends will eventually emerge from systematic analyses. This type of analysis could probably be applied to the fine structure proton scattering data performed at Duke University.

Although isolated resonances no longer appear in the cross section at larger bombarding energies, compound nuclear states still exist. They lead to statistical fluctuations of the scattering matrix. These give rise to fluctuations of the cross sections. The analysis of these fluctuations yields information on the level density of compound nuclear states. The present status of the theoretical calculation of the level density has been reviewed by S.M. Grimes. One of the remaining problems concerns the temperature dependence of the level density, which is of importance in astrophysics. Finite temperature Hartree-Fock theory may provide a guidance. One should however keep in mind that the Hartree-Fock Fermi gas model does not yield good results at zero temperature. The use of a radial- and temperature-dependent effective mass might be useful.

The calculation of average cross sections requires the energy average of products of fluctuating scattering matrix elements. H. Feshbach sketched how the pole expansion of the scattering matrix yields to the Hauser-Feshbach formula if one can use randomness assumptions for the residues at the resonance poles. The late P.A. Moldauer emphasized that the validity of these randomness assumptions is questionable, in particular because of correlations introduced by the unitarity of the scattering matrix; this has been recalled by J.E. Lynn. M.R. Zirnbauer described how this long standing problem of the calculation of the compound nucleus cross sections has recently been solved in the framework of a random matrix model in which statistical properties of the Hamiltonian are assumed to be given by the Gaussian orthogonal ensemble. The latter assumption is supported by the work of the Rochester group on nuclear spectroscopy. The final expression of the compound nucleus cross section is a threefold integral of a quantity which only depends upon the transmission coefficients. The Hauser-Feshbach formula is recovered in the limit where the sum of all transmission coefficients is much larger than unity.

J.E. Lynn recalled that the existence of a double-humped fission barrier leads to the necessity of distinguishing between two classes of compound states, each of these classes having its own statistical properties. The Hauser-Feshbach formula has to be modified to take into account the existence of these two classes of states of of the corresponding intermediate structure in subthreshold neutron-induced fission. The energy dependence of the fission cross section is able to yield information on the structure of the double-humped fission barrier.

H. Feshbach described recent improvements of the theory of multistep direct and multistep compound processes. The theory involves the eigenstates of optical-model Hamiltonians whose imaginary part is positive and thus corresponds to regeneration rather than to absorption. The scattering function associated with these regenerating potentials has poles near the real axis. Energy averages must therefore be introduced. The final result is close to the expression which had been used with success by the Milano-Oxford group for the analysis of preequilibrium cross section. H. Feshbach also described an appealing theoretical way for describing resonances which are associated with quasibound states that two deformed nuclei may form for a specific relative orientation between their symmetry axes. This quasibound state is not an eigenstate of the total angular momentum operator. Resonances of this type might occur in collisions between two very heavy ions, e.g. U + U .

The many interesting talks surveyed above illustrate the fundamental interest of neutron-nucleus collisions for reaching a better understanding of nuclear structure and of nuclear reactions. One should, however, keep in mind that an accurate knowledge and a good theoretical understanding of neutron-nucleus collisions are also of primary importance for technological or medical applications (H.H. Barschall) and for other fields of physics. For instance, neutron capture plays a fundamental role in the synthesis of heavy elements (R.R. Winters, J.W. Truran, D.N. Schramm).

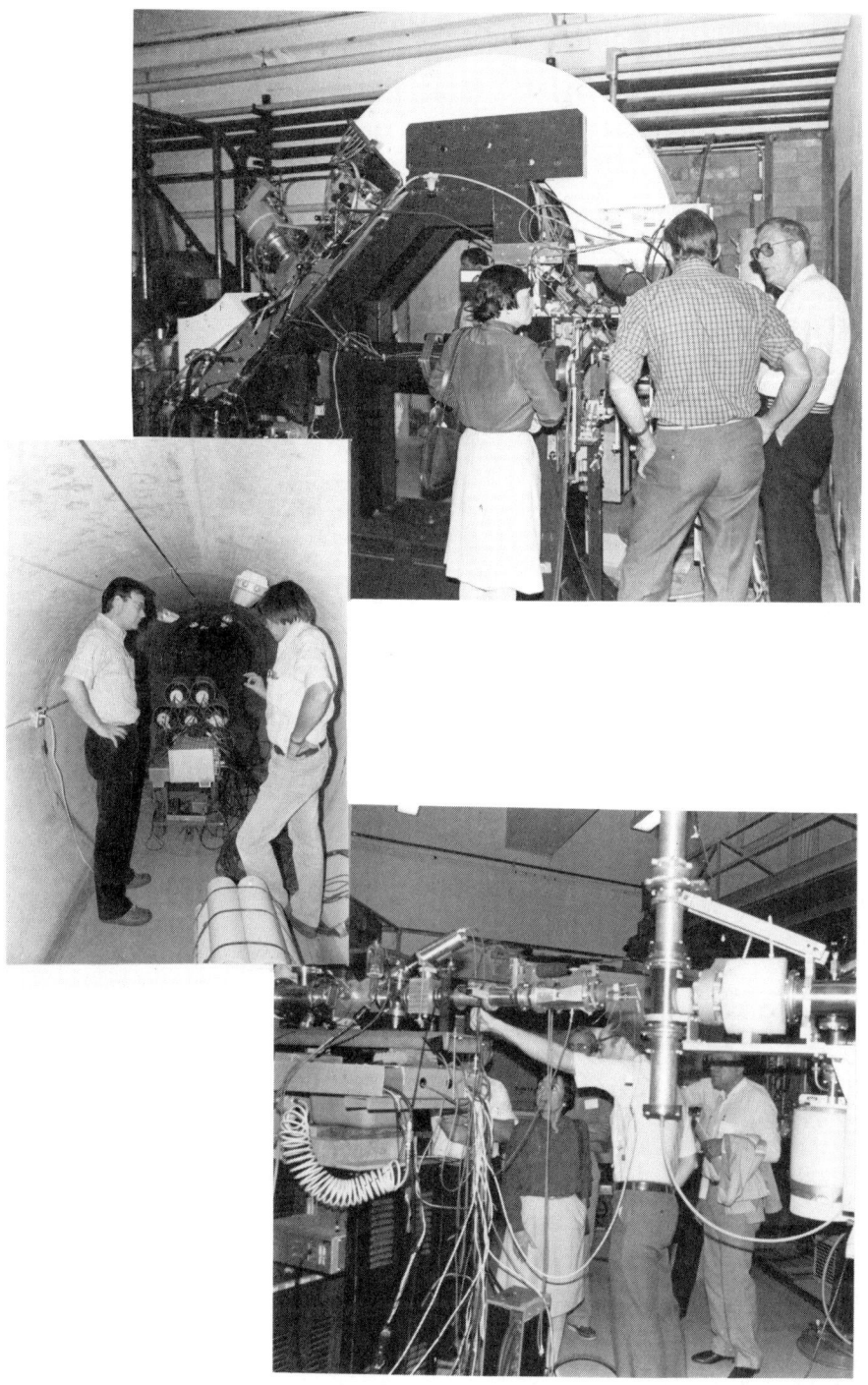

Author Index

Alarcon, R.   256
Annand, J.R.M.   322
Auchampaugh, G.F.   259
Aures, R.   137
Austin, S.M.   527

Barschall, H.H.   286
Bernard, V.   50
Bowman, C.D.   259
Brady, F.P.   137, 141, 143, 382
Brient, J.C.   320
Brown, B.A.   183, 256
Brown, V.R.   171

Carleton, R.F.   304, 308
Carr, J.A.   230
Castel, B.   304
Castenada, C.M.   141, 143
Clark, B.C.   123
Clegg, T.B.   297, 299
Coddens, G.P.   302
Condé, H.   403
Cooper, E.D.   123

Delaroche, J.P.   310, 324
de Swiniarski, R.   320
Dietrich, F.S.   299, 320, 322
Doll, P.   137
Drummond, J.R.   141, 143

Feshbach, H.   412
Finckh, E.   137
Finlay, R.W.   274, 322, 324
Ford, T.D.   141, 143
Friesel, D.L.   361

Girod, M.   318
Goodman, C.D.   375
Grangé, P.   50
Grimes, S.M.   463
Guss, P.P.   310, 312

Haight, R.C.   314
Hama, S.   123
Hansen, L.F.   314
Hansmeyer, J.   137
Haouat, G.   320
Harvey, J.A.   302, 304, 308
Heeringa, W.   137
Hiebert, J.C.   137
Hill, N.W.   302

Hodgson, P.E.   1
Hofmann, K.   137
Honore, G.M.   310
Howard, W.M.   511
Howell, C.R.   72, 310

Johnson, C.H.   304, 308, 446

Kaelbermann, S.G.   123
Kelly, J.J.   230
King, N.S.P.   141
Klages, H.O.   137
Koehler, R.   306
Kouzes, R.T.   256
Krupp, H.   137

Lagrauge, Ch.   314, 318, 320
Lejeune, A.   50
Lynn, J.E.   427

Madland, D.G.   318
Madsen, V.A.   26, 171
Mahoux, C.   540
Maier, Ch.   137
Martoff, J.   141
Martzolff, M.   50
Mathews, G.J.   511
Mc Eachern, B.   141
Mc Ellistrem, M.T.   208
Meigooni, A.S.   324
Mercer, R.L.   123
Mewissen, L.   306
Meyer, B.S.   515
Moore, W.H.   256

Nitz, W.   137

Olsson, N.   401
Orihara, H.   139
Osterfeld, F.   26

Patin, Y.   320
Petler, J.S.   322, 324
Petrovich, F.   90, 230
Plischke, P.   137
Pohl, B.A.   314
Poortmans, F.   306

Rapaport, J.   256
Rawitscher, G.H.   135
Romero, J.L.   141, 143

Salah, M.   302
Schramm, D.N.   515
Shepard, J.R.   107
Smith, A.B.   410

Taddeuci, T.N.   394
Takahashi, K.   511
Tornow, W.   312
Trostell, B.   401
Truran, J.W.   504

Van Parys, I.   306
Von Geramb, H.V.   14

Walter, R.L.   53, 310, 312
Wambach, J.   146
Wang, K.   141
Ward, R.A.   511
Webb, M.L.   141
Weigmann, H.   306
Wender, S.A.   259
Wilczynski, J.   137
Winters, R.R.   484
Wong, C.   314

Zafiratos, C.D.   327
Zeitnitz, B.   137
Zirnbauer, M.R.   481

AIP Conference Proceedings

		L.C. Number	ISBN
No. 1	Feedback and Dynamic Control of Plasmas	70-141596	0-88318-100-2
No. 2	Particles and Fields - 1971 (Rochester)	71-184662	0-88318-101-0
No. 3	Thermal Expansion - 1971 (Corning)	72-76970	0-88318-102-9
No. 4	Superconductivity in d- and f-Band Metals (Rochester, 1971)	74-18879	0-88318-103-7
No. 5	Magnetism and Magnetic Materials - 1971 (2 parts) (Chicago)	59-2468	0-88318-104-5
No. 6	Particle Physics (Irvine, 1971)	72-81239	0-88318-105-3
No. 7	Exploring the History of Nuclear Physics	72-81883	0-88318-106-1
No. 8	Experimental Meson Spectroscopy - 1972	72-88226	0-88318-107-X
No. 9	Cyclotrons - 1972 (Vancouver)	72-92798	0-88318-108-8
No. 10	Magnetism and Magnetic Materials - 1972	72-623469	0-88318-109-6
No. 11	Transport Phenomena - 1973 (Brown University Conference)	73-80682	0-88318-110-X
No. 12	Experiments on High Energy Particle Collisions - 1973 (Vanderbilt Conference)	73-81705	0-88318-111-8
No. 13	π-π Scattering - 1973 (Tallahassee Conference)	73-81704	0-88318-112-6
No. 14	Particles and Fields - 1973 (APS/DPF Berkeley)	73-91923	0-88318-113-4
No. 15	High Energy Collisions - 1973 (Stony Brook)	73-92324	0-88318-114-2
No. 16	Causality and Physical Theories (Wayne State University, 1973)	73-93420	0-88318-115-0
No. 17	Thermal Expansion - 1973 (lake of the Ozarks)	73-94415	0-88318-116-9
No. 18	Magnetism and Magnetic Materials - 1973 (2 parts) (Boston)	59-2468	0-88318-117-7
No. 19	Physics and the Energy Problem - 1974 (APS Chicago)	73-94416	0-88318-118-5
No. 20	Tetrahedrally Bonded Amorphous Semiconductors (Yorktown Heights, 1974)	74-80145	0-88318-119-3
No. 21	Experimental Meson Spectroscopy - 1974 (Boston)	74-82628	0-88318-120-7
No. 22	Neutrinos - 1974 (Philadelphia)	74-82413	0-88318-121-5
No. 23	Particles and Fields-1974 (APS/DPF Williamsburg)	74-27575	0-88318-122-3
No. 24	Magnetism and Magnetic Materials - 1974 (20th Annual Conference, San Francisco)	75-2647	0-88318-123-1
No. 25	Efficient Use of Energy (The APS Studies on the Technical Aspects of the More Efficient Use of Energy)	75-18227	0-88318-124-X
No. 26	High-Energy Physics and Nuclear Structure - 1975 (Santa Fe and Los Alamos)	75-26411	0-88318-125-8
No. 27	Topics in Statistical Mechanics and Biophysics: A Memorial to Julius L. Jackson (Wayne State University, 1975)	75-36309	0-88318-126-6
No. 28	Physics and Our World: A Symposium in Honor of Victor F. Weisskopf (M.I.T., 1974)	76-7207	0-88318-127-4

AIP Conference Proceedings

No. 29	Magnetism and Magnetic Materials - 1975 (21st Annual Conference, Philadelphia)	76-10931	0-88318-128-2
No. 30	Particle Searches and Discoveries - 1976 (Vanderbilt Conference)	76-19949	0-88318-129-0
No. 31	Structure and Excitations of Amorphous Solids (Williamsburg, VA., 1976)	76-22279	0-88318-130-4
No. 32	Materials Technology - 1976 (APS New York Meeting)	76-27967	0-88318-131-2
No. 33	Meson-Nuclear Physics - 1976 (Carnegie-Mellon Conference)	76-26811	0-88318-132-0
No. 34	Magnetism and Magnetic Materials - 1976 (Joint MMM-Intermag Conference, Pittsburgh)	76-47106	0-88318-133-9
No. 35	High Energy Physics with Polarized Beams and Targets (Argonne, 1976)	76-50181	0-88318-134-7
No. 36	Momentum Wave Functions - 1976 (Indiana University)	77-82145	0-88318-135-5
No. 37	Weak Interaction Physics - 1977 (Indiana University)	77-83344	0-88318-136-3
No. 38	Workshop on New Directions in Mossbauer Spectroscopy (Argonne, 1977)	77-90635	0-88318-137-1
No. 39	Physics Careers, Employment and Education (Penn State, 1977)	77-94053	0-88318-138-X
No. 40	Electrical Transport and Optical Properties of Inhomogeneous Media (Ohio State University, 1977)	78-54319	0-88318-139-8
No. 41	Nucleon-Nucleon Interactions - 1977 (Vancouver)	78-54249	0-88318-140-1
No. 42	Higher Energy Polarized Proton Beams (Ann Arbor, 1977)	78-55682	0-88318-141-X
No. 43	Particles and Fields - 1977 (APS/DPF, Argonne)	78-55683	0-88318-142-8
No. 44	Future Trends in Superconductive Electronics (Charlottesville, 1978)	77-9240	0-88318-143-6
No. 45	New Results in High Energy Physics - 1978 (Vanderbilt Conference)	78-67196	0-88318-144-4
No. 46	Topics in Nonlinear Dynamics (La Jolla Institute)	78-057870	0-88318-145-2
No. 47	Clustering Aspects of Nuclear Structure and Nuclear Reactions (Winnepeg, 1978)	78-64942	0-88318-146-0
No. 48	Current Trends in the Theory of Fields (Tallahassee, 1978)	78-72948	0-88318-147-9
No. 49	Cosmic Rays and Particle Physics - 1978 (Bartol Conference)	79-50489	0-88318-148-7
No. 50	Laser-Solid Interactions and Laser Processing - 1978 (Boston)	79-51564	0-88318-149-5
No. 51	High Energy Physics with Polarized Beams and Polarized Targets (Argonne, 1978)	79-64565	0-88318-150-9
No. 52	Long-Distance Neutrino Detection - 1978 (C.L. Cowan Memorial Symposium)	79-52078	0-88318-151-7
No. 53	Modulated Structures - 1979 (Kailua Kona, Hawaii)	79-53846	0-88318-152-5

AIP Conference Proceedings

No. 54	Meson-Nuclear Physics - 1979 (Houston)	79-53978	0-88318-153-3
No. 55	Quantum Chromodynamics (La Jolla, 1978)	79-54969	0-88318-154-1
No. 56	Particle Acceleration Mechanisms in Astrophysics (La Jolla, 1979)	79-55844	0-88318-155-X
No. 57	Nonlinear Dynamics and the Beam-Beam Interaction (Brookhaven, 1979)	79-57341	0-88318-156-8
No. 58	Inhomogeneous Superconductors - 1979 (Berkeley Springs, W.V.)	79-57620	0-88318-157-6
No. 59	Particles and Fields - 1979 (APS/DPF Montreal)	80-66631	0-88318-158-4
No. 60	History of the ZGS (Argonne, 1979)	80-67694	0-88318-159-2
No. 61	Aspects of the Kinetics and Dynamics of Surface Reactions (La Jolla Institute, 1979)	80-68004	0-88318-160-6
No. 62	High Energy e^+e^- Interactions (Vanderbilt, 1980)	80-53377	0-88318-161-4
No. 63	Supernovae Spectra (La Jolla, 1980)	80-70019	0-88318-162-2
No. 64	Laboratory EXAFS Facilities - 1980 (Univ. of Washington)	80-70579	0-88318-163-0
No. 65	Optics in Four Dimensions - 1980 (ICO, Ensenada)	80-70771	0-88318-164-9
No. 66	Physics in the Automotive Industry - 1980 (APS/AAPT Topical Conference)	80-70987	0-88318-165-7
No. 67	Experimental Meson Spectroscopy - 1980 (Sixth International Conference, Brookhaven)	80-71123	0-88318-166-5
No. 68	High Energy Physics - 1980 (XX International Conference, Madison)	81-65032	0-88318-167-3
No. 69	Polarization Phenomena in Nuclear Physics - 1980 (Fifth International Symposium, Santa Fe)	81-65107	0-88318-168-1
No. 70	Chemistry and Physics of Coal Utilization - 1980 (APS, Morgantown)	81-65106	0-88318-169-X
No. 71	Group Theory and its Applications in Physics - 1980 (Latin American School of Physics, Mexico City)	81-66132	0-88318-170-3
No. 72	Weak Interactions as a Probe of Unification (Virginia Polytechnic Institute - 1980)	81-67184	0-88318-171-1
No. 73	Tetrahedrally Bonded Amorphous Semiconductors (Carefree, Arizona, 1981)	81-67419	0-88318-172-X
No. 74	Perturbative Quantum Chromodynamics (Tallahassee, 1981)	81-70372	0-88318-173-8
No. 75	Low Energy X-ray Diagnostics - 1981 (Monterey)	81-69841	0-88318-174-6
No. 76	Nonlinear Properties of Internal Waves (La Jolla Institute, 1981)	81-71062	0-88318-175-4
No. 77	Gamma Ray Transients and Related Astrophysical Phenomena (La Jolla Institute, 1981)	81-71543	0-88318-176-2
No. 78	Shock Waves in Condensed Matter - 1981 (Menlo Park)	82-70014	0-88318-177-0
No. 79	Pion Production and Absorption in Nuclei - 1981 (Indiana University Cyclotron Facility)	82-70678	0-88318-178-9
No. 80	Polarized Proton Ion Sources (Ann Arbor, 1981)	82-71025	0-88318-179-7
No. 81	Particles and Fields - 1981: Testing the Standard Model (APS/DPF, Santa Cruz)	82-71156	0-88318-180-0

No. 82	Interpretation of Climate and Photochemical Models, Ozone and Temperature Measurements (La Jolla Institute, 1981)	82-071345	0-88318-181-9
No. 83	The Galactic Center (Cal. Inst. of Tech., 1982)	82-071635	0-88318-182-7
No. 84	Physics in the Steel Industry (APS.AISI, Lehigh University, 1981)	82-072033	0-88318-183-5
No 85	Proton-Antiproton Collider Physics - 1981 (Madison, Wisconsin)	82-072141	0-88318-184-3
No. 86	Momentum Wave Functions - 1982 (Adelaide, Australia)	82-072375	0-88318-185-1
No. 87	Physics of High Energy Particle Accelerators (Fermilab Summer School, 1981)	82-072421	0-88318-186-X
No. 88	Mathematical Methods in Hydrodynamics and Integrability in Dynamical Systems (La Jolla Institute, 1981)	82-072462	0-88318-187-8
No. 89	Neutron Scattering - 1981 (Argonne National Laboratory)	82-073094	0-88318-188-6
No. 90	Laser Techniques for Extreme Ultraviolet Spectroscopy (Boulder, 1982)	82-073205	0-88318-189-4
No. 91	Laser Acceleration of Particles (Los Alamos, 1982)	82-073361	0-88318-190-8
No. 92	The State of Particle Accelerators and High Energy Physics (Fermilab, 1981)	82-073861	0-88318-191-6
No. 93	Novel Results in Particle Physics (Vanderbilt, 1982)	82-73954	0-88318-192-4
No. 94	X-Ray and Atomic Inner-Shell Physics-1982 (International Conference, U. of Oregon)	82-74075	0-88318-193-2
No. 95	High Energy Spin Physics - 1982 (Brookhaven National Laboratory)	83-70154	0-88318-194-0
No. 96	Science Underground (Los Alamos, 1982)	83-70377	0-88318-195-9
No. 97	The Interaction Between Medium Energy Nucleons in Nuclei - 1982 (Indiana University)	83-70649	0-88318-196-7
No. 98	Particles and Fields - 1982 (APS/DPF University of Maryland)	83-70807	0-88318-197-5
No. 99	Neutrino Mass and Gauge Structure of Weak Interactions (Telemark, 1982)	83-71072	0-88318-198-3
No. 100	Excimer Lasers - 1983 (OSA, Lake Tahoe, Nevada)	83-71437	0-88318-199-1
No. 101	Positron-Electron Pairs in Astrophysics (Goddard Space Flight Center, 1983)	83-71926	0-88318-200-9
No. 102	Intense Medium Energy Sources of Strangeness (UC-Santa Cruz, 1983)	83-72261	0-88318-201-7
No. 103	Quantum Fluids and Solids - 1983 (Sanibel Island, Florida)	83-72440	0-88318-202-5
No. 104	Physics, Technology and the Nuclear Arms Race (APS Baltimore - 1983)	83-72533	0-88318-203-3
No. 105	Physics of High Energy Particle Accelerators (SLAC Summer School, 1982)	83-72986	0-88318-304-8

AIP Conference Proceedings

No. 106	Predictability of Fluid Motions (La Jolla Institute, 1983)	83-73641	0-88318-305-6
No. 107	Physics and Chemistry of Porous Media (Schlumberger-Doll Research, 1983)	83-73640	0-88318-306-4
No. 108	The Time Projection Chamber (TRIUMF, Vancouver, 1983)	83-83445	0-88318-307-2
No. 109	Random Walks and Their Applications in the Physical and Biological Sciences (NBS/La Jolla Institute, 1982)	84-70208	0-88318-308-0
No. 110	Hadron Substructure in Nuclear Physics (Indiana University, 1983)	84-70165	0-88318-309-9
No. 111	Production and Neutralization of Negative Ions and Beams (3rd Int'l Symposium, Brookhaven, 1983)	84-70379	0-88318-310-2
No. 112	Particles and Fields - 1983 (APS/DPF, Blacksburg, VA)	84-70378	0-88318-311-0
No. 113	Experimental Meson Spectroscopy - 1983 (Seventh International Conference, Brookhaven)	84-70910	0-88318-312-9
No. 114	Low Energy Tests of Conservation Laws in Particle Physics (Blacksburg, VA, 1983)	84-71157	0-88318-313-7
No. 115	High Energy Transients in Astrophysics (Santa Cruz, CA, 1983)	84-71205	0-88318-314-5
No. 116	Problems in Unification and Supergravity (La Jolla Institute, 1983)	84-71246	0-88318-315-3
No. 117	Polarized Proton Ion Sources (TRIUMF, Vancouver, 1983)	84-71235	0-88318-316-1
No. 118	Free Electron Generation of Extreme Ultraviolet Coherent Radiation (Brookhaven/OSA, 1983)	84-71539	0-88318-317-X
No. 119	Laser Techniques in the Extreme Ultraviolet (OSA, Boulder, Colorado, 1984)	84-72128	0-88318-318-8
No. 120	Optical Effects in Amorphous Semiconductors (Snowbird, Utah, 1984)	84-72419	0-88318-319-6
No. 121	High Energy e^+e^- Interactions (Vanderbilt, 1984)	84-72632	0-88318-320-X
No. 122	The Physics of VLSI (Xerox, Palo Alto, 1984)	84-72729	0-88318-321-8
No. 123	Intersections Between Particle and Nuclear Physics (Steamboat Springs, 1984)	84-72790	0-88318-322-6
No. 124	Neutron-Nucleus Collisions - A Probe of Nuclear Structure (Burr Oak State Park - 1984)	84-73216	0-88318-323-4
No. 125	Capture Gamma-Ray Spectroscopy and Related Topics - 1984 (International Symposium, Knoxville)		0-88318-324-2